Preslaughter handling and slaughter of meat animals

Preslaughter handling and slaughter of meat animals

edited by:

L. Faucitano

Wageningen Academic Publishers

Photo credits cover:

Main photo and sheep:	Claudia Terlouw, Université Clermont Auvergne, INRAE, France
Antelope, crocodile and ostrich:	Louw Hoffman, The University of Queensland, Australia
Horse:	Bert Driessen, Research, Control & Consult-Belgium, Belgium
Fish:	Bert Lambooij, WageningenUR Livestock Research, the Netherlands
Chicken:	Leslie Moffat, Eyes on Animals, the Netherlands
Pig:	Vanessa Pasquale, University of Guelph, Canada
Lama:	Genaro Miranda-de La Lama, University of Zaragoza, Spain

---

EAN: 9789086863723
e-EAN: 9789086869244
ISBN: 978-90-8686-372-3
e-ISBN: 978-90-8686-924-4
DOI: 10.3920/978-90-8686-924-4

First published, 2022

Table of contents

Introduction 9
References 11

Chapter 1 – Quantifying animal welfare preslaughter using behavioural, physiological and carcass and meat quality measures 12
C. Terlouw and C. Bourguet
Abstract 13
1.1 Stress and welfare 13
1.2 Studying and measuring stress 19
1.3 Consequences of stress reactions on carcass and meat quality 36
1.4 Conclusions and outlook 44
Acknowledgements 45
References 45

Chapter 2 – Cattle 62
C. Gallo, K. Schwartzkopf-Genswein and T. Gibson
Abstract 63
2.1 Pre-transport factors on farm 63
2.2 Transportation of cattle 75
2.3 Handling at the slaughter plant 81
2.4 Stunning and slaughter 86
2.5 Conclusions 100
References 100

Chapter 3 – Fish 118
M. Villarroel and E. Lambooij
Abstract 119
3.1 Introduction 119
3.2 Advances in the science of fish welfare 120
3.3 Feed withdrawal 121
3.4 Crowding 123
3.5 Emersion 125
3.6 Transport 125
3.7 Pre-slaughter stunning and killing with effects on unconsciousness and insensibility 127
3.8 Conclusions 140
References 141

Chapter 4 – Horses 150
B. Driessen, D. Marlin and J. Buyse
Abstract 151
4.1 Introduction 151
4.2 Pre-transport handling 153
4.3 Transport 158
4.4 Handling at the slaughter plant 165
4.5 Stunning and slaughter 168
4.6 Conclusions 170
References 171

Chapter 5 – Pigs 178
L. Faucitano and M. Raj
Abstract 179
5.1 Introduction 179
5.2 The farm of origin 180
5.3 Transport 189
5.4 Lairage 199
5.5 Stunning and slaughter 203
5.6 Conclusions 214
References 214

Chapter 6 – Poultry 232
M. Gerritzen, T. Shynkaruk, K. Buchynski, K. Schwean-Lardner and T.G. Crowe
Abstract 233
6.1 Introduction 233
6.2 Pre-pairing and handling for transport 234
6.3 Catching and loading 236
6.4 Transport 238
6.5 Lairage and unloading 247
6.6 Stunning methods 249
6.7 Bleeding 256
6.8 Conclusions 257
References 257

Chapter 7 – Sheep 266

M.S. Cockram and A. Velarde

Abstract 267

7.1 Introduction to pre-slaughter handling and transport 267

7.2 Pre-transport handling 269

7.3 Transport 273

7.4 Handling at the slaughter plant 282

7.5 Stunning and slaughter 287

7.6 Conclusions 297

References 298

Chapter 8 – Species destined for non-traditional meat production: 1. African game species, cervids, ostriches, crocodiles and kangaroos 312

T. Needham and L.C. Hoffman

Abstract 313

8.1 Introduction 313

8.2 Free-ranging African game species 315

8.3 Farmed cervids 324

8.4 Ostriches 331

8.5 Crocodiles 341

8.6 Kangaroos 342

8.7 Conclusions 343

References 343

Chapter 9 – Species destined for non-traditional meat production: 2. Goats and South American domestic camelids 348

G.C. Miranda-de la Lama, W. Mamani-Linares and L.X. Estévez-Moreno

Abstract 349

9.1 Introduction 349

9.2 Goats 350

9.3 South American domestic camelids 364

9.4 Conclusions 376

References 376

Chapter 10 – Avoiding live-animal transport to slaughter: mobile abattoirs 390
J. Hultgren
Abstract 391
10.1 Introduction 391
10.2 Industrialised slaughter 393
10.3 Transportation of meat versus animals 396
10.4 Mobile slaughter versus transportation 398
10.5 Mobile slaughter of fish 400
10.6 Mobile slaughter of birds 403
10.7 Mobile slaughter of mammals 405
10.8 Animal welfare 412
10.9 Feasibility and sustainability 414
10.10 Guidance and educational resources 418
10.11 Concluding remarks 420
Acknowledgements 420
References 421

Introduction

The practice of animal slaughter for meat consumption dates back to antiquity and can be traced to Roman civilisation and 15th-16th century France (Bello and Oyedemi, 2009). Animals were slaughtered in diverse places, such as backyards or butcher's sheds, until the first public slaughterhouse appeared in France (hence the term 'abattoir') as a unique institution and specific place where animals were slaughtered for human consumption in the early 19th century (Brantz, 2008; Vialles, 1994). This new establishment was part of a larger transition from a rural to industrial system, accompanied by increased urbanisation, technological developments, and concern about public hygiene (Brantz, 2008; Fitzgerald, 2010). Starting from the late 19th century, in a number of European countries, such as Italy, Denmark, Norway, Sweden, the Netherlands and Rumania, a slaughterhouse was required by law in all communities of more than 6,000 inhabitants (Oldfield, 1895). At that time, animals were killed by simply striking them with a blunt instrument, sometimes followed by exsanguination with a sharpened blade. This practice raised concerns among German animal protectionists, veterinarians and anti-Semitic societies who, among others, proposed the implementation of stricter inspection procedures and the slaughter of stunned and unconscious animals (Judd, 2003). A detailed history of the livestock and poultry slaughter industry is provided in Chapter 10 (Hultgren, 2022) in this book.

Laws for the protection of animals preventing cruelty, in terms of wilful or needless infliction of pain or spiteful neglect of the animal, existed in the Australian regions in 1901 (Jode *et al.*, 1906) and remained the dominant policy instrument for animal welfare until the 1980s. Since then, prevention of cruelty to animal acts have been revised regularly to keep pace with the time and ensure sufficient protection for animal welfare in many regions of the world. Based on the results of a questionnaire distributed by the World Organisation for Animal Health (OIE) in 2008, 83% of the OIE's member countries have a legislation regulating animal welfare during transport and 91% have a legislation addressed to the slaughter of animals (Stafford and Mellor, 2009).

Public concern for the welfare of farmed animals is changing the ways that meat products are being produced and marketed around the world. According to several recent consumer surveys in Europe and North America, humane transportation, proper handling and staff training, and humane slaughtering are deemed as very important criteria in terms of animal welfare by the large majority of respondents. These concerns about the welfare of animals from farm to slaughter are also growing in Australian, Asian and Latin American meat consumers (Faucitano *et al.*, 2017, 2022). In response to these requirements, food processors, supermarket and restaurant chains are driving changes in handling and slaughter practices by requiring that their meat suppliers adopt handling systems that consumers consider to be better for animal welfare and that slaughter plants pass regular animal welfare audits. These efforts are communicated to ethically conscious consumers through a number of food labelling and certification schemes (Faucitano *et al.*, 2017, 2022).

The day of slaughter is the last phase of the production period and consists of several stages, starting when animals leave their home pen and including transport, lairage, stunning and sticking. At each stage, animals destined for meat production are exposed to different stressors, including on-farm feed withdrawal, loading and transport, mixing, human interventions, unfamiliar noises and smells, and overcrowding that, both individually and/or additively, can affect their welfare and the quality of meat products. However, this last production phase tends to be overlooked given its short duration compared to the time spent by animals at the farm, although there is evidence showing that mistakes made in this phase may offset all efforts made by the production sector to improve performance and animal welfare. The application of maximum times for animal transportation to slaughter is particularly raising the interest of legislators and the public opinion as the time spent by animals contained in a truck is increasing due to the centralisation of the slaughtering industry, with more animals being killed in fewer larger plants, or emergency situations, e.g. the COVID-19 outbreak, which is presently stopping or slowing down slaughter operations at some plants and diverting shipments towards others with greater slaughter capacity in North America (Marchant-Forde and Boyle, 2020). The use of mobile abattoirs may help skip unnecessary handling and transport of live animals to slaughter, but the viability and profitability of these meat production systems are still being questioned (Hultgren, 2022 – Chapter 10).

The objective of the present book is to provide a science-based reference text which covers all events of the preslaughter period and slaughter and deal with their effects on the welfare of multiple species destined to meat production, including major species (cattle, fish, horses, pigs and poultry) and minor species for non-traditional meat production. Jointly with the chapter on mobile slaughtering (Hultgren, 2022 – Chapter 10), the chapters on non-traditional meat species (Needham and Hoffman, 2022 – Chapter 8 and Miranda-de la Lama *et al.*, 2022 – Chapter 9) specifically contribute to the uniqueness of this book content. Although these animals contribute to respond to the worldwide increasing demand for animal proteins and represent the major resource for agricultural economy in geographical locations characterised by adverse climate and scarcity of natural resources (Suman and McMillin, 2014), they have received significantly less attention when compared with conventional livestock species and the knowledge about the practices to ensure their welfare condition prior to and at slaughter is little, if none. An overview of indicators aimed at reliably quantifying the level of stress experienced by animal during the preslaughter phases is also included among the topics covered in this book (Terlouw and Bourguet, 2022 – Chapter 1).

This unique and comprehensive work is written by renowned experts and accomplished scientists from four continents. The book provides an in-depth scientific review and empirical assessment of the concepts, knowledge and technology practiced in the livestock and poultry industry today. This work will be of interest and relevance to animal scientists, academics, students, animal welfare and industry associations, policy makers, veterinarians, livestock and poultry chain stakeholders, and all those involved in and who care about the humane handling of meat species.

Luigi Faucitano
Editor

References

Bello, Y.O. and Oyedemi, D.T.A., 2009. The impact of abattoir activities and management in residential neighbourhoods: a case study of Ogbomoso, Nigeria. Journal of Social Science 19: 121-127.

Brantz, D., 2008. Animal bodies, human health, and the reform of slaughterhouses in nineteenth-century Berlin. In: Young Lee, P. (ed.) Meat, modernity and the rise of the slaughterhouse. University of New Hampshire Press, Durham, NH, USA, pp. 71-88.

Faucitano, L., Martelli, G., Nannoni, E. and Widowski, T.M., 2017. Fundamentals of animal welfare in meat animals and consumer attitudes to animal welfare. In: Purslow, P. (ed.) Meat quality aspects: from genes to ethics. Woodhead Publishing, London, UK, pp. 537-568.

Faucitano, L., Martelli, G., Nannoni, E. and Manteca, X., 2022 (in press). Fundamentals of animal welfare in meat animals and consumer attitudes to animal welfare. In: Purslow, P. (ed.) Meat quality aspects: from genes to ethics. 2nd edition, Woodhead Publishing, London, UK.

Fitzgerald, A.J., 2010. A social history of the slaughterhouse: from inception to contemporary implications. Human Ecology Review 17: 58-69.

Hultgren, J., 2022. Avoiding live-animal transport to slaughter: mobile abattoirs. In: Faucitano, L. (ed.) Preslaughter handling and slaughter of meat animals. Wageningen Academic Publishers, Wageningen, the Netherlands, pp. 390-434.

Jode, L., Martel, H. and Mallet, J., 1906. Les abattoirs publics. H. Dunod and E. Pinat Editeurs, Paris, France.

Judd, R., 2003. The politics of beef: animal advocacy and the kosher butchering. Jewish Social Studies 10: 117-150.

Marchant-Forde, J.N. and Boyle, L.A., 2020. COVID-19 effects on livestock production: a one welfare issue. Frontiers in Veterinary Science 7: 585787. https://doi.org/10.3389/fvets.2020.585787

Miranda-de la Lama, G.C., Mamani-Linares, W. and Estévez-Moreno, L.X., 2022. Species destined for non-traditional meat production: 2. Goats and South American domestic camelids. In: Faucitano, L. (ed.) Preslaughter handling and slaughter of meat animals. Wageningen Academic Publishers, Wageningen, the Netherlands, pp. 348-389.

Needham, T. and Hoffman, L.C., 2022. Species destined for non-traditional meat production: 1. African game species, cervids, ostriches, crocodiles and kangaroos. In: Faucitano, L. (ed.) Preslaughter handling and slaughter of meat animals. Wageningen Academic Publishers, Wageningen, the Netherlands, pp. 312-347.

Oldfield, J., 1895. The groaning creation. The Ideal Publishing Union, London, UK, 100 pp.

Stafford, K.J. and Mellor, D.J., 2009. The implementation of animal welfare standards by Member Countries of the World Organisation for Animal Health (OIE): analysis of an OIE questionnaire. Revue Scientifique et Technique 28: 1143-1164.

Suman, S.P. and McMillin, K.W., 2014. Contributions of non-traditional meat animals to global food security and agricultural economy. Animal Frontiers 4: 4-5.

Terlouw, C. and Bourguet, C., 2022. Quantifying animal welfare preslaughter using behavioural, physiological and carcass and meat quality measures. In: Faucitano, L. (ed.) Preslaughter handling and slaughter of meat animals. Wageningen Academic Publishers, Wageningen, the Netherlands, pp. 12-61.

Vialles, N., 1994. Animal to edible. Cambridge University Press, Cambridge, UK, 142 pp.

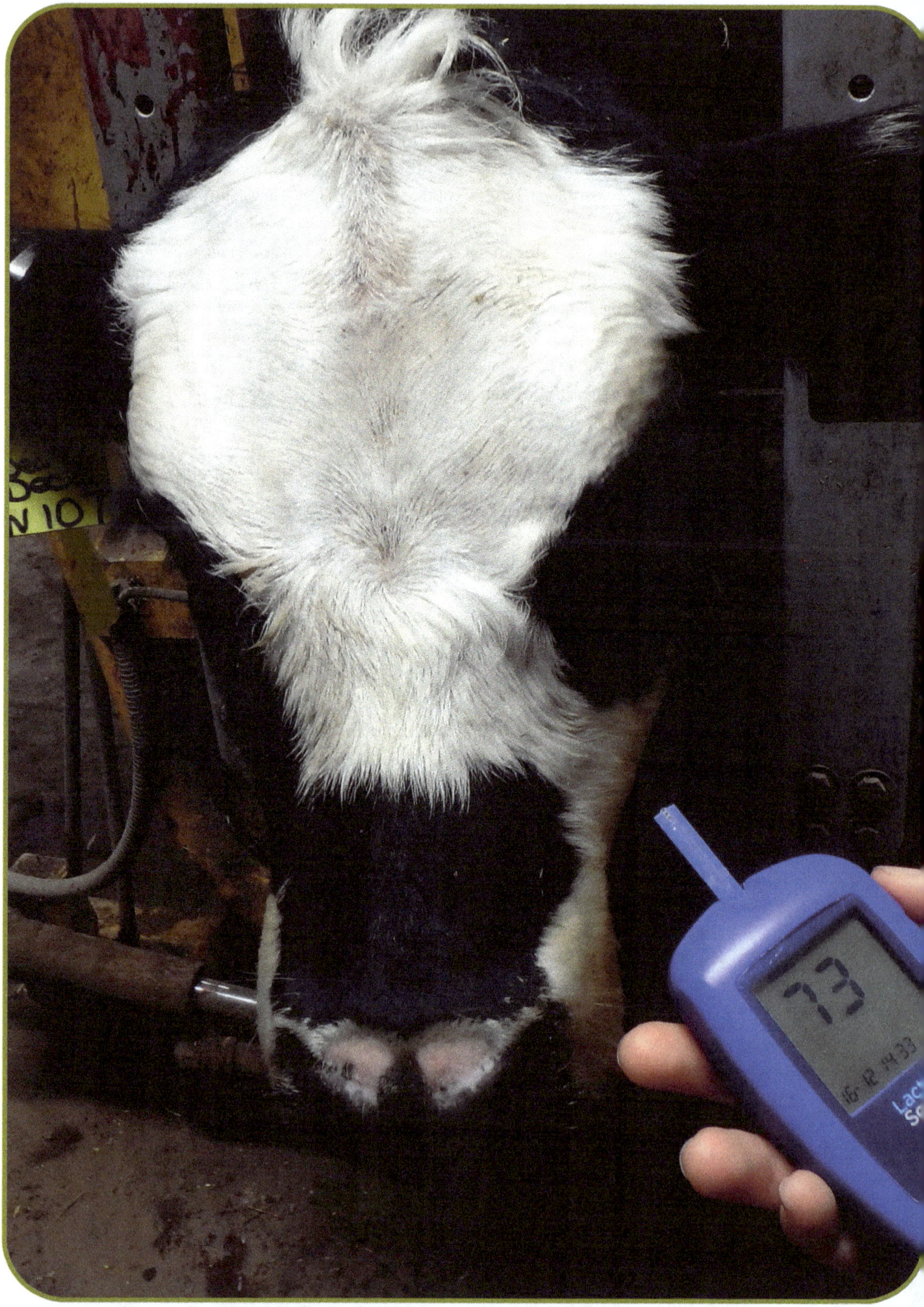

Photo: Daniela Meléndez, Agriculture & Agri-Food Canada, Lethbridge, Canada

Chapter 1

Quantifying animal welfare preslaughter using behavioural, physiological and carcass and meat quality measures

C. Terlouw[1*] *and C. Bourguet*[2]

[1]*UMR Herbivores, INRAE, VetAgro Sup, Theix, 63122 Saint-Genès Champanelle, France;* [2]*Bureau ETRE, 377 rue du Sou, Bravant, 63210 Olby, France; claudia.terlouw@inrae.fr*

Abstract

Animal welfare and stress refer to the subjective experience of the animal, which depends on its emotional and cognitive capacities to evaluate the situation it is in, to know its body state in terms of needs, and the higher-order, conscious, processing of the information. The origin of stress-inducing factors may be psychological, such as fear, or physical, such as food deprivation. In both cases, they influence the emotional status of the animal. Due to its subjective, therefore unmeasurable, character, the assessment of animal welfare is indirect. During the preslaughter period, this assessment relies on the physiological and behavioural stress reactions and environmental context. Physiological reactions are not specific to stress; they allow any adaptive response potentially requiring increased physical activity and vigilance. Combining physiological measurements and behavioural observations in relation to the environmental context allow interpreting these different types of information in terms of stress. Physiological and behavioural reactions further influence meat quality traits, such as post-mortem temperature and pH decline, which influence other traits, such as colour and tenderness, and carcass quality characteristics, such as bruising. Individuals show a degree of consistency in their responses to stress factors. The determination of the animal's stress reactivity using standardised tests before slaughter facilitates the interpretation of the physiological and behavioural observations made in the slaughter setting. Stress reactivity determined before slaughter may predict stress reactions at slaughter, and future meat quality. Summarising, the assessment of animal welfare and stress involves the interpretation of a range of measurements on the physiology and behaviour of the animal, the environmental context, post-mortem meat and carcass quality traits, and the characteristics of the animal itself, in terms of stress reactivity and rearing and genetic background.

Keywords: behaviour, blood, meat quality, stress

1.1 Stress and welfare

1.1.1 Can animals be stressed? Can they have emotions?

Stress has often been described as the state of the animal when it has difficulties to adapt to environmental or physical constraints (Broom, 1988; Fraser *et al.*, 1975). The question of adaptation is important, but animal stress is only an ethical problem if animals are capable to experience negative emotions, such as fear or frustration (Broom, 1991; Dantzer, 2002; Dawkins, 2008; Désiré *et al.*, 2002; Duncan, 1996). Can they? Studies on behaviour, physiology as well as on the anatomy and functioning of the brain indicate that this is likely (Boissy *et al.*, 2007a; Panksepp, 2005; Paul *et al.*, 2005). First, there are many neurological similarities between humans and non-human mammals. Particularly, the human brain contains a network of structures, called the limbic system, which processes emotions. Non-human mammalian brains also contain a limbic system, consisting of the same structures as in humans and involved in many behavioural reactions during stressful situations (Damasio, 1998; LeDoux, 2000; Panksepp, 2005; Box 1.1).

Box 1.1. Emotions and the limbic system.

Small, specific lesions in the limbic system modify the expression of emotions, sometimes very specifically. For example, the amygdala, a small nucleus in the lower parts of the forebrain and part of the limbic system, is involved in fear. Blocking its functioning stopped the immobility response of rats to fear-inducing stimuli, such as cat odour (Chen *et al.*, 2006; Li *et al.*, 2004). In Rhesus monkeys, blocking the amygdala reduced fear responses to snakes (Elorette *et al.*, 2020). Humans with amygdala damage do not exhibit physiological defence responses to threat (LeDoux, 2017). Despite the specific functions of specific structures of the limbic system, the processing of emotions is probably best described as the output of the integrative functioning of the limbic and other networks, including the cortex. For example, the experience of fear involves the processing of fear stimuli not only by the amygdala, but also by structures involved in memory and arousal, amongst others (LeDoux, 2017; LeDoux and Brown, 2017).

Second, humans and animals show similar behavioural and physiological responses to stress-inducing situations, like avoidance, immobilisation or aggressive responses to danger, and startle responses to sudden events, associated with increased heart rates, amongst others. It is further thought that animals may also experience positive emotions (Boissy *et al.*, 2007b; Finlayson *et al.*, 2016; Panksepp, 2005; Veissier *et al.*, 2009). For instance, young lambs receiving regular tactile contacts from humans showed behavioural and physiological signs of a calmed state while they were stroked (Coulon *et al.*, 2015).

Both negative and positive emotions help the animal to optimise its chances to survive and reproduce, that is, to adapt, because emotions are important drivers of motivation. Animals tend to avoid situations that result in negative emotions and are attracted to those that provide positive emotions. For example, animals avoid fear or pain-inducing situations and are attracted to pleasurable experiences, such as eating food when they are hungry or shelter when they need rest. The environment and the physical needs of the animal change constantly, consequently, its emotional state, that is, its collective negative and positive emotions, also.

The capacity of animals to experience emotions is best illustrated in a series of experiments inspired by studies on emotions in humans. In human research, analytical approaches indicate that the emotion elicited by a situation depends on several well-defined characteristics of the situation, such as suddenness, familiarity, predictability, pleasantness and conformity with expectations (Scherer, 2001). For example, in humans, a situation with a high degree of suddenness, and low familiarity, predictability, controllability and pleasantness evokes negative emotions, such as fear and anger (Scherer, 2001). It was shown that animals are also capable of recognising these characteristics (Veissier *et al.*, 2009). For instance, when a piece of textile fell rapidly and unexpectedly behind the trough from which sheep were eating, they showed startle responses, asymmetric ear postures and a sharp, short-lasting rise in heart rate, all indicative of a change in emotional state. When the object fell slowly, they did not show these reactions (Désiré *et al.*, 2004, 2006). When the appearance of the textile was predicted by a light signal, the sheep reacted less strongly (Greiveldinger *et al.*, 2007). Similarly, they looked more often at an object with their ears oriented forward when the object was unfamiliar rather than familiar (Désiré *et al.*, 2006). These results show that sheep are sensitive to suddenness, predictability and familiarity. Sheep can also form expectations. Sheep were trained to interrupt a light beam with their muzzle to receive a large food reward. When the amount was changed into a small reward, they adopted an asymmetric ear posture at the time of the reward delivery, increased locomotion and showed an increase in heart rate, indicative of an emotional response. Sheep responses to an increase in the reward were less pronounced (Greiveldinger *et al.*, 2011). It is likely that most or possibly all mammals are sensitive to the suddenness, unfamiliarity and predictability of their environment and, hence, are capable of a range of emotions, such as fear, anger, rage, despair, boredom, disgust, and happiness (Veissier *et al.*, 2009).

The above indicates that in the context of stress, while certain environmental and physical events do need adaptive responses from the animal, they are likely to provoke also negative emotions. It is important to realise that the state of stress of the animal depends only indirectly on the environment. It depends directly on the evaluation of the environment by the animal. The animal is stressed, that is, feels negative emotions, when it feels threatened, whether the threat is real or not. In order to adapt to this threat, the animal reacts behaviourally, with flight or defence reactions for example, associated with physiological changes, such as a faster heart rate and hormonal changes, which allow the increase in physical effort and vigilance (Terlouw, 2005; Veissier and Boissy, 2007).

1.1.2 Animal welfare and mental states, even more complex!

The concept of animal welfare is larger than of the concept of stress. While emotions are fluctuating and of short duration, the welfare depends on the global emotional state of the animal, as well as on its desires and preferences, considered over a longer time period (Boissy *et al.*, 2007b; Dawkins, 2003). Events with strong negative emotional impacts may have long-lasting effects on the global emotional state, and physiological and behavioural functioning, not only in humans, but also in animals. This was demonstrated in lambs that during rearing were subjected to repeated, unpredictable and uncontrollable aversive events, such as social separation, changes in the environment, noise, and handling. These lambs were subsequently trained in a choice test where they received a reward if they chose the correct bucket (e.g. the bucket placed on the left) and were punished if they chose the incorrect bucket (the one on the right). The final test consisted of placing the bucket in the middle, which is an ambiguous position. Lambs that had been subjected to repeated negative events during rearing chose the ambiguous position, 'middle', less often than their unstressed counterparts. This indicates that the way these lambs evaluated the situation had a negative or pessimistic bias compared to the unstressed controls (Destrez *et al.*, 2013).

Certain authors consider that the expression of self-awareness, memory, symbolic language, social understanding, and anticipating the behaviour of, or attributing desires to, conspecifics ('mindreading'), as well as having interests, matter for animal welfare (Jones, 2013; Krupenye and Call, 2019). The above considerations imply that at least some non-human mammals have complex, subjective, mental states. Today, many scientists believe that, like humans, non-human mammals construct an integrated mental image from the information perceived from the environment, the body and from the knowledge, they have through memories and other sources (Damasio, 2010). The construction of such an image needs a range of processes involved in the acquisition, storage and manipulation of information, referred to as cognition (Mendl and Paul, 2004; Box 1.2). The constructed mental image is foundational for the rich emotional, physiological and behavioural reactions of mammals to varying needs and a changing environment.

Box 1.2. Cognitive abilities of animals.

Emotional contagion refers to an emotional state in an observer as a direct result of perceiving the emotional state of another. Empathy refers to situations in which the observer has a similar emotional state as another being as a result of the perception of this being's situation. The processing underlying empathy is more complex than for emotional contagion. In contrast to emotional contagion, the distinction between self and other is maintained, and the emotional state remains focused on the other being rather than self-focused (Preston and De Waal, 2002). Examples of empathy are described for apes, elephants and dolphins (De Waal *et al.*, 2006). Emotional contagion has been described in other species (De Waal, 2008). For example, pigs, when joining another, stressed pig, showed behaviour that was different from when joining a non-stressed pig (Reimert *et al.*, 2015).

At least certain primates are capable of metacognition, that is, they are capable of judging their own state of knowledge. For example, Rhesus monkeys were trained to choose among four opaque tubes, one of which concealed food. On only half the trials, the monkeys observed the experimenter baiting the tube, so that they knew which tube contained the reward. They were allowed, however, to look down the length of each tube, before choosing. When they had seen the experimenter baiting, most monkeys chose without looking; when not, most often they looked in the tube first. This suggests that they discriminated between knowing and not knowing (Hampton *et al.*, 2004). Another experiment allowed dolphins performing a discrimination task, to choose the 'opt-out option'. Choosing this option, the dolphin received a greater reward than the reward given for an incorrect response, but lower than the reward given for a correct response. The dolphins seemed to choose this option when they had a bigger probability of failing the task (Smith *et al.*, 1995). Similar results were found in rats and pigeons (Adams and Santi, 2011; Foote and Crystal, 2007). However, it is believed that metacognition requires self-reflective consciousness and certain scientists question the existence of such capacities in non-primates, because more simple behavioural principles may also explain the observed behaviour, such as information seeking behaviour (Foote and Crystal, 2012; Kirk *et al.*, 2014; Smith *et al.*, 2008). Particularly, rats and bees accept to work for information on where food can be found, and to work harder if the food is more difficult to find (Foote and Crystal, 2012; Kirk *et al.*, 2014; Lehrer, 1993).

Helping behaviour is a social behaviour whereby an individual helps another irrespective of disadvantages to itself. Again, there are many examples in primates (Preston and De Waal, 2002). There is also recent evidence for helping behaviour in rats. Rats quickly learned to free a rat trapped in a wet area, and did only so if the wet rat showed signs of distress. When rats were forced to choose between opening the door to help a distressed cage mate and opening a different door to obtain a food reward, in most test trials, rats chose to help the cage mate before obtaining a food reward (Sato *et al.*, 2015).

Theory of mind is the ability to ascribe mental states, such as desires and beliefs, to others. While it is central to human social life, current research indicates that other mammals have also a theory of mind (Krupenye and Call, 2019; Van der Vaart and Hemelrijk, 2014). Most work was carried out on apes and monkeys, but dogs and corvidae have similar capacities. For example, Western scrub jays show caching behaviour: they bury food items for future consumption.

Conspecifics will steal such items if they see them being hidden, but cachers take measures to prevent such theft. When given worms to cache, with a visible competitor in an adjacent cage, they re-cached their worms in new sites once they were alone, but only if they had previous experience being pilferers themselves (Emery and Clayton, 2001).

As for emotions, the arguments for the existence of complex mental states in non-human mammals are based on analogies between humans and other mammals. In humans, the construction of complex mental states needs widespread functional connectivity between different brain areas, including the prefrontal cortex, necessary for higher order information processing (Salzman and Fusi, 2010). The question is whether non-human mammals have brain structures and connections allowing such complex processing. There are many similarities between humans and other mammals. Monkey and human brains contain essentially the same prefrontal cortex areas, but certain prefrontal cortex areas do not exist in rats for instance (Öngür and Price, 2000; Rolls, 2015). Despite these variations, in non-human mammals, areas of the prefrontal cortex make widespread connections with other cortical areas, and information processing subcortical areas, including the limbic system (Öngür and Price, 2000; Rolls, 2015). This suggests that all mammals have the necessary connections between the relevant brain areas to have at least some degree of higher-order information processing, allowing the construction by the brain of complex mental states that include emotions.

1.1.3 Can animals suffer? Are they conscious?

The above information indicates that non-human mammals may be capable to experience negative emotions and to construct complex mental states by higher-order processing of information. Scientists have argued, however, that these capacities are necessary, but not sufficient conditions for animal welfare matters to be an ethical problem (Mendl and Paul, 2004). Ethical questions related to animal welfare infer that animals are able to suffer and to experience pleasure. These scientists argue that the central question is whether animals have conscious knowledge of their sensations and emotions. The question has been debated for many years (Birch *et al.*, 2020; Dawkins, 2015; Mendl and Paul, 2004). Initially, the question was whether animals are conscious. Over the last ten years, the scientific understanding of human consciousness and the vision of animal consciousness have much evolved. Conceptual progression, clever experimental paradigms and new, sophisticated brain imagery techniques were major drivers (Lamme and Roelfsema, 2000; Laureys *et al.*, 2004; Tononi, 2012). In 2012, the Cambridge Declaration on Consciousness declared that humans are not the only conscious beings and that 'non-human animals, including all mammals and birds, and many other creatures, including octopuses' possess neurological substrates complex enough to support conscious experiences (Birch *et al.*, 2020). Todays' question is not *if*, but *how* animals are conscious. A recent paper considers the possible existence of different, multidimensional consciousness profiles, varying across species, including birds and fish (Birch *et al.*, 2020; Box 1.3). It can be argued on the same grounds that each *individual* has its own consciousness profile.

Box 1.3. Higher-order processing in birds and fish.

Like mammals, birds express complex cognitive behaviour. They can classify images according to their content or group them according to abstract concepts, such as 'similar' or 'different'. Pigeons, for example, distinguish photos containing humans from those that do not (Emery, 2006). Pigeons, hens and quails are also able to solve complex problems that require the application of a rule (Emery, 2006). The best performances are observed in corvids and parrots (Gossette *et al.*, 1966). Birds do not have a cortex, like mammals do, but their brain contains a structure called pallium, showing similarities in connectivity and functional organisation to the mammalian cortical structures (Jarvis *et al.*, 2005). The avian pallium receives information relative to the environment and the body state and is capable of higher order processing of information, explaining their cognitive abilities (Jarvis *et al.*, 2005; Nomura and Izawa, 2017).

Fish lack the cerebral cortical structures that are involved in consciousness in humans. Despite this, neuro-anatomical evidence and physiological and behavioural reactions suggest that fish experience pain, and other forms of negative emotions, including fear (Chandroo *et al.*, 2004; Mok and Munro, 1998). Behavioural observations indicate that teleost fish have a number of learning and memory mechanisms and basic cognitive capabilities, similar to those in mammals, birds and reptiles (Broglio *et al.*, 2005). For instance, trout that receive a painful stimulus change their behaviour by rubbing the painful site against available substrates (Sneddon, 2003, 2009). In goldfish, lesions in a specific area of the brain (the dorsomedial telencephalon) affected emotional pain-related behaviour, as the fish no longer avoided a painful stimulus (Portavella *et al.*, 2002). These scientists conclude that, although the fish brain lacks certain structures of the mammalian brain involved in higher-order processing of information, other structures may fulfil these functions. Other scientists claim however, that such results show that fish respond to noxious stimuli but do not necessarily experience pain (Rose *et al.*, 2014).

Despite these recent evolutions, sceptics maintain that unequivocal evidence of existence of higher-order mental states in animals is still wanting (Krupenye and Call, 2019; Box 1.2). Other scientists argue that materialistic approaches of science cannot elucidate the existence of consciousness (Beauregard *et al.*, 2014). They indicate that quantum mechanics have shown that atoms and subatomic particles are not really solid objects as they are essentially empty. Furthermore, they do not exist with certainty at definite spatial location at definite times. These scientists indicate that our vision of reality is exclusively materialistic and therefore incomplete. They argue that the mind is not derived from matter. Most scientists acknowledge the difficulty to provide formal proof for higher-order mental states in animals, as it is impossible to get access to the subjective state of another being, but they argue that, for the sake of precaution, we should assume that animals consciously perceive their inner mental state, including emotions (Mameli and Bortolotti, 2006; Mendl and Paul, 2004; Box 1.4).

Box 1.4. Are you sure that your neighbour has consciousness?

The question of the existence of higher-order states can be asked both for animals and humans. Most persons consider having emotional and complex mental states and believe that other persons have similar states. However, as these states are exclusively personal and subjective experiences, how to know that this is true? Can we prove that other humans are conscious? Would the world function in the same way if humans did not have any subjective experience, or no consciousness? This is a subject of much philosophical debate, opposing materialistic and non-materialist points of views, amongst others (Chalmers, 1996; Dennett, 1991; Hyslop and Jackson, 1975; Wittgenstein, 1958).

In conclusion, animal welfare and animal stress are subjective states, that is, they depend on the animal's conscious experience of its inner state. This inner state can only be measured indirectly, essentially by evaluating the behavioural and physiological expressions of these states, or the longer-term consequences of these expressions.

1.2 Studying and measuring stress

To measure and understand stress responses, it is useful to distinguish the potential causes of stress according to their origin: psychological or physical. Those of a psychological origin, such as fear, frustration, resignation and sadness, result from thwarted desires and preferences (Boissy, 1995; Destrez *et al.*, 2013). During the preslaughter period, stress of psychological origin is often caused by social disturbances, sudden events and the unfamiliarity of the different situations the animals encounter. They are due to the departure from the farm, the introduction into the unfamiliar environments of the lorry and the abattoir, the exposure to sounds and humans, and the disruption of the social group, due to separation and mixing of unfamiliar animals (Terlouw *et al.*, 2008).

Those of a physical origin inflict physical constraints on the functioning of the body, such as a state of hunger and thirst, fatigue, illness or tissue lesions. In the preslaughter context, they may result from food and water deprivation, negative events during transport and lairage, including physical compression, extreme temperatures, collisions with pen fittings or other animals, various interventions by humans, including the use of electric prod, and fighting between animals due to mixing, amongst others (Bourguet *et al.*, 2011b; Danziger, 2006). Despite the physical component of these stress factors, they influence also the emotional state of the animal. This was demonstrated by Bourguet *et al.* (2011a) who showed that cows and heifers that had been food deprived for 30 h, a stressor of physical origin, were more fearful to human presence and handling, and unexpected events (i.e. a light air blast on the nose), which are stressors of psychological origin. This further indicates that a given stress factor during the preslaughter period may exacerbate the effects of other stress factors on the levels of stress of the animal. Similarly, pain, which may be caused by fighting or bruising, for example, has a strong negative affective component; it is this negative experience which makes that painful situations are avoided (Villemure *et al.*, 2003; Box 1.5).

Box 1.5. Pain and the pain system.

Specialised neurons of the somatosensory system, called nociceptors, are activated by noxious stimuli, that is, stimuli that have the potential to cause injury. Nociceptors are found in the skin, joints, muscles and internal organs. They can respond to temperature (thermoreceptor), chemicals (chemoreceptors) and pressure (mechanoreceptors) (Millan, 1999). Other receptors are sensitive to the touch and are part of the non-nociceptive somatosensory system. Activation of nociceptors beyond a certain threshold causes the perception of pain (Millan, 1999). The perception of pain is associated with potential tissue damage and must be taken into account by the organism. Therefore, it is difficult to ignore pain (Eccleston and Crombez, 1999).

The nociceptors transform the harmful stimulus into a nerve message that is transmitted to other specialised neurons, called secondary nociceptors, in the dorsal horn of the spinal cord. The latter transmit their signals to various brain structures forming a network called the 'pain matrix' (Ingvar, 1999). This matrix includes structures of the brainstem, subcortical structures, and cortical areas. The pain experience results from the coordinated activity of the different regions of the brain that make up the pain matrix. Regardless of the type of pain, certain areas of the matrix are always activated during the perception of pain. Most pain matrix structures are, however, not pain specific. For example, most regions of the pain matrix also respond to tactile, auditory and visual stimuli in a similar way as to nociceptive stimuli. In addition, in humans, stimulation of structures of pain matrix, with a few exceptions, generally does not cause the experience of pain (Garcia-Larrea and Peyron, 2013; Mazzola *et al.*, 2012).

As indicated by the International Association for the Study of Pain (IASP; Raja *et al.*, 2020), the concept of pain refers to the situation where both dimensions, the sensory (the nature, intensity and duration of the stimulus) and motivational-affective (the unpleasantness) dimensions of pain are perceived. The pain sensation represents therefore a nociceptive sensation associated with a negative affective sensation. Different structures in the brain are associated with these different components.

1.2.1 Behavioural and physiological reactions to stress under controlled conditions

Reactions to specific stressors may be tested under controlled conditions (Table 1.1). These tests show that animals show a certain consistency in the way they react to stressors, referred to as the animal's stress reactivity (Bourguet *et al.*, 2015; Terlouw and Rybarczyk, 2008). Stress reactivity is considered high if an animal feels easily threatened and presents pronounced behavioural (whether overt or not) and/or physiological reactions. This consistency was illustrated by Boissy and Bouissou (1988, 1995; Table 1.1), who tested heifers in a series of tests confronting them with unfamiliarity, isolation, conflict between fear and motivation to feed, and social isolation. Heifers that were more reluctant to explore an unfamiliar object compared to their conspecifics were also more reluctant to feed near a fear-inducing stimulus and showed more signs of fear when they were socially isolated in an unfamiliar environment. This approach allowed the identification of several fear related behaviours, which differed according to the context (Table 1.2).

Individuals that were more fearful showed *higher* activity levels during the human exposure test, during restraint and during social isolation. However, they showed *lower* activity levels during the open field test, the novelty test and the surprise test. This shows that the behavioural fear response depends on the context. Fear induces defence and avoidance reactions, with defence reactions aiming to drive the threatening factor away from the animal, and avoidance reactions moving the animal away from the threat. The animal may also immobilise in order to remain unnoticed. Possibly, in the studies by Boissy and Bouissou (1988, 1995), the active responses expressed the motivation to escape, while the passive responses expressed the motivation to avoid approaching the threat or to remain unnoticed.

Other species show also consistent differences in stress reactivity. Gilts that accepted relatively easily to be removed from their home pen and to be separated from their social group, also accepted

Table 1.1. Examples of reactivity tests that examine different stress factors.

	Factors tested	Species	Reference
Individual tests[1,2]			
Novel object test	unfamiliarity	sheep, pigs	Boissy and Bouissou, 1988; Désiré *et al.*, 2004, 2006; Lawrence *et al.*, 1991; Olsson *et al.*, 1999; Terlouw and Rybarczyk, 2008
Walking through a corridor	unfamiliarity	heifers	Boissy and Bouissou, 1988
Surprise test	unfamiliarity, surprise	heifers, bulls, sheep, calves	Boissy and Bouissou, 1995; Bourguet *et al.*, 2015; Désiré *et al.*, 2004, 2006; Greiveldinger *et al.*, 2009; Lensink *et al.*, 2000
Open field test	open space	heifers, sheep	Boissy and Bouissou, 1995; Deiss *et al.*, 2009
Human exposure test	human passive presence, acceptance of being approached, handled and/or stroked	bulls, sheep, pigs, calves, goats	Bourguet *et al.*, 2015; Deiss *et al.*, 2009; Lawrence *et al.*, 1991; Lensink *et al.*, 2000; Lyons *et al.*, 1988; Terlouw and Porcher, 2005; Terlouw and Rybarczyk, 2008
Restraint in a crush	lack of freedom of movement	heifers	Boissy and Bouissou, 1995
Social isolation	absence of familiar conspecifics	heifers, sheep	Boissy and Bouissou, 1995; Bourguet *et al.*, 2010; Deiss *et al.*, 2009
Resident-intruder	territorial aggression/submission	pigs, rainbow trout	Olsson *et al.*, 1999; Øverli *et al.*, 2002
Group tests			
Food competition test	aggression, competition	pigs	Lawrence *et al.*, 1991; Terlouw and Rybarczyk, 2008
Human: passive presence or active approach	fear due to humans	sheep	Le Neindre *et al.*, 1993
Sorting out of cattle[3]	human presence and interventions, restraint, social separation	bulls, cows	Fordyce *et al.*, 1988a
Mixing of unfamiliar animals	aggression, competition	pigs	Terlouw and Rybarczyk, 2008; Terlouw *et al.*, 2005, 2009

[1] In individual tests, if a companion animal is not visible (behind a grid or fence, for instance), the reaction to social isolation influences the response.
[2] If the test is carried out in an unfamiliar environment, the reaction to unfamiliarity influences the response.
[3] Often part of normal stock procedures, using a crush, races and alleys.

Table 1.2. Behaviours indicative of high fear response in different tests (Boissy and Bouissou, 1988, 1995).

Test	High fear response
Novel object test	long latency to approach, long latency to exploration
Surprise test	strong startle, long latency to return to the surprise inducing object
Open field test	long latency to enter, high levels of standing immobile, high levels of head in an upright position
Human exposure test	moving away from human, low levels of immobility
Restraint	struggle, vocalisation
Social isolation	vocalisation, attempts to return to the familiar conspecifics
Walking through a corridor	refusal to walk, or in contrast, running

better to be driven through a corridor and were less reactive to the sudden approach of a human. In a novel object test, these same gilts touched and looked at the object less often than their more reactive counterparts. Furthermore, in a food competition test, these gilts won a smaller proportion of aggressive interactions, but had a greater feeding success (Lawrence *et al.*, 1991; Table 1.1). It is not clear whether the different profiles of these gilts were simply related to differences in fearfulness, or more generally in reactivity levels, irrespective of the stress factor. In sheep, certain individuals showed relatively high levels of locomotion, vigilance and bleating in several tests, including an open field test, social isolation, and exposure to human presence, compared to their less reactive group members (Deiss *et al.*, 2009; Table 1.1). Similar results were found in goats, fowl and fish (Castanheira *et al.*, 2017; Jones *et al.*, 1994; Lyons *et al.*, 1988). For instance, fish that were more reactive to confinement were also more active in the presence of an intruder and did not feed in isolation, in contrast to their less reactive counterparts (Øverli *et al.*, 2002).

Stress reactivity depends at least partly on genetic background and rearing conditions. In a surprise test comparing young bulls of different breeds, each bull was isolated in a corridor and an umbrella was presented. The umbrella was first closed, and subsequently suddenly opened. Strictly speaking, in addition to surprise, this test combines social isolation, and novelty due to the unfamiliar test environment and the unfamiliarity of the umbrella (Table 1.1). Compared to Angus, Blond d'Aquitaine bulls looked less at the closed umbrella and remained at a greater distance from the opened umbrella. Throughout the test they expressed more often startle responses and had faster heat rates. Other tests also showed breed differences. During a 'human exposure' test, the Blond d'Aquitaine bulls expressed more vigilance. These results indicate that the Blond d'Aquitaine bulls were more reactive to various stressful situations, probably due to greater fearfulness (Bourguet *et al.*, 2015). Other species also show breed differences. During a human exposure test, Duroc pigs approached and touched the human significantly more often than Large Whites. This breed difference was specific for the motivation to touch humans, because these same pigs did not differ behaviourally or physiologically during a novel object test (Terlouw and Rybarczyk, 2008; Table 1.1). Romanov sheep eliminated more, ate less and avoided a human more than Merinos sheep (Le Neindre *et al.*, 1993). Concerning rearing conditions, pigs and calves that had positive contacts with a human or the farmer during rearing, approached more easily other humans and were easier to handle, respectively (Lensink *et al.*, 2000; Terlouw *et al.*, 2005). Free-range born pigs were less aggressive during resident-intruder tests and established more quickly a clearly defined dominance/subordinate relationship. Possibly, during conflicts, the pigs reared in a confined space had not learned to withdraw and showed less avoidance (Barton-Gade, 2008; Olsson *et al.*, 1999; Terlouw *et al.*, 2009). These pigs also differed in non-social tests, such as the novel object test (Olsson *et al.*, 1999; Table 1.1).

1.2.2 Behavioural stress responses in the slaughter context

In abattoirs, fear is a common cause of stress and may be expressed by the refusal to go forward, walking backwards, immobilisation or standing still despite interventions by operators. Cattle may press other animals against walls or gates while walking backwards and may turn their head if the size of the single-file corridor does not allow turning of the body (Bourguet *et al.*, 2011b; Table 1.3). Fear may also cause escape attempts, resulting in running, slipping and/or falling (Table

Table 1.3. Examples of animal based observations in an abattoir and possible causes in terms of stress for cattle (Bourguet *et al.*, 2011b; Grandin, 2019).

Animal based observation	Cause
Retreat, attempted retreat (animals turn their heads backwards), refusal to move forward, refusal to enter certain places including the stunning area	fear of what animal perceive or expect to encounter if moving forward
Slips, falls[1]	fear, pain[1]
Vigilance (immobile, poll of the head above the shoulder, ears immobile or moving back and forward, fixed gaze directed towards a point in the environment)	perception of a potential threat or stressful event, possibly causing fear
Compressed animal (the animal has a curved back because it lacks space due to the close presence of other animals and/or space limiting objects, such as barriers or walls)	fear, pain
Tremor (rhythmic shaking of the body)	fear, pain, cold
Accelerated and/or deep breathing	fear, pain, heat stress
Vocalisation[2] of an isolated animal	fear of social isolation
Vocalisation[2] during driving or other interventions	pain due to use of the goad or due to too strong compression of the animal in the stunning box
Vocalisation[2] during lairage	at least in calves[3]: fear due to an unfamiliar environment and/or conspecifics
Kick	fear or pain if observed during driving, pain if observed during restraint in the stunning box

[1] Fear and pain can be the cause, but can also be the result of slipping and especially falling.
[2] Vocalisation of non-isolated cattle is considered a social expression, and not related to stress (Grandin, 2001).
[3] Similar interpretations can be made for pigs (Friel *et al.*, 2019).

1.3). Other frequent causes of stress in the preslaughter context are social separation or isolation, which may cause vocalisation, particularly in cattle and sheep (Deiss *et al.*, 2009; Grandin, 2001; Table 1.3). The responses to physical causes of stress, such as thirst, hunger and fatigue include attempts to replenish the physical needs, such as trying to get access to the water bowl, or to lie down (Jarvis *et al.*, 1996a,b; Knowles *et al.*, 1999). Generally, there is no searching for food, as food is often not available in abattoirs (Table 1.4). Known expressions of hunger are increased activity and competition, difficult handling, and vocalisation (Bourguet *et al.*, 2011a; Brown *et al.*, 1999; Dalla Costa *et al.*, 2016; Thomas *et al.*, 2001; Vieira *et al.*, 2008). In fish, low oxygen levels or high unionised ammonia levels provoked escape reactions (Danley *et al.*, 2005; Pickering *et al.*, 1991).

During the pre-slaughter and slaughter period various factors may cause pain. According to the IASP, pain is: 'An unpleasant sensory and emotional experience associated with, or resembling that associated with, actual or potential tissue damage.' The IASP adds: 'Verbal description is only one of several behaviours to express pain; inability to communicate does not negate the possibility that a human or a non-human animal experiences pain' (Raja *et al.*, 2020; Box 1.5). Examples of potentially painful events include falling, slipping, adverse interactions between animals, such as being bitten, being head-butted, or being mounted, hitting structures, pen fittings, gates and other equipment, human interventions to drive animals forwards and unsuccessful stuns (Jarvis *et al.*, 1996b). Acute painful stimuli may cause avoidance reactions, aggression, agitation, and, at least in pigs and cattle, vocalisation (Cordeiro *et al.*, 2018; Grandin, 2001; Marx *et al.*, 2003; Prunier *et al.*, 2013). It causes kicking in cattle (Bourguet *et al.*, 2011b; Table 1.3). In birds, painful stimulation may induce passive immobility or active avoidance behaviour, such as jumping, and occasionally, calling (Gentle, 1992). Wing flapping in birds is a sign of pain and discomfort and may be caused at shackling by poor handling and compression of the birds' hock due to tight fitting shackles, or as a result of pre-stun shocks when birds enter the water bath stunner (Gerritzen *et al.*, 2022 – Chapter 6 of this book; Gregory and Bell, 1987; Sparrey and Kettlewell, 1994; Terlouw *et al.*, 2008). Long-lasting shackling may also be very painful for the birds, causing wing flapping and sometimes broken bones (Gentle and Tilston, 2000; Gerritzen *et al.*, 2022 – Chapter 6 of this book; Gregory and Wilkins, 1990). Depending on the body site concerned, pain may also cause abnormal postures, abnormal or avoidance of walking (Prunier *et al.*, 2013). Certain reactions, such as lack of activity and prostration, are not overt and more difficult to identify (Prunier *et al.*, 2013). Bruising is an indirect indicator of pain.

Certain stress-inducing characteristics of the abattoir can also be identified by evaluating the environment itself. The environmental context may cause lack of physical comfort or fear (Table 1.4). Detailed observations involving watching the animals, while taking into consideration the context, need careful analysis to understand the causes of the behaviour, to recognise the problem, to identify the cause, and find solutions.

As indicated above, vocalisations may be simply social exchanges, or reactions to pain or hunger, depending on the situation and the species and age of the animal (Grandin, 2001; Marx *et al.*, 2003; Thomas *et al.*, 2001; Table 1.3). It is important to realise that the visual, olfactory and auditory capacities of animals are different from those of humans. For example, the eyes of farm animals adapt more slowly to changes in light conditions and light contrasts may be interpreted as obstacles (Willson *et al.*, 2021; Box 1.6).

Box 1.6. Vision in farm animals.

Vision relies on a highly complex organised system. The retina is a sophisticated image processor, containing various cell types that form microcircuits that have different functions, encoding different aspects of the visual scene. The signals processed by these different retinal circuits are relayed to specific brain regions (Zhang *et al.*, 2017). Form, orientation, colour, motion, and depth are processed in discrete areas of the cerebral cortex. The number of functionally discrete areas of the visual cortex varies between species.

Mammals and birds have two types of photoreceptors in the retina, rods and cones; they contain photosensitive pigments. Rods are found predominantly in the periphery of the retina, whereas the cones are located mostly in the centre. In mammals, there are many more rods than cones in the retina. The cones allow vision of fine details (called visual acuity) and colour discrimination, but do not function under low light conditions. The rods cannot distinguish colours and have poor resolution, but they mediate vision under low light conditions. At intermediate levels of illumination, rods and cones function simultaneously (Jonas *et al.*, 1992).

As they have different characteristics, the study of amounts and distribution of rods and cones delivers important information on visual capacities. In the human retina, depending on the location, there are 30,000 to 150,000 rods/mm^2, and 6,000 to 180,000 cones/mm^2. As the cones are confined to a limited area, overall, humans have 20 times

Table 1.4. Examples of environment-based observations, possible negative consequences for animal stress and proposed solutions (Bourguet *et al.*, 2011b; Grandin, 2019).

Environment-based observations	Risks in terms of stress	Possible solutions
During unloading at slaughter		
Outdoor unloading facilities	Fear of novel environment and odours	Integration of the unloading area as much as possible into the lairage area, with a roof and high solid walls
Absence of equipment for injured animals after transport	Pain	Provision of pens easily and quickly accessible from the unloading area
During lairage		
Mixing of unfamiliar animals	Risk of aggressive behaviour, resulting in pain, fear and fatigue	Use of individual stalls or maintain only familiar conspecifics in groups
Large groups of animals	Risk of aggressive behaviour, resulting in pain, fear and fatigue	Decrease group size with more but smaller pens
Use of individual stalls with no front gate	Need to make animals to walk backwards: resulting in fear and discomfort	Installation of front gates
Absence of food[1]	Hunger, increased reactivity to fear-inducing factors	Improved planning to reduce lairage durations, or provision of food
Absence of functional drinkers available and accessible for each animal[1]	Thirst, risk of increased reactivity to stress factors	Regular checks and maintenance of drinkers, avoidance of excessive densities of the animals in pens. Adjusted height of the drinkers according to the species, the age of the animals and the type of drinkers
Lack of space	Animals unable to lie down: discomfort, fatigue, and thirst due to low accessibility of drinkers	Written indications of the maximum number of animals allowed in each pen and compliance with the indication
Inappropriate temperatures	Cold or heat stress, discomfort	Cold: install windbreakers, improve thermic isolation of the stalls or area, add litter or rubber mats Heat: use humidifiers, add openings to ventilate the stalls
Bad air quality, presence of ammonia	Discomfort	Improve the ventilation system and/or add openings to ventilate the stalls, increase the frequency and quality of cleaning of the stalls
Noise	Fear and discomfort	Limit noise pollution: Installation of rubber stoppers on the doors and sound insulation on the hydraulic systems, lubrication of cylinders
During handling		
Light contrasts: (1) bright light (e.g. when the truck is opened on a sunny day), (2) darkness (e.g. when the animal has to enter a dark abattoir on a sunny day) or (3) shadows on the ground (Willson *et al.*, 2021)	The animal is (1) dazzled, (2) perceives only darkness or (3) perceives shadows as obstacles which cause fear when forced to move forward	Allowance of the necessary time to the animals, so they can adapt to the light conditions if they change (indoors/outdoors). Sharp shadows should be avoided. Installation of artificial lights where needed. Use of warm colour light (above 3,000 K)

>>>

Table 1.4. Continued.

Environment-based observations	Risks in terms of stress	Possible solutions
Irregular or insufficient light indoors	Incapacity of the animal to see ahead which will cause fear reactions	Light should be sufficient and homogeneous throughout the abattoir; for example, neon lights above and in the same direction of the animal corridors
Light reflections	Dazzling effect: fear	The operator may place her/himself at the place of the animals (eyes at the same height) to see if water or metallic objects are reflecting light (sun, artificial lights). Remove sources, paint reflecting objects, or block view
Races equipped with openwork barriers or solid walls but too low (Figure 1A, B and C).	The animal can see objects or humans outside the race, which may cause fear	Installation of solid sides or other means that prevent animals from seeing out
Individual corridors	Fear due to social isolation, risk of injury	Limitation of individual sections of corridors as much as possible in order to allow walking of the animals in shoulder-to-shoulder or flank contact
Straight corridor and/or sharp angles	Fear due to a lack of exit, the animal perceiving a dead end	Avoiding use of straight corridors and sharp angles, but rather curved corridors
Excessive downward slopes	Fear of slipping/falling	Use of gentle (<20°) slopes with non-slipping floor
Slippery floor	Risk of slipping and falling: fear, pain	Regular cleaning of the floors. Reparation of floors to make them non-slip
Floor with heterogeneous or irregular aspects (Figure 1D and E)	The animal interprets irregularities as obstacles	Reparation of floors to give them a regular aspect
Unstable or moving floor, or changes in floor level (Figure 1B)	Fear of falling	Provision of facilities with solid and stable floors
Air blast	Suddenness inducing startle and fear responses including backward movements	Move the air exhausts of the pneumatic systems out of areas where animals move (in the attic for example)
Inappropriate use of the electric goad (continuously on the same animal, on an animal that cannot move forward or that has collapsed, or on sensitive parts of the body (head, genitals, etc.)[2]	Pain, fear	Identification of the reason for the excessive use and correct the cause; improve equipment and conception, organisation of the abattoir, remind operators of the rules, and/or offer appropriate training
Metal doors, ungreased cylinders, uninsulated hydraulic systems	Fear and discomfort	Limit noise pollution: Installation of rubber stoppers on the doors and sound insulation on the hydraulic systems, lubrication of cylinders

[1] Council Regulation (EC) No 1099/2009 stipulates that operators take the necessary measures to ensure that animals:
(a) are provided with physical comfort and protection, in particular by being kept clean in adequate thermal conditions and prevented from falling or slipping;
(b) are protected from injury;
(c) are handled and housed taking into consideration their normal behaviour;
(d) do not show signs of avoidable pain or fear or exhibit abnormal behaviour;
(e) do not suffer from prolonged withdrawal of feed or water;
(f) are prevented from avoidable interaction with other animals that could harm their welfare.

[2] Council Regulation (EC) No 1099/2009 stipulates that the use of instruments, which administer electric shocks, shall be avoided as far as possible. In any case, such instruments shall only be used for adult bovine animals and adult pigs which refuse to move, and only when they have room ahead of them in which to move. The shocks shall last no longer than one second, be adequately spaced and shall only be applied to the muscles of the hindquarters. Shocks shall not be used repeatedly if the animal fails to respond.

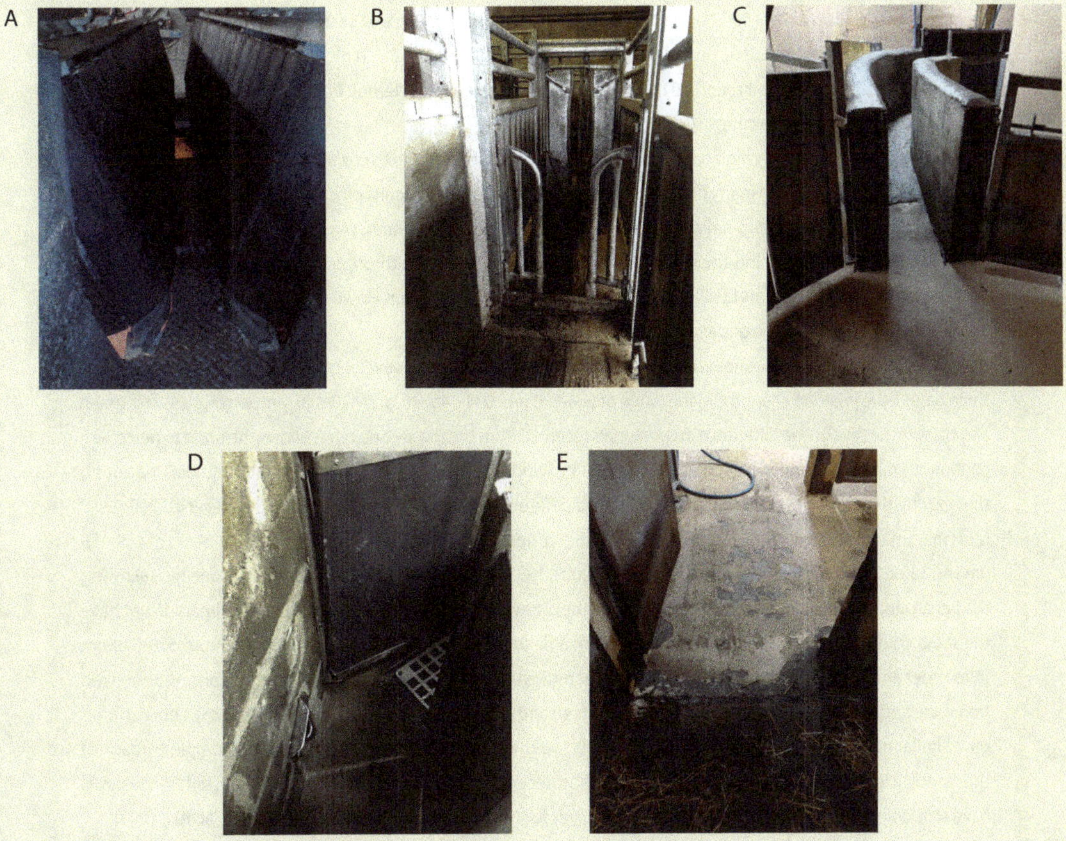

igure 1.1. (A) Entrance of a pig restrainer. The negative points are that the area is too dark; the environment outside the restrainer visible causing too many distractions for the animal; the false-floor is lacking in the restrainer, while it is known that the absence f a floor causes fear; and metal floors are inappropriate because they may vibrate and be noisy. (B) Cattle identification crush. he negative points are that the environment may be difficult to interpret for the animal because there are too many objects nd shapes; the walls are too low allowing vision of the outside environment distracting the animals; there is insufficient light, ith dark areas; the floor is non-homogeneous and there is a step at the entrance of the crush which may present obstacles for e animal; and the one-way backstop is little efficient. (C) Sheep entrance of a single file race towards the slaughter area. The egative points are that the reduction in the size of the entrance of the race is too abrupt; the environment outside of the race too light, possibly distracting the animal; and the aspect of the floor changes which the animal may perceive as an obstacle.) Entrance of a pig restraining box. The negative points are that the corner is too sharp; the aspect of the floor changes which e animal may perceive as an obstacle; metal floors are inappropriate because they may vibrate and be noisy; and the concrete oor does not have a non-slip surface. (E) Sheep entrance to the slaughter area. The negative points are the complexity of the nvironment, which may be difficult to interpret for the animal; the presence of the water hose and the changing aspect of the oor, which may present obstacles for the animal. (Photo courtesy of C. Bourguet, Bureau ETRE, France)

more rods than cones (Jonas *et al.*, 1992). These values differ according to species. In pigs, highest rod and cone densities were 153,000 and 39,000 /mm^2, respectively, and in sheep, 270,000 and 28,000/mm^2, respectively (Chandler *et al.*, 1999; Shinozaki *et al.*, 2010). Rod to cone ratios ranged from 3:1 to 16:1 in pigs and from 6:1 to 20:1 in sheep (Chandler *et al.*, 1999; Shinozaki *et al.*, 2010). In birds, the ratios are very different. Common birds of various species had 55,000 rods and 175,000 cones/mm^2 (McNeil *et al.*, 2005).

Human acuity is exceptionally high compared with most animal species (Caves *et al.*, 2018). Visual acuity depends on the number of cones, but also many other factors, such as by how retinal circuits process the information (Caves *et al.*, 2018). Furthermore, many mammalian species (with the exception of primates, pigs, and rodents) have a tapetum, a reflective layer located behind the retina. By reflecting visible light back through the retina, it increases the light available to the photoreceptors facilitating vision under low light conditions. However, the tapetum may cause blur under daylight conditions (Maggs *et al.*, 2008).

Humans have three types of cones which allow the detection of different wavelengths, that is, colours, of light. The three types have peak wavelengths in the range of 564-580 nm, 534-545 nm, and 420-440 nm, perceived as red, green and blue, respectively. The difference in the signals received from the three cone types allows the brain to perceive a continuous range of colours. The context dependency of colour perception is shown by laboratory studies indicating that exactly the same wavelength may be perceived differently depending on its surround (Purves *et al.*, 2004).

Most mammals have only two types of cones. Pigs, sheep, goats and cows have cones with spectral peaks of 439-455 nm and 552-556 nm (Jacobs *et al.*, 1998; Neitz and Jacobs, 1989). Birds have four types of cones, with one type allowing the perception of very short wave lengths, towards the ultraviolet. Overall, this means that farm animals do not have the same colour perception as humans. Sheep were able to distinguish colours of wavelengths from 520-640 nm (Alexander and Stevens, 1979). A study on colour discrimination in cattle found that 5 out 8 heifers distinguished red from blue and only three distinguished blue from green and green from red (Gilbert and Arave, 1986). Other studies gave similar results (Dabrowska *et al.*, 1981; Riol *et al.*, 1989). However, such studies do not give a complete image of the complex interactions between outputs of the photoreceptors (Jacobs, 1992, 1993). In addition, different colours have also different brightness that may be used as cues for the animals, rather than wavelength (Jacobs, 1993).

The eye needs to adapt to changes in light conditions. The adjustment of the aperture of the pupil is a fast response to adjust the amount of light that reaches the retina. However, its contribution to the overall light adaptation response is very small. An important mechanism is the intra-cellular response of the rods to incoming photons, which is adjusted to the light conditions of the environment. Furthermore, the retina contains a vast network of horizontal cells of which the activity depends on the brightness of the surrounding illumination. These horizontal cells are thought to influence the responsiveness of the photoreceptors to incoming photons (Miller and Tredici, 1992).

The above mechanisms play an important role during light adaptation: when suddenly the environment becomes much brighter, the eye has to adapt quickly to increased background illumination and be capable to discern objects in this brighter background. In humans, light adaptation occurs within seconds (Miller and Tredici, 1992). Dark adaptation, where the eye needs to adjust to very low light following bright light conditions, is in contrast a much slower process. Although the above cellular and neural adjustments occur within seconds, they are not sufficient. For complete dark adaptation, rods and cones need to regenerate their photosensitive pigments. In humans, the first phase of dark adaptation, based on the regeneration of pigments of cones, takes 5 to 8 min. The recovery of rod sensitivity needs much more time: about 40-50 min for completion (Reuter, 2011).

The pupillary light reflex, that is, the rapid constriction of the pupil in response to bright light, takes less than a second in humans, but several seconds in horses, cattle and sheep (Hachol *et al.*, 2006; Maggs *et al.*, 2008). In addition, as most non-human animals have a lower cone/rod ration than humans, they have also a less efficient dark adaptation (Wouters *et al.*, 1980). Little is known of the reactivity of retinal intracellular and neural mechanisms involved in the adaptation to surrounding illumination in animals. However, behavioural responses of animals to changing light conditions suggest that the retinal responses are relatively slow (Grandin, 2019; Willson *et al.*, 2021).

The organisation of the eye is adapted to the habitat and characteristics of the species. For example, herbivores have horizontally elongated pupils. The pupils remain more or less horizontally orientated through a rotation movement of the eyeball, including when the animals are grazing (Banks *et al.*, 2015). In sheep and probably other herbivores, the horizontal pupils, the lateral placement of the eyes, and the specific distribution of cones, rods and the tapetum, allow the creation of sharp images of horizontal contours ahead and behind, creating a horizontally panoramic view that facilitates detection of predators from various directions, including when grazing at low light conditions, and forward locomotion across uneven terrain (Banks *et al.*, 2015; Piggins and Phillips, 1996; Shinozaki *et al.*, 2010).

Summarising, 'seeing' is a perception and not a faithful rendition of the external world in all its factual detail. The visual system takes great liberties with the information it receives from the external environment and uses a number of shortcuts to process the image rapidly into information relevant to the species so that can be used to guide behaviours (Gilger, 2010). Hence, what is perceived by a normal, human eye does not necessarily reflect what is perceived by the eye of the various farm animal species.

Similarly, hissing or high frequency sounds may seem unimportant or simply remain undetected by humans (Heffner and Heffner, 1992; Phillips, 2010; Box 1.7), whereas they may induce fear-responses in animals (Grandin, 2019).

Box 1.7. Sound perception in farm animals.

Cattle perceive sounds up to around 35 kHz compared to about 20 kHz for humans (Adamczyk *et al.*, 2015; Heffner and Heffner, 1992). The human ear is most sensitive to sounds ranging between 500 to 4 kHz (Weeks *et al.*, 2009). The low frequencies hearing limit of cattle is similar to that of humans (about 25-30 Hz; Adamczyk *et al.*, 2015; Phillips, 2010). Thresholds for discomfort for cattle were noted at 70-85 dB, with physical damage to the ear occurring at 110 dB (Adamczyk *et al.*, 2015; Phillips, 2010; Weeks *et al.*, 2009). The auditory range of sheep is 125 to 40 kHz and for pigs between 42 and 40 kHz (Adamczyk *et al.*, 2015; Brouček, 2014; Weeks, 2008). Pigs show stress responses to sounds above 70 dB (Goumon and Faucitano, 2017). Loud (around 100 dB) and sudden noise induce stress responses in animals during the slaughter period. For example, the shot of the captive bolt device used to stun cattle produces sounds of 110 dB and is perceived by the other animals in the slaughter corridor resulting in stress (Gregory *et al.*, 2007).

1.2.3 Physiological stress responses that can be measured in the preslaughter and slaughter context

1.2.3.1 Metabolic indicators

The preslaughter period is a physically and psychologically demanding situation, which needs energy expenditure to allow increased physical action and vigilance. In addition, food deprivation is generally practised to avoid full intestines at slaughter because during carcass dressing, it reduces the risk to open accidentally the filled digestive tract, which would lead to microbial contamination of the carcass. In pigs, food deprivation reduces further travel sickness, during transport, expressed as sniffing, foaming and chomping and retching and vomiting (Bradshaw *et al.*, 1996; Faucitano and Raj, 2022 – Chapter 5 of this book; Rioja-Lang *et al.*, 2019). The food deprivation, physical activity and psychological stress all influence the metabolism of the animals.

The two major energy substrates of the body are glucose and free fatty acids (FFA). Glucose is a monosaccharide; it is stored as a multi-branched polysaccharide, called glycogen, in skeletal muscle and the liver. FFAs are elongated hydrocarbon chains with a terminal carboxylate group and are stored in body fat. Glucose and FFA need to be degraded via specific pathways to yield energy. The first step of glucose breakdown is called glycolysis and does not require oxygen. It yields pyruvate, which, if sufficient oxygen is present, enters the mitochondrion for a subsequent process called oxidative phospohorylation (Robergs *et al.*, 2004). Both glycolysis and oxidative phospohorylation yield adenosine triphosphate (ATP), an energy carrying organic compound necessary for processes that would not occur automatically such as metabolic reactions, transport of substances across membranes, nerve impulse propagation and muscle contraction. Their efficiency is very different: glycolysis yields 2 ATP per glucose, much less than the oxidative phosphorylation that yields about 30 ATP (Robergs *et al.*, 2004).

FFA are broken down via β-oxidation and oxidative phospohorylation in the mitochondria to yield ATP. The amount of ATP the FFA breakdown yields depends on the length of the chain of the fatty acid, but FFA are very rich in energy. For instance, palmitic acid, a common FFA in the body, contains 16 carbon atoms, and yields 106 ATP (Darvey, 1998).

The availability of glucose and FFA in the plasma and in organs are tuned together. Under basal circumstances, glucose is the primary energy source for skeletal muscle. Food deprivation diminishes plasma glucose levels and glycogen contents in the liver and muscle (Knowles *et al.*, 1999). When glucose levels are relatively low or when needs are high, plasma FFA levels, and their uptake and oxidation by the muscles increase (Romijn *et al.*, 1995). The elevation of plasma FFA levels further stimulates glycogenolysis (breaking down of glycogen into glucose) and hepatic gluconeogenesis (formation of glucose from lactate, pyruvate, glycerol and amino acids), thus enhancing availability of glucose (Lam *et al.*, 2003; Staehr *et al.*, 2003). The maintenance of glucose levels is of major importance for the functioning of the brain, which uses essentially glucose as energy source, contrary to other organs (Kuo *et al.*, 2015). Ruminants are capable of forming β-hydroxybutyrate from volatile fatty acids that are produced by rumen fermentation and use it as alternative source of energy.

Due to the increased effort, plasma glucose and FFA levels change during the preslaughter period (Table 1.5). For example, during transport, plasma levels of FFA and glucose increased (Jarvis *et al.*, 1996a; Knowles *et al.*, 1999; McVeigh and Tarrant, 1982; Warriss *et al.*, 1995). Following long transports, a decrease in glucose may also be observed (Navarro *et al.*, 2019). In this case, the consumption of glucose outweighed the capacity of replenishment, probably due to depletion of liver and muscle glycogen reserves. In agreement, durations of more than 14 h of transport and food deprivation led to lower glycogen contents in the muscle (Knowles *et al.*, 1999). Following transport by road or sea, food deprived cattle had higher plasma β-hydroxybutyrate levels (Table 1.5; Navarro *et al.*, 2019; Warriss *et al.*, 1995). The increase in β-hydroxybutyrate may reflect increased needs,

able 1.5. Indicators related to physical condition in meat animals.

Variable	Cause
;lucose (plasma) ree fatty acids (plasma) .actate (whole blood, plasma, muscles, liver)	physical effort, food deprivation, stress
Jrea (plasma) otal proteins (plasma) \lbumin (plasma) Haematocrit/packed cell volume (plasma) 3-hydroxybutyrate (plasma)	food/water deprivation
Creatine kinase (plasma, serum, muscle)	physical effort
3-endorphins (plasma)	stress
3-endorphins (pain matrix)	pain

associated or not with reduced glucose availability. In keeping with this, the increased levels of β-hydroxybutyrate were accompanied by increased levels of glucose and FFA in one study (Warriss *et al.*, 1995) and by a decrease in glucose levels in the other (Navarro *et al.*, 2019).

The increased production of ATP in the skeletal muscle is necessary to allow muscle contraction. This involves binding of myosin heads to strings of actin, which is an energy-needing process. The energy is delivered by the hydrolysation at the myosin head of ATP into adenosine-di-phosphate (ADP). Creatine kinase (CK) plays an essential role in maintaining ATP levels by catalysing the transfer from creatine phosphate to the ADP to yield ATP. Plasma CK levels increase following vigorous or long-lasting physical effort, due to damage of or a transient increase in the permeability of the muscle fibre membranes (Baird *et al.*, 2012; Ferraz *et al.*, 2010; Warriss *et al.*, 1995). The time of CK release into and clearance from plasma depends on the level of training (or daily exercise), type, intensity and duration of exercise, age and gender (Brancaccio *et al.*, 2007). For example, post-exercise plasma CK levels were higher in men than women (Sewright *et al.*, 2008). Peak plasma CK values depend on the type of exercise. Eight hours after strength training (such as lifting light or moderate weights), they are about two-fold above baseline (Brancaccio *et al.*, 2007). Eccentric exercise (a slow muscle elongation exercise) causes a pronounced increase only between 2 and 7 days after exercise, due to the muscle damage associated with the exercise (Brancaccio *et al.*, 2007). After prolonged exercise, total serum CK activity is markedly elevated for 24 h after the exercise bout when subjects rest and remain elevated for longer in the absence of rest (Brancaccio *et al.*, 2007).

Blood CK levels were increased following transport in pigs and cattle (Table 1.5; Knowles *et al.*, 1999; Śmiecińska *et al.*, 2011; Warriss *et al.*, 1995) and following thermal stress in poultry (Debut *et al.*, 2005; Mitchell and Sandercock, 1995; Mitchell *et al.*, 1992). Compared to cattle arriving directly from the farm, plasma CK activity in blood collected at exsanguination was significantly greater in cattle from markets, probably caused by additional handling and transport (Jarvis *et al.*, 1996a). Pigs transported in a top rear compartment of the lorry or for a longer duration (12 or 18 h) had higher levels of serum CK than those transported in a top front compartment or for a shorter duration (6 h; Sommavilla *et al.*, 2017). Plasma CK levels were higher in exsanguination blood in abattoirs considered particularly stressful for pigs and after transport and slaughter described as having a low welfare index (Sardi *et al.*, 2020; Warriss *et al.*, 1994). In certain experiments, serum CK levels following transport were higher in winter than in summer (Correa *et al.*, 2013, 2014). Other experiments found the opposite (Sommavilla *et al.*, 2017). Lower increases during the warmer season may be explained by lower activity levels due to high ambient temperatures (Nybo *et al.*, 2013).

Lactate is a product of glycolysis when levels of oxygen in the cells are low. Under normal conditions, in humans, lactate is produced at a rate of about 15-30 mmol/kg/day. Blood lactate levels are maintained between 0.5 and 1.0 mmol/l, which reflects the balance between lactate production and consumption (Allen and Holm, 2008). In the exercising muscle, oxygen levels are generally insufficient, and consequently, plasma lactate levels may rise to levels above 20 mmol/l. The effort-related rise is only transitory because the body has the capacity for rapid lactate disposal, which at least in humans can rise to 500 mmol/h. In the resting state, the liver and kidneys account for disposal of around 60 and 25-30% of circulating lactate by converting lactate into glucose (Cohen and Woods, 1983). During exercise at higher ambient temperatures, lactate production may be greater (Claremont *et al.*, 1975; Morrissey *et al.*, 2019), but opposite effects have also been reported (Marino *et al.*, 2001).

The above indicates that plasma lactate levels depend on many factors, such as the oxidative capacity of the muscles, the level of daily exercise of the animal, the ambient temperature and the capacity of the liver and kidneys to convert lactate into glucose (Thomas *et al.*, 2004). However, overall, during transport and slaughter, plasma lactate levels rise due the physical activity associated with this period (Brandt and Aaslyng, 2015; Table 1.5). Greater levels of lactate were observed in pigs loaded on steep ramps or using electrical prods, and following longer durations of transport, compared to controls (Faucitano and Lambooij, 2019). Pigs showing more often rearing, walking backwards and being more often involved in jams while attempting to enter a space had higher lactate levels (Edwards *et al.*, 2010). Seasonal effects were also observed and seemed to be associated with seasonal effects on activity levels. Specifically, one study found greater (Correa *et al.*, 2013), and another smaller lactate increases during transport and lairage in summer compared to winter (Correa *et al.*, 2014). These differences reflected higher and lower activity levels in summer compared to winter, respectively (Correa *et al.*, 2013, 2014).

Plasma lactate levels and body temperature both rise during exercise (Claremont *et al.*, 1975). Accordingly, ocular temperature, measured by infrared thermography and being an indicator of core temperature, was positively correlated, albeit weakly, with plasma lactate levels (Weschenfelder *et al.*, 2013). However, in transported pigs, negative correlations were found between plasma lactate levels on the one, and temperature in the gastrointestinal tract on the other hand (Sommavilla *et al.*, 2017). The reason for this negative association remains to be explained.

1.2.3.2 Body fluids

Body fluids are often considered according to their fluid compartment. The intracellular fluid compartment refers to all fluids enclosed in cells by their plasma membranes. It represents about two thirds of all body fluids. Extracellular fluid represents one third and is constituted of the fluid component of the blood and the interstitial fluid that surrounds all cells. Dehydration causes increases in blood osmolality, in plasma levels of urea, a waste product of protein breakdown, in total proteins, in albumin, one of the most abundant proteins in the blood and involved in carrying certain hormones, and in packed cell volume (PCV), a measurement of the percentage of blood that is made up of cells (Jarvis *et al.*, 1996a; Knowles *et al.*, 1999; Navarro *et al.*, 2019; Warriss *et al.*, 1995). Normal reference PCV values are 24-46, 36-43 and 27-45% for cows, pigs and sheep, respectively (Latimer, 2011; Weiss and Wardrop, 2010). Haematocrits (red blood cells) account for nearly all the cells in the blood. Under stressful conditions, such as during transport and restraint in the box prior to religious bleeding, the spleen may release haematocrits into the blood stream, which may be another cause of an increase in PCV (Bourguet *et al.*, 2011b; Liotta *et al.*, 2007; Warriss *et al.*, 1995). It is useful to combine several indicators, physiological and behavioural, to evaluate a possible state of dehydration. Only plasma total protein concentration in blood collected at exsanguination was significantly greater in cattle arrived from markets than in those from farms (Jarvis *et al.*, 1996a). Cattle from distant markets had higher PCV and plasma total protein concentration than those from markets at shorter distances, indicating additional effects of transport duration on dehydration. The cattle from markets spent more time drinking during the initial 3 h in the lairage pen, in agreement with their state of dehydration (Jarvis *et al.*, 1996a). In pigs, plasma albumin levels were greater after unloading than before loading (Averos *et al.*, 2007). The clinical status of the animal must be taken into account as various diseases may also influence haematological indicators (Abutarbush, 2015).

The combination of food and water deprivation may result in decreased water in extracellular space, different body compartments and in Na excretion in urine in sheep and cattle, indicative of an imbalance of the normal water and electrolyte distribution (Cole, 2000; Hogan *et al.*, 2007).

1.2.3.3 Nervous system and hormones

The changes in metabolism described above are associated with other physiological responses that allow an increase in availability of energy substrates, efficient gas exchanges and increased blood circulation to the organs that are most needed for the increased activity and vigilance needed in demanding situations. These responses involve changes in neurological and endocrine systems. They may be induced by psychological stress, without any overt behavioural reaction, but also by physical activity, without any psychological stress. The combined physiological responses help sustain the increased physical effort and vigilance needed in stressful situations. Certain responses occur very fast, other changes occur progressively over the time of the duration of the stressor or of the increased activity. Generally, the physiological responses are more or less proportional to the stress level of the animal, until the system reaches a certain ceiling threshold.

The autonomic nervous system is a neurological system involved in vital functions, consisting of a sympathetic and a parasympathetic branch, which have opposite effects. Their balance has a direct influence on the activity of many organs, the sympathetic branch stimulating physically active responses, while the parasympathetic stimulates body functions associated with resting and digesting. During psychological stress and physical activity, the sympathetic/parasympathetic balance shifts towards increased sympathetic activity. It is a fast responding system. Increased sympathetic activity causes the release of catecholamines (i.e. adrenaline, noradrenaline and dopamine) from the medulla of the adrenal glands. However, most of the noradrenaline and dopamine present in the blood stream has diffused from the noradrenergic nerve endings of the sympathetic nervous system into the blood (Goldstein and Holmes, 2008). The role of dopamine in acute stress responses is not well understood. Increased activity of the sympathetic nerves to the heart and increased adrenaline levels accelerates heart rate and increases stroke volume, while blood pressure rises due to vasoconstriction, providing better perfusion of the vital organs and muscles. Sympathetic activity stimulates the dilation of bronchioles, hence facilitating gas exchanges in the lungs and decreases the variability between inter-beat intervals of the heart, resulting in lower heart rate variability (Shaffer and Ginsberg, 2017). Low heart rate variability can therefore be used as an indicator of acute stress (Després *et al.*, 2003). Increased sympathetic activity further relaxes accommodation and dilates the pupils, letting more light into the eyes, and constricts the blood vessels of the skin, which limits bleeding from wounds and makes more blood available for organs. It also stimulates structures in the brain stem, particularly the reticular formation, responsible for vigilance. The shift towards the stronger sympathetic and lower parasympathetic input continues as long as the threat is perceived. Removal of the stress is associated with a relatively rapid restoration of the pre-stress balance.

Stress causes further the activation of the hypothalamic-pituitary-adrenal axis (HPA). This axis is controlled by neurons in the hypothalamus of the brain, secreting the corticotropin releasing hormone. This hormone stimulates the secretion of adrenocorticotropic hormone (ACTH) into the bloodstream from the anterior lobe of the pituitary. The ACTH subsequently reaches the adrenal cortex via the blood stream where it stimulates the secretion of glucocorticoids (i.e. cortisol in most mammals and corticosterone in rodents, birds, reptiles and amphibians). The HPA is a relatively slowly responding system, often reaching (context-dependent) maximal values after 30 to 60 min after onset of the stress (Herman *et al.*, 2016). Plasma cortisol has a half-life of 66 min at normal hormone levels (Weitzman *et al.*, 1971). Consequently, increased cortisol may still be observed after removal of the stress. When pre-stress levels are restored it is considered that the animals have completely recovered from the stressful event (Knowles *et al.*, 1999).

Catecholamines and glucocorticoids act in concert. The presence of glucocorticoids is necessary to allow the effect of catecholamines on the vasoconstriction of blood vessels (Barrett *et al.*, 2016). Increased circulating glucocorticoid and catecholamine concentrations contribute to mobilise energy substrates, glucose and FFA, in the blood. More specifically, adrenaline stimulates lipolysis and glycogenolysis in the liver and muscle (Qvisth *et al.*, 2006; Watt *et al.*, 2001). Hence, during stress, the combined increase in cortisol and adrenaline make lipids available from adipose tissue and glucose through glycogenolysis for increased muscle activity, while energy storage is inhibited. Glucocorticoids block further insulin-mediated glucose uptake by the muscles reserving part of the blood glucose for brain functioning (Bingham *et al.*, 2002; Kuo *et al.*, 2015).

Given their prominent roles in adaptive responses, the autonomic nervous system and HPA axis are strongly influenced by pre-slaughter procedures. Birds being caught and carried upside down (Duncan *et al.*, 1986; Kannan and Mench, 1996) and adult cattle, calves, sheep and pigs after transport, loading and unloading showed increased heart rates and glucocorticoids (Baldock and

Sibly, 1990; Bradshaw *et al.*, 1996; Kenny and Tarrant, 1987a,b; Knowles *et al.*, 1999; Śmiecińska *et al.*, 2011; Van de Water *et al.*, 2003). Increased plasma levels of adrenaline and glucocorticoids have been reported in fighting pigs (Fernandez *et al.*, 1994; Terlouw *et al.*, 2005) and in fish, under conditions of crowding, handling, struggling or hypoxia (Barton, 2002; Mazeaud *et al.*, 1977; Ramsay *et al.*, 2006). Following a long (14 to 31 h) transport, cattle that lay down in lairage had higher plasma glucocorticoids levels than those that remained standing (Knowles *et al.*, 1999), indicating a relationship with level of fatigue. Low welfare levels of transport and slaughter in pigs were associated with higher cortisol and adrenaline levels (Sardi *et al.*, 2020). Introduction into the stunning box or the restrainer in cattle and pigs were associated with faster heart rates (Chevillon, 2000; Knowles, 1999; Reiche *et al.*, 2019).

The measurement of physiological indicators of pain is relatively complex. Pain induces physiological changes that are similar to non-painful responses to stress (Prunier *et al.*, 2013). The pain perceived can be modulated by β-endorphins in humans or in animals (Box 1.8). Some reports mention the possibility that β-endorphins are secreted following bleeding without prior stunning (EFSA, 2004). Today, there are little arguments to support this. Although a study in sheep found an increase in the level of β-endorphins in the blood following a slow haemorrhage, compared to a control group, this increase was gradual until reaching a maximum value only after 45 min (Smith *et al.*, 1986). Furthermore, a later cattle study reported that blood endorphin levels did not increase after bleeding without stunning (Zulkifli *et al.*, 2014). Overall, there is much doubt on a role of circulating endorphins in acute pain perception (Box 1.8).

Box 1.8. The role of endorphins in pain.

In humans and animals, the secretion of endogenous opioid peptides, called β-endorphins, can reduce the pain experienced. Endorphins act on opiate receptors and have analgesic action at all levels of the pain system: at the level of the wound, the spinal cord and the brain (Luan *et al.*, 2017).

The role of β-endorphins in the central nervous system in pain perception has been clearly established (Millan *et al.*, 1987). The concentration of β-endorphins in the brain increases rapidly after noxious stimulation, which demonstrates its role as a regulatory mechanism (Young *et al.*, 1993). Micro-injections of morphine (a synthetic opioid) into certain areas of the pain matrix decrease, and of agents blocking the effects of β-endorphins increase, pain-related behaviours in rats (Basbaum and Fields, 1984).

The role of β-endorphins present in the blood in pain perception is less clear. The pituitary gland secretes β-endorphins into the bloodstream in response to stress or pain. In production animals, stressful or painful procedures, such as shearing, electro-immobilisation or tail-docking in sheep, provoked β-endorphins secretion in blood (Jephcott *et al.*, 1987; Shutt *et al.*, 1987). However, these increases in plasma β-endorphins are believed to be a response to stress rather than acute pain (Bach *et al.*, 1987). Circulating endorphins do not seem to reach the central nervous system and the level of pain perceived is generally not correlated with plasma β-endorphin levels (Basbaum and Fields, 1984; Johansen *et al.*, 2003). Basal circulating β-endorphin levels may, however, be involved in chronic pain perception (Bruehl *et al.*, 2012; Falcone *et al.*, 1988).

Finally, the immune system produces β-endorphins to reduce inflammatory (non-acute) pain, resulting from healing processes. These β-endorphins are produced and secreted by immune cells locally present in inflamed tissues, for example in wounds, and act on primary nociceptors (Stein, 2013). The endorphin receptors are produced by the nociceptor itself, at the level of the dorsal root, and transported to the wound site (Mousa *et al.*, 2001).

The secretion of endorphins in certain brain structures are in contrast associated with pain perception (Box 1.8). However, in induced haemorrhages in sheep, the concentration of β-endorphins in the cerebrospinal fluid was not significantly different from that of non-haemorrhaged sheep (Smith *et al.*, 1986). The possibility of an increase in the brain or the increase of other molecules of this opioid family, such as enkephalin or dynorphin, remains to be studied.

1.2.3.4 Sampling

Saliva and blood samples may be collected during the pre-slaughter period or just after stunning to evaluate blood constituents (Table 1.5 and 1.6). Blood samples may be taken directly from the jugular vein (venepuncture) in the case of cattle and sheep. For pigs, blood can be sampled from the ear vein or using an indwelling jugular catheter. In pigs and sheep, bleeding takes place immediately after stunning leaving no time allowance for sampling, but blood can be sampled from the bleeding wound. This is less precise as the sample obtained is a mixture of arterial and venous blood, which differ in levels of plasma constituents. In addition, the stunning procedure, particularly electrical stunning may influence hormone levels, such as catecholamines (Shaw and Tume, 1990). After slaughter, urine may be sampled from the bladder to estimate catecholamine (adrenaline, noradrenaline and dopamine) and glucocorticoids (cortisol, corticosterone) levels (Bourguet *et al.*, 2015). Plasma hormone levels reflect the physiological status on the moment of sampling, while because of their slower diffusion, the hormones levels in the saliva and urine reflect the physiological status of the animal over a longer period (Elfering *et al.*, 2003; Jung *et al.*, 2014).

1.3 Consequences of stress reactions on carcass and meat quality

1.3.1 Carcass quality

Certain animal stress indicators can be collected at the levels of the carcass. Food and/or water deprivation obviously result in weight loss (Knowles, 1999; Nicol and Scott, 1990). After slaughter, the carcass can be inspected for different types of skin damage, caused by human-animal, animal-animal and animal-facility interactions (Strappini *et al.*, 2013). Bruises result from damaged blood vessels due to a blow on the skin, and scratches are surface wounds. They may be caused by the use of sticks, collisions with equipment or other animals, or by a vertical sliding door closed while an animal is underneath, striking its back or shoulders (Bourguet *et al.*, 2011a; Sanchez-Hidalgo *et al.*, 2020; Strappini *et al.*, 2013). Lacerations and tears penetrate the skin and are caused by sharp objects, such as hoofs and teeth of conspecifics, or sharp pen fittings. The part of the carcass may also be distinguished, such as the head and shoulders, ears, middle part and hind part. Certain events cause more specifically lesions on specific parts of the body. For example, fighting pig direct their bites most often, but not exclusively, to the front parts (head, should, ears) and vertical sliding doors impact the back and shoulders of cattle (Bottacini *et al.*, 2018; Strappini *et al.*, 2013). Pigs mounting other pigs scratch with their forelegs the back of the mounted pig (Faucitano, 2001). During the

able 1.6. Examples of physiological responses to stress and meat quality aspects that are influenced by stress, relative to the reslaughter context (Section 1.2.3, 1.3.1 and 1.3.2).

Variable indicative of stress	Interpretation in terms of effects of stress
Ante-mortem	
Fast heart rate Low heart rate variability	Immediate response to stress of psychological origin, and/or to physical effort
High adrenaline and noradrenaline levels (plasma, saliva, urine)	The substances are secreted in the blood in response to stress, and progressively enter saliva and urine
High cortisol levels (plasma, saliva, urine)	Progressive response to stress of psychological origin, and/or to physical effort. The substances are secreted in the blood, and progressively enter saliva and urine
At bleeding	
High plasma adrenaline and noradrenaline levels	Immediate response to stress of psychological origin, and/or to physical effort
High plasma cortisol levels	Progressive response to stress of psychological origin, and/or to physical effort
Post-mortem	
High urinary adrenaline and noradrenaline levels[1] High urinary cortisol levels[1]	Indicative of ante-mortem stress of psychological origin, and/or to physical effort
Low early post-mortem pH[1] Light meat colour[2] High drip loss[3]	Indicative of stress just before stunning/bleeding
High pHu[2] Dark meat colour[2] Low drip loss[3] Low muscle glycogen content[4]	Indicative of stress during the pre-slaughter period
Carcass lesions	Aggressive behaviour, interventions (slaps, sticks) by humans, impacts from equipment
Decreased tenderness or juiciness	Stress during the preslaughter period. The exact mechanisms vary according to the moment of stress and the species. For pigs, the underlying mechanism is to a certain extent explained by rate and extent of pH decline. For cattle, the underlying mechanisms are not well known

[1] Generally measured between 30 and 60 min post-mortem.
[2] Generally measured around 24 h post-mortem.
[3] Generally measured over several days post-mortem.
[4] Glycogen is progressively degraded over the first 24 h post-mortem. The interpretation depends on the moment of sampling.

resolution of a bruise, generally, its colour changes from red or purple to blue, then to brown, yellow or green (Bohnert *et al.*, 2000; Northcutt *et al.*, 2000). Characteristics of tissue injured, mechanism of injury, length, duration of force, and depth of injury amongst other influence the speed of this process (Bohnert *et al.*, 2000; Northcutt *et al.*, 2000). Generally, bright red bruises have occurred more recently than darker coloured bruises (Faucitano, 2001), although in fowl, the changes may go in the opposite direction in certain muscles (Northcutt *et al.*, 2000). A precise timing using visual colour assessment is difficult, however. A detailed analysis of skin lesions caused by biting or mounting found that recent wounds (<7 h) were darker and redder than older wounds (>7 h) when assessing the colours spectrophotometrically, but visual assessment of the colours did not allow this differentiation accurately (Vitali *et al.*, 2017).

Skin damage may occur at varying moments of the slaughter period, during transport, lairage and in the stunning box, but may also already have been present before leaving the farm. If injures are present irrespectively of the transporter or farm, it may be assumed that they have occurred at the abattoir and animals need to be observed during their trajectory through the abattoir to identify the risk zones. In cattle, bruising was largely related to animal-facility and human-animal interactions (Strappini *et al.*, 2013). However, animal-animal interactions can lead to very severe conditions with fallen cattle being stamped on by other animals (Bourguet *et al.*, 2011b; Strappini *et al.*, 2013). In pigs, increased on-farm fasting time, mixing of unfamiliar groups, longer loading times, increased human interventions during loading and greater space allowance during transport were associated with increased amounts of scratches and lacerations (Correa *et al.*, 2013; Guàrdia *et al.*, 2009). Body weight or age, number of falls and slips due to poor handling or poor equipment and overall level of activity may also influence degree of skin damage in cattle and pigs (Correa *et al.*, 2010; Grandin, 1980; Mpakama *et al.*, 2014; Tarrant *et al.*, 1992). Electrical goads do not mark the skin, unless they are vigorously applied on the animal, but may indirectly lead to increased skin damage due to the increased speed of pigs, frequency of slips, and panic-like reactions that make them collide with walls and other objects (Correa *et al.*, 2010; McGlone *et al.*, 2004; Rabaste *et al.*, 2007; Strappini *et al.*, 2013). Similar observations were made for cattle (Jarvis *et al.*, 1995). In pigs, increased mounting and associated damage was caused by overcrowding (Faucitano, 2001). Seasonal effects are sometimes observed with higher degrees of skin damage during certain seasons, but the effects seem generally indirect, and related to other factors, such as different weights, handling practices, transport conditions or activity levels, which vary according to season (Correa *et al.*, 2013; Dalla Costa *et al.*, 2007). In cattle, interventions that potentially cause bruises often leave no visible bruise (Jarvis *et al.*, 1996; Strappini *et al.*, 2013). This indicates that the indicator is relatively insensitive and that negative events may have occurred without leaving marks on the skin.

In birds, the most frequent injuries seen are bruises, broken limbs and damaged wings (Nicol and Scott, 1990). Damaged wings are often caused by wing flapping, which may cause bruising, dislocation, broken wings, and haemorrhage resulting in red wing tips after slaughter (Gregory, 1994; Gregory *et al.*, 1989; Knowles and Broom, 1990; Lambooij *et al.*, 2010; Mitchell and De Boom, 1986; Rao *et al.*, 2013). Wing flapping and struggling may result from manual catching, lifting, holding, and carrying the birds inverted by their legs. Wing flapping may also occur after placing them in the transport crates or after shackling them on the slaughter line (Gerritzen *et al.*, 2022 – Chapter 6 of this book). Joint dislocations and haemorrhaging may result from carrying birds by one leg, amongst others, during loading (Cockram *et al.*, 2020; Nicol and Scott, 1990). Significantly more fractures occurred during pre-slaughter handling at the abattoirs than during catching/transportation (Kittelsen *et al.*, 2015). The authors stress that despite their lower prevalence, damage occurring before arrival at the abattoir are of particular importance, due to the longer duration of suffering (Kittelsen *et al.*, 2015). Factors that influenced the risk of wing injuries were the handlers, weight of the birds, number of birds per cage, loading speed, season and time of day (Cockram *et al.*, 2020). Care must be taken with interpretation of the data obtained on the carcass, because damage caused by stunning, whether electrical or by CO_2, is not a welfare issue as the animals are unconscious when it occurs, provided that the stunning process is carried out correctly. Thus, CO_2 stunning may cause wing fractures, due to struggling, wing flapping and muscular spasms experienced during the last phase of the stunning in unconscious birds (Abeyesinghe *et al.*, 2007; McKeegan *et al* 2007a,b). High current intensity stunning increases the incidence of red wing tips, bones fractures, dislocations, and haemorrhaging (Gregory and Wilkins, 1989). Similar effects may be observed in pigs where electrical stunning may induce bone fractures in the thoracic vertebrae and in the scapula where it joins the humerus (Channon *et al.*, 2003).

1.3.2 Meat quality

Pre-slaughter physiological and behavioural reactions of the animal influence ante- and post-mortem muscle energy metabolism with potentially very negative effects on the quality of meat (Bendall, 1973; Bourguet *et al.*, 2015; Hambrecht *et al.*, 2005; Rocha *et al.*, 2016; Terlouw and Rybarczyk, 2008). Following bleeding of the animal, biochemical reactions continue in the muscle cells for several hours. These reactions are necessary for the conversion of muscle into meat. The reactions are not identical to those occurring in the living animal. Since blood no longer circulates, glucose and oxygen are not delivered to the muscle. The reactions use therefore glycogen stored locally in the muscle as energy source. Due to the absence of oxygen, glycogen is catabolised anaerobically, causing a net production of hydrogen ions (H^+), amongst others. Due to the absence of blood circulation, these ions accumulate in the muscle resulting in a pH decline (Hambrecht *et al.*, 2005; Robergs *et al.*, 2004). This decline is initially fast, then slows to reach a stable value of around 5.5, reached approximately 24 h post-mortem, referred to as the ultimate pH (pHu). The rate and extent of the pH decline may have significant effects on the quality of the meat that is produced (Figure 1.2).

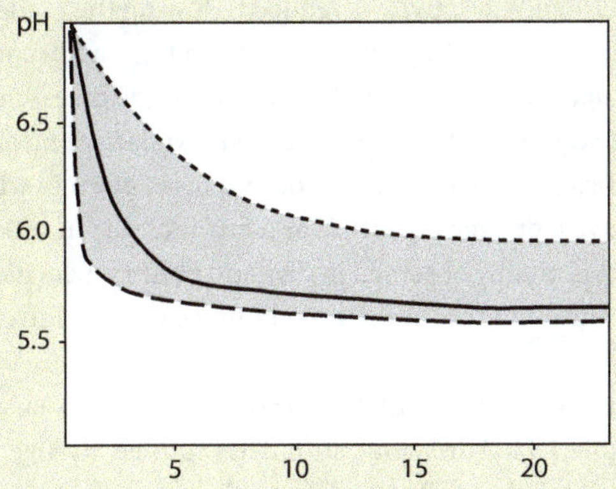

igure 1.2. The post-mortem pH decline in muscles is initially fast then slows to reach a stable value around 5.4-5.6. Preslaughter :ress may cause a fast pH decline (broken line) or a pH decline of insufficient amplitude (dotted line). The resulting meats are f lesser quality.

The effects of stress reactions on post-mortem muscle metabolism and, consequently, on meat quality, can be distinguished in two broad categories. As muscle glycogen stores fuel the post-mortem metabolic reactions resulting in acidification, low glycogen levels at the moment of slaughter may result in relatively high pHu values. Low muscle glycogen levels may occur if the day or hours before slaughter physical activity and psychological stress were high and have consumed glycogen stores from the muscles. For instance, food deprivation, long transports and physical interactions between mixed unfamiliar animals lead to depletion of muscle glycogen and increased pHu values (Knowles *et al.*, 1999; McVeigh and Tarrant, 1982; Warriss, 1990). At high pHu value, the physical state of the proteins is above their iso-electric point and associate with more water. The meat appears dark because the fibres are more tightly packed and its surface does not scatter light to the same extent as the more open surface of meat with a lower pHu value (Seideman *et al.*, 1984). Extreme cases characterised by a pHu value >5.9 are called dark cutting (DFD: dark, firm, dry) meat (Figure 1.2). Although high pHu value has positive effects on water-holding capacity and tenderness, flavour and shelf life are degraded at values above 6.1 (Bidner *et al.*, 2004). DFD meat can occur in all major meat species (Ponnampalam *et al.*, 2017). The effects are only visible if glycogen levels reach a certain, low threshold (55 μmol glucose equivalents/g fresh muscle tissue; Terlouw *et al.*, 2021). Basal glycogen levels are generally much higher, which means that only longer-lasting or very high stress levels will influence pHu values.

If on the other hand stress reactions take place in the minutes preceding slaughter, whole body and particularly muscle metabolic activity is high at the moment of death. This high metabolic activity continues after death of the animal, causing a faster than normal pH decline during the early post-mortem period (Figure 1.2). Studies found that stress or exercise immediately before slaughter, due to the use of an electric prod (D'Souza *et al.*, 1998), intense physical activity (Rosenvold and Andersen, 2003), or fighting (Karlsson and Lundström, 1992) resulted in higher muscle temperature and lactic acid content and faster pH value decline. As the carcass remains relatively warm due to increased heat production, the denaturation of muscle proteins is accelerated (Aalhus *et al.*, 1998; Bendall, 1973; Hambrecht *et al.*, 2005). This leads to light coloured meats with low water-holding capacity (high drip loss). Proteolytic activity is reduced, slowing the tenderisation process (Barbut *et al.*, 2008). During cooking, the denatured proteins precipitate and become less soluble adding to the toughness (Hegarty *et al.*, 1963; Minelli *et al.*, 1995). Extreme cases, characterised by a pH <6.0 at 45 min post-mortem are designated as PSE (pale, soft, exudative) meat (Bendall and Swatland, 1988). A fast early post-mortem pH value decline occurs more often in white muscle fibres as they are fast contracting and have a greater potential for a fast increase in catabolism compared to red fibres. Therefore, this defect is mainly observed in pigs and poultry, who have muscles relatively rich in white fibres, but it exists also in other species, including cattle (Aalhus *et al.*, 1998).

Particularly the combination of physical activity and adrenaline (Christensen and Galbo, 1983) may influence meat quality (Table 1.7). Adrenaline stimulates muscle glycogen breakdown specifically in the exercising muscle, while it has little effect when the muscle is at rest (Febbraio *et al.*, 1998). In abattoirs, the muscles of the animals are mostly not at rest as driving the animals through the alleys and races of the abattoir are inherent to the slaughter procedure. If animals are, in addition, psychologically stressed (e.g. avoidance and escape responses), they may be even more active. The psychological stress and physical activation cause increases in adrenaline levels. If this occurs just

Table 1.7. Examples of correlations among behavioural and physiological responses, during the slaughter period or in reactivity tests conducted during rearing, and meat quality.

Variable 1	Variable 2	r (p) values	Animal type	Reference
Vocalisations in response to use of the prod in the abattoir handling area	Lactate (exsanguination blood)	0.23 (0.006)	Slaughter pigs	Edwards *et al.*, 2010
Occurrences of jamming of pigs during driving in the abattoir		0.24 (0.005)		
Creatine kinase (CK) (exsanguination blood)	Adrenaline (urine, post-mortem)	0.28 (<0.05)	Slaughter pigs	Foury *et al.*, 2011
CK (exsanguination blood)	pH 24 h post-mortem (*Semimembranosus* muscle)	0.17 (<0.05)		
Adrenaline (urine, post-mortem)		0.44 (<0.001)		
Noradrenaline (urine, post-mortem)		0.31 (<0.05)		
Ocular infrared temperature	Lactate (blood sample in restrainer)	0.20 (0.001)	Slaughter pigs	Weschenfelder *et al.*, 2013
	pH 1 h post-mortem (*Semimembranosus* muscle)	-0.20 (0.02)		
Lactate (exsanguination blood)	pH 1 h post-mortem (*Semimembranosus* muscle)	-0.37 (<0.001)	Slaughter pigs	Rocha *et al.*, 2015
Lactate (blood sample after lairage)	pH 24 h post-mortem (*Semimembranosus* muscle)	0.46 (<0.001)		
Gastrointestinal tract temperature from loading to lairage	CK (exsanguination blood)	0.41 (<0.001)	Slaughter pigs	Sommavilla *et al.*, 2017
Cortisol (exsanguination blood)	CK (exsanguination blood)	0.45 (<0.001)		
CK (exsanguination blood)	pH 24 h post-mortem (*Longissimus* muscle)	0.41 (<0.001)		
Fighting during pre-slaughter mixing	pH 24 h post-mortem (*Adductor* muscle)	0.59 (0.01)	Large White and Duroc slaughter pigs	Terlouw and Rybarczyk, 2008
	pH 24 h post-mortem (*Longissimus* muscle)	0.33 (<0.05)	Large White pigs	Terlouw *et al.*, 2005
Heart rate before slaughter	pH 40 min post-mortem (*Longissimus* muscle)	0.72 (<0.02)	Duroc pigs slaughtered with low stress	Terlouw and Rybarczyk, 2008
		0.69 (<0.02)	Large White pigs slaughtered with low stress	
Adrenaline (urine, post-mortem)	Meat lightness (*Longissimus* muscle)	-0.65 (<0.01)	Nguni cattle	Muchenje *et al.*, 2009
Noradrenaline (urine, post-mortem)		-0.52 (<0.05)		
Dopamine (urine, post-mortem)		0.53 (<0.05)		
		-0.52 (<0.05)	Bonsmara cattle	
Behavioural score after transportation to the slaughterhouse (calm vs restless)	Meat toughness (*Longissimus* muscle; Warner-Bratzler Shear Force)	0.34 (<0.05)	Crossbred steers	Gruber *et al.*, 2010
Plasma lactate (exsanguination blood)		0.26 (0.002)		

>>>

Table 1.7. Continued.

Variable 1	Variable 2	r (p) values	Animal type	Reference
Noradrenaline (urine, post-mortem)	pH 1 h post-mortem (*Longissimus* muscle)	0.57 (<0.001)	Young bulls	Reiche *et al.*, 2019
Heart rate during the first minutes of transport	Time needed to be driven into the abattoir	0.84 (0.003)	Normand cows (supplementary stress slaughter conditions)	Bourguet *et al.*, 2010
Time needed to be driven into the abattoir	pH 3 h post-mortem (*Semitendinosus* muscle)	-0.60 (0.03)		
Heart rate at arrival at the abattoir		-0.64 (0.04)	Normand cows (low stress slaughter conditions)	
Heart rate at arrival at the abattoir	pH 40 min post-mortem (*Semitendinosus* muscle)	-0.58 (<0.001)	Young bulls (Angus, Limousin and Blond d'Aquitaine)	Bourguet *et al.*, 2015
Correlations between reactions during reactivity tests conducted during rearing and meat quality				
N° of contacts with a novel object	pH 24 h post-mortem (*Adductor* muscle)	0.83 (0.001)	Large White and Duroc pigs	Terlouw and Rybarczyk, 2008
N° of contacts with an unfamiliar person	Temperature, 5 min post-mortem (*Longissimus* muscle)	-0.86 (0.003)	Large White pigs	
Plasma adrenaline levels during restraint	Meat toughness (*Longissimus* muscle; Warner-Bratzler Shear Force)	0.22 (0.006)	Crossbred steers	Gruber *et al.*, 2010
Heart rate response to the presence of the closed umbrella	Meat tenderness (*Longissimus* muscle; sensory analysis)	-0.59 (0.01)	Young bulls (Angus, Limousin and Blond d'Aquitaine)	Terlouw *et al.*, 2021

before slaughter, it may result in a fast early post-mortem pH value decline, while if it occurs during the hours preceding slaughter, it may result in high pHu value (Table 1.3). Correlations may also be found for noradrenaline as it is released in concert with adrenaline (Reiche *et al.*, 2019). For example, Foury *et al.* (2011) found that pigs that had been more active (higher plasma CK level at exsanguination) and/or had higher adrenaline or noradrenaline levels had lower preslaughter muscle glycogen content and higher meat pHu value (Table 1.7). Muchenje *et al.* (2009) found for cattle of the Nguni breed a negative correlation between adrenaline and noradrenaline levels and lightness of the meat (Table 1.7). In birds, increased physical activity and food deprivation were associated with lower hepatic and muscle glycogen levels (Warriss *et al.*, 1999). Pigs with higher lactate levels after lairage, indicative of greater activity and/or stress levels, had lower pHu value in the muscles of the thigh (Rocha *et al.*, 2015; Table 1.7). Muscles that are used for locomotion, during fighting, for mounting other animals or when being mounted, have a greater probability to be influenced by negative slaughter conditions compared to other muscles (Terlouw and Rybarczyk, 2008).

The effects of stress are not limited to the rate and extent of pH value decline and associated meat quality traits, such as colour and water holding capacity. They may also influence sensory traits. In cattle, the use of the electric prod, or back and forwards driving of the animals just before slaughter caused decreases in tenderness and juiciness (Reiche *et al.*, 2019; Warner *et al.*, 2007). The effects are proportional to the degree of stress as preslaughter behavioural, and physiological stress indicators and beef tenderness showed negative correlations (Gruber *et al.*, 2010; Table 1.7). The relationships between rate and extent of pH value decline and tenderness have been described for pigs, poultry and fish; in beef and lamb, they also exist, but are more complex (Dransfield, 1994; El Rammouz *et al.*, 2004; Huff-Lonergan *et al.*, 2002; Gagaoua *et al.*, 2015; Hwang and Thompson, 2001). It is also important to realise that meat quality is determined not only by metabolic, but also by many other, interdependent, processes (Ouali *et al.*, 2006). For instance, tenderness of the *Longissimus* muscle in cattle was influenced by interconnected pathways involved in muscle contraction, ATP metabolism, muscle structure, oxidative stress and chaperone-mediated protein folding (Gagaoua *et al.*, 2021). Preslaughter stress likely influences all these processes. For example, meat from cows slaughtered following stressful handling presented higher level of oxidation during storage (Delosière *et al.*, 2020).

As stress reactions to the pre- and slaughter context influence meat quality, stress reactivity is a characteristic of the animal that influences its potential for the production of meat and carcasses of high quality. For example, more reactive cattle had higher bruising scores (Fordyce *et al.*, 1988b; Gallo *et al.*, 2022 – Chapter 2 of this book). Pigs that fought more during a food competition test, fought also more when they were mixed with other, unfamiliar pigs during the slaughter period, and therefore had higher ultimate pH and darker meat. The effects on pHu and meat colour were proportional to the level of fighting (Terlouw and Rybarczyk, 2008; Terlouw *et al.*, 2005, 2021; Table 1.7). Normand cows that had faster heart rates when they were removed from their conspecifics during a test conducted three weeks before slaughter, had higher heart rates at the start of transport, where more difficult to drive into the abattoir, and had higher urinary catecholamine levels at slaughter. After slaughter, their *Semitendinosus* muscle, a muscle of the thigh, presented a faster pH value decline and higher temperature, indicative of a faster post-mortem metabolism, compared to their less reactive counterparts (Bourguet *et al.*, 2010; Figure 1.3; Table 1.7). When the closed umbrella was presented to the young bulls described above (Section 1.2.2), their heart rates increased (Bourguet *et al.*, 2015). Following slaughter 3 weeks later, bulls with a greater heart rate increase during presentation of the umbrella had a faster early post-mortem pH value decline in the *Longissimus* muscle (loin muscle). The meat of these bulls was further tougher during sensory

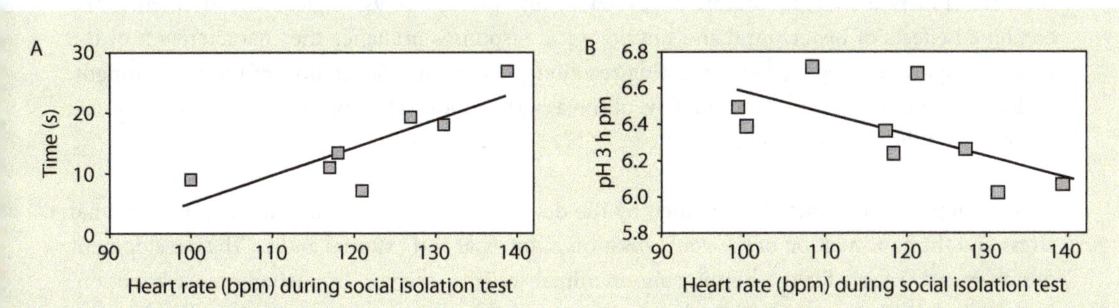

igure 1.3. Correlations between heart rate when the distance between the cow and her group members was increased and \) time need to introduce her into the abattoir (*r*=0.81; *P*=0.03) and (B) the pH of the *Semintendinosus* muscle 3 h post-mortem '=0.65; *P*=0.06) (Bourguet *et al.*, 2010).

testing (Terlouw *et al.*, 2021; Table 1.7). In the above experiments, the cows and bulls that were more reactive to unfamiliar situations during the tests were probably more stressed by the unfamiliar preslaughter situation, leading to faster ante- and post-mortem muscle metabolism. Other studies showed that a more excitable temperament had negative effects on weight gain during rearing and various meat quality traits, including pH and temperature values decline (Behrends *et al.*, 2009; Café *et al.*, 2011; Gruber *et al.*, 2010; King *et al.*, 2006; Table 1.7). Similar effects were observed in pigs (Terlouw and Rybarczyk, 2008; Table 1.7).

Finally, the influence of preslaughter stress on muscle traits vary according to muscle, breeds, rearing system, feeding regime, physical condition and species amongst others. The effects of stress on different muscles depend on the extent to which the muscles are used in the behavioural response, and their fibre composition, that is, the relative amounts of slow and fast fibres. Compared to slow fibres, fast fibres can be activated rapidly but have fewer mitochondria and therefore, if the demand is high, for a given amount of glucose less ATP is produced. In addition, slow and fast contracting fibres do not have the same glycogen contents. Differences in muscle traits are not necessarily caused by different stress levels. For instance, slow contracting muscle fibres are darker than fast contracting fibres. Certain species, such as pigs and poultry, have more fast contracting fibres and therefore lighter meat than other species, such as cattle and sheep. For the same reason, in pigs, the *Longissimus* muscle has a lighter colour than the *Adductor* muscle, a muscle of the thigh (Terlouw and Rybarczyk, 2008). Thus, the simple observation of lighter coloured meat of an animal cannot be interpreted as an indicator of high preslaughter stress, as other underlying causes are also possible.

1.4 Conclusions and outlook

The animal's state of welfare depends on its conscious, subjective experience of emotions. Today, much progress is needed to understand the mechanisms underlying conscious experiences, including in humans. Negative emotions may be indirectly measured, using behavioural and physiological approaches. Many measurements and indicators of stress are relevant for the preslaughter and slaughter period and they can be obtained before or after death of the animal. Physiological variables change due to stress factors of psychological or physical origin, but they are only partly specific. Behavioural indicators are more specific. They need to be combined with observations on the environmental context, and are necessary to interpret the physiological measurements. The combined effects of behavioural and physiological responses influence the characteristics of the meat. Analytical strategies, such as testing reactions to very specific features of the environment, including suddenness, or unfamiliarity, allow a better understanding of which aspects of the environment the animals react to.

Further improvements may be obtained by the development of remote measurements of animal stress and the continuation of the generalisation of internal and external audits. The development of non-invasive tools for the monitoring of animal welfare receives increasingly more attention. Such tools should aim to identify high risk situations. Many different events may be at the origin of high-risk situations; therefore, most efficient would be to monitor the resulting stress state of the animal. Video images can be analysed using algorithms to detect pain or changes in posture and certain behaviours as indicators of illness under experimental conditions or during rearing, but a standardised recording of such images under abattoir conditions seems still complex (Jorquera-Chavez *et al.*, 2019; Mahmoud *et al.*, 2018). Remotely measured increases in body temperature during slaughter may be a candidate as it is indicative of increased activity and/or stress, as well as a predictor of reduced meat quality (Jorquera-Chavez *et al.*, 2019; Rocha *et al.*, 2019; Weschenfelder *et al.*, 2013). Blood lactate content measured using a hand-held analyser gives similar information, although if used non-invasively on the exsanguination blood, the information is only obtained after the slaughter procedure (Rocha *et al.*, 2015). Increases in heart rate may also be measured remotely, but the animal has to be restrained to be fitted with the system (Jorquera-Chavez *et al.*, 2019). Such tools cannot replace the regular checks carried out by slaughterhouse personnel and veterinarians and should be considered as an additional support. The increasingly frequent use by abattoirs of internal and external audits is further very beneficial for the rapid identification of critical aspects of its organisation and processes and the immediate implementation of corrective actions.

Acknowledgements

The authors thank the abattoirs for allowing them taking pictures to present examples of concrete slaughterhouse situations.

References

Aalhus, J.L., Best, D.R., Murray, A.C. and Jones, S.D.M., 1998. A comparison of the quality characteristics of pale, soft and exudative beef and pork. Journal of Muscle Foods 9: 267-280.

Abeyesinghe, S.M., McKeegan, D.E.F., Mcleman, M.A., Lowe, J.C., Demmers, T.G.M., White, R.P., Kranen, R.W., Van Bemmel, H., Lankhaar, J.A.C. and Wathes, C.M., 2007. Controlled atmosphere stunning of broiler chickens. I. Effects on behaviour, physiology and meat quality in a pilot scale system at a processing plant. British Poultry Science 48: 406-423.

Abutarbush, S.M., 2015. Hematological and serum biochemical findings in clinical cases of cattle naturally infected with lumpy skin disease. Journal of Infection in Developing Countries 9: 283-288.

Adamczyk, K., Gorecka-Bruzda, A., Nowicki, J., Gumulka, M., Molik, E., Schwarz, T., Earley, B. and Klocek, C., 2015. Perception of environment in farm animals – a review. Annals of Animal Science 15: 565-589.

Adams, A. and Santi, A., 2011. Pigeons exhibit higher accuracy for chosen memory tests than for forced memory tests in duration matching-to-sample. Learning & Behavior 39: 1-11.

Alexander, G. and Stevens, D., 1979. Discrimination of colours and grey shades by Merino ewes: tests using coloured lambs. Applied Animal Ethology 5: 215-231.

Allen, S.E. and Holm, J.L., 2008. Lactate: physiology and clinical utility. Journal of Veterinary Emergency and Critical Care 18: 123-132.

Averos, X., Herranz, A., Sanchez, R., Comella, J.X. and Gosalvez, L.F., 2007. Serum stress parameters in pigs transported to slaughter under commercial conditions in different seasons. Veterinarni Medicina 52: 333-342.

Bach, F.W., Knigge, U. and Warberg, J., 1987. The inhibition of stress-induced beta-endorphin secretion by histamine-receptor antagonists. Anesthesiology 67: 605-605.

Baird, M.F., Graham, S.M., Baker, J.S. and Bickerstaff, G.F., 2012. Creatine-kinase and exercise-related muscle damage implications for muscle performance and recovery. Journal of Nutrition and Metabolism, Article ID 960363. https://doi.org/10.1155/2012/960363

Baldock, N.M. and Sibly, R.M., 1990. Effects of handling and transportation on the heart rate and behaviour of sheep. Applied Animal Behaviour Science 28: 15-39.

Banks, M.S., Sprague, W.W., Schmoll, J., Parnell, J.A. and Love, G.D., 2015. Why do animal eyes have pupils of different shapes? Science Advances 1: e1500391. https://doi.org/10.1126/sciadv.1500391

Barbut, S., Sosnicki, A.A., Lonergan, S.M., Knapp, T., Ciobanu, D.C., Gatcliffe, L J., Huff-Lonergan, E. and Wilson, E.W., 2008. Progress in reducing the pale, soft and exudative (PSE) problem in pork and poultry meat. Meat Science 79: 46-63.

Barrett, B., Barma, S.M., Boitano, S. and Brooks, H.L., 2016. Ganong's review of medical physiology. McGraw Hill Education, New York, NY, USA, 750 pp.

Barton, B.A., 2002. Stress in fishes: a diversity of responses with particular reference to changes in circulating corticosteroids. Integrative and Comparative Biology 42: 517-525.

Barton-Gade, P., 2008. Effect of rearing system and mixing at loading on transport and lairage behaviour and meat quality: comparison of outdoor and conventionally raised pigs. Animal 2: 902-911.

Basbaum, A.I. and Fields, H.L., 1984. Endogenous pain control-systems – brain-stem spinal pathways and endorphin circuitry. Annual Review of Neuroscience 7: 309-338.

Beauregard, M., Schwartz, G.E., Miller, L., Dossey, L., Moreira-Almeida, A., Schlitz, M., Sheldrake, R. and Tart, C., 2014. Manifesto for a post-materialist science. Explore: The Journal of Science and Healing 10: 272-274.

Behrends, S.M., Miller, R.K., Rouquette, F. M., Randel, R.D., Warrington, B.G., Forbes, T.D.A., Welsh, T.H., Lippke, H., Behrends, J. M., Carstens, G.E. and Holloway, J.W., 2009. Relationship of temperament, growth, carcass characteristics and tenderness in beef steers. Meat Science 81: 433-438.

Bendall, J.R., 1973. Postmortem changes in muscle. In: Bourne, G.H. (ed.) The structure and function of muscle. Academic Press, New York, NY, USA, pp. 244-309.

Bendall, J.R. and Swatland, H.J., 1988. A review of the relationships of pH with physical aspects of pork quality. Meat Science 24: 85-126.

Bidner, B.S., Ellis, M., Brewer, M.S., Campion, D., Wilson, E.R. and McKeith, F.K., 2004. Effect of ultimate pH on the quality characteristics of pork. Journal of Muscle Foods 15: 139-154.

Bingham, E.M., Hopkins, D., Smith, D., Pernet, A., Hallett, W., Reed, L., Marsden, P.K. and Amiel, S.A., 2002. The role of insulin in human brain glucose metabolism – an (18) fluoro-deoxyglucose positron emission tomography study. Diabetes 51: 3384-3390.

Birch, J., Schnell, A.K. and Clayton, N.S., 2020. Dimensions of animal consciousness. Trends in Cognitive Sciences 24: 789-801.

Bohnert, M., Baumgartner, R. and Pollak, S., 2000. Spectrophotometric evaluation of the colour of intra- and subcutaneous bruises. International Journal of Legal Medicine 113: 343-348.

Boissy, A., 1995. Fear and fearfulness in animals. The Quarterly Review of Biology 70: 165-191.

Boissy, A. and Bouissou, M.F., 1988. Effects of early handling on heifers' subsequent reactivity to humans and to unfamiliar situations. Applied Animal Behaviour Science 20: 259-273.

Boissy, A. and Bouissou, M.F., 1995. Assessment of individual differences in behavioural reactions of heifers exposed to various fear-eliciting situations. Applied Animal Behaviour Science 46: 17-31.

Boissy, A., Arnould, C., Chaillou, E., Désiré, L., Duvaux-Ponter, C., Greiveldinger, L., Leterrier, C., Richard, S., Roussel, S., Saint-Dizier, H., Meunier, S., M. C., Valance, D. and Veissier, I., 2007a. Emotions and cognition: a new approach to animal welfare. Animal Welfare 16 (Suppl. 1): 37-43.

Boissy, A., Manteuffel, G., Jensen, M. B., Moe, R. O., Spruijt, B., Keeling, L. J., Winckler, C., Forkman, B., Dimitrov, I., Langbein, J., Bakken, M., Veissier, I. and Aubert, A., 2007b. Assessment of positive emotions in animals to improve their welfare. Physiology & Behavior 92: 375-397.

Bottacini, M., Scollo, A., Edwards, S.A., Contiero, B., Veloci, M., Pace, V. and Gottardo, F., 2018. Skin lesion monitoring at slaughter on heavy pigs, (170 kg): welfare indicators and ham defects. PLoS One 13: e0207115.

Bourguet, C., Deiss, V., Boissy, A. and Terlouw, E.M.C., 2015. Young Blond d'Aquitaine, Angus and Limousin bulls differ in emotional reactivity: relationships with animal traits, stress reactions at slaughter and post-mortem muscle metabolism. Applied Animal Behaviour Science 164: 41-55.

Bourguet, C., Deiss, V., Boissy, A., Andanson, S. and Terlouw, E.M.C., 2011a. Effects of feed deprivation on behavioral reactivity and physiological status in Holstein cattle. Journal of Animal Science 89: 3272-3285.

Bourguet, C., Deiss, V., Gobert, M., Durand, D., Boissy, A. and Terlouw, E.M.C., 2010. Characterising the emotional reactivity of cows to understand and predict their stress reactions to the slaughter procedure. Applied Animal Behaviour Science 125: 9-21.

Bourguet, C., Deiss, V., Tannugi, C.C. and Terlouw, E.M.C., 2011b. Behavioural and physiological reactions of cattle in a commercial abattoir: relationships with organisational aspects of the abattoir and animal characteristics. Meat Science 88: 158-168.

Bradshaw, R.H., Parrott, R.F., Forsling, M.L., Goode, J.A., Lloyd, D.M., Rodway, R.G. and Broom, D.M., 1996. Stress and travel sickness in pigs: effects of road transport on plasma concentrations of cortisol, beta-endorphin and lysine vasopressin. Animal Science 63: 507-516.

Brancaccio, P., Maffulli, N. and Limongelli, F.M., 2007. Creatine kinase monitoring in sport medicine. British Medical Bulletin 81-82: 209-230.

Brandt, P. and Aaslyng, M.D., 2015. Welfare measurements of finishing pigs on the day of slaughter: a review. Meat Science 103: 13-23.

Broglio, C., Gomez, A., Duran, E., Ocana, F.M., Jimenez-Moya, F., Rodriguez, F. and Salas, C., 2005. Hallmarks of a common forebrain vertebrate plan: specialized pallial areas for spatial, temporal and emotional memory in actinopterygian fish. Brain Research Bulletin 66: 277-281.

Broom, D.M., 1988. The scientific assessment of animal welfare. Applied Animal Behaviour Science 20: 5-19.

Broom, D.M., 1991. Assessing welfare and suffering. Behavioural Processes 25: 117-123.

Brouček, J., 2014. Effect of noise on performance, stress, and behaviour of animals. Slovak Journal of Animal Science 47: 111-123.

Brown, S.N., Knowles, T.G., Edwards, J.E. and Warriss, P.D., 1999. Relationship between food deprivation before transport and aggression in pigs held in lairage before slaughter. Veterinary Record 145: 630-634.

Bruehl, S., Burns, J.W., Chung, O.Y. and Chont, M., 2012. What do plasma beta-endorphin levels reveal about endogenous opioid analgesic function? European Journal of Pain 16: 370-380.

Café, L.M., Robinson, D.L., Ferguson, D.M., McIntyre, B.L., Geesink, G.H. and Greenwood, P.L., 2011. Cattle temperament: persistence of assessments and associations with productivity, efficiency, carcass and meat quality traits. Journal of Animal Science 89: 1452-1465.

Castanheira, M.F., Conceicao, L.E.C., Millot, S., Rey, S., Begout, M.L., Damsgard, B., Kristiansen, T., Hoglund, E., Overli, O. and Martins, C.I.M., 2017. Coping styles in farmed fish: consequences for aquaculture. Reviews in Aquaculture 9: 23-41.

Caves, E.M., Brandley, N.C. and Johnsen, S., 2018. Visual acuity and the evolution of signals. Trends in Ecology & Evolution 33: 358-372.

Chalmers, D.J., 1996. Facing up to the problem of consciousness. Journal of Consciousness Studies 2: 200-219.

Chandler, M., Smith, P., Samuelson, D. and MacKay, E., 1999. Photoreceptor density of the domestic pig retina. Veterinary Ophthalmology 2: 179-184.

Chandroo, K.P., Duncan, I.J.H. and Moccia, R.D., 2004. Can fish suffer?: perspectives on sentience, pain, fear and stress. Applied Animal Behaviour Science 86: 225-250.

Channon, H.A., Payne, A.M. and Warner, R.D., 2003. Effect of stun duration and current level applied during head to back and head only electrical stunning of pigs on pork quality compared with pigs stunned with CO_2. Meat Science 65: 1325-1333.

Chen, S.W., Shemyakin, A. and Wiedenmayer, C.P., 2006. The role of the amygdala and olfaction in unconditioned fear in developing rats. The Journal of Neuroscience 26: 233-240.

Chevillon, P., 2000. Bien-être des porcs lors des opérations de pre-abattage et d'anesthésie. Proceedings of the 1st International Virtual Conference on Pork Quality, Concórdia, Brazil, p. 159-173.

Christensen, N.J. and Galbo, H., 1983. Sympathetic nervous activity during exercise. Annual Reviews of Physiology 45: 139-153.

Claremont, A.D., Nagle, F., Reddan, W.D. and Brooks, G.A., 1975. Comparison of metabolic, temperature, heart-rate and ventilatory responses to exercise at extreme ambient temperatures (0 degrees and 35 degrees C.). Medicine and Science in Sports and Exercise 7: 150-154.

Cockram, M.S., Dulal, K.J., Stryhn, H. and Revie, C.W., 2020. Rearing and handling injuries in broiler chickens and risk factors for wing injuries during loading. Canadian Journal of Animal Science 100: 402-410.

Cohen, R.D. and Woods, H.F., 1983. Lactic acidosis revisited. Diabetes 32: 181-191.

Cole, N.A., 2000. Changes in postprandial plasma and extracellular and ruminal fluid volumes in wethers fed or unfed for 72 hours. Journal of Animal Science 78: 216-223.

Cordeiro, A.F.D.S., Nääs, I.D.A., Baracho, M.D.S., Jacob, F.G. and De Moura, D.J., 2018. The use of vocalization signals to estimate the level of pain in piglets. Engenharia Agricola 38: 486-490.

Correa, J.A., Gonyou, H.W., Torrey, S., Widowski, T., Bergeron, R., Crowe, T.G., Laforest, J.P. and Faucitano, L., 2013. Welfare and carcass and meat quality of pigs being transported for 2 hours using two vehicle types during two seasons of the year. Canadian Journal of Animal Science 93: 43-55.

Correa, J.A., Gonyou, H.W., Torrey, S., Widowski, T., Bergeron, R., Crowe, T.G., Laforest, J.P. and Faucitano, L., 2014. Welfare of pigs being transported for long distance using a pot-belly trailer during winter and summer. Animals 4: 200-213.

Correa, J.A., Torrey, S., Devillers, N., Laforest, J.P., Gonyou, H.W. and Faucitano, L., 2010. Effects of different moving devices at loading on stress response and meat quality in pigs. Journal of Animal Science 88: 4086-4093.

Coulon, M., Nowak, R., Peyrat, J., Chandeze, H., Boissy, A. and Boivin, X., 2015. Do lambs perceive regular human stroking as pleasant? Behavior and heart rate variability analyses. Plos One 10. https://doi.org/10.1371/journal.pone.0118617

Dabrowska, B., Harmata, W., Lenkiewicz, Z., Schiffer, Z. and Wojtusiak, R.J., 1981. Colour perception in cows. Behavioural Processes 6: 1-10.

Dalla Costa, F.A., Devillers, N., Paranhos da Costa, M.J.R. and Faucitano, L., 2016. Effects of applying preslaughter feed withdrawal at the abattoir on behaviour, blood parameters and meat quality in pigs. Meat Science 119: 89-94.

Dalla Costa, O.A., Faucitano, L., Coldebella, A., Ludke, J.V., Peloso, J.V., Dalla Roza, D. and Paranhos da Costa, M.J.R., 2007. Effects of the season of the year, truck type and location on truck on skin bruises and meat quality in pigs. Livestock Science 107: 29-36.

Damasio, A.R., 1998. Emotion in the perspective of an integrated nervous system. Brain Research Reviews 26: 83-86.

Damasio, A.R., 2010. Self comes to mind: constructing the conscious brain. Pantheon Books, New York, NY, USA, 367 pp.

Danley, M.L., Kenney, P.B., Mazik, P.M., Kiser, R. and Hankins, J.A., 2005. Effects of carbon dioxide exposure on intensively cultured rainbow trout Oneorhynchus mykiss: physiological responses and fillet attributes. Journal of the World Aquaculture Society 36: 249-261.

Dantzer, R., 2002. Can farm animal welfare be understood without taking into account the issues of emotion and cognition? Journal of Animal Science 80: E1-9.

Danziger, N., 2006. Bases neurologiques de l'affect douloureux. Revue Neurologique 162: 395-399.

Darvey, I.G., 1998. How does the ratio of ATP yield from the complete oxidation of palmitic acid to that of glucose compare with the relative energy contents of fat and carbohydrate? Biochemical Education 26: 22-23.

Dawkins, M., 2015. Animal welfare and the paradox of animal consciousness. Advances in the Study of Behavior 47: 5-38.

Dawkins, M.S., 2003. Behaviour as a tool in the assessment of animal welfare. Zoology 106: 383-387.

Dawkins, M.S., 2008. The science of animal suffering. Ethology 114: 937-945.

De Waal, F.B.M., 2008. Putting the altruism back into altruism: the evolution of empathy. The Annual Review of Psychology 59: 279-300.

De Waal, F.B.M., Wright, R., Korsgaard, C.M., Kitcher, P. and Singer, P., 2006. Morally evolved primate social instincts, human morality, and the rise and fall of 'veneer theory'. In: Macedo, S. and Ober, J. (eds.) Primates and philosophers: how morality evolved. Princeton University Press, Princeton, NJ, USA, pp. 1-80.

Debut, M., Berri, C., Arnould, C., Guemene, D., Sante-Lhoutellier, V., Sellier, N., Baeza, E., Jehl, N., Jego, Y., Beaumont, C. and Le Bihan-Duval, E., 2005. Behavioural and physiological responses of three chicken breeds to pre-slaughter shackling and acute heat stress. British Poultry Science 46: 527-535.

Deiss, V., Temple, D., Ligout, S., Racine, C., Bouix, J., Terlouw, C. and Boissy, A., 2009. Can emotional reactivity predict stress responses at slaughter in sheep? Applied Animal Behaviour Science 119: 193-202.

Delosière, M., Thomas, A., Terlouw, C.E.M., Gruffat, D., Habeanu, M. and Durand, D., 2020. Dataset of lipids, antioxidative status and color attributes in cows meat from slaughter to storage: impacts of diet supplementations and pre-slaughter stress. Data In Brief 30: 105409. https://doi.org/10.1016/j.dib.2020.105409

Dennett, D.C., 1991. Consciousness explained. Back Bay Books Little, Brown and Company, New York, NY, USA, 492 pp.

Désiré, L., Boissy, A. and Veissier, I., 2002. Emotions in farm animals: a new approach to animal welfare in applied ethology. Behavioural Processes 60: 165-180.

Désiré, L., Veissier, I., Despres, G. and Boissy, A., 2004. On the way to assess emotions in animals: do lambs (*Ovis aries*) evaluate an event through its suddenness, novelty, or unpredictability? Journal of Comparative Psychology 118: 363-374.

Désiré, L., Veissier, I., Despres, G., Delval, E., Toporenko, G. and Boissy, A., 2006. Appraisal process in sheep (*Ovis aries*): interactive effect of suddenness and unfamiliarity on cardiac and behavioral responses. Journal of Comparative Psychology 120: 280-287.

Després, G., Boissy, A., Desire, L., LeNeindre, P. and Veissier, I., 2003. Validation of the measure of sympatho-vagal effect in lambs through autonomic blockades and heart rate variability indexes. Journal of Animal and Veterinary Advances 2: 615-619.

Destrez, A., Deiss, V., Lévy, F., Calandreau, L., Lee, C., Chaillou-Sagon, E. and Boissy, A., 2013. Chronic stress induces pessimistic-like judgment and learning deficits in sheep. Applied Animal Behaviour Science 148: 28-36.

Dransfield, E., 1994. Tenderness of meat, poultry and fish. In: Pearson, A.M. and Dutson, T.R. (eds.) Quality attributes and their measurement in meat. Poultry and fish products. Springer, Boston, MA, USA, pp. 289-315.

D'Souza, D.N., Warner, R.D., Dunshea, F.R. and Leury, B.J., 1998. Effect of on-farm and pre-slaughter handing of pigs on meat quality. Australian Journal of Agricultural Research 49: 1022-1025.

Duncan, I.J., 1996. Animal welfare defined in terms of feelings. Acta Agriculturae Scandinavica. Section A. Animal Science. Supplement 27: 29-35.

Duncan, I.J.H., Slee, G.S., Kettlewell, P., Berry, P. and Carlisle, A.J., 1986. Comparison of the stressfulness of harvesting broiler chickens by machine and by hand. British Poultry Science 27: 109-114.

Eccleston, C. and Crombez, G., 1999. Pain demands attention: a cognitive-affective model of the interruptive function of pain. Psychological Bulletin 125: 356-366.

Edwards, L.N., Grandin, T., Engle, T.E., Porter, S.P., Ritter, M.J., Sosnicki, A.A. and Anderson, D.B., 2010. Use of exsanguination blood lactate to assess the quality of pre-slaughter pig handling. Meat Science 86: 384-390.

EFSA (European Food Safety Authority), 2004. Opinion of the Scientific Panel on Animal Health and Welfare (AHAW) on a request from the Commission related to welfare aspects of the main systems of stunning and killing the main commercial species of animals. EFSA Journal 45: 1-29.

El Rammouz, R., Babilé, R. and Fernandez, X., 2004. Effect of ultimate pH on the physicochemical and biochemical characteristics of turkey breast muscle showing normal rate of post-mortem pH fall. Poultry Science 83: 1750-1757.

Elfering, A., Grebner, S., Semmer, N.K., Byland, C. and Gerber, H., 2003. Two urinary catecholamine measurement indices for applied stress research: effects of time and temperature until freezing. Human Factors 45: 563-574.

Elorette, C., Aguilar, B.L., Novak, V., Forcelli, P.A. and Malkova, L., 2020. Dysregulation of behavioral and autonomic responses to emotional and social stimuli following bidirectional pharmacological manipulation of the basolateral amygdala in macaques. Neuropharmacology 179: 108275. https://doi.org/10.1016/j.neuropharm.2020.108275

Emery, N.J., 2006. Cognitive ornithology: the evolution of avian intelligence. Philosophical Transactions of the Royal Society B 361: 23-43.

Emery, N.J. and Clayton, N.S., 2001. Effects of experience and social context on prospective caching strategies by scrub jays. Nature 414: 443-446.

European Commission (EC), 2009. Council Regulation (EC) No 1099/2009 of 24 September 2009 on the protection of animals at the time of killing. Official Journal of the European Union L 303: 1-30. Available at: http://data.europa.eu/eli/reg/2009/1099/oj.

Falcone, C., Specchia, G., Rondanelli, R., Guasti, L., Corsico, G., Codega, S. and Montemartini, C., 1988. Correlation between beta-endorphin plasma-levels and anginal symptoms in patients with coronary-artery disease. Journals of the American College of Cardiology 11: 719-723.

Faucitano, L., 2001. Causes of skin damage to pig carcasses. Canadian Journal of Animal Science 81: 39-45.

Faucitano, L. and Lambooij, E., 2019. Transport of pigs. In: Grandin, T. (ed.) Livestock handling and transport. CABI Publ., Wallingford, UK, pp. 307-327.

Febbraio, M.A., Lambert, D.L., Starkie, R.L., Proietto, J. and Hargreaves, M., 1998. Effect of epinephrine in trained men. Journal of Applied Physiology 84: 465-470.

Fernandez, X., Meunier-Salaun, M.-C. and Mormede, P., 1994. Agonistic behavior, plasma stress hormones, and metabolites in response to dyadic encounters in domestic pigs: Interrelationships and effect of dominance status. Physiology & Behavior 56: 841-847.

Ferraz, G.C., Soares, O.A.B., Foz, N.S.B., Pereira, M.C. and Queiroz-Neto, A., 2010. The workload and plasma ion concentration in a training match session of high-goal (elite) polo ponies. Equine Veterinary Journal 42: 191-195.

Finlayson, K., Lampe, J.F., Hintze, S., Wurbel, H. and Melotti, L., 2016. Facial indicators of positive emotions in rats. Plos One 11: 0166446. https://doi.org/10.1371/journal.pone.0166446

Foote, A.L. and Crystal, J.D., 2007. Metacognition in the rat. Current Biology 17: 551-555.

Foote, A.L. and Crystal, J.D., 2012. 'Play it again': a new method for testing metacognition in animals. Animal Cognition 15: 187-199.

Fordyce G., Whythes J.R., Shorthose W.R., Underwood D.W. and Shepherd R.K., 1988b. Cattle temperaments in extensive beef herds in Northern Queensland 2. Effect of temperament on carcass and meat quality. Australian Journal of Experimental Agriculture 28: 689-693.

Fordyce, G., Dodt, R. M. and Wythes, J., 1988a. Cattle temperaments in extensive beef herds in northern Queensland. 1. Factors affecting temperament. Australian Journal of Experimental Agriculture 28: 683-687.

Foury, A., Lebret, B., Chevillon, P., Vautier, A., Terlouw, C. and Mormede, P., 2011. Alternative rearing systems in pigs: consequences on stress indicators at slaughter and meat quality. Animal 5: 1620-1625.

Fraser, D., Ritchie, J.S.D. and Faser, A.F., 1975. The term 'stress' in a veterinary context. British Veterinary Journal 131: 653-662.

Friel, M., Kunc, H.P., Griffin, K., Asher, L. and Collins, L.M., 2019. Positive and negative contexts predict duration of pig vocalisations. Scientific Reports 9: 2062.

Gagaoua, M., Terlouw, E.C., Mullen, A.M., Franco, D., Warner, R.D., Lorenzo, J.M., Purslow, P.P., Gerrard, D., Hopkins, D.L. and Troy, D., 2021. Molecular signatures of beef tenderness: underlying mechanisms based on integromics of protein biomarkers from multi-platform proteomics studies. Meat Science 172: 108311. https://doi.org/10.1016/j.meatsci.2020.108311.

Gagaoua, M., Terlouw, E.M.C., Micol, D., Boudjellal, A., Hocquette, J.F. and Picard, B., 2015. understanding early post-mortem biochemical processes underlying meat color and pH decline in the longissimus thoracis muscle of Young Blond d'Aquitaine bulls using protein biomarkers. Journal of Agriculture and Food Chemistry 63: 6799-6809.

Gallo, C., Schwartzkopf-Genswein, K. and Gibson, T., 2022. Cattle. In: Faucitano, L. (ed.) Preslaughter handling and slaughter of meat animals. Wageningen Academic Publishers, Wageningen, the Netherlands, pp. 62-116.

Garcia-Larrea, L. and Peyron, R., 2013. Pain matrices and neuropathic pain matrices: a review. Pain 154 (Suppl. 1): S29-S43.

Gentle, M.J., 1992. Pain in birds. Animal Welfare 1: 235-247.

Gentle, M.J. and Tilston, V.L., 2000. Nociceptors in the legs of poultry: implications for potential pain in pre-slaughter shackling. Animal Welfare 9: 227-236.

Gerritzen, M., Shynkaruk, T., Buchynski, K., Schwean-Lardner, K. and Crowe, T.G., 2022. Poultry. In: Faucitano, L. (ed.) Preslaughter handling and slaughter of meat animals. Wageningen Academic Publishers, Wageningen, the Netherlands, pp. 232-265.

Gilbert Jr., B. and Arave, C., 1986. Ability of cattle to distinguish among different wavelengths of light. Journal of Dairy Science 69: 825-832.

Gilger, B., 2010. Equine ophthalmology. Elsevier Health Sciences, St-Louis, MI, USA, 514 pp.

Goldstein, D.S. and Holmes, C., 2008. Neuronal source of plasma dopamine. Clinical Chemistry 54: 1864-1871.

Gossette, R.L., Gossette, M.F. and Riddell, W., 1966. Comparisons of successive discrimination reversal performances among closely and remotely related avian species. Animal Behaviour 14: 560-564.

Goumon, S. and Faucitano, L., 2017. Influence of loading handling and facilities on subsequent response to pre-slaughter stress in pigs. Livestock Science 200: 6-13.

Grandin, T., 1980. Bruises and carcass damage. International Journal for the Study of Animal Problems 1: 121-137.

Grandin, T., 2001. Cattle vocalizations are associated with handling and equipment problems at beef slaughter plants. Applied Animal Behaviour Science 71: 191-201.

Grandin, T., 2019. Recommended animal handling guidelines and audit guide: a systematic approach to animal welfare. North American American Meat Institute, Washington, DC, USA, 126 pp.

Gregory, N.G. and Bell, J.C., 1987. Duration of wing flapping in chickens shackled before slaughter. Veterinary Record 121: 567-569.

Gregory, N.G. and Wilkins, L.J., 1989. Broken bones in domestic-fowl – handling and processing damage in end-of-lay battery hens. British Poultry Science 30: 555-562.

Gregory, N.G. and Wilkins, L.J., 1990. Broken bones in chickens – effect of stunning and processing in broilers. British Poultry Science 31: 53-58.

Gregory, N.G., 1994. Pathology and handling of poultry at the slaughterhouse. Worlds Poultry Science Journal 50: 66-67.

Gregory, N.G., Austin, S.D. and Wilkins, L.J., 1989. Relationship between wing flapping at shackling and red wingtips in chicken carcases. Veterinary Record 124: 62-62.

Gregory, N.G., Lee, C.J. and Widdicombe, J.P., 2007. Depth of concussion in cattle shot by penetrating captive bolt. Meat Science 77: 499-503.

Greiveldinger, L., Veissier, I. and Boissy, A., 2007. Emotional experience in sheep: predictability of a sudden event lowers subsequent emotional responses. Physiology & Behavior 92: 675-683.

Greiveldinger, L., Veissier, I. and Boissy, A., 2009. Behavioural and physiological responses of lambs to controllable vs. uncontrollable aversive events. Psychoneuroendocrinology 34: 805-814.

Greiveldinger, L., Veissier, I. and Boissy, A., 2011. The ability of lambs to form expectations and the emotional consequences of a discrepancy from their expectations. Psychoneuroendocrinology 36: 806-815.

Gruber, S.L., Tatum, J.D., Engle, T.E., Chapman, P.L., Belk, K.E. and Smith, G.C., 2010. Relationships of behavioral and physiological symptoms of preslaughter stress to beef longissimus muscle tenderness. Journal of Animal Science 88: 1148-1159.

Guàrdia, M., Estany, J., Balasch, S., Oliver, M., Gispert, M. and Diestre, A., 2009. Risk assessment of skin damage due to pre-slaughter conditions and RYR1 gene in pigs. Meat Science 81: 745-751.

Hachol, A., Szczepanowska-Nowak, W., Kasprzak, H., Zawojska, I., Dudzinski, A., Kinasz, R. and Wygledowska-Promienska, D., 2006. Measurement of pupil reactivity using fast pupillometry. Physiological Measurement 28: 61.

Hambrecht, E., Eissen, J.J., Newman, D.J., Smits, C.H.M., den Hartog, L.A. and Verstegen, M.W.A., 2005. Negative effects of stress immediately before slaughter on pork quality are aggravated by suboptimal transport and lairage conditions. Journal of Animal Science 83: 440-448.

Hampton, R.R., Zivin, A. and Murray, E.A., 2004. Rhesus monkeys (*Macaca mulatta*) discriminate between knowing and not knowing and collect information as needed before acting. Animal Cognition 7: 239-246.

Heffner, H.E. and Heffner, R.S., 1992. Auditory perception. In: Phillips, C. and Piggins, D. (eds.) Farm animals and the environment. CAB International, Oxon, UK, 430 pp.

Hegarty, G.R., Bratzler, L.J. and Pearson, A.M., 1963. Studies on the emulsifying properties of some intracellular beef muscle proteins. Journal of Food Science 28: 663-668.

Herman, J.P., McKlveen, J.M., Ghosal, S., Kopp, B., Wulsin, A., Makinson, R., Scheimann, J. and Myers, B., 2016. Regulation of the hypothalamic-pituitary-adrenocortical stress response. Comparative Physiology 6: 603-621.

Hogan, J.P., Petherick, J.C. and Phillips, C.J.C., 2007. The physiological and metabolic impacts on sheep and cattle of feed and water deprivation before and during transport. Nutrition Research Reviews 20: 17-28.

Huff-Lonergan, E., Baas, T.J., Malek, M., Dekkers, J.C., Prusa, K. and Rothschild, M.F., 2002. Correlations among selected pork quality traits. Journal of Animal Science 80: 617-627.

Hwang, I.H. and Thompson, J.M., 2001. The interaction between pH and temperature decline early postmortem on the calpain system and objective tenderness in electrically stimulated beef longissimus dorsi muscle. Meat Science 58: 167-174.

Hyslop, A. and Jackson, F., 1975. A reply to Don Locke. Australasian Journal of Philosophy 53: 68-69.

Ingvar, M., 1999. Pain and functional imaging. Philosophical Transactions of the Royal Society London series B, Biological Sciences 354: 1347-1358.

Jacobs, G.H., 1992. Data and interpretation in comparative color vision. Behavioral and Brain Sciences 15: 40-41.

Jacobs, G.H., 1993. The distribution and nature of colour vision among the mammals. Biological Reviews of the Cambridge Philosophical Society 68: 413-471.

Jacobs, G.H., Deegan, J.F. and Neitz, J., 1998. Photopigment basis for dichromatic color vision in cows, goats, and sheep. Visual Neuroscience 15: 581-584.

Jarvis, A., Selkirk, L. and Cockram, M., 1995. The influence of source, sex class and pre-slaughter handling on the bruising of cattle at two slaughterhouses. Livestock Production Science 43: 215-224.

Jarvis, A.M., Harrington, D.W.J. and Cockram, M.S., 1996a. Effect of source and lairage on some behavioural and biochemical measurements of feed restriction and dehydration in cattle at a slaughterhouse. Applied Animal Behaviour Science 50: 83-94.

Jarvis, A.M., Messer, C.D.A. and Cockram, M.S., 1996b. Handling, bruising and dehydration of cattle at the time of slaughter. Animal Welfare 5: 259-270.

Jarvis, E.D., Gunturkun, O., Bruce, L., Csillag, A., Karten, H., Kuenzel, W., Medina, L., Paxinos, G., Perkel, D.J., Shimizu, T., Striedter, G. and Wild, J.M., 2005. Avian brains and a new understanding of vertebrate brain evolution. Nature Reviews Neurosciences 6: 151-159.

Jephcott, E.H., McMillen, I.C., Rushen, J.P. and Thorburn, G.D., 1987. A comparison of the effects of electroimmobilisation and, or, shearing procedures on ovine plasma concentrations of beta-endorphin/beta-lipoprotein and cortisol. Research in Veterinary Science 43: 97-100.

Johansen, O., Brox, J. and Flaten, M.A., 2003. Placebo and nocebo responses, cortisol, and circulating beta-endorphin. Psychosomatic Medicine 65: 786-790.

Jonas, J.B., Schneider, U. and Naumann, G.O.H., 1992. Count and density of human retinal photoreceptors. Graefes Archive for Clinical and Experimental Ophthalmology 230: 505-510.

Jones, R.B., Mills, A.D., Faure, J.-M. and Williams, J.B., 1994. Restraint, fear, and distress in Japanese quail genetically selected for long or short tonic immobility reactions. Physiology & Behavior 56: 529-534.

Jones, R.C., 2013. Science, sentience, and animal welfare. Biology & Philosophy 28: 1-30.

Jorquera-Chavez, M., Fuentes, S., Dunshea, F.R., Jongman, E.C. and Warner, R.D., 2019. Computer vision and remote sensing to assess physiological responses of cattle to pre-slaughter stress, and its impact on beef quality: a review. Meat Science 156: 11-22.

Jung, C., Greco, S., Nguyen, H.H.T., Ho, J.T., Lewis, J.G., Torpy, D.J. and Inder, W.J., 2014. Plasma, salivary and urinary cortisol levels following physiological and stress doses of hydrocortisone in normal volunteers. BMC Endocrine Disorders 14: 1-10.

Kannan, G. and Mench, J.A., 1996. Influence of different handling methods and crating periods on plasma corticosterone concentrations in broilers. British Poultry Science 37: 21-31.

Karlsson, A. and Lundström, K., 1992. Meat quality in pigs reared in groups kept as a unit during the fattening period and slaughter. Animal Production 54: 421-426.

Kenny, F.J. and Tarrant, P.V., 1987a. The physiological and behavioural responses of crossbred Friesian steers to short-haul transport by road. Livestock Production Science 17: 63-75.

Kenny, F.J. and Tarrant, P.V., 1987b. The reaction of young bulls to short-haul road transport. Applied Animal Behaviour Science 17: 209-227.

King, D.A., Schuehle Pfeiffer, C.E., Randel, R.D., Welsh, T.H., Jr., Oliphint, R.A., Baird, B.E., Curley, K.O., Jr., Vann, R.C., Hale, D.S. and Savell, J.W., 2006. Influence of animal temperament and stress responsiveness on the carcass quality and beef tenderness of feedlot cattle. Meat Science 74: 546-556.

Kirk, C.R., McMillan, N. and Roberts, W.A., 2014. Rats respond for information: metacognition in a rodent? Journal of Experimental Psychology 40: 249-259.

Kittelsen, K., Granquist, E., Vasdal, G., Tolo, E. and Moe, R., 2015. Effects of catching and transportation versus pre-slaughter handling at the abattoir on the prevalence of wing fractures in broilers. Animal Welfare 24: 387-389.

Knowles, T. and Broom, D., 1990. The handling and transport of broilers and spent hens. Applied Animal Behaviour Science 28: 75-91.

Knowles, T.G., 1999. A review of the road transport of cattle. Veterinary Record 144: 197-201.

Knowles, T.G., Warriss, P.D., Brown, S.N. and Edwards, J.E., 1999. Effects on cattle of transportation by road for up to 31 hours. Veterinary Record 145: 575-582.

Krupenye, C. and Call, J., 2019. Theory of mind in animals: current and future directions. Wires Cognitive Science 10: 6, e1503. https://doi.org/10.1002/wcs.1503

Kuo, T., McQueen, A., Chen, T.C. and Wang, J.C., 2015. Regulation of glucose homeostasis by glucocorticoids. Advances in Experimental Medicine and Biology 872: 99-126.

Lam, T.K., Carpentier, A., Lewis, G.F., Van de Werve, G., Fantus, I. G. and Giacca, A., 2003. Mechanisms of the free fatty acid-induced increase in hepatic glucose production. American Journal of Physiology-Endocrinology and Metabolism 284: E863-E873.

Lambooij, E., Reimert, H.G.M. and Hindle, V.A., 2010. Evaluation of head-only electrical stunning for practical application: assessment of neural and meat quality parameters. Poultry Science 89: 2551-2558.

Lamme, V.A. and Roelfsema, P.R., 2000. The distinct modes of vision offered by feedforward and recurrent processing. Trends in Neurosciences 23: 571-579.

Latimer, K.S., 2011. Duncan & Prasse's veterinary laboratory medicine: clinical pathology. Wiley-Blackwell, Ames, IA, USA, 515 pp.

Laureys, S., Faymonville, M.E., De Tiege, X., Peigneux, P., Berre, J., Moonen, G., Goldman, S. and Maque, P., 2004. Brain function in the vegetative state. Advances in Experimental Medicine and Biology 550: 229-238.

Lawrence, A.B., Terlouw, E.M.C. and Illius, A.W., 1991. Individual differences in behavioural responses of pigs exposed to non-social and social challenges. Applied Animal Behaviour Science 30: 73-86.

Le Neindre, P., Poindron, P., Trillat, G. and Orgeur, P., 1993. Influence of breed on reactivity of sheep to humans. Genetics Selection Evolution 25: 447-458.

LeDoux, J.E., 2000. Emotion circuits in the brain. Annual Review of Neuroscience 23: 155-184.

LeDoux, J.E., 2017. Semantics, surplus meaning, and the science of fear. Trends in Cognitive Sciences 21: 303-306.

LeDoux, J.E. and Brown, R., 2017. A higher-order theory of emotional consciousness. Proceedings of the National Academy of Sciences 114: E2016-E2025.

Lehrer, M., 1993. Why do bees turn back and look. Journal of Comparative Physiology 172: 549-563.

Lensink, B.J., Boivin, X., Pradel, P., Le Neindre, P. and Veissier, I., 2000. Reducing veal calves' reactivity to people by providing additional human contact. Journal of Animal Science 78: 1213-1218.

Li, C.I., Maglinao, T.L. and Takahashi, L.K., 2004. Medial amygdala modulation of predator odor-induced unconditioned fear in the rat. Behavioral Neuroscience 118: 324-332.

Liotta, L., Nanni Costa, L., Chiofalo, B., Ravarotto, L. and Chiofalo, V., 2007. Effect of lairage duration on some blood constituents and beef quality in bulls after long journey. Italian Journal of Animal Science 6: 375-384.

Luan, Y.H., Wang, D., Yu, Q. and Chai, X.Q., 2017. Action of β-endorphin and nonsteroidal anti-inflammatory drugs, and the possible effects of nonsteroidal anti-inflammatory drugs on β-endorphin. Journal of Clinical Anesthesia 37: 123-128.

Lyons, D.M., Price, E.O. and Moberg, G.P., 1988. Individual differences in temperament of domestic dairy goats: constancy and change. Animal Behaviour 36: 1323-1333.

Maggs, D., Miller, P. and Ofri, R., 2008. Slatter's fundamentals of veterinary ophthalmology. Elsevier Health Sciences, St-Louis, MI, USA, 478 pp.

Mahmoud, M., Lu, Y., Hou, X., McLennan, K. and Robinson, P., 2018. Estimation of pain in sheep using computer vision. In: Moore, R.J. (ed.) Handbook of pain and palliative care. Springer, Champ, Switzerland, pp. 145-157.

Mameli, M. and Bortolotti, L., 2006. Animal rights, animal minds, and human mindreading. Journal of Medical Ethics 32: 84-89.

Marino, F.E., Mbambo, Z., Kortekaas, E., Wilson, G., Lambert, M.I., Noakes, T.D. and Dennis, S.C., 2001. Influence of ambient temperature on plasma ammonia and lactate accumulation during prolonged submaximal and self-paced running. European Journal of Applied Physiology 86: 71-78.

Marx, G., Horn, T., Thielebein, J., Knubel, B. and von Borell, E., 2003. Analysis of pain-related vocalization in young pigs. Journal of Sound and Vibration 266: 687-698.

Mazeaud, M.M., Mazeaud, F. and Donaldson, E.M., 1977. Primary and secondary effects of stress in fish – some new data with a general review. Transactions of the American Fisheries Society 106: 201-212.

Mazzola, L., Isnard, J., Peyron, R. and Mauguière, F., 2012. Stimulation of the human cortex and the experience of pain: Wilder Penfield's observations revisited. Brain 135: 631-640.

McGlone, J.J., McPherson, R.L. and Anderson, D.L., 2004. Moving devices for finishing pigs: efficacy of electric prod, board, paddle, or flag. The Professional Animal Scientist 20: 518-523.

McKeegan, D.E., Abeyesinghe, S.M., McLeman, M.A., Lowe, J.C., Demmers, T.G., White, R.P., Kranen, R.W., Van Bemmel, H., Lankhaar, J.A. and Wathes, C.M., 2007a. Controlled atmosphere stunning of broiler chickens. II. Effects on behaviour, physiology and meat quality in a commercial processing plant. British Poultry Science 48: 430-442.

McKeegan, D.E.F., McIntyre, J.A., Demmers, T.G.M., Lowe, J.C., Wathes, C.M., Van Den Broek, P.L.C, Coenen, A.M.L and Gentle, M., 2007b. Physiological and behavioural responses of broilers to controlled atmosphere stunning: implications for welfare. Animal Welfare 16: 409-426.

McNeil, R., McSween, A. and Lachapelle, P., 2005. Comparison of the retinal structure and function in four bird species as a function of the time they start singing in the morning. Brain, Behavior and Evolution 65: 202-214.

McVeigh, J.M. and Tarrant, P.V., 1982. Glycogen-content and repletion rates in beef muscle, effect of feeding and fasting. Journal of Nutrition 112: 1306-1314.

Mendl, M. and Paul, E.S., 2004. Consciousness, emotion and animal welfare: insights from cognitive science. Animal Welfare 13: S17-S25.

Millan, M.J., 1999. The induction of pain: an integrative review. Progress in Neurobiology 57: 1-164.

Millan, M.J., Czlonkowski, A., Millan, M.H. and Herz, A., 1987. Activation of periaqueductal gray pools of beta-endorphin by analgesic electrical-stimulation in freely moving rats. Brain Research 407: 199-203.

Miller, R.E. and Tredici, T.J., 1992. Night vision manual for the flight surgeon. Armstrong Lab Brooks, Arlington, TX, USA, 101 pp.

Minelli, G., Culioli, J., Vignon, X. and Monin, G., 1995. Postmortem changes in the mechanical properties and ultrastructure of the longissimus in two porcine breeds. Journal of Muscle Foods 6: 313-326.

Mitchell, J.R. and De Boom, H.P.A., 1986. Traumatic avulsion of the proximal femoral articular-cartilage as a cause of hip dislocation in broiler-chickens. Journal of the South African Veterinary Association 57: 133-137.

Mitchell, M.A. and Sandercock, D.A., 1995. Creatine kinase isoenzyme profiles in the plasma of the domestic fowl (*Gallus domesticus*): effects of acute heat stress. Research in Veterinary Science 59: 30-34.

Mitchell, M.A., Kettlewell, P.J. and Maxwell, M.H., 1992. Indicators of physiological stress in broiler chickens during road transportation. Animal Welfare 1: 91-103.

Mok, E.Y.M. and Munro, A.D., 1998. Effects of dopaminergic drugs on locomotor activity in teleost fish of the genus *Oreochromis* (Cichlidae): involvement of the telencephalon. Physiology & Behavior 64: 227-234.

Morrissey, M.C., Kisiolek, J.N., Ragland, T.J., Willingham, B.D., Hunt, R.L., Hickner, R.C. and Ormsbee, M.J., 2019. The effect of cold ambient temperature and preceding active warm-up on lactate kinetics in female cyclists and triathletes. Applied Physiology Nutrition and Metabolism 44: 1043-1051.

Mousa, S.A., Zhang, Q., Sitte, N., Ji, R.R. and Stein, C., 2001. Beta-endorphin-containing memory-cells and mu-opioid receptors undergo transport to peripheral inflamed tissue. Journal of Neuroimmunology 115: 71-78.

Mpakama, T., Chulayo, A.Y. and Muchenje, V., 2014. Bruising in slaughter cattle and its relationship with creatine kinase levels and beef quality as affected by animal related factors. Asian-Australasian Journal of Animal Sciences 27: 717-725.

Muchenje, V., Dzama, K., Chimonyo, M., Strydom, P.E. and Raats, J.G., 2009. Relationship between pre-slaughter stress responsiveness and beef quality in three cattle breeds. Meat Science 81: 653-657.

Navarro, G., Bravo, V., Gallo, C. and Phillips, C.J.C., 2019. Physiological and behavioural responses of cattle to high and low space, feed and water allowances during long distance transport in the south of Chile. Animals 9: 229. https://doi.org/10.3390/ani9050229

Neitz, J. and Jacobs, G.H., 1989. Spectral sensitivity of cones in an ungulate. Visual Neuroscience 2: 97-100.

Nicol, C.J. and Scott, G.B., 1990. Pre-slaughter handling and transport of broiler-chickens. Applied Animal Behaviour Science 28: 57-73.

Nomura, T. and Izawa, E., 2017. Avian brains: insights from development, behaviors and evolution. Development, Growth & Differentiation 59: 244-257.

Northcutt, J.R., Buhr, R.J. and Rowland, G.N., 2000. Relationship of broiler bruise age to appearance and tissue histological characteristics. Journal of Applied Poultry Research 9: 13-20.

Nybo, L., Girard, O., Mohr, M., Knez, W., Voss, S. and Racinais, S., 2013. Markers of muscle damage and performance recovery after exercise in the heat. Medicine & Science in Sports & Exercise 45: 860-868.

Olsson, I.A.S., De Jonge, F.H., Schuurman, T. and Helmond, F.A., 1999. Poor rearing conditions and social stress in pigs: repeated social challenge and the effect on behavioural and physiological responses to stressors. Behavioural Processes 46: 201-215.

Öngür, D. and Price, J.L., 2000. The organization of networks within the orbital and medial prefrontal cortex of rats, monkeys and humans. Cerebral Cortex 10: 206-219.

Ouali, A., Herrera-Mendez, C.H., Coulis, G., Becila, S., Boudjellal, A., Aubry, L. and Sentandreu, M.A., 2006. Revisiting the conversion of muscle into meat and the underlying mechanisms. Meat Science 74: 44-58.

Øverli, O., Pottinger, T.G., Carrick, T.R., Overli, E. and Winberg, S., 2002. Differences in behaviour between rainbow trout selected for high- and low-stress responsiveness. Journal of Experimental Biology 205: 391-395.

Panksepp, J., 2005. Affective consciousness: core emotional feelings in animals and humans: neurobiology of animal consciousness. Consciousness and Cognition 14: 30-80.

Paul, E.S., Harding, E.J. and Mendl, M., 2005. Measuring emotional processes in animals: the utility of a cognitive approach. Neuroscience & Biobehavioral Reviews 29: 469-491.

Phillips, C.J.C., 2010. Principles of cattle production. CABI, Wallingford, UK, 233 pp.

Pickering, A.D., Pottinger, T.G., Sumpter, J.P., Carragher, J.F. and Lebail, P.Y., 1991. Effects of acute and chronic stress on the levels of circulating growth-hormone in the rainbow-trout, *Oncorhynchus-mykiss*. General and Comparative Endocrinology 83: 86-93.

Piggins, D. and Phillips, C., 1996. The eye of the domesticated sheep with implications for vision. Animal Science 62: 301-308.

Ponnampalam, E.N., Hopkins, D.L., Bruce, H., Li, D., Baldi, G. and El-din Bekhit, A., 2017. Causes and contributing factors to 'dark cutting' meat: current trends and future directions: a review. Comprehensive Reviews in Food Science and Food Safety 16: 400-430.

Portavella, M., Vargas, J.P., Torres, B. and Salas, C., 2002. The effects of telencephalic pallial lesions on spatial, temporal, and emotional learning in goldfish. Brain Research Bulletin 57: 397-399.

Preston, S.D. and De Waal, F.B., 2002. The communication of emotions and the possibility of empathy in animals. In: Monroe, K.R., Post, S.G., Underwood, L., Schloss, J. and Hurlbut, W. (eds.) Altruism & altruistic love: science, philosophy and religion in dialogue. Oxford University Press, Oxford, UK, pp. 284-308.

Prunier, A., Mounier, L., Le Neindre, P., Leterrier, C., Mormède, P., Paulmier, V., Prunet, P., Terlouw, C. and Guatteo, R., 2013. Identifying and monitoring pain in farm animals: a review. Animal 7: 998-1010.

Purves, D., Augustine, G., Fitzpatrick, D., Hall, W., LaMantia, A., McNamara, J. and White, L., 2004. Neuroscience. Sinauer Associates Inc., Sunderland, MA, USA, 960 pp.

Qvisth, V., Hagström-Toft, E., Enoksson, S., Moberg, E., Arner, P. and Bolinder, J., 2006. Human skeletal muscle lipolysis is more responsive to epinephrine than to norepinephrine stimulation *in vivo*. Journal of Clinical Endocrinology and Metabolism 91: 665-670.

Rabaste, C., Faucitano, L., Saucier, L., Foury, D., Mormède, P., Correa, J.A., Giguère, A. and Bergeron, R., 2007. The effects of handling and group size on welfare of pigs in lairage and its influence on stomach weight, carcass microbial contamination and meat quality variation. Canadian Journal of Animal Science 87: 3-12.

Raja, S.N., Carr, D.B., Cohen, M., Finnerup, N.B., Flor, H., Gibson, S., Keefe, F.J., Mogil, J.S., Ringkamp, M. and Sluka, K.A., 2020. The revised international association for the study of pain definition of pain: concepts, challenges, and compromises. Pain 161: 1976-1982.

Ramsay, J.M., Feist, G.W., Varga, Z.M., Westerfield, M., Kent, M.L. and Schreck, C.B., 2006. Whole-body cortisol is an indicator of crowding stress in adult zebrafish, *Danio rerio*. Aquaculture 258: 565-574.

Rao, M.A., Knowles, T.G. and Wotton, S.B., 2013. The effect of pre-stun shocks in electrical water-bath stunners on carcase and meat quality in broilers. Animal Welfare 22: 79-84.

Reiche, A.M., Oberson, J.L., Silacci, P., Messadene-Chelali, J., Hess, H.D., Dohme-Meier, F., Dufey, P.A. and Terlouw, E.M.C., 2019. Pre-slaughter stress and horn status influence physiology and meat quality of young bulls. Meat Science 158: 107892. https://doi.org/10.1016/j.meatsci.2019.107892

Reimert, I., Bolhuis, J.E., Kemp, B. and Rodenburg, T.B., 2015. Emotions on the loose: emotional contagion and the role of oxytocin in pigs. Animal Cognition 18: 517-532.

Reuter, T., 2011. Fifty years of dark adaptation 1961-2011. Vision research 51: 2243-2262.

Rioja-Lang, F.C., Brown, J.A., Brockhoff, E.J. and Faucitano, L., 2019. A review of swine transportation research on priority welfare issues: a Canadian perspective. Frontiers in Veterinary Science 6: 36. https://doi.org/10.3389/fvets.2019.00036

Riol, J., Sanchez, J., Eguren, V. and Gaudioso, V., 1989. Colour perception in fighting cattle. Applied Animal Behaviour Science 23: 199-206.

Robergs, R.A., Ghiasvand, F. and Parker, D., 2004. Biochemistry of exercise-induced metabolic acidosis. American Journal of Physiology-Regulatory, Integrative and Comparative Physiology 287: R502-R516.

Rocha, L.M., Devillers, N., Maldague, X., Kabemba, F.Z., Fleuret, J., Guay, F. and Faucitano, L., 2019. Validation of anatomical sites for the measurement of infrared body surface temperature variation in response to handling and transport. Animals 9: 425. https://doi.org/10.3390/ani9070425

Rocha, L.M., Dionne, A., Saucier, L., Nannoni, E. and Faucitano, L., 2015. Hand-held lactate analyzer as a tool for the real-time measurement of physical fatigue before slaughter and pork quality prediction. Animal 9: 707-714.

Rocha, L.M., Velarde, A., Dalmau, A., Saucier, L. and Faucitano, L., 2016. Can the monitoring of animal welfare parameters predict pork meat quality variation through the supply chain (from farm to slaughter)? Journal of Animal Science 94: 359-376.

Rolls, E.T., 2015. Limbic systems for emotion and for memory, but no single limbic system. Cortex 62: 119-157.

Romijn, J., Coyle, E., Sidossis, L., Zhang, X. and Wolfe, R., 1995. Relationship between fatty acid delivery and fatty acid oxidation during strenuous exercise. Journal of Applied Physiology 79: 1939-1945.

Rose, J.D., Arlinghaus, R., Cooke, S.J., Diggles, B.K., Sawynok, W., Stevens, E. and Wynne, C.D., 2014. Can fish really feel pain? Fish and Fisheries 15: 97-133.

Rosenvold, K. and Andersen, H.J., 2003. The signification of pre-slaughter stress and diet on colour and colour stability of pork. Meat Science 63: 199-209.

Salzman, C.D. and Fusi, S., 2010. Emotion, cognition, and mental state representation in amygdala and prefrontal cortex. Annual Review of Neuroscience 33: 173-202.

Sanchez-Hidalgo, M., Bravo, V. and Gallo, C., 2020. Behavior and health indicators to assess cull cow's welfare in livestock markets. Frontiers in Veterinary Science 7: 471. https://doi.org/10.3389/fvets.2020.00471

Sardi, L., Gastaldo, A., Borciani, M., Bertolini, A., Musi, V., Martelli, G., Cavallini, D., Rubini, G. and Nannoni, E., 2020. Identification of possible pre-slaughter indicators to predict stress and meat quality: a study on heavy pigs. Animals 10: 945. https://doi.org/10.3390/ani10060945

Sato, N., Tan, L., Tate, K. and Okada, M., 2015. Rats demonstrate helping behavior toward a soaked conspecific. Animal Cognition 18: 1049-1049.

Scherer, K.R., 2001. Appraisal considered as a process of multilevel sequential checking, In: Scherer, K.R., Schorr, A. and Johnstone, T. (eds.) Appraisal processes in emotion: theory, methods, research. Oxford University Press, New York, NY, USA, pp. 92-120.

Seideman, S.C., Cross, H.R., Smith, G.C. and Durland, P.R., 1984. Factors associated with fresh meat color: a review. Journal of Food Quality 6: 211-237.

Sewright, K.A., Hubal, M.J., Kearns, A., Holbrook, M.T. and Clarkson, P.M., 2008. Sex differences in response to maximal eccentric exercise. Medicine and Science in Sports and Exercise 40: 242-251.

Shaffer, F. and Ginsberg, J.P., 2017. An overview of heart rate variability metrics and norms. Frontiers in Public Health 5: 258. https://doi.org/10.3389/fpubh.2017.00258

Shaw, F.D. and Tume, R.K., 1990. Beta-endorphin and cortisol concentrations in plasma of cattle. Australian Veterinary Journal 67: 423-424.

Shinozaki, A., Hosaka, Y., Imagawa, T. and Uehara, M., 2010. Topography of ganglion cells and photoreceptors in the sheep retina. Journal of Comparative Neurology 518: 2305-2315.

Shutt, D.A., Fell, L.R., Connell, R., Bell, A.K., Wallace, C.A. and Smith, A.I., 1987. Stress-induced changes in plasma-concentrations of immunoreactive beta-endorphin and cortisol in response to routine surgical-procedures in lambs. Australian Journal of Biological Science 40: 97-103.

Śmiecińska, K., Denaburski, J. and Sobotka, W., 2011. Slaughter value, meat quality, creatine kinase activity and cortisol levels in the blood serum of growing-finishing pigs slaughtered immediately after transport and after a rest period. Polish Journal of Veterinary Science 14: 47-54.

Smith, J.D., Beran, M.J., Couchman, J.J. and Coutinho, M.V.C., 2008. The comparative study of metacognition: sharper paradigms, safer inferences. Psychonomic Bulletin & Review 15: 679-691.

Smith, J.D., Schull, J., Strote, J., McGee, K., Egnor, R. and Erb, L., 1995. The uncertain response in the bottlenosed dolphin (*Tursiops truncatus*). Journal of Experimental Psychology 124: 391-408.

Smith, R., Owens, P.C., Lovelock, M., Chan, E.C. and Falconer, J., 1986. Acute hemorrhagic stress in conscious sheep elevates immunoreactive beta-endorphin in plasma but not in cerebrospinal fluid. Endocrinology 118: 2572-2576.

Sneddon, L.U., 2003. The evidence for pain in fish: the use of morphine as an analgesic. Applied Animal Behaviour Science 83: 153-162.

Sneddon, L.U., 2009. Pain perception in fish: indicators and endpoints. Ilar Journal 50: 338-342.

Sommavilla, R., Faucitano, L., Gonyou, H., Seddon, Y., Bergeron, R., Widowski, T., Crowe, T., Connor, L., Scheeren, M.B., Goumon, S. and Brown, J., 2017. Season, transport duration and trailer compartment effects on blood stress indicators in pigs: relationship to environmental, behavioral and other physiological factors, and pork quality traits. Animals 7: 8. https://doi.org/10.3390/ani7020008

Sparrey, J.M. and Kettlewell, P.J., 1994. Shackling of poultry – is it a welfare problem. Worlds Poultry Science Journal 50: 167-176.

Staehr, P., Hother-Nielsen, O., Landau, B.R., Chandramouli, V., Holst, J.J. and Beck-Nielsen, H., 2003. Effects of free fatty acids per se on glucose production, gluconeogenesis, and glycogenolysis. Diabetes 52: 260-267.

Stein, C., 2013. Targeting pain and inflammation by peripherally acting opioids. Frontiers in Pharmacology 4: 123. https://doi.org/10.3389/fphar.2013.00123

Strappini, A.C., Metz, J.H.M., Gallo, C., Frankena, K., Vargas, R., De Freslon, I. and Kemp, B., 2013. Bruises in culled cows: when, where and how are they inflicted? Animal 7: 485-491.

Tarrant, P.V., Kenny, F.J., Harrington, D. and Murphy, M., 1992. Long-distance transportation of steers to slaughter – effect of stocking density on physiology, behavior and carcass quality. Livestock Production Science 30: 223-238.

Terlouw, C., 2005. Stress reactions at slaughter and meat quality in pigs: genetic background and prior experience: a brief review of recent findings. Livestock Production Science 94: 125-135.

Terlouw, C., Berne, A. and Astruc, T., 2009. Effect of rearing and slaughter conditions on behaviour, physiology and meat quality of Large White and Duroc-sired pigs. Livestock Science 122: 199-213.

Terlouw, E.M.C. and Porcher, J., 2005. Repeated handling of pigs during rearing. I. Refusal of contact by the handler and reactivity to familiar and unfamiliar humans. Journal of Animal Science 83: 1653-1663.

Terlouw, E.M.C. and Rybarczyk, P., 2008. Explaining and predicting differences in meat quality through stress reactions at slaughter: the case of Large White and Duroc pigs. Meat Science 79: 795-805.

Terlouw, E.M.C., Arnould, C., Auperin, B., Berri, C., Le Bihan-Duval, E., Deiss, V., Lefèvre, F., Lensink, B.J. and Mounier, L., 2008. Pre-slaughter conditions, animal stress and welfare: current status and possible future research. Animal 2: 1501-1517.

Terlouw, E.M.C., Picard, B., Deiss, V., Berri, C., Hocquette, J.F., Lebret, B., Lefevre, F., Hamill, R. and Gagaoua, M., 2021. Understanding the determination of meat quality using biochemical characteristics of the muscle: stress at slaughter and other missing keys. Foods 10: 84. https://doi.org/10.3390/foods10010084

Terlouw, E.M.C., Porcher, J. and Fernandez, X., 2005. Repeated handling of pigs during rearing. II. Effect of reactivity to humans on aggression during mixing and on meat quality. Journal of Animal Science 83: 1664-1672.

Thomas, C., Sirvent, P., Perrey, S., Raynaud, E. and Mercier, J., 2004. Relationships between maximal muscle oxidative capacity and blood lactate removal after supramaximal exercise and fatigue indexes in humans. Journal of Applied Physiology 97: 2132-2138.

Thomas, T.J., Weary, D.M. and Appleby, M.C., 2001. Newborn and 5-week-old calves vocalize in response to milk deprivation. Applied Animal Behaviour Science 74: 165-173.

Tononi, G., 2012. Integrated information theory of consciousness: an updated account. Archives Italiennes de Biologie 150: 56-90.

Van de Water, G., Heylen, T., Swinnen, K. and Geers, R., 2003. The impact of vertical vibrations on the welfare of calves. Deutsche Tierärztliche Wochenschrift 110: 111-114.

Van der Vaart, E. and Hemelrijk, C.K., 2014. 'Theory of mind' in animals: ways to make progress. Synthese 191: 335-354.

Veissier, I. and Boissy, A., 2007. Stress and welfare: two complementary concepts that are intrinsically related to the animal's point of view. Physiology & Behavior 92: 429-433.

Veissier, I., Boissy, A., Desire, L. and Greiveldinger, L., 2009. Animals' emotions: studies in sheep using appraisal theories. Animal Welfare 18: 347-354.

Vieira, A.D., Guesdon, V., De Passille, A.M., von Keyserlingk, M.A.G. and Weary, D.M., 2008. Behavioural indicators of hunger in dairy calves. Applied Animal Behaviour Science 109: 180-189.

Villemure, C., Slotnick, B.M. and Bushnell, M.C., 2003. Effects of odours on pain perception: deciphering the roles of emotion and attention. Pain 106: 101-108.

Vitali, M., Conte, S., Lessard, M., Deschêne, K., Benoit-Biancamano, M., Celeste, C., Martelli, G., Sardi, L., Guay, F. and Faucitano, L., 2017. Use of the spectrophotometric color method for the determination of the age of skin lesions on the pig carcass and its relationship with gene expression and histological and histochemical parameters. Journal of Animal Science 95: 3873-3884.

Warner, R.D., Ferguson, D.M., Cottrell, J.J. and Knee, B.W., 2007. Acute stress induced by the preslaughter use of electric prodders causes tougher beef meat. Australian Journal of Experimental Agriculture 47: 782-788.

Warriss, P., Wilkins, L. and Knowles, T.G. 1999. The influence of ante-mortem handling on poultry meat quality. Proceedings of Poultry Science Symposium Series, pp. 217-230.

Warriss, P.D., 1990. The handling of cattle pre-slaughter and its effects on carcass and meat quality. Applied Animal Behaviour Science 28: 171-186.

Warriss, P.D., Brown, S.N., Adams, S.J.M. and Corlett, I.K., 1994. Relationships between subjective and objective assessments of stress at slaughter and meat quality in pigs. Meat Science 38: 329-340.

Warriss, P.D., Brown, S.N., Knowles, T.G., Kestin, S.C., Edwards, J.E., Dolan, S.K. and Phillips, A.J., 1995. Effects on cattle of transport by road for up to 15 hours. Veterinary Record 136: 319-323.

Watt, M.J., Howlett, K.F., Febbraio, M.A., Spriet, L.L. and Hargreaves, M., 2001. Adrenaline increases skeletal muscle glycogenolysis, pyruvate dehydrogenase activation and carbohydrate oxidation during moderate exercise in humans. Journal of Physiology 534: 269-278.

Weeks, C., 2008. A review of welfare in cattle, sheep and pig lairages, with emphasis on stocking rates, ventilation and noise. Animal Welfare 17: 275-284.

Weeks, C.A., Brown, S.N., Lane, S., Heasman, L., Benson, T. and Warriss, P.D., 2009. Noise levels in lairages for cattle, sheep and pigs in abattoirs in England and Wales. Veterinary Record 165: 308-314.

Weiss, D.J. and Wardrop, K.J., 2010. Schalm's veterinary hematology. Wiley-Blackwell, Ames, IA, 1206 pp.

Weitzman, E.D., Fukushima, D., Nogeire, C., Roffwarg, H., Gallagher, T.F. and Hellman, L., 1971. Twenty-four hour pattern of the episodic secretion of cortisol in normal subjects. The Journal of Clinical Endocrinology and Metabolism 33: 14-22.

Weschenfelder, A.V., Saucier, L., Maldague, X., Rocha, L.M., Schaefer, A.L. and Faucitano, L., 2013. Use of infrared ocular thermography to assess physiological conditions of pigs prior to slaughter and predict pork quality variation. Meat Science 95: 616-620.

Willson, D.W., Baier, F.S. and Grandin, T., 2021. An observational field study on the effects of changes in shadow contrasts and noise on cattle movement in a small abattoir. Meat Science 179: 108539. https://doi.org/10.1016/j.meatsci.2021.108539

Wittgenstein, L., 1958. Philosophical investigations. Basil Blackwell, Oxford, UK, 250 pp.

Wouters, L., De Moor, A. and Moens, Y., 1980. Rod and cone components in the electroretinogram of the horse. Zentralblatt für Veterinärmedizin Reihe A 27: 330-338.

Young, R.F., Bach, F.W., Van Norman, A.S. and Yaksh, T.L., 1993. Release of beta-endorphin and methionine-enkephalin into cerebrospinal fluid during deep brain stimulation for chronic pain – effects of stimulation locus and site of sampling. Journal of Neurosurgery 79: 816-825.

Zhang, C., Kolodkin, A.L., Wong, R.O. and James, R.E. 2017. Establishing wiring specificity in visual system circuits: from the retina to the brain. Annual Review of Neuroscience 40: 395-424.

Zulkifli, I., Goh, Y.M., Norbaiyah, B., Sazili, A.Q., Lotfi, M., Soleimani, A.F. and Small, A.H., 2014. Changes in blood parameters and electroencephalogram of cattle as affected by different stunning and slaughter methods in cattle. Animal Production Science 54: 187-193.

Photo: Karen Schwarzkopf-Genswein, Agriculture & Agri-Food Canada, Lethbridge, Canada

Chapter 2
Cattle

C. Gallo[1*]*, K. Schwartzkopf-Genswein*[2] *and T. Gibson*[3]

[1]*Animal Welfare Programme, Faculty of Veterinary Science, Universidad Austral de Chile, Casilla 567, Valdivia, Chile;* [2]*Agriculture and Agri-Food Canada, Lethbridge Research and Development Centre, 5403 1*[st] *Ave S, Lethbridge, T1J 4B1, Canada;* [3]*Pathobiology and Population Sciences, Royal Veterinary College, Royal College Street, London, NW1 0TU, United Kingdom; cgallo@uach.cl*

Abstract

Animal welfare, food safety and quality are major issues in the production of food animals. A combination of factors including animal (i.e. genetics, temperament, age and sex), stockperson/driver/slaughterhouse personnel (i.e. experience, training, attitudes and empathy) and environment (i.e. climate, facilities, production systems and equipment) can increase or reduce the physical, physiological and emotional stress imposed on cattle at all stages prior to and at slaughter. These stages include handling on the farm, loading, transport, unloading at the slaughter plant, lairage, stunning and/or slaughter. In this chapter we discuss these factors and their relationship to cattle welfare. In most cases, reduced welfare during any of the pre-slaughter stages can negatively impact the quantity and quality of the meat produced. More specifically, carcass condemnations, reduced carcass yield, trimming of bruised meat and quality defects like dark cutting meat all contribute to discounts that are passed onto the producer. It is of utmost importance that cattle producers, stockpersons, livestock transporters and slaughterhouse personnel can assure high standards of animal welfare from farm to slaughter, as demanded by today's consumers.

Keywords: animal welfare, cattle, meat quality, preslaughter, slaughter

2.1 Pre-transport factors on farm

Animal handling practices prior to transport and slaughter can directly impact animal stress and welfare as well as meat quantity and quality. The number and frequency of stressors cattle are exposed to on farm will depend on factors, such as the rearing and fattening conditions, facilities, handling, food and water availability/deprivation and on the quality of the human-animal interactions. The responses of farm animals to these various extrinsic factors will in turn depend on intrinsic factors, such as genotype, temperament, age, sex and physiological state, and the experiences of the animals. Provided that the adverse effects are reduced as much as possible on farm, the probability that animal welfare during transport, lairage and slaughter are compromised is reduced (Simova *et al.*, 2016).

2.1.1 Production system

The response of cattle to handling and transport stress will vary according to their daily life experience in the production system of the farm of origin (extensive vs intensive) and will have an influence on the animal stress and behaviour during all subsequent production phases (Grandin, 1993).

Extensive production systems provide more natural environmental conditions for cattle, allowing animals to perform most of their normal behaviour patterns; these patterns may be restricted only for short periods of time when they are gathered and penned. However, in these conditions animals may experience climatic extremes and concomitant 'natural disasters' (Petherick, 2005). In many extensive systems cattle are gathered to be driven to the corrals only once or twice a year, involving less frequent human-animal interactions compared to intensive systems. This can make cattle fearful and thus aggressive during handling at loading and during subsequent preslaughter stages (Le Neindre *et al.*, 1996). When animals are inspected less frequently, it is also more difficult to identify sick and injured animals in the early stages, which has consequences for welfare, in terms of the duration of pain and suffering through the pre-slaughter period (Petherick, 2005; Weary *et al.*, 2009). Moreover, when animals are handled in extensive systems it is often for procedures like veterinary treatment, identification or other painful husbandry practices, which may be aversive experiences that animals will remember. It is well documented that repeated exposure to and experience with stimuli will reduce the fear of them, provided that the experience is not aversive (Grandin, 1993, 1998a,b).

Increased meat consumption has resulted in increased production intensification on pasture in some countries. Rotational grazing is considered a promising alternative for the intensification of pasture systems, as it combines environmental benefits with improved cattle welfare. This system typically involves more frequent human-animal interactions during paddock changes that can reduce animal reactivity. For example, Ceballos *et al.* (2016) found that frequent handling in rotationally grazed cattle reduced their reactivity through consecutive measurements of flight speed and composite reactivity score, and improved their temperament over time. The more frequent human-animal interactions during paddock changes in rotational grazing, may hence imply having calmer animals during pre-transport handling.

The use of open outdoor and indoor feedlots for housing large numbers of cattle to be fattened is increasing in many parts of the world. The deviation of feedlots away from the more 'natural' environment offered by extensive systems has led to increased consumer concerns for cattle welfare (Petherick, 2005). When considering animal welfare, livestock producers tend to emphasise the importance of health and accessibility to resources, such as food, water and shelter (Phillips *et al.*, 2009). However, consumers place an additional emphasis on freedom to move and the ability to display natural behaviours when considering animal welfare (Vanhonacker *et al.*, 2008). Within the feedlot environment, provision of resources, such as sufficient space, appropriate flooring, bedding, ventilation, shade, protection from inclement weather conditions, and enrichment (grooming and foraging, straw) are important to ensure positive welfare and growth outcomes for cattle (Cozzi *et al.*, 2009; Grandin, 2016; Ishiwata *et al.*, 2006; Salvin *et al.*, 2020), and increase their resilience during the preslaughter stages (Park *et al.*, 2020). Park *et al.* (2020) reviewed the relationship between housing and welfare status and demonstrated that smaller space allowances can negatively impact welfare (i.e. a decrease in lying, ruminating, and positive social behaviour, reduced feed efficiency and productivity, greater concentrations of stress hormones), whereas larger space allowances can positively impact welfare (e.g. reduced frequency of abnormal behaviours, greater body weights, and lower stress hormone concentrations). Overall, feedlot environments that provided between 3.0 and 4.5 m^2 per animal of around 500 kg liveweight showed better welfare and productive outcomes than those that provided 1.5 m^2. Using much higher space allowances (6, 12 and 24 m^2/animal), in an outdoor feedlot in Brazil, Braga *et al.* (2018) found that calm Nellore bulls showed better performance regardless of the space allowance in the pens. These authors concluded that this was probably because the reactive bulls faced more difficulties to adapt to the feedlot conditions, suffering with chronic stress. Aggressive behaviours have shown to increase when fattening bulls are housed with insufficient space and barren environments (Grandin, 2016) resulting in increased stress and bruised carcasses. Suitable housing systems should ensure that an animal can move, lie or stand freely on a non-slippery floor (Park *et al.*, 2020). Inadequate floor conditions can affect cattle health by causing swollen joints and leg injuries (Cozzi *et al.*, 2005; Gottardo *et al.*, 2003) that may affect their fitness for transport and increase their risk of falling and being trampled by other animals.

2.1.2 Human and facility factors

One of the most effective means of minimising preslaughter stress is to ensure that the animal handling facilities on farm enable smooth and efficient movement of livestock and that the stockpersons understand the principles of animal handling (Grandin 2014a; Grandin and Deesing, 2008). These recommendations and those from the OIE (2019) include good, even lighting, as it is easier to move cattle from dimly to more brightly illuminated areas. Cattle should be moved in groups, rather than individually. Gently curved races facilitate movement by preventing animals from seeing a dead-end and thus baulking. The use of solid, high walls in handling facilities prevents visual distraction of the animals and therefore makes handling easier. Restraint of animals is an important welfare issue on farms and considerable efforts have gone into designing crushes and head-bails that will hold animals securely with minimal risks of injury to themselves and stockpersons (Grandin, 1997a,b, 1998a,b). Rubber stops should be installed on gates and other devices like head holders to reduce the sound produced by metal to metal contact and reduce fear of cattle. Inadequate handling facilities and distractions on farms and cattle markets require more effort from humans to move the animals (Grandin, 2014a).

Direct measures of the quality of handling include the time required for an animal to move a certain distance, the number of vocalisations, slips, falls, turns, baulks, stops and other behavioural measurements during movement. Whereas, indirect measures are based on the effort the handler makes to move the animal, usually expressed as the numbers of shouts, pushes and hits. When cattle are being handled, stockpersons can be assessed on the percentage of cattle moved with an electric prod, cattle falling while exiting from the squeeze chute, stumbling, vocalisation and being miscaught in the squeeze chute (Grandin, 2016). The higher the former indicators, the higher the risk of poor welfare for the animals when moving them. The use of numerical scoring of welfare indicators to assess cattle handling has been effective for improving cattle handling and welfare at slaughter plants (Grandin, 2010, 2012) and can be used in feedlots (Grandin, 2016; Paranhos Da Costa *et al.*, 2012) in order to pick up facility and handler problems and then correct them.

Infrastructure and handling problems (Figure 2.1A en 2.1B) examined at cattle markets usually include similar animal indicators like refusal to move, baulking, slipping, falling, and striking an animal (Bravo *et al.*, 2019; Gregory *et al.*, 2009a; Sánchez-Hidalgo *et al.*, 2020; Weeks *et al.*, 2002). Bravo *et al.* (2019) calculated a movement index based on several behavioural indicators (i.e. slips,

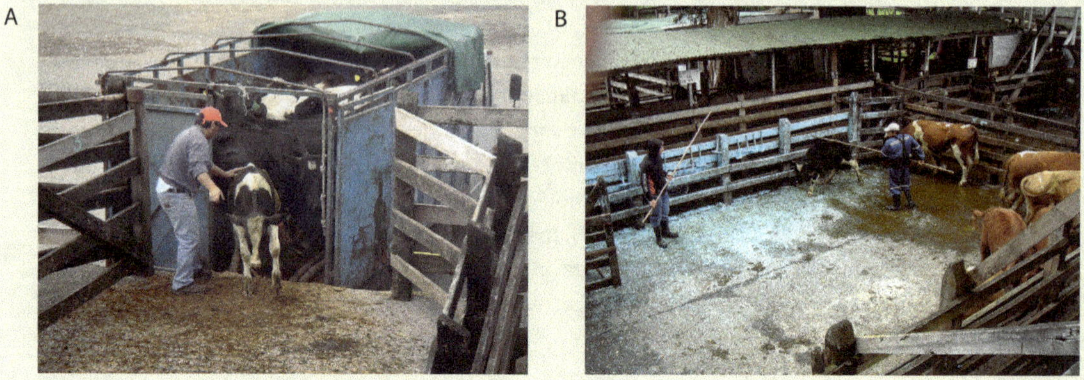

Figure 2.1. Infrastructure and handling problems at a livestock market during unloading: (A) wrong parking of the truck agains the ramp leaving spaces, too high ramp for the height of truck, person standing in front of the cattle being unloaded and only one truck door open; (B) moving cattle by pricking and hitting with sticks (C. Gallo, Universidad Austral de Chile, Chile).

falls, baulks, jumps and returns) observed during the driving of calves in Chilean cattle markets and found that the presence of inadequate lighting, and slippery flooring were associated with an increased movement index, indicator of reduced welfare. Poor handling, such as hitting, poking with a handling tool, kicking or flexing/pulling of the tail, have all been associated with an increase in movement index. Injuries due to striking or excessive and inappropriate use of handling devices on farm and cattle markets result in animal fear and pain as well as carcass bruises that need to be trimmed at the slaughterhouse (Strappini *et al.*, 2010, 2012; Weeks *et al.*, 2002).

Increased physical activity and fatigue are common when cattle are mustered on extensive fattening systems and handled before transport on farm. The mixing and separating of groups increase the number of social interactions, which may result in psychological stress and physical exhaustion. Mixing unfamiliar animals results in fighting, particularly in young bulls, until a new dominance hierarchy is established (Warriss, 2000; Warriss *et al.*, 1984). Animal to animal aggression during mixing can result in lesions and also increase the incidence of dark cutters (Warriss, 2000). For this reason, whenever possible, animals should be kept in stable social groups and avoiding mixing before transport. If social groups are to be split or animals removed for sale, then cattle should be sorted 1-2 weeks before weighing and selling (Schwartzkopf-Genswein *et al.*, 2012). Minka and Ayo (2007) found that cattle that exhibit higher levels of behavioural activities (weighed subjectively according to their possible effects on welfare, like falls, slips, aggressions/fights between animals, mounts and jumps), took longer to load (4.0 vs 1.3 min per animal). They also found a significant correlation (r=0.58 to 1.0) between the behavioural activities before loading and the number of injuries sustained by the cattle during transport.

An animal's response to handling and restraint depends on genetics, but also on previous experiences (Grandin 1993, 1998a). Two studies compared the stress of beef cattle raised on pasture and unaccustomed to handling to that of dairy cattle that were habituated to frequent handling and found that beef cattle had greater heart rates and blood cortisol levels during handling (Lay *et al.*, 1992a,b). The welfare of beef cattle can be severely affected by the quality of the stockmanship and particularly by a negative human-animal relationship (HAR), which is a key component of human and animal welfare (Hemsworth, 2003; Hemsworth and Coleman, 2011). The stockpersons' behaviour is a major variable determining an animals' fear of or confidence in human beings and, hence, the quality of the HAR. High quality HAR requires a certain level of positive human contact, and this is most likely in husbandry systems that involve regular, intense and long-term contact with humans (Waiblinger *et al.*, 2006). An animal's first experience with handling and restraint should be positive, as their memories of a negative handling event can last a long time (Grandin and Shivley, 2015). Animals learn to avoid conditioned stimuli associated with aversive events and, through this kind of learning (conditioning), their behavioural reactions may be regulated by the nature of previous experiences when interacting with humans (Hemsworth *et al.*, 2002; Waiblinger *et al.*, 2006).

Aversive or rough handling practices (frequent or not) has a negative influence on cattle behaviour, making animals fearful of humans (Hemsworth and Coleman, 2011). From the stockpersons' point of view, fearful animals are often more difficult to handle and manage making their job more difficult and increasing their risk of injury or death (Le Neindre *et al.*, 1996; Petherick 2005; Waiblinger *et al.*, 2006). This, in turn, exacerbates problems encountered during routine handling procedures and decreases job satisfaction, motivation, commitment and self-esteem (Hemsworth and Coleman, 2011). The development of a positive HAR, defined as low levels of fear or high levels of confidence in humans, can be beneficial for cattle and stockpersons. Ceballos *et al.* (2018) evaluated the potential impacts of training in good cattle handling practices on stockperson's attitudes and behaviour, and on cattle welfare in Brazilian farms keeping Nellore cattle in extensive conditions, during vaccination. They observed that farms with non-trained stockpersons had the poorest quality of handling, as well as more undesirable animal behaviours during handling. It is advisable that there should be repeated positive humane handling during rearing and immediately prior to transport, in order to minimise aversive reactions during loading and transportation (Probst *et al.*, 2013).

When cattle are sold through livestock markets, additional stressors include new environments, mixing with animals from other origins and categories (Gregory, 2008; Gregory *et al.*, 2009a) and handling by inexperienced stockpersons in a noisy and harassing environment (Bravo *et al.*, 2019; Romero *et al.*, 2013). Studies have also shown an association between handlers' attitudes and empathy towards animals and the quality of HAR at auction markets (Herrán *et al.*, 2017; Leon *et al.*, 2020).

Improving the skills of stockpersons in adopting welfare friendly handling practices is a key factor in achieving good animal welfare (Hemsworth and Coleman, 2011). Hemsworth (2003) gives examples of methods to select suitable stockpersons. Training stockpersons can be an effective and practical strategy to promote positive human-animal interactions on beef cattle farms and livestock markets improving the quality of life of both animals and workers (Coleman and Hemsworth, 2014).

2.1.3 Genotype, sex and temperament of cattle

Temperament is the behavioural response of an animal to close handling by humans. Good temperament means that animals can be approached and handled safely and easily during routine management procedures on or off the farm (Burrow, 1997). Nervous or reactive animals are not desired, because they represent a high risk for causing injury to handlers and themselves during routine handling and management (Grandin and Shivley, 2015). Currently, two simple and safe temperament evaluation methods, i.e. flight speed and crush score, are being used by the beef cattle industry to select for calmer animals (Terlouw and Bourguet, 2022 – Chapter 1 of this book). Slower flight speeds (measured objectively as m/s or subjectively as 1 = walk; 2 = trot; 3 = run, and 4 = jumping out of the crush) and lower crush scores (scores 1 to 5) are associated with calmer temperaments, which makes the animals less vulnerable to injury when handled on and off the farm (Grandin, 1993). These behavioural measurements of temperament are positively correlated with physiological measures of stress like catecholamines and serum cortisol concentration in cattle (Curley *et al.*, 2006; De Freslon *et al.*, 2014; King *et al.*, 2006). In fact, Braga *et al.* (2018) found that reactive Nellore and crossbred bulls, compared to intermediate and calm bulls of the same breed, had a significantly greater weight and larger cortical area of the adrenal glands, which is consistent with chronic stress.

Both the crush-test and flight-speed scoring are adequate tools to detect individual differences in beef cattle temperament under field conditions and have been shown to be repeatable and heritable (Benhajali *et al.*, 2010; Curley *et al.*, 2006; Hoppe *et al.*, 2010). They can be easily integrated in the routine weighing process and, if done at weaning, it may prevent additional workload for cattle handling and further stress for the animals in future preslaughter stages (Hoppe *et al.*, 2010). Animal temperament has been shown to be moderately heritable in beef cattle and therefore could respond to selection (Burrow, 1997). According to Haskell *et al.* (2014) the unweighted mean and range of heritabilities (irrespective of the models used) for temperament traits are in the same range [chute scores/response to restraint: 0.24 (0.03-0.67); flight speed: 0.36 (0.05-0.7); and docility: 0.26 (0.0-0.61)]. However, heritability estimates do vary between breeds, and are generally higher for *Bos indicus* breeds and crosses than for *Bos taurus* breeds; a mean of repeated measurements has usually higher heritability (Haskell *et al.*, 2014). Temperament traits are consistent across evaluations. However, the magnitude of the values of flight speed and crush score along time may decrease, i.e. animals become less fearful and reactive (Braga *et al.*, 2018; King *et al.*, 2006) or increase (animals become more fearful and reactive; Petherick *et al.*, 2002), which could depend on how fearful the animals perceived the repeated handling for measurements in each particular study.

In studies in which the rearing/fattening conditions are standardised, breed differences in temperament can be found (Haskell *et al.*, 2014). More reactive breeds have greater blood cortisol concentrations and exhibit more agitated behaviour after handling, transport, or restraint. For instance, Zavy *et al.* (1992) found that *B. indicus* beef calves had greater baseline cortisol concentrations (32.60±0.66 ng/ml) compared to *B. taurus* (25.81±0.76). Voisinet *et al.* (1997a) found that animals that had Brahman breeding had a higher mean temperament rating (3.45±0.09) or were significantly more excitable than animals that had no Brahman influence (1.80±0.10). Differences in the reactivity have also been found within *B. indicus* breeds. Mach *et al.* (2008)

observed a higher reactivity in Nellore cross cattle which made them more difficult to handle and increased the risk of them running into fences, thereby increasing muscular fatigue, falls in the truck and having reduced meat quality. However, Piovezan *et al.* (2013) reported that *B. indicus* Nellore cattle were calmer than Guzerat and Gyr. Within *B. taurus* breeds, cattle from continental breeds and their crosses (i.e. Chianina, Charolais, Limousin, Simmental, Friesian) have in general higher temperament scores than Hereford and Angus breeds and their crosses (Hoppe *et al.*, 2010; Morris *et al.*, 1994). Higher temperament scores imply more excitable cattle, and a reduced ability to cope with stressful situations which makes handling during the preslaughter stages more difficult. No studies comparing the temperament of dairy and beef cattle reared and handled together under the same standardised conditions are available.

Bulls exhibit in general more aggressive and reactive behaviours than steers, which in turn are calmer than bulls in the chute test (Grandin, 1993). Some studies have found that heifers have a higher mean temperament rating than steers (Hoppe *et al.*, 2010; Voisinet *et al.*, 1997a), whereas others have found no difference between male and female calves (Benhajali *et al.*, 2010). Café *et al.* (2011) found that heifers had greater crush scores than steers in Brahmans, but not in Angus.

Cattle temperament has an important role in their individual capacity to adapt to the feedlot environment with more reactive animals suffering chronic stress (Braga *et al.*, 2018; Gellatly *et al.*, 2020). Reactive cattle will probably experience greater amounts of fear when they are suddenly subjected to new experiences, like handling for loading, than animals with a calm temperament (Grandin and Shivley, 2015). Cattle that have had excessively calm or reactive temperaments may experience greater stress (as shown by higher blood cortisol level) at loading than those with moderate temperaments and will have more difficulties to cope with pre-transport handling (Burdick *et al.*, 2011). Carefully acclimating young animals to gentle handling and restraint can decrease temperament scores and can help improve both productivity and welfare by reducing fear stress (Probst *et al.*, 2013). An improvement of beef cattle temperament across successive handlings has been evidenced mostly in studies conducted with *B. indicus* and *B. taurus* animals housed in feedlots (Titto *et al.*, 2010; Turner *et al.*, 2011), but also on pasture (Ceballos *et al.*, 2018). Beef cattle not accustomed to handling had greater blood cortisol concentrations after restraint compared to dairy cattle accustomed to handling (Bulitta *et al.*, 2015). These studies suggest that animals with little previous exposure to people or having extreme temperaments, likely experience more stress and greater potential for negative outcomes during handling on farm and during the further stages of the preslaughter chain.

Cattle with calm temperament also result in benefits across a comprehensive range of beef production traits. Voisinet *et al.* (1997a) reported that calm *B. indicus* cross and *B. taurus* cattle had higher average daily liveweight gains than more excitable animals of the same breed. Petherick *et al.* (2002) recorded better body condition score and better feed conversion efficiency in good compared to poor temperament lot-fed *B. indicus* steers, although they found only a trend for a difference in average daily weight gain and no significant difference in carcass weight, dressing percentage or fat values. Café *et al.* (2011) found that *B. indicus* cattle with a calmer temperament, as measured by reduced flight speed and crush score, had superior performance in average daily weight gain, carcass weight, loin area and fat cover. Associations between temperament and growth rates, and carcass traits were found to be weaker in Angus than in Brahmans, probably due to a greater variation between measurements of flight speed and crush score observed in Brahmans (Café *et al.*, 2011). Bourguet *et al.* (2015) found that a calm response of Limousin and Aberdeen Angus cattle in the crush score test was associated with greater average daily gain during fattening. Temperament of young *B. indicus* bulls (Braga *et al.*, 2018) and heifers (Olson *et al.*, 2019) assessed at the end of the feedlot period showed that calm animals had better performance than the intermediate and reactive ones, and also produced heavier carcasses. Most of the above studies refer to feedlot cattle, but on pasture, calm Hereford and Bramford steers also had higher average daily liveweight gain than reactive steers of the same breeds (Del Campo *et al.*, 2010). Differences in productive traits would be based on the fact that poor temperament reduces dry matter intake and average daily weight gains through behavioural rather than metabolic mechanisms (Café *et al.*, 2011).

Similar to other species, in cattle stress is a risk factor for the occurrence of poor meat quality, i.e. higher ultimate pH (pHu; Ferguson and Warner, 2008), and therefore there is a risk that meat quality from more reactive animals is poorer. A number of studies have shown a relationship between temperament and meat quality. Voisinet *et al.* (1997b) reported that *B. indicus* cross feedlot cattle with excitable temperaments had tougher meat (higher shear force values) and a higher incidence of borderline dark cutters (visually assessed) than calmer animals. Café *et al.* (2011) reported that increasing flight speeds resulted in a higher pHu and darker meat colour, as well as increased shear force, compression and cooking loss, all considered detrimental for eating quality. Beef from calm Hereford and Bramford steers grazing pasture was also found more tender than that from reactive steers of the same breeds (Del Campo *et al.*, 2010). However, other studies have failed to find a clear association between temperament and meat quality. In lot fed *B. indicus* cattle, Petherick *et al.* (2002) neither found an effect of temperament on pH nor on eating quality. King *et al.* (2006) recorded that temperament categories (calm, intermediate, excitable) in Angus and Bonsmara cross cattle did not affect marbling score or meat colour. There appears to be no phenotypic association between meat quality and temperament in frequently handled *B. taurus* animals (Turner *et al.*, 2011). Bourguet *et al.* (2015) found that a calm response of Limousin and Aberdeen Angus cattle in the crush score test did not appear to relate to meat quality variation. In Red and Black Friesian steers finished on pasture, De Freslon *et al.* (2014) found that behavioural parameters related to temperament were not statistically associated with carcass bruising scores and meat pHu values. Moreover, Coombes *et al.* (2014) measured the effect of temperament on muscle glycogen concentration and on pHu in *B. indicus* and *B. taurus* and found no effect, although animals with high flight speed mobilised more glycogen before slaughter.

In conclusion, as handling temperament and other behavioural traits clearly have economic value and animals that respond poorly to handling suffer negative emotional and physical experiences that may result in reduced welfare, it seems important to improve temperament. The relationship between temperament and productive traits, like weight gain, carcass yield and yield grades in cattle seems also clear and positive. However, the relationship between temperament, stress, pH, and meat tenderness is not straight-forward. This is probably due to the fact that the effects of stress on muscle physiology depend on a number of other factors, such as preslaughter handling practices and even post-mortem meat processing procedures. Hence, selection against stress susceptible animals, and not only culling of cattle with the most reactive temperaments, can improve productivity and the safety and welfare of cattle, as well as the safety for human handlers.

2.1.4 Other animal factors and fitness for transport

Animal age, weight class, temperament, previous experience and health, all play an important role in an animal's ability to cope with the multitude of challenges cattle encounter within the transport continuum (Edwards-Callaway *et al.*, 2019; Schwartzkopf-Genswein and Grandin, 2019). Perhaps the most relevant and socially sensitive factor is the animal's overall condition or 'fitness' at the time of loading onto the transport vehicle (Grandin, 2001). This is because the probability of negative welfare outcomes (during or after transportation) has been shown to increase dramatically with reduced fitness (Edwards-Callaway *et al.*, 2019; Schwartzkopf-Genswein and Grandin, 2019). For example, a greater incidence of lameness, downer cattle and death was observed in cull cows and calves compared to feeder and fat cattle following transports of varying length (400 to 2,560 km) and environmental conditions (-42 to 45 °C) (González *et al.*, 2012b; Knowles, 1995). Cull cows also had greater liveweight losses than feeder and fat cattle (7.2 vs 4.3 and 2.3%, respectively) shipped similar distances (800-900 km; González *et al.*, 2012a), which provides further evidence of increased susceptibility to transport stress.

The increased risk of poor welfare outcomes in culls and calves has been explained by their advanced or young age which is associated with the reduced ability to control body temperature, particularly during inclement weather (Knowles *et al.*, 1997; Knowles, 1999), low energy reserves related to low body condition scores (<2 on a 9 point scale ranging from 1-3 = under conditioned or thin to 7-9 = over conditioned; Ahola *et al.*, 2011; Nicholson *et al.*, 2013) and reduced overall health, i.e. cull cows (Edwards-Callaway *et al.*, 2019; Gallo and Strappini, 2016; Figure 2.2). A recent descriptive study has documented that a significant portion (75%) of dairy cull cows transported for slaughter were lame (31%), had mastitis (20%) or severe wounds (22%). However, few of them were considered unfit for transport (Dahl-Pedersen *et al.*, 2018a). Similarly, Rezac *et al.* (2014) found that the prevalence of liver abscesses (18.5%) and rumen and lung lesions (10 and 10.3%, respectively) was substantial in cull cows (87% Holstein) at slaughter. Stojkov *et al.* (2020) reported that around 30% of the cows sold at livestock markets in British Columbia (Canada) had poor fitness for transport because they were thin, severely lame, had udder engorgement or inflammation, among other defects. Similar figures have been reported by Sánchez-Hidalgo *et al.* (2019, 2020) for cows sold at livestock markets or directly to the slaughterhouse in Chile. These cow's conditions on arrival at the slaughter plant result in carcass condemnations (Sánchez-Hidalgo *et al.*, 2019). To ensure the welfare of cattle during transport and limit economic losses, it is therefore recommended to use specific fitness-for-transport guidelines (Canadian Food Inspection Agency, 2021; European Union, 2017) that would improve on farm culling decisions, to train all stakeholders to consistently evaluate cattle fitness for transport, and to make a greater use of alternative options, such as direct transport-to-slaughter, emergency slaughter and euthanasia for severely compromised animals (Stojkov *et al.*, 2020).

Calves are also at higher risk of illness due to the fact that their immune systems and adrenal glands are not fully developed until 2-4 months of age (Barrington 2001; Knowles *et al.*, 1997) reducing their ability to mount a sufficient immune response to pathogens and cope with transport stressors. It is

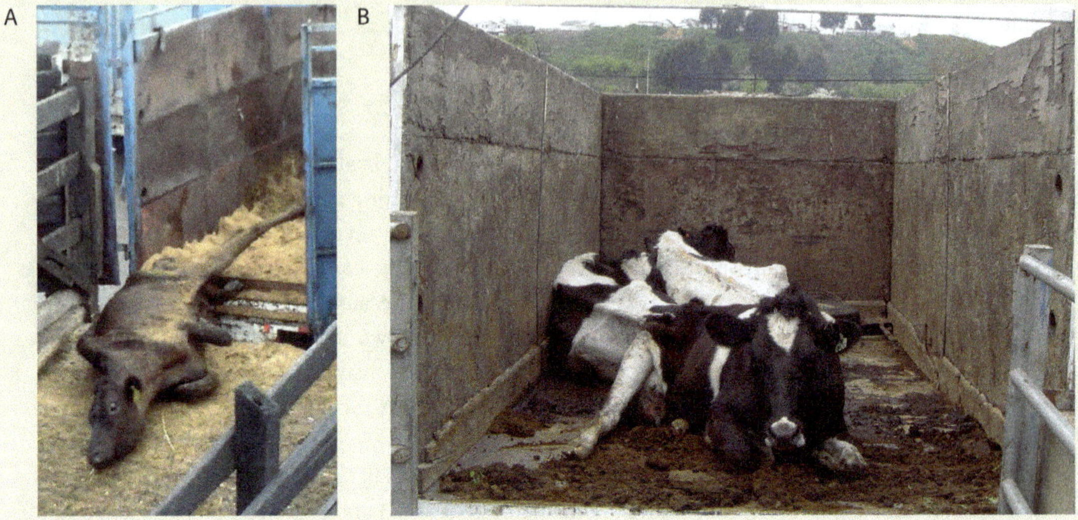

Figure 2.2. Assuring fitness before transport is decisive: culled dairy cows arriving non-ambulatory at (A) a livestock market and (B) a slaughterhouse (C. Gallo, Universidad Austral de Chile, Chile).

well known that the combined effects of pre-transport management (i.e. weaning, which includes separation stress and a change from milk to a solid diet, vaccination, castration, dehorning, ear tagging, handling, and mixing) conducted within a few days of transporting calves off the ranch is associated with the increased incidence of bovine respiratory disease (BRD) for the first 45 days in the feedlot (Eicher, 2001; Van Engen and Coetzee, 2018). This disease is the most prevalent and costly illness of feedlot calves to date, having significant implications for both antimicrobial resistance and poor calf welfare. In addition to pre-transport management stressors, calves are usually transported and sold through an auction market requiring they be unloaded/reloaded as well as comingled with unfamiliar animals, further increasing their exposure to pathogens and the likelihood of contracting BRD (González *et al.*, 2012b; Van Engen and Coetzee, 2018). To mitigate the risk of poor welfare outcomes it is recommended that calves not be transported before they are at least one month of age (Knowles *et al.*, 2014). In contrast, feeder and fat cattle are generally better able to cope with the stressors of transport due to their greater body condition score, mature immune system, ability to thermoregulate and overall condition concomitant with fit animals (González *et al.*, 2012c; Knowles, 1999; Knowles *et al.*, 1997). However, hoof lesions can be frequent in cattle finished in feedlots (Cozzi *et al.*, 2005; Gottardo *et al.*, 2003). Lameness and leg injuries are indicators of poor welfare and should be assessed in all types of housing systems before loading, in order to determine if animals are fit enough for transport or have to be transported separately. When lame or injured animals are sent for slaughter the risks of becoming worse and/or falling during transport are higher (Dahl-Pedersen *et al.*, 2018b). Improving understanding of lameness, detection and scoring of it is important to reduce pain in animals and also reduce economic losses (Tucker *et al.*, 2015).

Animal age and type, and breed are inherently associated with fitness as in the case of young dairy type calves that are unweaned and younger (0-30 days old) at the time of transport compared to beef calves that are typically weaned at 4 to 6 months of age before they are transported (Gonzalez *et al.*, 2012b) or cull cows that are typically removed from the breeding herd and shipped to an auction for marketing. One study found that the odds of being unfit (at an auction or abattoir) was 15.5 times greater in dairy than in beef cows (Heuston, 2017).

Although definitions for unfit and compromised animals exist (OIE, 2019), regulations vary by country and the identification of these animals is still based on the subjective visual evaluation and judgement of the individual(s) loading the cattle. Consequently, the potential for poor loading decisions remains high, particularly in cattle that do not exhibit obvious signs/conditions of compromise or being unfit. Transporting unfit cattle is not only inconsistent with animal welfare requirements by causing additional pain and suffering to animals, but can also lead to economic losses due to mortality and carcass condemnations (Cockram, 2019). Until recently, slaughterhouse records of carcass and organ condemnations in cattle were used to report prevalence of illnesses without an animal welfare approach. However, a broader vision, which includes the evaluation of post-mortem findings at slaughter as animal welfare indicators during the preslaughter period, has been proposed in a number of studies (Losada-Espinosa *et al.*, 2018; Miranda-de la Lama *et al.*, 2012; Romero *et al.*, 2012; Sánchez-Hidalgo *et al.*, 2019; Valkova *et al.*, 2021; Velarde and Dalmau, 2012).

2.1.5 Feed withdrawal

It is frequent that feed is withdrawn from cattle since the moment they are gathered from grazing fields or feedlot pens to be put in yards or corrals in preparation for transport directly to slaughter or through livestock markets. Although animals should never be deprived of water, this element is not always available in the corrals pre-transport and certainly not in vehicles during short distance transport. Feed and water deprivation before transport reduces digesta load in the gastrointestinal tract and decreases the risks of carcass contamination during slaughter. It also reduces fouling of other animals, the trucks and roads over which they pass and allows a more accurate prediction of carcass weight (Hogan *et al.*, 2007). In the case of feedlot cattle savings, feed and trucking savings (less liveweight transported), may be also considered (Janloo *et al.*, 1998). However, with more cattle being sold on a carcass weight instead of a liveweight basis and including payments for carcass quality, negative effects of feed withdrawal on carcass yield, fat cover, marbling grades and meat pHu values need to be considered (Ferguson and Warner, 2008). Long fasting times pre-transport can be also critical from an animal welfare point of view, because animals may suffer from hunger and thirst, increasing stress. The recommendation of any maximum curfew time on farm will depend on the subsequent time off food and water during transport and lairage at the slaughter plant (Hogan *et al.*, 2007). Therefore, fasting time on farm should be closely connected with the expected duration of transport and the expected slaughter time, as these factors will all add-up over time and increase stress and liveweight losses in cattle.

In young cattle entering feedlots, feed and water withdrawal for more than 24 h also appears to severely negatively impact the length of time that it takes the animals to regain live weight (Hogan *et al.*, 2007). In weaned beef calves (mean 146 kg) live weight decreased 10 kg per calf after 24 h of fasting (including 3 h transport) and was consistent with a significant increase in the concentration of betahydroxybutirate (0.26 to 0.35 mmol/l) after fasting, indicating that even with the short time off feed, the calves had to make use of their body reserves (Bravo *et al.*, 2018). In South America it is common to sell weaned calves for fattening through livestock markets, being submitted to over 24 h periods without feed and water (Bravo *et al.*, 2019; Gallo and Tadich, 2008) which could have consequences on the welfare and productivity of the calves during recovery periods. Selling cattle through auction markets not only increases fasting times, but also handling and stress, because animals have to be loaded, transported and unloaded several times and exposed to many stressors (Gregory, 2008). Although most of these calves are not sent directly to slaughter after these procedures, the effects of the increased fasting time (and increased handling stress) would apply if they are sent to slaughter; moreover this also applies to fat cattle being sold through the market for slaughter.

The main factors which influence the initial liveweight loss during feed deprivation are the amount and quality of feed consumed and the recent water consumption. Most of the live weight lost during the first 24 h corresponds to gut contents, with animals on high-roughage pasture having larger weight of gut contents than those fed grain diets, but losing weight more slowly (Bass and Duganzich, 1980). Fasting steers confined in farm pens for 3 and 16 h, without water and food, compared to homologous steers of the same farm transported for 3 and 16 h, at the same time and under the same conditions, caused less liveweight losses and fewer changes in blood variables related to stress, like cortisol, packed cell volume (PCV), leucocyte count and glucose (Tadich *et al.*, 2003).

For finished cattle producers, carcass weight loss is more important than liveweight loss. Carcass weight loss during feed withdrawal is mainly due to tissue dehydration and fat mobilisation (Warriss, 2000). Because animals necessarily have to be transported to a slaughterhouse to study the effects of fasting on carcass weight, fasting and transport losses may be confounded. Bass and Duganzich (1980) found that losses in carcass yield after 48 h fasting reached up to 8%. Economic losses due to reduced carcass weight and increase of dark cutters carcass penalties to the cattle feeder and packing plant from feed withdrawal on farm 24 h before sending cattle to slaughter overshadow the potential savings in feed and trucking (Janloo *et al.*, 1998). Results indicative of dehydration, as well as impaired meat quality, have been reported in cows that were deprived of water and feed for 18 up to 36 h (Vogel *et al.*, 2011). Gallo and Gatica (1995) found a reduction of 11.5% in live weight and 6% in carcass weight after 60 h of food deprivation in fat steers. In many parts of the world, like in Australia and South America, the total period off-feed in cattle for slaughter can be extended to more than 60 h due to long distance transport and sale through auction markets (Gallo and Tadich, 2008; Hogan *et al.*, 2007).

Fasting is also likely to reduce muscle glycogen concentration (MGC), because of energy expenditure due to physical exercise (i.e. mustering, mixing and agonistic acts, like butting, pushing and mounting) or psychological stress, which may in turn increase muscle pHu values and compromise key meat quality traits, such as colour, tenderness and shelf life (Ferguson and Warner, 2008; Warriss, 2000). The normal MGC in cattle ranges between 75 and 120 mmol/kg and if this concentration reaches a critical threshold of 45-57 mmol/kg, the normal meat pHu value of 5.5-5.6 will not be attained (Immonen *et al.*, 2000). A lower initial concentration of glycogen occurs in pasture raised cattle compared to those supplemented with grain (Apaoblaza and Gallo, 2014; Apaoblaza *et al.*, 2017) which may result in a higher risk of increased muscle pH values in cattle from extensive production systems under similar conditions of handling and fasting. Apaoblaza and Gallo (2014) determined through biopsies the live changes in MGC of housed steers fed on hay only or hay plus energy concentrate for 30 days and then determined the effect of 24 h food deprivation (fasting) on farm on the same variable. These authors found a 27 and 22% decrease in MGC in the steers fed only hay and in those supplemented with a concentrate, respectively, reaching after fasting 35.8 and 46.9 mmol/kg also, respectively, values close to the critical threshold. Depending on the extent of glycogen depletion, a substantial period of rest with access to water and feed of reasonable nutritional value may be needed before slaughter if dark, firm and dry (DFD) meat is to be avoided (Hogan *et al.*, 2007). However, this is usually not viable at a slaughter plant. In order to protect cattle from reaching the critical glycogen concentration during the inevitable stress of the further preslaughter stages, like transport and lairage, and the concomitant time off food, the initial concentration of glycogen can be increased on farm by supplementing pasture fattened cattle with grain for at least 3 weeks before slaughter (Apaoblaza and Gallo, 2014) and by reducing the time cattle are deprived of food during all preslaughter procedures (Amtmann *et al.*, 2006; Apaoblaza *et al.*, 2017; Ferguson *et al.*, 2001).

2.2 Transportation of cattle

Cattle are typically transported between 1 and 7 times during their lives to various production (ranch, feedlot), marketing (assembly yard, auction market) or processing (abattoir) facilities. They can be moved by hoof, road, rail, ship or air. However, in developed countries, road transport is the most common mode of transport due to its ease, flexibility and efficiency of relocating cattle (FAO, 2001). Therefore, this section will focus on the impacts of road transportation on indicators of cattle stress and well-being during the transport continuum which encompasses loading/unloading, waiting for departure, in transit, rest stops and stationary periods and waiting to off-load.

2.2.1 Human and facility factors during loading

Human (handler) interactions with animals have an obvious role in facilitating either positive or negative responses in cattle. Handler qualities, such as training in the use of good handling techniques and tools, experience with applying these techniques and understanding cattle behaviour as well as positive attitudes towards animal care and welfare, cannot be over emphasised. These qualities have the potential to reduce fear, excessive running and injury in all types of cattle, but particularly in animals unaccustomed to people/handling (Figure 2.3) or with poor temperament as previously discussed. Unfortunately, rough handling, characterised by excessive yelling, running, and prodding by stock personnel lacking experience and training or having poor attitude, is still common and is frequently implicated in increasing animal stress.

Handler quality has steadily improved at commercial abattoirs as a consequence of in-house welfare audits that focus on ensuring good handling practice (Grandin, 2017). However, similar on-farm and auction market audits are rare, increasing the risk of poor handling which is highly dependent upon the attitudes and practices of individual producers and transporters loading/unloading and

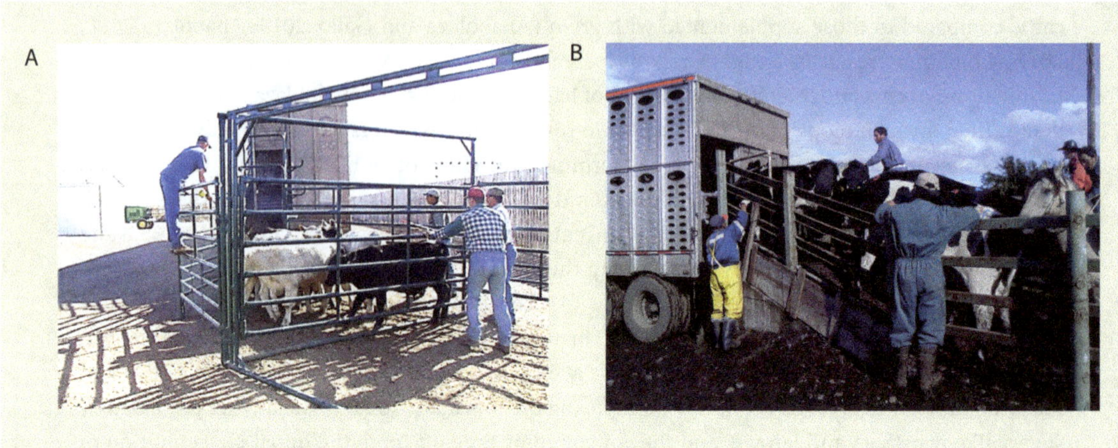

Figure 2.3. Ranch calves being loaded onto a truck for transport to a local feedlot in Canada. (A) Loading is a stressful event fc calves that are not typically handled beforehand. (B) Dairy cows used to interact with people are calmer to be handled durin loading (photo A: K. Schwartzkopf-Genswein, AAFC, Canada; photo B: C. Gallo, AWP, Chile).

hauling the cattle. Previous studies have reported that loading and unloading resulted in greater heart rate and plasma cortisol concentrations (Booth-McLean *et al.*, 2007; Fell and Shutt 1986; Grigor *et al.*, 2001; Warriss *et al.*, 1995) than transport (Pettiford *et al.*, 2008). One study compared the stress of loading with unloading and found a larger increase in plasma cortisol, creatine kinase (CK) and lactate concentrations in loaded compared to unloaded cattle (María *et al.*, 2004). It should be noted that these earlier studies did not include an evaluation of handling quality, animal temperament or experience making it difficult to separate the impacts of human from animal stress factors, including novelty and physical exertion of negotiating ramps onto or within the transport vehicle. Brunel *et al.* (2018) conducted one of the few studies assessing handling quality (rough vs gentle) in bulls (*B. indicus)* prior to being transported for 24, 48 or 72 h on their stress, immune, fatigue and energy reserve responses. Overall, roughly handled bulls were found to have greater blood cortisol concentrations and reduced immune function (i.e. elevated neutrophils and lower lymphocytes numbers) than gently handled cattle. More specifically, greater blood CK levels, indicator of muscle damage, and fructosamine, indicator of feed deprivation, were observed in rough compared to gently handled bulls after 72 h of transport. This study supports the fact that multiple stress factors contribute to greater physiological changes implying greater risk of reduced welfare.

There is a significant body of literature discussing the relationship between proper handling, loading facility design and animal stress during handling. Optimal loading facilities (ramp surface, angle, tread length, etc.) and handling guidelines that reduce baulking during loading/unloading have been well documented (EFSA, 2011; Faucitano and Pedernera, 2016; Grandin 1997a,b, 1998a,b, 2001, 2007). To ease handling and limit the occurrence of lesions the ramp angle should not be steeper than 20° for calves or 26° for adult cattle. When the slope is steeper than 10°, ramps should be fitted with foot battens or cleats in order to improve animal stability on its feet while walking on the ramp reducing the risk of slips and falls (EFSA, 2011; Faucitano and Pedernera, 2016). To ease cattle handling through the ramp a 20 cm spacing between the cleats is recommended (Grandin,

2014b), It should be noted that the recommendations for facility design have largely been based on common sense, years of visual observations and experience evaluating handling efficiency (i.e. ease of moving animals, and handling related injuries, slipping and falling or deaths) rather than assessing measured indicators of stress associated with varying facility configurations. In addition, most studies assessing the impacts of facility design on animal welfare are related to carcass bruising rather than the stress of moving animals through them and will be discussed in more detail in a later section.

2.2.2 Driver skills and experience

The quality of transport experienced by the cattle is associated with a combination of driver's experience and driving quality. Driver experience refers to the number of years and loads of live cattle a hauler has transported during their career and is typically associated with the increased ability to manage risk of poor welfare outcomes. These risks include, but are not limited to selecting optimal transport times or routes to reduce the effects of extreme weather and or poor road conditions, minimising delays and managing trailer microclimate through the use of bedding, boarding or reducing stationary periods (Schwartzkopf-Genswein and Grandin, 2019). Driving quality refers to driving skill that can either increase or decrease poor animal outcomes including speed, cornering, stopping and breaking. Research on the effects these factors have on animal stress during transport are currently lacking. Survey data from Spain (Miranda-de la Lama *et al.*, 2014) and the US and Canada (Woods and Grandin, 2008) show that driver fatigue and lack of driving skills were causes of road accidents, especially during cattle transportation.

In one of the few studies evaluating the effects of driver experience, González *et al.* (2012b) found cattle transported by drivers with ≥5 years of cattle hauling experience had reduced incidence of lameness, becoming non-ambulatory and death than drivers with <5 years of experience. Driver experience and training is recognised to improve animal condition during transport and at unloading and many transport companies now require their drivers successfully complete training (e.g. US Transport Quality Assurance) which includes knowledge of animal behaviour, driving skill, handling techniques, proper loading density, vehicle maintenance, emergency preparedness, etc.

2.2.3 Transport duration

Transport duration is the most studied of all the transport related stressors. This is likely a function of its relationship to weight loss and production economics as well as public concern. Collectively, these studies provide a significant body of evidence showing the greater the duration of transport, the greater the likelihood of reduced welfare (Broom, 2008; Knowles *et al.*, 2014). The main contributors of reduced welfare have been identified as water and tissue loss, depleted energy reserves, dehydration, fatigue and injury (Schwartzkopf-Genswein and Grandin, 2019; Van Engen and Coetzee, 2018). Reported live weight loss, also known as shrink, ranges between 0 and 16% and is the result of faeces, urine, evaporation, and respiration losses within the first 12 h of transport (Barnes *et al.*, 2004; Coffey *et al.*, 2001; Jones *et al.*, 1988, 1990), while weight loss after that is related to tissue loss resulting from mobilisation of energy reserves (Van Engen and Coetzee, 2018). Weight losses exceeding 10% of the animal´s body weight have been attributed to such factors as energy required for maintaining balance, reduced ability to lie down, feed and water deprivation and exposure to extreme ambient temperatures for extended periods of time which all increase the potential for poor welfare outcomes. This has been confirmed by several cattle studies documenting that shrink losses of ≥10% resulted in the mobilisation of body reserves and dehydration (Knowles, 1999; Tarrant *et al.*, 1992; Warriss *et al.*, 1995) and increased the risk of cattle mortality (González *et al.*, 2012a). Mobilisation of energy reserves as assessed by blood non-esterified fatty acids (NEFA), glycogen and fructosamine increase in cattle transported ≥24 h in comparison to non-transported or cattle transported <12 h (Knowles *et al.*, 1999; Tarrant *et al.*, 1992; Warriss *et al.*, 1995). Increased dehydration as assessed by blood PCV, total protein and albumin, and osmolality (Terlouw and Bourguet, 2022 – Chapter 1 of this book) has been also observed in cattle transported >24 h compared to pre-transport levels (Tarrant *et al.*, 1992; Warriss *et al.*, 1995). Signs of feed deprivation and dehydration are not surprising given that feed and water are typically not provided on board trailers in most countries. Muscle fatigue is evaluated using blood metabolites, such as lactate, CK, and glycogen (Terlouw and Bourguet, 2022 – Chapter 1 of this book) and several studies have reported these metabolites increase as transport durations increase from, for example, from 12 to 36 h (Meléndez *et al.*, 2020) or from 2 to 11 h (Gebresenbet *et al.*, 2012). The relationship between transport duration and injury (other than bruising, such as lacerations, broken horns, etc.) has not been well documented in the literature, although it is likely that opportunities for injury and bruising increase the longer the cattle remain on the truck. Some studies have documented a positive association between transport duration and risk of illness, such as BRD, once cattle are placed in the feedlot (Cernicchiaro *et al.*, 2012; Hay *et al.*, 2014). Although several studies recommended that transport duration not exceed 24-30 h (González *et al.*, 2012b,c; Knowles, 1999; Tarrant *et al.*, 1992), they lack comparisons across varying animal types and fitness. Regulations governing the duration of livestock transport differ among countries. According to the Canadian Health of Animal Regulations (Canadian Food Inspection Agency, 2020), the maximum allowable transport time for adult cattle is 36 h and then cattle must be rested for a minimum of 8 h. Regulations for the European Union indicate that journeys should not exceed 8 h for adult cattle, but can be extended to 14 h providing the vehicle meets additional requirements at which point a minimum rest period of 1 h (with water) must be provided and then they can be transported for an added 14 h (European Union, 2004). The US has a '28-h Law' stating that cattle cannot be confined on a vehicle for more than 28 consecutive hours without being unloaded for access to feed, water and rest for at least 5 h (USDA, 1994). The regulations indicated above are for adult cattle however, each country, with the exception of the US, has different requirements for young calves (under 6 months of age) that allow less time in transit before a rest period is required. Other transport duration guidelines can be found in Miranda-de la Lama *et al.* (2014).

2.2.4 Space allowance

The amount of space provided to cattle during transport, also known as loading density, is yet another factor that can influence the quality of cattle transport. Loading density is conventionally expressed as unit of body weight (kg) per available space (m^2) and space availability as m^2 per animal (Schwartzkopf-Genswein and Grandin, 2019). However, researchers more commonly use an allometric coefficient (k-value) calculated as the observed loading density/(body weight$^{0.667}$) which removes the need to consider animal weight when making comparisons across studies (Petherick and Phillips, 2009).

Loading density can be influenced by several factors, including animal age, weight and frame size, fat cover (or body condition score), fitness, or presence of horns, the number of axles on tractors and trailers (the greater the number of axles the more weight can be hauled), weather conditions, road regulations, and the total distance they are shipped. However, most of these factors have not been studied and therefore there is little science-based information from which to improve loading density recommendations tailored to specific scenarios. Cattle tend to be loaded more densely than recommended because hauling costs are calculated using both transport distance and truck weight, meaning that the closer the trailer is to maximum allowable weight the more cost efficient the transport will be. This has greater implications for lighter weight animals (i.e. calves) that are loaded into larger compartments, such as the middle compartments of the bottom and top decks of the trailer, since more calves can be loaded within those compartments or the trailer before maximum weight limits are reached (González *et al.*, 2012c). Loading density was found to be 56% greater in the front compartment of the top deck 'nose' compared to the bottom compartments (average k-value = 0.029 vs 0.017) which means that cattle within the same load may be transported with either too much (low density) or too little (high density) space which are associated with increased incidence of shrink, injury or death (González *et al.*, 2012c). In general, studies assessing loading density have reported greater blood cortisol and CK concentrations, bruising, losses of balance and falling, as well as inhibited exploratory behaviours, movement and choice of standing orientation at densities >550 kg/m^2 compared to <550 kg/m^2 (Tarrant *et al.*, 1988, 1992). Several studies recommend that loading densities exceeding 0.035 k-value (corresponding to 1.75 m^2/animal or 550 kg/m^2) should be avoided to mitigate poor welfare outcomes (González *et al.*, 2012c; Petherick and Phillips, 2009; Tarrant *et al.*, 1992).

2.2.5 Resting periods

By definition the provision of a 'rest period', in which cattle are offloaded so they can eat, drink and lie down, should benefit all animals, and particularly those that have been transported for extended periods of time (>24 h) under adverse conditions (high loading density, extreme temperature, no feed or water, rough driving, etc.; Cooke *et al.*, 2013; Flint *et al.*, 2014; Ross *et al.*, 2016).

Despite its potential for reducing transport stress, relatively few studies have assessed the impacts of rest periods on cattle welfare. One of the main criticisms of providing a rest has been the need to offload/and reload the cattle, which is known to increase stress responses, injury, mixing with unfamiliar animals, exposure to a novel environment, and extension of the entire transport period (Swanson and Morrow-Tesch, 2001). Consequently, two main questions still need to be addressed (1) do rested cattle (mid-journey) have reduced indicators of stress and improved welfare than non-rested cattle and (2) if there is a benefit, what length of rest (proportionate to the transport duration) is most beneficial? Studies assessing varying rest stop durations from 0 (no rest) to 15 h have reported variable results. Cooke *et al.* (2013) showed that calves that were not rested had 10.1% more shrink, and greater blood NEFA and cortisol concentrations compared to rested calves. However, another study found that cattle rested for ≥10 h (following 20 h of continuous transport in a stock trailer) did not prevent increased short- and long-term stress (as assessed by blood cortisol concentration) or improve average daily gain for the first 25 days after transport when compared to unrested cattle (Marti *et al.*, 2017). Two commercial scale trials reported similar findings where, with the exception of blood NEFA concentration, indicator of food deprivation, the level of dehydration, fatigue or reduced immune function did not differ between calves given 4, 8 or 12 h of rest (compared to no rest) following 12 or 36 h of transport (Meléndez *et al.*, 2020) or calves preconditioned (or not) and sourced from either an auction market or ranch given 8 h of rest (compared to no rest) following 36 h of transport (Meléndez *et al.*, 2021). In fact, greater transport distances and lack of conditioning had much greater effects on increasing indicators of stress, such as haptoglobin and NEFA concentrations, shrink and lying percentage as well as reduced body weight, average daily weight gain and dry matter intake than lack of rest (Meléndez *et al.*, 2020, 2021). Cattle were only transported an additional 4 h after their assigned rest in both studies. Collectively, the research suggests that fitness for travel may have more profound benefits than other strategies in mitigating transport stress. This is not to imply that rest periods are not beneficial for certain types of cattle or under extreme environmental conditions and continued research is needed that further defines the optimal length and quality of the rest period.

2.2.6 Environmental factors

As homeotherms, cattle have the ability to adapt to ambient temperature changes gradually over days or weeks. However, abrupt changes in ambient conditions are much more challenging and can result in greater negative impacts during transportation particularly when they exceed the thermal neutral zone for cattle. Environmental conditions within the trailer, also referred to as microclimate, can be affected by several factors including airflow patterns, ambient temperature and humidity, solar loading, presence of bedding, as well as animal density, respiration, sweat, and excretions (Schwartzkopf-Genswein and Grandin, 2019). The majority of livestock trailers use passive ventilation to manage heat and humidity build up as they are not climate controlled. This is accomplished using perforations of various sizes and configurations along the walls of the trailer in addition to roof hatches that can be opened or closed. Ventilation can be further controlled through the use of boarding whereby solid pieces of plywood or plastic are placed over the perforations so that air flow can be fully or partially obstructed. Studies have shown that passive ventilation is greatest when a trailer is in motion and lowest when it is stationary (Bryan, 2013). In fact, one strategy to reduce poor animal outcomes during hot conditions is to ensure the trucks remain moving, if possible (Figure 2.4). It is not surprising that most studies assessing the effects of trailer microclimate on cattle welfare have reported significant associations with extreme ambient temperatures. For example, death rates were observed to increase when ambient temperatures dipped below -15 °C, while becoming non-ambulatory increased when temperatures rose above 30 °C (González *et al.*, 2012b). Another study reported that summer temperatures had a greater impact on trailer microclimate than loading density (Goldhawk *et al.*, 2014a,b). These findings suggest that calves and cull cows may be more affected by temperature extremes during transportation due to their reduced ability to thermoregulate and explain why the provision of ample dry bedding and boarding are particularly important for these cattle. Most countries have regulations enforcing the use of bedding during cold weather transport (Faucitano and Pedernera, 2016).

Figure 2.4. Loads of fat cattle being transported from Canada to the USA for slaughter. Trucks must cross the Canada/USA border where trucks may remain stationary for varying periods of time. During periods of hot weather there is significant risk of heat stress due to lack of ventilation as a result of being stationary (K. Schwartzkopf-Genswein, AAFC, Canada).

2.3 Handling at the slaughter plant

The different events involved during pre-slaughter handling of cattle at a slaughter plant, such as unloading and driving to lairage pens, environmental conditions and duration of lairage, feed and water deprivation, are all stressing factors that affect animal welfare and can also affect meat quality (Ferguson and Warner, 2008; Gallo *et al.*, 2016). Because consumers demand higher animal welfare standards, there is a pressure to adhere to good welfare protocols and this has led to improvements in welfare practices at slaughter plants (Grandin 2010, 2020). Assessing welfare at slaughter can also have a significant economic impact due to the links between animal welfare and meat quantity and quality (Gallo and Huertas, 2016). Several welfare protocols have been proposed and discussed to be used at the slaughter plants (Grandin 2007, 2020; Velarde and Dalmau, 2012; Wigham *et al.*, 2018). Although there is no consensus on one specific protocol that should be used, the most important aspect is that welfare problems at the slaughter plants need to be identified, measured, monitored and managed in order to make improvements.

2.3.1 Unloading

The welfare of animals at arrival at the slaughter plant will depend on factors at the premises, but also on factors previous to arrival, like the fitness for travel, the handling during loading and the transport duration and conditions, which have been discussed earlier in this chapter. Similarly to loading, the quality of unloading is strongly influenced by the handling methods, but also the category of cattle (Bulitta *et al.*, 2015). According to the behavioural indicators observed in the animals, unloading is usually easier than loading cattle (Bravo *et al.*, 2019; María *et al.*, 2004). If unloading facilities, like ramps, are according to recommendations, in terms of height (ideally less than 20 cm difference between ramp and truck height), maximum slope of 20°, solid fences and non-slippery floor (Grandin, 2007), just by opening the truck doors the animals will naturally leave the truck, unless they are injured or sick. Monitoring mobility during unloading allows for the detection of lame and injured cattle, which is common in culled cows (Grandin, 2007; Sánchez-Hidalgo *et al.*, 2019), but also in feedlot cattle (Edwards-Callaway *et al.*, 2017). Lameness scoring is also useful to detect which producers, dealers or transporters have delivered high percentages of affected cattle (Grandin, 2020). During unloading, the same behavioural indicators of cattle used on farm or at livestock markets can be used for assessing welfare, such as slips, falls, baulks, returns and vocalisations (Grandin, 2017). The behaviour of stockpersons can be assessed by monitoring improper handling methods, like shouting, kicking, hitting animals or twisting their tails, and the percentage use of electric goads.

The health conditions of the animals should be assessed upon their arrival for any animal welfare and health problems. Injured or sick animals, requiring immediate slaughter, should be killed humanely and without delay, in accordance with the recommendations of the OIE (2019). Appropriate handling and euthanasia of non-ambulatory or disabled animals is critical. These animals, usually culled cows, are prone to suffering pain associated with lameness and other diseases and should be unloaded avoiding further welfare problems (Gallo and Strappini 2016; Grandin, 2001, 2007). Non ambulatory and fractious animals should be checked by the veterinary inspector and slaughtered (or euthanised, depending on the general compromise of health) as soon as possible (emergency slaughter) after arrival.

2.3.2 Moving cattle

Stress and lesions due to moving cattle between the unloading point and the lairage pens, as well as from the lairage pens to the stunning point, can be minimised by ensuring that the animal handling facilities are designed, constructed and maintained to favour a smooth and efficient movement of livestock (Grandin and Shivley, 2015). Distractions that may cause approaching animals to stop, baulk or turn back should be designed out of new facilities or removed from existing ones (Grandin, 2010, 2020). The lairage area should be designed to allow a one-way flow of animals from unloading to the point of slaughter. Passageways and races should have solid sides and floors should be well drained and not slippery, with appropriate lighting, adequate to facilitate animal movement (Grandin, 2007; Grandin and Shivley, 2015). Moving cattle through narrow raceways, with the presence of right angles and protruding objects, such as door hinges, metalwork or nails can result in injury and pain for the animals and in bruises in the carcasses. Poor design or maintenance of structures, inadequate lighting or presence of distractors, such as objects on the floor, drainages or change of type of flooring on the passageway, sounds of metal or others, can result in animals baulking or refusing to move, making the handler´s job more difficult. It can also lead to the overuse of movement aids by handlers, such as electric cattle prods, sticks, slapping and shouting, and cause agitation and distress in both cattle and stockpersons. A common problem in races for cattle is the guillotine door being dropped on the back of animals, which leads to bruises on the loin and back of the cattle (Strappini *et al.*, 2013; Figure 2.5). To avoid this type of bruise, the guillotine door bottom edges should be cushioned and the doors handled correctly by the personnel.

Poor handling practices need to be monitored continuously and most used welfare indicators at this level are behavioural and related to the ease of movement through passageways and races, such as the percentage of electric prod use, falls and vocalisations of cattle, and wilful acts of abuse by stockpersons (Grandin, 2010). Cattle vocalise after adverse events, such as electric prod use, slipping or excessive pressure by restraining equipment. The percentage of cattle slipping, and falling while being driven through the race from lairage to stunning reflect difficulties during handling and are a useful tool to assess the effects of improvements resulting from handler training or facility design (Grandin, 2017). It should be noted that different sources (farm or market) and types of cattle (breed, category, size, temperament) may react differently to similar slaughter procedures and this could have consequences for subsequent handling procedures and meat quality (Bourguet *et al.*, 2011; Jarvis *et al.*, 1995; Strappini *et al.*, 2012). For instance, Bourguet *et al.* (2011) reported in their study at a slaughterhouse in France that all animals were not subjected to the same procedures due to organisational aspects (i.e. bulls were slaughtered first because of being more reactive and females remained longer in lairage) and that within a same procedure, like being put in the rotary box for Halal slaughter, kicking was more frequent in Blond d'Aquitaine than Charolais males. Similarly, in Chile, steers and heifers for export (higher value) are slaughtered first in the line and cows (lower value) at the end. Regarding animal differences, Muñoz *et al.* (2012) found that in the stunning box for captive bolt shot, behaviours like vocalising, falling and struggling were more frequent in heifers than in steers or cows.

Low stress livestock handling at slaughter plants has the benefit of improving both animal welfare and meat quality. Links between handling at slaughter and meat quality (i.e. altered post-mortem

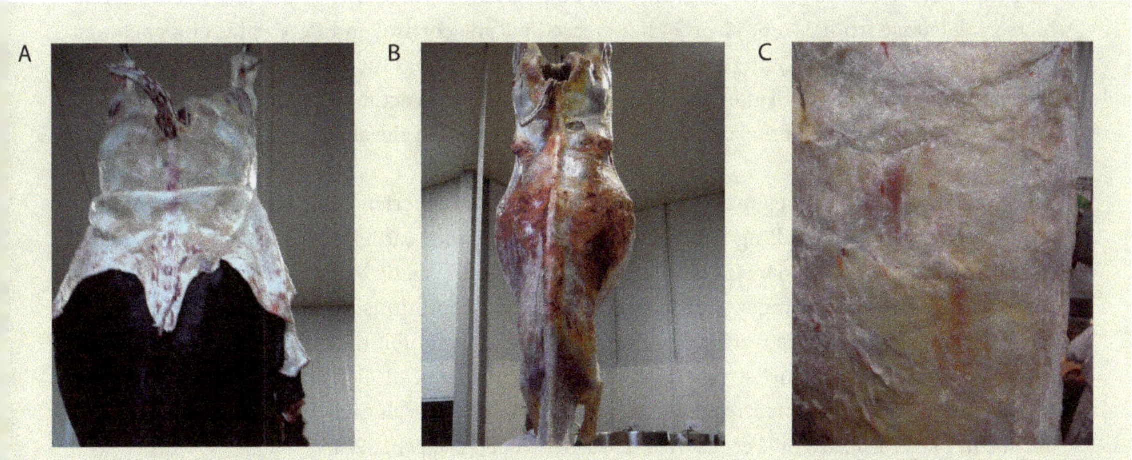

igure 2.5. Consequences of poor handling are visible on cattle carcasses after dehiding: (A) bruise due to hitting on the back ith guillotine door; (B) bruised carcass due to trampling on fallen cattle during transport; (C) bruises due to pricking with electric oads (C. Gallo, Universidad Austral de Chile, Chile).

muscle pH value decline) have been reported (Bourguet *et al.*, 2015; Warner *et al.*, 2007) and may also include bruises due to inadequate human-animal interactions like prodding or hitting animals with the driving instruments (Jarvis *et al.*, 1995; Strappini *et al.*, 2013; Figure 2.5). The use of procedures that cause pain to the animals, such as pricking or hitting cattle with sticks, is forbidden (OIE, 2019). The use of electric goads is permissible in certain situations for cattle, but an inappropriate application of this tool (too frequent or too long application or too high voltage) can also produce pain and leave marks on the carcass (Gallo *et al.*, 2016). The stress caused by the use of electric goads prior to slaughter has been shown to negatively affect water-holding capacity and consumer acceptability of loin meat (Warner *et al.*, 2007).

To minimise stress and improve welfare during unloading and moving animals within the slaughter plant, the handling crews should be composed by personnel in sufficient number and trained in humane cattle handling principles and OIE standards (Coleman and Hemsworth, 2014; Hemsworth *et al.*, 2002; OIE, 2019).

2.3.3 Lairage conditions and duration

The lairage pens should provide animals a reasonably comfortable place to stay for some time, so that veterinary inspection can be performed and animals can 'rest' before being slaughtered, although resting is difficult considering that this is an unfamiliar environment for the animals.

During lairage the provision of water may help cattle rehydrate, improving welfare and reducing carcass yield losses. However, not all animals will drink readily. Jarvis *et al.* (1996a) observed that after arrival, cattle were more thirsty and tired if they had a long transport journey and originated from livestock markets. On the contrary, after a short transport journey, steers and cows show more agonistic behaviours during lairage, lie down less and approach water troughs less than after long transport journeys (C. Gallo, unpublished data). However, Tadich *et al.* (2005) showed that increasing lairage from 3 to 6, 12 or 24 h increased plasma cortisol and PCV values independent of the transport duration (3 or 16 h), whereas blood CK level was increased at arrival and did not return to normal values until after 24 h lairage. Hence there is no actual recovery or beneficial effect on the stress and welfare of the animals by a long lairage time at the abattoir.

Cattle arriving from livestock markets to the slaughterhouse experience additional periods of food and water restriction, handling and transport compared with cattle sent directly from farms to slaughter (Jarvis *et al.*, 1996a; Strappini *et al.*,2012). Exposing cattle to various adverse conditions prior to slaughter, like hunger, thirst, adverse climatic conditions, unfamiliar animals, noise, muscular fatigue, and space restrictions during lairage, may inhibit their ability to rest or recover from handling and transport stress (Grandin, 2007; Jarvis *et al.*, 1996a,b; Weeks, 2008), and can deplete muscle glycogen stores, resulting in elevation of meat pH values and alteration of the normal decline in pH post-mortem (Ferguson and Warner, 2008;Teke *et al.*, 2014).

Recommendations during lairage include having protection from adverse climatic conditions (roofing in cold climates and shade in hot climates) and in general reducing physical and psychological stress in cattle. Lairages can be relatively noisy compared with most farm environments, due to the fact that animals are being continually moved in and out of pens, with the necessary opening and closing of gates and doors, arrival of new animals, people moving around and the noise produced by the functioning of slaughter equipment (Bourguet *et al.*, 2011; Weeks, 2008). Cattle situated near a noisy environment exhibited more movement than cattle held in pens in quieter locations (Eldridge *et al.*, 1989), which impairs resting and could cause faster depletion of muscle glycogen stores and alter normal pH value decline.

Ruminants have a wide range of thermoneutrality from about 10 to 30 °C and therefore are able to tolerate well most lairage conditions (Weeks, 2008); for comfort and good welfare animals should not be observed panting. In South American countries, like Brazil, and sometimes Argentina and Chile, where ambient temperatures can exceed 30 °C, sprinklers are used to increase comfort during lairage in cattle. Regarding ammonia levels, Weeks (2008) reported that in some cattle lairages in the UK, the mean daytime ammonia levels in summer were as high as 18 ppm, which is close to the recommended maximum of 25 ppm, thereby indicating that ventilation rates were insufficient and that fan assisted ventilation may be needed. Considering that for humans, ammonia levels should be kept under 3 ppm, ammonia levels should be monitored more carefully.

When confined in lairage pens, each animal should have room to stand up and lie down, and to turn around (OIE, 2019). According to Kline *et al.* (2019), this space is equivalent to 1.86 m^2 per 544 kg liveweight beef cattle and approximately an additional 0.093 m^2 (1.0 ft^2) per 45.36 kg (100 lb) over the 544.31kg (1,200 lb) base liveweight should be accounted for in case of heavier cattle. A similar relationship could be also used for somehow lighter cattle for calculating pen stocking densities at the slaughter facility.

Mixing animals from different origins during lairage will increase stress and reduce glycogen reserves due to greater physical activity or emotional stress, all leading to high meat pHu values, especially in bulls as they show more aggressive behaviours (i.e. mounting, butting and teasing) than steers (Warriss *et al.*, 1984). Therefore, bulls or active groups in lairage should be separated and/or slaughtered as soon as possible (Raj *et al.*, 1991; Weeks, 2008).

Time in lairage should be minimised and should not exceed 12 h according to OIE standards (OIE, 2019). For animal welfare reasons, the slaughter order should prioritise (shorter lairage) culled cows, because they are generally thinner, older and more fragile, often suffering from conditions, such as mastitis and lameness (Gallo and Strappini, 2016; Sánchez-Hidalgo *et al.*, 2019). The duration of fasting in lairage has to be added to the fasting during previous preslaughter stages on farm, during transport and eventually during their stay at livestock markets. Hence, if animals are not to be slaughtered within the 12 h period, suitable feed should be available along with drinking water, which should be available at all times, in order to limit reductions in welfare, carcass yield, bruises and meat quality (Gallo and Gatica, 1995, Gallo *et al.*, 2003).

The longer cattle spend in lairage, the greater the risk of carcass weight loss, bruising and dark cutters (Amtmann *et al.*, 2006; Ferguson and Warner, 2008; Strappini *et al.*, 2013), particularly if they had a prolonged transport previously (Gallo *et al.*, 2003). As the duration of feed and water deprivation extends beyond 48 h, tissue catabolism and dehydration increase their contribution to liveweight loss and carcass yields will be negatively affected (Ferguson and Warner, 2008; Gallo *et al.*, 2003).

In cattle, stress prior to slaughter can deplete muscle glycogen stores, resulting in alteration of the normal decline in pH value post-mortem and a higher meat pHu value (Apaoblaza *et al.*, 2017; Ferguson and Warner, 2008; Teke *et al.*, 2014). However, reported effects of lairage duration on meat quality in cattle may vary between studies and countries in the world because of differences in production systems (e.g. pasture fed vs grain fed), transport duration and handling conditions, as well as cattle temperament. Del Campo *et al.* (2010) and Del Campo Gigena *et al.* (2021) in Uruguay found that carcasses from cattle transported for 3.5 h with 15 h lairage had a better rate of pH value decline and more tender meat than cattle with 3 h lairage, whereas Gallo *et al.* (2003) in Chile found that the risks of reaching higher meat pHu value become greater when steers transported for 3 and 16 h remained in lairage for more than 3 h. Coincidently, Alende *et al.* (2014) in Argentina reported that after up to 8 h transport, a lairage time of 14 h compared to 2 h impaired both physiological indicators of stress and beef quality in the same animals. Steers transported for 3 to 24 h and kept in lairage for 24 h had 9.4 greater odds of having dark cutters (muscle pHu value >5.8) than those that were transported for the same time, but remained in lairage for 3 h (Amtmann *et al.*, 2006). The increase in dark cutters can be mainly attributed to a decrease in muscle glycogen concentration due to the concurrent effects of longer fasting times (Amtmann *et al.*, 2006: Apaoblaza and Gallo, 2014; Ferguson *et al.*, 2001).

Many events that lead to physical injury and bruises can happen during the time animals wait in lairage. Strappini *et al.* (2013) found that 91% of 1,792 potential bruising events recorded through videos of 52 cows during loading, transport, unloading, lairage and stunning, occurred during the 19 h lairage they were submitted to and were mostly due to social interactions between animals and handling. Similar findings were reported by Romero *et al.* (2013), who added loading density and stops during transportation as risk factors for bruises in cattle. Cattle passing through auction markets before arriving at the slaughterhouse have a higher prevalence of bruises than cattle sent directly from farm to abattoir (McNally and Warriss, 1996: Romero *et al.*, 2013; Strappini *et al.*, 2010). The shape of the bruises in cattle passing through the markets indicate that these lesions are mostly caused by sticks used to hit or prick cattle (McNally and Warriss, 1996; Strappini *et al.*, 2012). Bruising, besides indicating poor handling conditions and impaired welfare, results in economic losses due to trimming, reduced yield and carcass downgrading (Gallo and Huertas, 2016; Warriss, 2000).

2.4 Stunning and slaughter

For the majority of cattle, the final destination of their productive lives is the slaughterhouse. During the slaughter process there is significant potential for compromised welfare. All aspects of the slaughter process are all interrelated and can contribute to either positive or negative end of life conditions.

Key factors that relate to the performance and humaneness of the slaughter process from entry into the stun/slaughter restraint are:

- human/animal interactions prior to and during stunning/slaughter;
- level of restraint during the stunning and slaughter process;
- stunning method and how it is performed (for stunned slaughter);
- performance of the ventral neck incision (for non-stunned slaughter);
- performance of exsanguination and the stun to stick interval (stunned slaughter);
- monitoring of stun and bleeding performance;
- monitoring on the bleeding rail.

In addition, stress and distress prior to and during the slaughter process can have a significant influence on the ability to correctly perform stunning and slaughter. Agitated animals in the stun and slaughter pen are more difficult to restrain, complicate accurate placement of the stunning equipment, and make performance of a ventral neck incision during non-stunned slaughter difficult. All these factors can increase the time the animal spends in the stunning pen, lead to inadequate stunning and bleeding, which have the potential to compromise animal welfare.

2.4.1 Restraint for slaughter

An essential requirement for the slaughter of cattle with or without stunning is adequate restraint. Restraint during the stunning and slaughter process is one of the greatest determinants of the success or otherwise of the process and animal welfare. The objective of the different systems is to separate and restrain the animal in the pen so that stunning and slaughter can be performed effectively, safely, while minimising stress and pain in the animals (Anil, 2012). Separation and adequate restraint allows the controlled performance of stunning and/or slaughter (Gregory, 2007). Common to all restraint systems are some basic design and operation principles, which can significantly influence animal behaviour and the ability to perform stunning and slaughter: (1) flooring surfaces in raceways and pens should be non-slip in both dry and wet conditions; (2) ideally raceways/entrance ways should be curved and avoid sharp corners to prevent baulking and allow easy access into the pen; (3) there should be no protruding obstructions that can cause the animals to baulk or cause injury when entering or being restrained in the pen; (4) there should be sufficient lighting to minimise baulking; (5) reflections and variations in surface materials should be reduced to minimise baulking; (6) gates and restraining equipment should be engineered to minimise noise that could cause agitation/distress to the animals; (7) in mechanised stunning/slaughter pens all moving restraining equipment that contacts the animal, should move in a smooth, steady motion. Sudden jerky motions can cause agitation and distress; (8) the pressure applied by the device should not be excessive. The application of excessive pressure could cause struggling, vocalisations, pain and distress; (9) stunning or slaughter without stunning should occur immediately after the animal is fully restrained; (10) solid panels should be used to prevent entering animals from seeing people or moving equipment; (11) animals must be irrecoverably unconsciousness and insensible before release from the pen; and (12) the pen should be designed to allow the safe and effective ejection of the carcase from the device, without putting the operators safety at risk (adapted from Grandin, 2013).

2.4.2 Restraint during stunned slaughter

There are a variety of restraint systems employed for stunned slaughter. Some systems are highly mechanised (central track conveyor restrainer), while others either incorporate stunning in their design (electrical stunning) or only limit movement and prevent escape prior to the act of stunning (free standing stunning pens). Each system has particular strengths and weaknesses associated with their designs but also with incorrect operation.

Free standing stunning pens are generally found in low- to medium-throughput slaughterhouses and are generally used with cartridge powered captive bolt stunning. The animals enter the pen followed by closure of the backdoor. Refusal to enter the pen can often be due to the front wall of the pen appearing to be a dead end to the animal. This can be reduced with small openings on the front wall, generally at eye level, which give sufficient view ahead to reduce baulking. The pen is designed to hold the animal and restrict movement, most pens have sufficient height to prevent cattle from escaping with the stunning operator positioned on a raised platform to the side of the pen out of immediate level eyesight of the animal. Some pens have sloped or hinged floors that help to expel the carcass after stunning and opening of the side door. When performing stunning the operator leans in from the top and stuns the animal. In this system the animal is minimally restrained and movement prior to and during stunning is a significant issue, this is particularly so with sloped floors, where the animal is often moving to maintain posture. This can result in the incorrect placement and angulation of the captive bolt gun, resulting in incomplete stunning. Some pens have rotating lights set above the pen or other distractions, designed to attract the animal´s attention and allow placement of the captive bolt shot. Others have static head raisers that typically extend from the front wall and encourage the animal to raise its head when propelled to move forwards when entering the pen or via a rump pusher (Gregory, 2007).

Some free-standing systems use additional mechanical restraint which typically include rump pushers, neck yokes and chin lifts. These are required when operating pneumatic captive bolt with free-standing cattle to allow accurate placement of the shot. Head-restraints have been reported to improve captive bolt stunning accuracy, which can improve welfare (Atkinson *et al.*, 2013; Ewbank *et al.*, 1992; Grandin 2002; von Wenzlawowicz *et al.*, 2012). A study in Colombian slaughterhouses found that when water buffalos were stunned without head or body restraint systems, an average of 2.8 shots (range 1 to 8) were required to stun the animal. However, when head and body restraints were in operation this figure decreased to 2.45 (range 2 to 3) (De la Cruz *et al.*, 2018; De la Cruz-Cruz *et al.*, 2014). Interestingly, Ewbank *et al.* (1992) reported that the use of head-restraints during captive bolt stunning for cattle can reduce line speed and result in more distress than shooting without head-restraint. However, it has been suggested that provision of simultaneous body restraint with head-restraint can reduce animal struggling (Gregory, 2007).

Conveyor restrainers are most commonly used in high-throughput slaughterhouses. These types of restrainers generally have a central track conveyor which supports, suspends and restrains the animal on the ventral surface of their abdomen, lifting them off their feet. Side walls provide further restraint and support, and some systems have a hold down bar or rack. Central track conveyors move the animals forward without the need for relying on animal locomotion. This reduces the issues associated with the misuse of movement aids. However, these are sometimes required if the walk on ramp and belly rail are not correctly designed or adjusted for the size of the animal. With central track conveyors it is important that the side walls are sufficiently high, are adjustable to account for different sized animals and there are no gaps above or below that allow the animals to see distractions (light, operators, processing etc.), as this could cause baulking on entry and agitation when restrained. There should be no objects or protrusions under the conveyor which could make contact with the suspended legs and there should be a solid false floor to avoid perception of the 'visible cliff' effect (Grandin, 2013).

In some developing countries (e.g. Bolivia, some sub-Saharan African countries and China), some small- and medium-sized slaughterhouses still immobilise cattle during stunned and non-stunned slaughter using a practice known as 'puntilla' (Limon *et al.*, 2012). Puntilla involves the insertion of a sharp knife into the back of the neck at the foramen magnum in an attempt to sever the spinal cord and is illegal in most countries. The aim is to induce immediate collapse and immobilisation allowing the animal to be bled without risk of injury to the operator. Research on cattle has found that the stab is often ineffective, as 91% of animals show signs of brain and spinal cord function, thus suggesting that they were still conscious and able to experience pain and distress (Limon *et al.*, 2010). Based on the potential for reduced animal welfare this practice should be discontinued as a method of immobilising animals during slaughter. However, it is still practice in some least developed countries.

2.4.3 Stunning systems

Cattle are generally stunned prior to the act of slaughter. Stunning, when performed correctly, is designed to render the animal immediately unconscious and insensible to pain and distress. It is also practiced to improve slaughterhouse worker safety and to improve line efficiency. The most commonly used stunning methods for cattle are captive bolt and electrical stunning systems. However, free bullet slaughter is also used for on-farm slaughter of large animals.

2.4.3.1 Penetrating and non-penetrating captive bolt

There are two captive bolt systems: non-penetrating, which has a retractable knocker head (can have a flat mushroom head) that strikes the cranium, but does not enter the cranium, and penetrating, which has a retractable steel bolt that strikes and enters the cranial vault and brain (Figure 2.6). For use in cattle these guns are either powered by blank cartridges (gun powder) or compressed air.

Both devices administer a concussive blow, which involves the transferring of the kinetic energy of the rapidly moving bolt to the cranium that results in focal and diffuse damage throughout the brain and disrupting cerebral function (Gibson *et al.*, 2012). Focal damage is seen as direct physical damage at the injury site (e.g. bolt penetration), while diffuse injury is caused by a combination of shock waves, coup and contrecoup forces, shear forces, cerebral haemorrhage and neuron depolarisation. This wide scale injury and damage to specific brain structures prevents normal function and consciousness (Schwenk *et al.*, 2016; Terlouw *et al.*, 2008). Focal damage associated with the non-penetrating captive bolt is generally limited to the underlying brain tissues

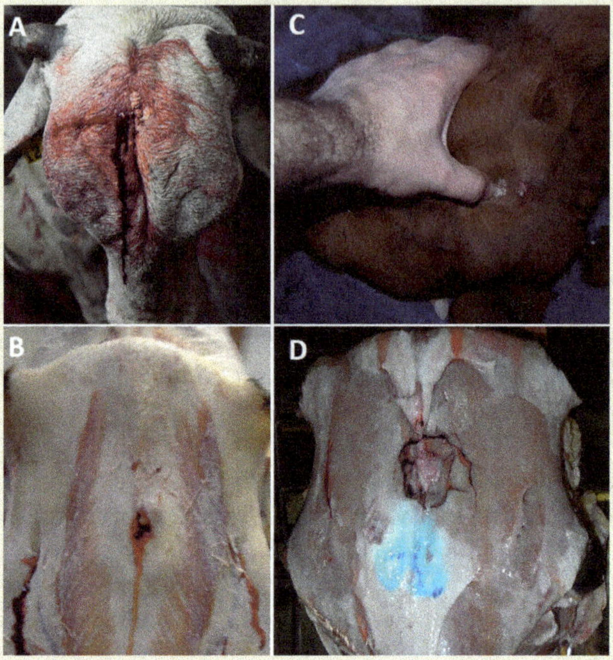

Figure 2.6. Cattle shot with penetrating and non-penetrating captive bolt. Animals shot with a penetrating captive bolt ((A) pre-skinning and (B) post-skinning): the entry position of the bolt into the cranial vault is denoted by the circular hole in the skull with no other associated fractures. Animals shot with non-penetrating captive bolt ((C) pre-skinning and (D) post-skinning): note the depressed fractures in both images (photo A and C: T. Gibson, Royal Veterinary College, UK; photo B and D: C. Gallo, AWP, Chile)

in immediate proximity to the bolt impact site. Meanwhile, penetrating captive bolt produces extensive focal damage along the bolt penetration path. Previous studies have found that it is the kinetic energy imparted to the cranium by the bolt (penetrating and non-penetrating) that produces unconsciousness and insensibility (Daly and Whittington, 1989a), while the actual physical damage by the bolt (penetrating) to specific brain structures is responsible for producing irrecoverable insensibility (Gibson *et al.*, 2012).

The performance of captive bolt stunning in rendering animals unconscious and insensible can be affected by the following factors: (1) selection of the appropriate captive bolt gun/cartridge combination for the animals stage of production; (2) correct placement of the captive bolt shot (cross over of lines drawn between the opposing eyes and horns); (3) variations between animals in skull morphology; (4) experience of the stunner operator; (5) storage of the stunner and cartridges in damp conditions; (6) behaviour of the animals; (7) variations in cartridge fill and quality; and (8) regular maintenance of the stunner, including removing excessive carbon build-up, regular rotation of the rubber buffers (recuperator sleeves), and replacement of damaged firing pins and bolts (Grandin 1980, 1994, 2002, 2012; Gibson *et al.*, 2015a; Gregory and Shaw, 2000; Grist and Wotton, 2021; Grist *et al.*, 2019, 2020; Večerek *et al.*, 2021).

Selection of the correct captive bolt gun/cartridge combination has a significant influence on the effectiveness of stunning. Many captive bolt guns are designed for specific species or stages of production, and the kinetic energy delivered can vary substantially between models and within guns with different powerloads. The kinetic energy delivered to the head during captive bolt stunning is affected to a much greater extent by variation in the velocity of the captive bolt as opposed to the mass of the bolt (Gibson *et al.*, 2015a). It is often reported that the minimum required velocity for steers, heifers and cull cows is generally no less than 55 m/s (Daly and Whittington, 1989b; Daly *et al.*, 1987; Grandin, 2013; Gregory, 2005). In contrast, young and mature bulls are often quoted as requiring a minimum velocity of 70 m/s (Grandin, 2013; Gregory, 2007). Many conventional captive bolt guns designed for cattle are able to reach the minimum velocity for animals up to cull cows, but fail to have sufficient velocity to reliably stun large cattle and mature bulls (Gibson *et al.*, 2015a). Several manufacturers have designed specific captive bolt guns with extended bolts and higher powered powerloads (Magnum XL with 6 gr cartridges; Accles and Shelvoke Ltd, Sutton Coldfield, West Midlands, UK) and the Schermer KL (Karl Schermer GmbH & Co., Ettlingen, Germany) for use in very large cattle (De la Cruz *et al.*, 2018).

Pneumatically powered penetrating and non-penetrating captive bolt stunning is often used in high-throughput slaughterhouses for large cattle. The velocity and resulting kinetic energy of pneumatically powered captive bolt guns is generally greater than that achieved with cartridge powered devices. Atkinson (2016) examined behavioural signs and reflexes of cattle in an audit of Swedish cattle slaughterhouses and found that pneumatic penetrating captive bolt stunning in one slaughterhouse produced adequate stuns in 96% of bulls compared to only 64% with a cartridge powered penetrating captive bolt gun. For successful pneumatic captive bolt stunning the following are required: adequate head restraint, correct gun maintenance, a consistent and good quality supply of compressed air and accurate shot placement. Work by Oliveira *et al.* (2017, 2018a) has examined the role of airline pressure on velocity, and the behavioural signs and reflexes associated with incomplete concussion following pneumatic penetrating and non-penetrating captive bolt stunning in different classes of Brazilian beef cattle (Nellore and Nellore hybrids). Based on behavioural responses these studies suggest that only airline pressures of 190 psi (1,310 kPa) and above should be used for stunning cattle with a pneumatically powered penetrating captive bolt (Oliveira *et al.*, 2017, 2018a,b). In addition to airline pressure and velocity, increased penetrating bolt length has been demonstrated to increase induced brain trauma in finishing cattle (Kline *et al.*, 2019; Wagner *et al.*, 2019), but also the prevalence of post-stun clonic convulsive kicking (Martin *et al.*, 2018).

Non-penetrating captive bolt stunning is used and accepted by some Muslim communities for use in Halal slaughter of cattle. The requirement is that the device does not penetrate or fracture the cranium (Nakyinsige *et al.*, 2013). However, non-penetrating stunning is recognised as being less effective in inducing reliable unconsciousness in large cattle (EFSA, 2004; Gerritzen and Gibson, 2016; Gibson *et al.*, 2019; Oliveira *et al.*, 2018a,b). Gibson *et al.* (2019) reported that the pneumatic penetrating captive bolt was 100% effective in inducing unconsciousness in mature bulls as assessed with electroencephalographic (EEG) parameters, whereas non-penetrating captive bolts were only effective in 82% of cattle. Oliveira *et al.* (2018b) reported that pneumatic non-penetrating captive bolts failed to produce macroscopic lesions in the brainstems of cattle. These results demonstrate a risk to animal welfare associated with non-penetrating stunning of mature animals in commercial conditions. In the European Union, use of the non-penetrating captive bolt is only permitted for ruminants less than 10 kg live weight and cannot be used on mature cattle (European Union, 2009).

2.4.3.2 Free bullet

Free bullet shooting is not commonly used in the commercial slaughter of cattle, but is typically used in Northern Europe for on-farm slaughter of large animals (breeding animals and water buffalo) for human consumption, where captive bolts can be ineffective. Unlike the captive bolt, the free bullet shooting is considered to be a killing method when performed correctly requiring no secondary procedure(s) provided the bullet is fired into the cranium to destroy the brain (Hultgren, 2022 – Chapter 10 of this book). However, for meat quality and animal welfare it is important that animals are bled after shooting to ensure unconsciousness.

Rifles and pistols are most commonly used to kill animals (Gibson *et al.*, 2015b; Machado *et al.*, 2013; Thomson *et al.*, 2013). There are a wide range of weapons available for this use including those specifically designed to deliver single shot (Cash .32 Humane Killer; Hultgren, 2022 – Chapter 10 of this book) to military grade weapons (Schwenk *et al.*, 2016). Animals are generally shot at close range, however, for extensively reared, stressed or dangerous livestock it can be more appropriate to shoot from a distance. It is also advisable to use a silencer to minimise noise and hence disturbances to other animals. The gun and ammunition combination should be appropriate for the shooting distance and stage of production of the animal. The shooting position for cattle is the same as for the captive bolt with the goal of maximising damage to the structures of the brainstem (midbrain, pons and medulla) (Gerritzen and Gibson, 2016).

It is important to ensure that the firearm operators are trained and licenced, and the ammunition is designed and constructed such that they do not exit the head of the animals. The use of hallow point or non-jacketed rounds can reduce the risk of ricochets or bullets exiting the animal as they are designed to deform and fragment on impact (Gibson *et al.*, 2015a). This has the added advantage of increasing surface area and severity of the wound (Farjo and Miclau 1997; Hollerman *et al.*, 1990; Thomson *et al.*, 2013).

2.4.3.3 Electrical stunning

Cattle electrical stunning systems vary from those used for other species. The stunning systems are often incorporated within the restraint, stunning can be head-only or head-to-body, and placement of the stunning electrodes and application of the current is often automated. For both head-only and head-to-body stunning, electrodes are often incorporated within the neck yoke and the chin is raised to contact a fixed nose plate electrode. However, for head-only electrical stunning of calves, hand-held scissor tongs are also used. The stunning systems used for cattle are generally 50 Hz constant current, with voltage ranging in response to the electrical impedance of the head to ensure near immediate and consistent current flow. In most systems a low voltage impedance sensing circuit is often incorporated to determine when the animals head is in position based on its impedance before switching to the stunning voltage. The stunning voltage is designed to break down the electrical impedance of the head allowing sufficient electrical current through the animal's brain, to cause depolarisation of cortical neurons resulting in synchronised neuronal activity, an epileptiform EEG trace and unconsciousness. This epileptiform activity is caused by the release of the excitatory neurotransmitters, glutamate, aspartate and gamma-aminobutyric acid (GABA; Cook *et al.*, 1992, 1995). The brain cannot function normally until the neurons are repolarised and resume their normal, desynchronised cycle of activity (Terlouw and Gibson, 2016). In the European Union the minimum electrical stunning current for head-only is 1.26 and 1.28A for cattle less than <6 and ≥6 months of age, respectively (European Union, 2009).

In head-to-body systems there are generally two to three cycles, which operate sequentially. The first cycle is represented by the head-only electrical stun, the second is the chest stun with a pair of additional brisket electrode that passes electrical current across the heart to induce cardiac dysfunction or arrest. This results in the reduction or cessation of blood circulation and causes cerebral hypoxemia, making unconsciousness irreversible. The third cycle is electro-immobilisation, which is performed after stunning. Its use is not widespread, but it is used in some countries (e.g. New Zealand and Australia) with high throughput slaughterhouses. This involves a low voltage high frequency stimulus which prevents clonic convulsions, improving operator safety and allowing performance of exsanguination. Some concerns about this method include the potential to mask signs of incomplete stunning or recovery of consciousness. Consequently, electro-immobilisation is not permitted in some countries (e.g. UK and EU). A further low-frequency electrical stimulation can be applied to the bleeding carcass to improve bleeding and meat quality (Gregory, 2007).

Potential welfare issues associated with electrical stunning of cattle include:

- Pain and distress associated with excessive restraint. However, such a restraint is needed to ensure a correct contact with stunning electrodes.
- Delays in entry into the restraint to the onset of electrical stunning. This will result in a stressed, agitated animal and could complicate positioning of electrodes and performance of stunning.
- Incorrect electrode positioning and poor electrical contact, resulting in insufficient electrical current passing through the brain. In some cases, this could result in pre-stun electrical shocks. This may also be associated with variations in animal size and shape and the presence of horns.
- Delays in the switching from the impedance sensing to stunning voltage could result in prolonged distress from restraint.
- Incorrect maintenance of stunning electrodes, such as carbon build up on electrodes or loose connections, producing variations in the current flow and stunning performance.
- Inappropriate electrical parameters and exposure time for the animal type, potentially resulting in failure to induce unconsciousness or a short duration of induced unconsciousness.
- In head-to-body systems, failure to induce cardiac arrest or dysfunction, potentially producing a situation in which animals could recover after the stun or prior to and during exsanguination.
- In head-only systems, short duration of unconsciousness and/or failure to immediately bleed the animal could result in some animals recovering consciousness prior to, during or after exsanguination (neck incision or chest stick).
- Animals should be monitored for signs of successful or incomplete stunning, this is especially important when a reversible method, such as head-only is used.

More recently there has been new experimental research into single pulse ultra-high electrical current stunning of cattle. This involves a single short pulse (50 ms) of at least 5,000 V at 70 A or greater, which is designed to produce a state of reversible unconsciousness. Initial results with this method suggest that stun duration is more prolonged than conventional electrical stunning with little to no post-stun seizure activity (Fuseini, 2019; Robins *et al.*, 2014). However, work is still on going with this method and it is not commercially available currently.

2.4.4 Assessment of stunning

Immediately after stunning it is essential that operators are able to safely monitor the effectiveness or otherwise of the method and apply a secondary stunning procedure, if required. After penetrating, non-penetrating captive bolt stunning and free bullet shooting effectiveness is assessed by a combination of behavioural, brainstem mediated and spinal reflexes (Table 2.1). None of these indices by themselves are direct indicators of consciousness/unconsciousness, but they should be used in combination to determine the effectiveness of the stun. If any of these behaviours or reflexes are present (e.g. the animal is breathing rhythmically), the animal may be incompletely concussed and there is a risk that the animal could regain consciousness. In this situation, it is advisable to re-stun.

When assessing the effectiveness of electrical stunning it is important to inspect the equipment, stunning parameters and electric stun records to make sure it is delivering the required electrical current, to check that the electrodes are being applied in the appropriate position and to ensure that the animals are being bleed within the prescribed period (stun-to-stick interval). In parallel it is essential to observe the behaviour and reflexes of the animal. Many of these are the same signs as for captive bolt stunning (Table 2.1). Normally, when an animal is electrically stunned by the head-only method, the animal immediately collapses if it is not supported in a restraining conveyor or pen.

Table 2.1. Signs of effective captive bolt stunning and free bullet shooting of cattle.

Signs	Description
Immediate collapse	Animal immediately collapses and does not attempt to regain posture
Immediate loss of normal rhythmic breathing	Immediate onset of apnoea with no rhythmic pattern of respiratory movement, noise or expiration of air via the mouth or nostrils
Tonic seizure	Muscles in the back and legs are in spasm and contracted
Absence of eyeball rotation	The eyes are not rotated and are fixed forwards
Absence of spontaneous blinking	Absence of spontaneous unprovoked blinking activity
Absence of corneal reflex	No blinking response when the exposed cornea is stimulated
Absence of palpebral reflex	No response to stimulation of the eyelashes, inner or outer eye canthus
Absence of coordinated muscle tone	Absence of coordinated muscle tone (e.g. relaxed jaw, protruding tongue when suspended on the bleeding rail)
Absence of nystagmus	Spontaneous rapid involuntary movement of the eyeballs
Absence of vocalisations	No grunting, bellowing or mooing. Different to noises associated with exhalation of gases during movement of the carcass post-mortem
Coordinated body movements	Intentional coordinated movements of the body and head to regain posture or in response to further stimulation (e.g. shackling, insertion of the knife)

During the electroepileptic phase the body is tense and tonic (rigid), and breathing is absent. The tonic phase is followed by clonic phase, which is characterised by convulsive kicking. It is during the tonic and clonic phases that the animal is unconscious. At the end of this phase, the body relaxes, rapid rhythmic breathing returns, and the animal will attempt to stand. The animal must be bled and killed before this last phase begins.

In animals that are successfully stunned with head-to-body electrical stunning, after the clonic phase the body will remain relaxed, with no recovery of consciousness. Sometimes animals that have experienced a cardiac arrest, whilst stunning can present with very shallow slow non-functional breaths. This does not indicate recovery of consciousness or return of true rhythmic breathing. In these cases, operators should check the other behavioural/brainstem indices (Table 2.1) to ensure unconsciousness.

2.4.5 Exsanguination

Regardless of the stunning method employed, the act of cattle slaughter for human consumption is a two-step process. After the act of stunning, cattle require rapid exsanguination (thoracic stab and/or ventral neck incision) to prevent the potential recovery of consciousness. This is particularly important with reversible stunning methods, when there is variation in shot position (captive bolt) and when less reliable methods are employed (non-penetrating captive bolt and head-only electrical stunning).

Exsanguination in cattle must be performed immediately after stunning to prevent potential recovery of consciousness with reversible stunning methods or in cases of incomplete stunning. In commercial stunned slaughter, the majority of cattle are bled via a chest stick, which severs the brachiocephalic artery, the first branch of the aortic arch. The severance of the brachiocephalic artery eliminates the supply of oxygenated blood to the brain via the common carotid and vertebral arteries, resulting in the disruption of oxygenated blood to the brain. For some stunned and all non-stunned slaughter, cattle are bled with a ventral neck incision, which involves the cutting of the soft tissues of the ventral aspect of the neck. Ventral neck incision slaughter is discussed further below.

2.4.6 Non-stunned slaughter

The issues associated with non-stunned slaughter of cattle are complex and multifaceted, including theology, human rights, law, societal views and animal welfare. From the animal welfare perspective, the principal concerns are: (1) pain and distress associated with restraint (Berg and Jakobsson, 2007; Daly *et al.*, 1988; Grandin, 1998c; Grandin and Regenstein, 1994; Gregory *et al.*, 2012a; Velarde *et al.*, 2014); (2) pain from the neck cut and stimulation of nociceptors in the wound (Gibson *et al.*, 2009a,b; Zulkifli *et al.*, 2014); (3) distress associated with delays in the time to loss of consciousness (Daly *et al.*, 1988; Gibson *et al.*, 2015c; Gregory *et al.*, 2010); and (4) aspiration of blood into the respiratory tract (Grandin and Regenstein, 1994; Gregory *et al.*, 2010).

2.4.6.1 Restraint for non-stunned slaughter of cattle

During non-stunned slaughter it is essential for welfare and religious requirements that the animals are sufficiently restrained. Much of the issues associated with restraint for non-stunned slaughter are similar to those for stunned slaughter. The issues that are specific to non-stunned slaughter are orientation of the animals, body restraint and restraint of the neck and head for incision. A variety of restraining methods are employed for non-stunned slaughter of cattle. In developed and developing countries, mechanised upright and rotating restraint devices (dorsal, inverted 180° or lateral recumbency, 90°) are often used, while in under developed countries non-mechanical devices (e.g. modified Mark I box, which confines the animal upright in a pen where a forelimb and hindlimb are restrained with ropes, the side door to the box is opened and the animal moves towards the perceived route of escape, but is instead tripped over due to the robed limbs into lateral recumbency onto a slopped concrete slab (Jones, 2011), manual casting on to lateral or sternal recumbency, or immobilisation with 'puntilla' are employed.

Manual methods of restraint often involve the casting of conscious animals on the ground or on a concrete slab into lateral recumbency using ropes to secure their legs. This can result in significant pain and discomfort, especially where animals fall onto concrete slabs. The lateral positioning of the animal makes completion of the neck cut difficult, requiring the twisting of the neck to allow access to the ventral surface. This can cause further distress and in some cases is not possible due to the presence of horns (De la Cruz *et al.*, 2018). The practice of hind leg shackling and hoisting of conscious cattle, despite being prohibited in many countries (e.g. Canada, UK, EU, Australia), is still practiced in countries like China, Turkey and Brazil. Hoisting of fully conscious cattle and nose clamping to allow stretching of the neck for performance of the neck cut causes significant pain and distress (Grandin, 2009) and as a method of restraint should be discontinued.

Mechanised rotating and upright restraint systems (Figure 2.7) are principally used in medium to large-scale slaughterhouses in developed countries. The slaughter of un-stunned cattle in dorsal recumbency (inverted) is prohibited in many countries (Grandin, 1994), because of the stress involved in inverting the animal (Dunn, 1990; Grandin and Regenstein, 1994; Shragge and Price, 2004). A number of bodies have stated that the inversion of cattle for slaughter without stunning could result in welfare compromise compared to upright restraint (Blokhuis *et al.*, 2004; FAWC, 1995; Wathes, 2012). The welfare concerns with rotating systems relate to unnatural posture, abdominal pressure on visceral tissues, and stress during inversion. Inversion has been shown to induce a greater stress response (as showed by increased plasma cortisol concentration and vocalisations) in cattle compared to upright and lateral restraint (Bourguet *et al.*, 2011; Dunn, 1990; Tagawa *et al.*, 1994). Lambooij *et al.* (2012) reported that, independent of the angle (90°, 120° and 180°), inversion caused a significant increase in heart rate, suggesting a stress mediated sympathetic response. Lateral and dorsal recumbency has been reported to significantly impair arterial oxygenation (Tagawa *et al.*, 1994; Wagner *et al.*, 1990). Being in lateral and dorsal recumbency can result in reduced lung volumes, due to compression of the thoracic cavity by the weight of the rumen and abdominal viscera pressing against the diaphragm. However, Von Holleben *et al.* (2010) suggested that lateral restraint can reduce the pressure of the rumen depressing on the internal organs and diaphragm, compared to full 180° inversion. The advantage of the 180° rotating systems is that they often provide

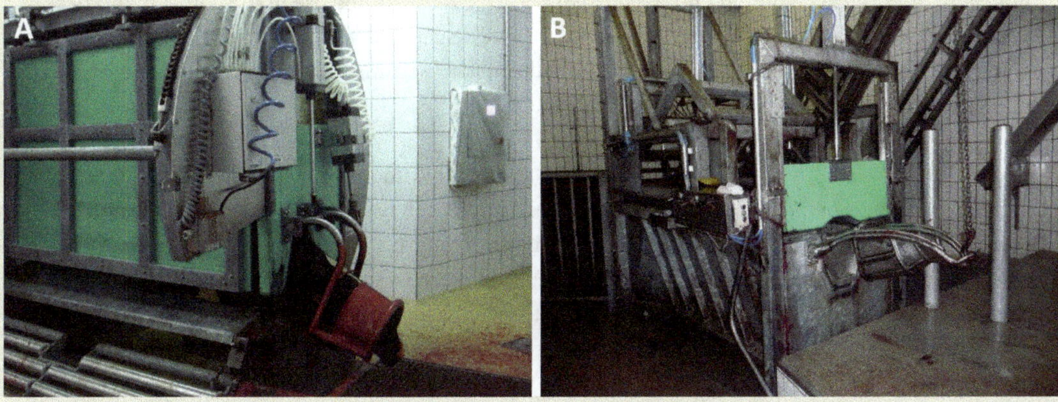

Figure 2.7. Restraint pens used for non-stunned slaughter of cattle: (A) rotating design that rotates the animal 180° prior to performance of the neck cut and (B) upright design that uses a back pusher, deflecting plates, neck yoke and chin lift to restrain the animal in an upright position for the neck cut. Other upright systems often include a belly plate, which is absent in this design (T. Gibson, Royal Veterinary College, UK).

good presentation of the ventral surface of the neck for the neck incision. However, a criticism of inverted restraint positions is that blood and rumen contents can contact the cranial and caudal aspects of the wound surface. This could result in carcass contamination, aspiration of liquid into the respiratory tract (Blokhuis *et al.*, 2004; Grandin and Regenstein, 1994), and possible further stimulation of the wound site which could cause pain and distress (Gibson *et al.*, 2009a; Zulkifi *et al.*, 2014).

Upright restraint systems confine cattle in an upright position for slaughter. This can either be free standing or straddling a centre track conveyor restrainer in sternal recumbency, where the feet are off the floor and the conveyor and walls support the body weight via the brisket, abdomen and flanks (Grandin, 1990). The head is secured and stretched by a chin-lift into position for the ventral neck cut, either mechanically (hydraulic or electrical) or manually for calves. Many systems, such as the Cincinnati or American Society for the Prevention of Cruelty to Animals (ASPCA) pen, have a back-pusher/tailgate and belly plate that further restrains the animal. A common complaint of upright systems is that animals can be over restrained. Poorly designed upright restraints can apply excessive pressure (Berg and Jakobsson, 2007; Grandin and Regenstein, 1994). For example, excessive pressure of the back-pusher, lifting of animals with the belly plate (problematic with smaller or young stock) and excessive neck tension (hyperextension) with the chin-lift, could all cause discomfort, pain and suffering. Grandin (1998c) reported that in two separate slaughterhouses 3 and 8% of cattle vocalised due to excessive pressure of the back-pusher and chin-lift respectively. Similarly, Velarde and Cenci-Goga (2010) reported that in spot visits 63% of cattle restrained in upright pens showed struggling compared to 37% in modern inverted pens, potentially due to excessive pressure from the chin-lift, back-pusher and belly plate. Ideally chin-lifts and back-pushers should be designed with pressure limiting devices when possible (Grandin, 1992; Grandin and Regenstein, 1994), and should be operated with the concept of 'optimal pressure' required to firmly restrain the animal, but not cause discomfort (Grandin and Regenstein, 1994). Even with the use of the chin-lift and neck yoke, the ventral cutting surface of the neck in upright pens is less well presented than for animals restrained in the inverted position. The cut is made upwards against the ventral aspect of the neck, making the cut more awkward (Gregory, 2005).

At the end of a EU study, Mirabito *et al.* (2015) concluded that upright and inverted restraint systems have their own strengths and weaknesses, in terms of welfare, religious acceptability and worker safety. This often related to their design and operation, and not just the orientation of restraint. Irrespective of the system or slaughter method, all restraints, including those of the highest standards, are open to operator abuse, if used incorrectly, and can result in poor animal welfare (Grandin and Regenstein, 1994).

2.4.6.2 Time to unconsciousness

During slaughter without stunning, there is a period of time following the neck incision, and before the onset of cerebral hypoxia and unconsciousness/insensibility, during which the animal is both conscious and sensible to pain and distress (Gibson *et al.*, 2009a,b). There is considerable species variation in this period of time, which is well documented to be the greatest and most variable in cattle (ranging from 2 to 385 s; Bager *et al.*, 1992; Blackmore, 1979, 1984; Blackmore and Newhook, 1981; Bourguet *et al.*, 2011; Daly *et al.*, 1988; Gibson *et al.*, 2015c; Gregory and Wotton, 1984; Gregory *et al.*, 2010; Groβ, 1979; Lambooij *et al.*, 2012; Levinger 1961; Nangeroni and Kennett, 1964; Neves *et al.*, 2016; Newhook and Blackmore, 1982).

The delayed time to unconsciousness in cattle compared to other species has been attributed to continued blood supply via the vertebral arteries and the formation of false aneurysms (carotid occlusion) on the severed ends of the carotid arteries (cephalic and cardiac ends) (Gibson *et al.*, 2015c; Gregory *et al.*, 2008). Cerebral perfusion in cattle is primarily supplied by a mixture of both carotid and vertebral arterial blood. Cattle are unique to other species in that their vertebral blood flows to the entire brain as well as anterior structures (Baldwin, 1971). During slaughter by ventral neck incision only the common carotid arteries are severed, leaving the vertebral arteries intact within the vertebrae. In cattle, the vertebral arteries communicate directly to the rete mirabile by the basi-occipital plexus as well as communicating via the occipito-vertebral anastomosis to the common carotids (Baldwin and Bell, 1963).

A false aneurysm is thought to form when the severed carotid artery retracts within its surrounding connective tissue sheath. Blood from the severed vessel impregnates and swells the adventitia (Gregory *et al.*, 2006). The swelling of the adventitia impedes and can block the flow of blood from the artery (Gregory *et al.*, 2010, 2012a). There is a welfare concern in animals have not been stunned prior to neck incision because the false aneurysm can obstruct the flow of blood, and if blood continues to flow to the brain via the vertebral arteries (which are not cut during slaughter), there can be delays in loss of consciousness (Anil *et al.*, 1995a,b; Bager *et al.*, 1988; Blackman *et al.*, 1986; Blackmore, 1984; Gregory *et al.*, 2012a,b; Newhook and Blackmore, 1982). Some authors have suggested that, following incision of the common carotids, blood would flow via the path of least-resistance (including from the vertebral arteries) to the severed cephalic ends of the jugular veins and carotid arteries (Levinger, 1995; Rosen, 2004). However, this would be dependent on the absence of occlusion of the cut cephalic ends of these blood vessels. The prevalence of carotid occlusion in finished cattle (n=387) and calves (n=189) after stunning and ventral neck incision slaughter was reported as 16 and 25%, respectively (Gregory *et al.*, 2006). This suggests that carotid occlusion is not a rare or isolated occurrence in cattle, especially in calves.

Work with non-stunned Halal cattle has shown that performing a high neck cut (in relation to the first two trachea rings or first cervical vertebra) compared to the standard low neck cut significantly reduced the time to final collapse (13.5±1.3 s vs 18.9±1.1 s) and the proportion of animals that took over 20 s to final collapse (Gibson *et al.*, 2015c). Similarly, Bozzo *et al.* (2020) and Gregory *et al.* (2012a) reported reduced false aneurysm rates in Halal-slaughtered cattle cut in the high neck cut position. Taken together these results suggest that the high neck cut could significantly improve exsanguination and time to loss of consciousness.

2.4.6.3 Pain during non-stunned slaughter

The act of ventral neck incision involves the transection of skin, muscle, trachea, oesophagus, sensory nerves and connective tissues. All these tissues are highly innervated and, when cut, would be expected to activate nociceptive pathways leading to perceived pain by the conscious animal prior to the onset of hypoxia-induced insensibility. Gibson *et al.* (2009a,b) and Zulkifi *et al.* (2014) have shown that the tissue damage associated with the neck cut causes pain (as assessed by EEG indices of noxiousness) during the slaughter of cattle without stunning. However, this response could be eliminated with the application of a post-cut stun (Gibson *et al.*, 2009c).

2.4.6.4 Aspiration of blood

There are concerns that slaughter without stunning could prolong animal consciousness/sensibility resulting in aspiration of blood into the respiratory tract (Grandin and Regenstein, 1994; Gregory *et al.*, 2009b), causing irritation and distress. Aspiration of blood occurs during both inverted and upright restraint (Blokhuis *et al.*, 2004; Gregory *et al.*, 2009b). Severing the vagus nerves was previously thought to prevent lower respiratory tract irritation. However, stimulation of the glottis (innervated by the cranial laryngeal nerve) by haemorrhaged blood entering the lower respiratory tract (innervated by a collateral spinal afferent pathway) could result in animal discomfort, distress, and suffering (Gregory *et al.*, 2009b).

2.4.7 Meat quality and method of slaughter

There is conflicting information on the effects of stunning and slaughter methods on carcass and meat quality in cattle. This is further confused by much of the work in this area being focused on other species, and of the limited studies on cattle, subsequent review authors have often reinterpreted the original findings. Of the empirical studies, work by Önenç and Kaya (2004) reported that sensory attributes of meat quality were significantly improved with captive bolt and electrically stunned cattle compared to those slaughtered without stunning. Meanwhile, there were no significant differences in colour or water-holding capacity. Sazili *et al.* (2013) did report some differences between pre-, post- and non-stunned slaughter, however, they characterise the final results as been comparable in meat quality across the treatments. Anil *et al.* (2006) reported similar findings when comparing captive bolt and non-stunned cattle, also reporting that total blood loss, blood loss as a percentage of live weight and rate were not significantly different between the two methods. These findings were further supported by a similar South African study (Agbeniga and Webb, 2012).

2.5 Conclusions

Animal handling practices on farm, during transport and slaughter can directly impact animal stress and welfare as well as meat quantity and quality. The number and frequency of stressors cattle are exposed to on farm may be reduced by designing, building and maintaining appropriate facilities for cattle, by breeding/selecting cattle with calm temperament and by training stockpersons for gentle handling and knowledge in animal behaviour. Although cattle are exposed to a multitude of stressors during transport, all with the potential to reduce coping and welfare, the majority are transported without experiencing severe welfare issues. This does not imply that improvements in transport management practices are not warranted as there are still obvious places where changes are required. There is growing evidence to support that animal fitness is a main contributing factor for successful transports. Consequently, focus on education and management of vulnerable cattle should take precedence. In addition, continued work on developing objective criteria for determining fitness for travel would be welcomed by producers and regulatory agents alike. Likewise, studies assessing interactions between animals of varying age, weight, condition and temperament and preslaughter procedures, human and environmental factors are needed and will benefit from the use of new technologies and biomarkers of stress and welfare. This continued effort will lead to an improved understanding of what the most significant stressors are, where and when they occur and what mitigations strategies can be employed to alleviate them. It is likely that the mitigation of preslaughter stress is highly dependent on the type of animal and situation and therefore targeted rather than general recommendations will be more beneficial and necessary. Welfare problems at the slaughter plants during handling and stunning need to be identified and regularly measured, monitored and managed in order to make improvements. The use of behavioural indicators of welfare in the cattle, as well as associated productive and economic indicators regarding meat quantity and quality is recommended. A constant feedback to slaughterhouse managers/operators, transporters and producers about the final outcome of their pre-slaughter procedures on product yields, and on the technological and ethical quality of the meat is urgently needed.

References

Agbeniga, B. and Webb, E.C., 2012. Effect of slaughter technique on bleed-out, blood in the trachea and blood splash in the lungs of cattle. South African Journal of Animal Science 42: 524-529.

Ahola, J.K., Foster, H.A., Van Overbeke, D.L., Jensen, K.S., Wilson, R.L., Glaze, Jr, J.B., Fife, T.E., Gray, C.W., Nash, S.A., Panting, R.R. and Rimbey, N.R., 2011. Survey of quality defects in market beef and dairy cows and bulls sold through livestock auction markets in the Western United States. Journal of Animal Science 89: 1471-1483.

Alende, M., Volpi Lagreca, G., Pordomingo, JA, Pighín, D., Grigioni, G., Carduza, F., Pazos, A., Babinec, F. and Sancho, A.M., 2014. Effects of transport, lairage and ageing time on stress indicators and on instrumental and sensory quality of beef from steers. Archivos de Medicina Veterinaria 46: 217-227.

Amtmann, V., Gallo, C., Van Schaik, G. and Tadich, N., 2006. Relaciones entre el manejo antemortem, variables sanguíneas indicadoras de estrés y pH de la canal en novillos. Archivos de Medicina Veterinaria 38: 259-264.

Anil, M.H., 2012. Religious slaughter: a current controversial animal welfare issue. Animal Frontiers 2: 64-67.

Anil, M.H., McKinstry, J.L., Gregory, N.G., Wotton, S.B. and Symonds, H., 1995b. Welfare of calves – 2. Increase in vertebral artery blood flow following exsanguination by neck sticking and evaluation of chest sticking as an alternative slaughter method. Meat Science 41: 113-123.

Anil, M.H., McKinstry, J.L., Wotton, S.B. and Gregory, N.G., 1995a. Welfare of calves – 1. Investigations into some aspects of calf slaughter. Meat Science 41: 101-112.

Anil, M.H., Yesildere, T., Aksu, H., Matur, E., McKinstry, J.L., Weaver, H.R., Erdogan, O., Hughes, S. and Mason, C., 2006. Comparison of Halal slaughter with captive bolt stunning and neck cutting in cattle: exsanguination and quality parameters. Animal Welfare 15: 325-330.

Apaoblaza, A. and Gallo, C., 2014. Live changes in muscle glycogen concentration of steers due to feeding and fasting as determined through serial biopsies of the Longissimus dorsi muscle. Chilean Journal of Agricultural Research 74: 55-58.

Apaoblaza, A., Strobel, P., Ramírez-Reveco, A., Jeréz-Timaure, N., Monti, G. and Gallo, C., 2017. Effect of season, supplementation and fasting on glycolytic potential and activity of AMP-activated protein kinase, glycogen phosphorylase and glycogen debranching enzyme in grass-fed steers as determined in Longissimus lumborum muscle. Livestock Science 202: 101-108.

Atkinson, S., 2016. Assessment of cattle and pig welfare at stunning in commercial abattoirs. PhD Thesis, Swedish University of Agricultural Science, Uppsala, Sweden.

Atkinson, S., Velarde, A. and Algers, B., 2013. Assessment of stun quality at commercial slaughter in cattle shot with captive bolt. Animal Welfare 22: 473-481.

Bager, F., Braggins, T.J., Devine, C.E., Graafhuis, A.E., Mellor, D.J., Tavener, A. and Upsdell, M.P., 1992. Onset of insensibility at slaughter in calves – effects of electroplectic seizure and exsanguination on spontaneous electrocortical activity and indexes of cerebral metabolism. Research in Veterinary Science 52: 162-173.

Bager, F., Devine, C.E. and Gilbert, K.V., 1988 Jugular blood flow in calves after head-only electrical stunning and throat-cutting. Meat Science 22: 237-243.

Baldwin, B.A. and Bell, F.R., 1963. The anatomy of the cerebral circulation of the sheep and ox. The dynamic distribution of the blood supplied by the carotid and vertebral arteries to cranial regions. Journal of Anatomy 97: 203-215.

Baldwin, B.A., 1971. Anatomical and physiological factors involved in slaughter by section of the carotid arteries. In: Ewer, T.K. (ed.) Humane killing and slaughterhouse techniques. The Universities Federation for Animal Welfare, Wheathampstead, UK, pp. 34-43.

Barnes, K., Smith, S. and Lalman, D., 2004. Managing shrink and weighing conditions in beef cattle. Factsheet ANSI-3257. Oklahoma State University Cooperative Extension Service. Available at: https://tinyurl.com/t5wp3st2.

Barrington, G.M., 2001. Bovine neonatal immunology. Veterinary Clinics of North America Food Animal Practice 17: 463-476.

Bass, J.J. and Duganzich, D.M., 1980. A note on the effect of starvation on the bovine alimentary tract and its contents. Animal Production 31: 111-113.

Benhajali, H., Boivin, X., Sapa, J., Pellegrini, P., Boulesteix, P., Lajudie, P. and Phocaset, F., 2010. Assessment of different on-farm measures of beef cattle temperament for use in genetic evaluation. Journal of Animal Science 88: 3529-3537.

Berg, C. and Jakobsson, T., 2007. Post-cut stunning at religious slaughter. Svensk Veterinärtidning 59: 21-28.

Blackman, N.L., Cheetham, K. and Blackmore, D.K., 1986. Differences in blood-supply to the cerebral-cortex between sheep and calves during slaughter. Research in Veterinary Science 40: 252-254.

Blackmore, D.K. and Newhook, J.C., 1981. Insensibility during slaughter of pigs in comparison to other domestic stock. New Zealand Veterinary Journal 29: 219-222.

Blackmore, D.K., 1979. Non-penetrative percussion stunning of sheep and calves. Veterinary Record 105: 372-375.

Blackmore, D.K., 1984. Differences in behaviour between sheep and cattle during slaughter. Research in Veterinary Science 37: 223-226.

Blokhuis, H.J., Roth, B., Holst, S., Kestin, S., Raj, M., Terlouw, C., Velarde, A. and von Wenzlawowicz, M., 2004. Welfare aspects of animal stunning and killing methods. Scientific Report of the Scientific Panel for Animal Health and Welfare on request from the Commission related to welfare aspects of animal stunning and killing methods. European Food Safety Authority, 24 pp.

Booth-McLean, M.E., Schwartzkopf-Genswein, K.S., Brown, F.A., Holmes C.L., Schaefer, A.L., McAllister, T.A. and Mears, G.J., 2007. Physiological and behavioural responses to short-haul transport by stock trailer in finished steers. Canadian Journal of Animal Science 87: 291-297.

Bourguet, C., Deiss, V., Boissy, A. and Terlouw, E.C., 2015. Young Blond d'Aquitaine, Angus and Limousin bulls differ in emotional reactivity: relationships with animal traits, stress reactions at slaughter and post-mortem muscle metabolism. Applied Animal Behaviour Science 164: 41-55.

Bourguet, C., Deiss, V., Tannugi, C. and Terlouw, E.C., 2011. Behavioural and physiological reactions of cattle in a commercial abattoir: relationships with organisational aspects of the abattoir and animal characteristics. Meat Science 88: 158-168.

Bozzo, G., Bonerba, E., Barrasso, R., Roma, R., Luposella, F., Zizzo, N. and Tantillo, G., 2020. Evaluation of the occurrence of false aneurysms during halal slaughtering and consequences on the animal's state of consciousness. Animals 10: 183. https://doi.org/10.3390/ani10071183

Braga, J.S., Faucitano, L., Macitelli, F., Sant'Anna, A.C., Méthot, S. and Paranhos da Costa, M.J.R., 2018. Temperament effects on performance and adaptability of Nellore young bulls to the feedlot environment. Livestock Science 216: 88-93.

Bravo, V., Gallo, C. and Acosta-Jamett, G., 2018. Effects of short transport and prolonged fasting in beef calves. Animals 8: 170. https://doi.org/10.3390/ani8100170

Bravo, V.M., Knowles, T.G. and Gallo, C., 2019. Factors affecting the welfare of calves in auction markets. Animals 9: 333. https://doi.org/10.3390/ani9060333

Broom, D.M., 2008. The welfare of livestock during road transport. In: Appleby, V. Cussen, I., Garcés, I., Lambert, J. and Turner, M.C. (eds.) Long distance transport and welfare of farm animals. CABI, Wallingford, UK, p. 157-181.

Brunel, H., Dallago, B., De Almeida, A., De Assis, A. and De Bento Calzada, R., 2018. Hemato-biochemical profile of meat cattle submitted to different types of pre-loading handling and transport times. International Journal of Veterinary Science and Medicine 6: 90-96.

Bryan, M., 2013. Trailer micro-climate during long-distance transport of finished beef cattle for the summer months in North America. MSc Thesis, University of Saskatchewan, Saskatoon, Canada.

Bulitta, F.S., Aradom, S. and Gebresenbet, G., 2015. Effect of transport time of up to 12 hours on welfare of cows and bulls. Journal of Service Science and Management 8: 161-182.

Burdick, N.C., Carroll, J.A., Randel, R.D., Willard, S.T., Vann, R.C., Chase, Jr. C.C., Lawhon, S.D., Hulbert, L.E. and Welsh, Jr. T.H., 2011. Influence of temperament and transportation on physiological and endocrinological parameters in bulls. Livestock Science 139: 213-221.

Burrow, H.M., 1997. Measurements of temperament and their relationships with performance traits of beef cattle. Animal Breeding Abstracts 65: 477-495.

Café, L.M., Robinson, D.L., Ferguson, D.M., McIntyre, B.L., Geesink, G.H. and Greenwood, P.L., 2011. Cattle temperament: persistence of assessments and associations with productivity, efficiency, carcass and meat quality traits. Journal of Animal Science 89: 1452-1465.

Canadian Food Inspection Agency, 2020. Health of animals regulations: Part XII: transport of animals-regulatory amendment – interpretive guidance for regulated parties. Available at: https://tinyurl.com/3fn66zn6.

Canadian Food Inspection Agency, 2021. Livestock transport in Canada. Available at: https://tinyurl.com/3zm7r4e2.

Ceballos, M.C., Góis, K.C.R., Sant'Anna, A.C. and Paranhos da Costa, M.J.R., 2016. Frequent handling of grazing beef cattle maintained under the rotational stocking method improves temperament over time. Animal Production Science 58: 307. https://doi.org/10.1071/AN16025

Ceballos, M.C., Sant'Anna, A., Boivin, X., De Oliveira Costa, F., Carvalhal, M.V. de L. and Paranhos da Costa, M.J., 2018. Impact of good practices of handling training on beef cattle welfare and stockpeople attitudes and behaviours. Livestock Science 216: 24-31.

Cernicchiaro, N., White, B.J., Renter, D.G., Babcock, A.H., Kelly, L. and Slattery, R., 2012. Associations between the distance travelled from sale barns to commercial feedlots in the United States and overall performance, risk of respiratory disease, and cumulative mortality in feeder cattle during 1997 to 2009. Journal of Animal Science 90: 1929-1939.

Cockram, M., 2019. Fitness of animals for transport to slaughter. Canadian Veterinary Journal 60: 423-429.

Coffey, K.P., Coblentz, W.K., Humphry, J.B. and Brazle, F.K., 2001. Review: basic principles and economics of transportation shrink in beef. The Professional Animal Scientist 17: 247-255.

Coleman, G.J. and Hemsworth, P.H., 2014. Training to improve stockperson beliefs and behaviour towards livestock enhances welfare and productivity. Revue Scientifique e Technique Office International Epizootiologie 33: 131-137.

Cook, C.J., Devine, C.E., Gilbert, K.V., Smith, D.D. and Maasland, S.A., 1995. The effect of electrical head-only stun duration on electroencephalographic-measured seizure and brain amino-acid neurotransmitter release. Meat Science: 40: 137-147.

Cook, C.J., Devine, C.E., Tavener, A. and Gilbert, K.V., 1992. Contribution of amino acid transmitters to epileptiform activity and reflex suppression in electrically head stunned sheep. Research in Veterinary Science 52: 48-56.

Cooke, R.F., Cappellozza, B.I., Guarnieri Filho, T.A. and Bohnert, D.W., 2013. Effects of flunixin meglumine administration on physiological and performance responses of transported feeder cattle. Journal of Animal Science 91: 5500-5506.

Coombes, S.V., Gardner, G.E., Pethick, D.W. and McGilchrist, P., 2014. The impact of beef cattle temperament assessed using flight speed on muscle glycogen, muscle lactate and plasma lactate concentrations at slaughter. Meat Science 98: 815-821.

Cozzi, G., Brscic, M. and Gottardo, F., 2009. Main critical factors affecting the welfare of beef cattle and veal calves raised under intensive rearing systems in Italy: a review. Italian Journal of Animal Science 8 (Suppl. 1): 67-80.

Cozzi, G., Ricci, R., Dorigo, M. and Zanet, D., 2005. Growth performance, cleanliness and lameness of finishing Charolais bulls housed in littered pens of different design. Italian Journal of Animal Science 4 (Suppl. 1): 251-253.

Curley, Jr. K.O., Paschal, J.C., Welsh, Jr. T.H. and Randel, R.D., 2006. Exit velocity as a measure of cattle temperament is repeatable and associated with serum concentration of cortisol in Brahman bulls. Journal of Animal Science 84: 3100-3103.

Dahl-Pedersen, K., Herskin, M.S., Houe, H. and Thomsen, P.T., 2018b. Risk factors for deterioration of the clinical condition of cull dairy cows during transport to slaughter. Frontiers in Veterinary Science 5: 297. https://doi.org/10.3389/fvets.2018.00297

Dahl-Pedersen, K., Herskin, M.S., Houe. H. and Thomsen, P.T. 2018a. A descriptive study of the clinical condition of cull dairy cows before transport to slaughter. Livestock Science 218: 108-113.

Daly, C.C. and Whittington, P.E., 1989a. Investigation into the principal determinants of effective captive bolt stunning of sheep. Research in Veterinary Science 46: 406-408.

Daly, C.C. and Whittington, P.E., 1989b. A survey of commercial practices used in the stunning of cattle. Report to the RSPCA and MAFF, UK. Bristol University, Langford, UK, 26 pp.

Daly, C.C., Gregory, N.G. and Wotton, S.B., 1987. Captive bolt stunning of cattle – effects on brain-function and role of bolt velocity. British Veterinary Journal 143: 574-580.

Daly, C.C., Kallweit, E. and Ellendorf, F., 1988. Cortical function in cattle during slaughter – conventional captive bolt stunning followed by exsanguination compared with shechita slaughter. Veterinary Record 122: 325-329.

De Freslon, I., Strappini, A.C., Soto-Gamboa, M. and Gallo, C., 2014. Characterization of behavioural reactivity in steers during handling and its relationship with blood cortisol, bruising and meat pH. Archivos de Medicina Veterinaria 46: 229-237.

De la Cruz, L., Gibson, T.J., Guerrero-Legarreta, I., Napolitano, F., Mora-Medina, P. and Mota-Rojas, D., 2018. The welfare of water buffaloes during the slaughter process: a review. Livestock Science 212: 22-33.

De la Cruz-Cruz, L.A., Mota-Rojas, D., Guerreo-Legarreta, I., Hernandez-Gonzalez, R., Roldan-Santiago, P. and Berdugo-Gutiérrez, J.A., 2014. Errores y consecuencias del manejo pre sacrificio de los búfalos de agua. 4° Encuentro Nacional de Criadores de Búfalos. Montería, Colombia.

Del Campo Gigena, M., Soares de Lima, J.M., Brito, G., Manteca, X., Hernández, P. and Montossi, F., 2021. Effect of finishing diet and lairage time on steers welfare in Uruguay. Animals 11: 1329. https://doi.org/10.3390/ani11051329

Del Campo, M., Brito, G., Soares de Lima, J., Hernández, P. and Montossi, F., 2010. Finishing diet, temperament and lairage time effects on carcass and meat quality traits in steers. Meat Science 86: 908-914.

Dunn, C.S., 1990. Stress reactions of cattle undergoing ritual slaughter using two methods of restraint. Veterinary Record 126: 522-525.

Edwards-Callaway, L.N., Calvo-Lorenzo, M.S., Scanga, J.A. and Grandin, T., 2017. Mobility scoring of finished cattle. Veterinary Clinics Food Animals 33: 235-250.

Edwards-Callaway, L.N., Walker, J. and Tucker, C.B., 2019. Culling decisions and dairy cattle welfare during transport to slaughter in the United States. Frontiers in Veterinary Science 5: 343. https://doi.org/10.3389/fvets.2018.00343

EFSA (European Food Safety Authority), 2004. Opinion of the Scientific Panel on Animal Health and Welfare on a request from the Commission related to welfare aspects of the main systems of stunning and killing the main commercial species of animals. EFSA Journal 45: 1-29.

EFSA (European Food Safety Authority), 2011. Scientific opinion concerning the welfare of animals during transport EFSA panel on animal health and welfare (AHAW). EFSA Journal 9: 1966. https://doi.org/10.2903/j.efsa.2011.1966

Eicher, S.D., 2001. Transport of cattle in the dairy industry: current research and future directions. Journal of Animal Science 84 (Suppl. E): E19-E23.

Eldridge, G.A., Warner, R.D., Winfield, C.G., and Vowles, W.J., 1989. Pre-slaughter management and marketing systems for cattle in relation to improving meat yield, meat quality and animal welfare. Department of Agriculture and Rural Affairs, Melbourne, Australia, 56 pp.

European Union, 2004. Council Regulation (EC) No 1/2005 of 22 December 2004 on the protection of animals during transport and related operations and amending Directives 64/432/EEC and 93/119/EC and Regulation (EC) No 1255/97. Official Journal of the European Union L3: 1-44. Available at: http://data.europa.eu/eli/reg/2005/1/oj.

European Union, 2009. Council Regulation (EC) No 1099/2009 of 24 September 2009 on the protection of animals at the time of killing. Official Journal of the European Union L 303: 1-30.

European Union, 2017. Animal transport guides: guide to good practices for the transport of cattle. Available at: www.animaltransportguides.eu.

Ewbank, R., Parker, M.J. and Mason, C.W., 1992. Reactions of cattle to head-restraint at stunning: a practical dilemma. Animal Welfare 1: 55-63.

FAO (Food and Agriculture Organization of the United Nations), 2001. Guidelines for humane handling, transport and slaughter of livestock. FAO, Rome, Italy. Available at: https://tinyurl.com/jnw8xbvt.

Farjo, L.A. and Miclau, T., 1997. Ballistics and mechanisms of tissue wounding. Injury Journal 28: 12-17.

Faucitano, L. and Pedernera, C., 2016. Reception and unloading. In: Raj, M. and Velarde, A. (eds.) Animal welfare at slaughter: a practitioner guide. 5M Publishing, Sheffield, UK, pp. 33-50.

FAWC, 1995. Report on the welfare of livestock when slaughtered by religious methods. Farm Animal Welfare Council, London, UK.

Fell, L.R. and Shutt, D.A., 1986. Adrenal response of calves to transport stress as measured by salivary cortisol. Canadian Journal of Animal Science 66: 637-641

Ferguson, D.M. and Warner, R., 2008. Have we underestimated the impact of pre-slaughter stress on meat quality in ruminants? Meat Science 80: 12-19.

Ferguson, D.M., Bruce, H.L., Thompson, J.M., Egan, A.F., Perry, D. and Shorthose, W.R., 2001. Factors affecting beef palatability – farm gate to chilled carcass. Australian Journal of Experimental Agriculture 41: 879-891.

Flint, H.E., Schwartzkopf-Genswein, K.S., Bateman, K.G. and Haley, D.B., 2014. Characteristics of loads of cattle stopping for feed, water and rest during long-distance transport in Canada. Animals 4: 62-81.

Fuseini, A. 2019. The development and practical implementation of single pulse ultra-high current for humane and halal compliant slaughter of cattle. PhD thesis. University of Bristol, Langford, UK.

Gallo, C. and Gatica, C., 1995. Efectos del tiempo de ayuno, sobre el peso vivo, de la canal y algunos órganos en novillos. Archivos de Medicina Veterinaria 27: 69-78.

Gallo, C. and Strappini, A., 2016. Ensuring the welfare of culled dairy cows during transport and slaughter. In: Webster, J. (ed.) Achieving sustainable production of milk. Burleigh Dodds Series in Agricultural Science, Cambridge, UK, pp. 1-20.

Gallo, C. and Tadich, T.A., 2008. South America. In: Appleby, M.C., Cussen, V., Garcés, L., Lambert, L. and Turner, J. (eds.) Long distance transport and welfare of farm animals. CABI, Wallingford, UK, pp. 261-287.

Gallo, C., Faucitano, L. and Gerritzen, M., 2016. Effects of preslaughter handling and stunning on meat quality. In: Velarde, A. and Raj, M. (eds.) Animal welfare at slaughter. 5M Publishing Ltd., Sheffield, UK, pp. 251-269.

Gallo, C., Lizondo, G. and Knowles T., 2003. The effects of journey and lairage time on steers transported to slaughter in Chile. Veterinary Record 152: 361-364.

Gallo, C.B. and Huertas, S.M., 2016. Main animal welfare problems in ruminant livestock during preslaughter operations: a South American view. Animal 10: 342-348.

Gebresenbet, G. Wikner, I. Bobobee, E.Y.H. Maria, G. and Villarroel, M., 2012. Effect of transport time and handling on physiological responses of cattle. Journal of Agriculture Science and Technology 2: 800-810.

Gellatly, D., Cyrillo, J.N.S.G., Sant'Anna, A.C., Valente, T.S., Schwartzkopf-Genswein, K. and Paranhos da Costa, M.J.R., 2020. Flight speed as an indicator of Zebu cattle growth performance and suitability to the feedlot environment. Animal Production Science 60: 560-566.

Gerritzen, M.A. and Gibson, T.J., 2016. Animal welfare at depopulation strategies during disease control actions. In: Velarde, A. and Raj, M. (eds.) Animal welfare at slaughter. 5M Publishing Ltd., Sheffield, UK, p. 203-218.

Gibson, T.J., Bedford, E.M., Chancellor, N.M. and Limon, G., 2015b. Pathophysiology of free-bullet slaughter of horses and ponies. Meat Science 108: 120-124.

Gibson, T.J., Dadios, N. and Gregory, N.G., 2015c. Effect of neck cut position on time to collapse in halal slaughtered cattle without stunning. Meat Science 110: 310-314.

Gibson, T.J., Johnson, C.B., Murrell, J.C., Chambers, J.P., Stafford, K.J. and Mellor, D.J., 2009b. Components of electroencephalographic responses to slaughter in halothane-anaesthetised calves: effects of cutting neck tissues compared with major blood vessels. New Zealand Veterinary Journal 57: 84-89.

Gibson, T.J., Johnson, C.B., Murrell, J.C., Hulls, C.M., Mitchinson, S.L., Stafford, K.J., Johnstone, A.C. and Mellor, D.J., 2009a. Electroencephalographic responses of halothane-anaesthetised calves to slaughter by ventral-neck incision without prior stunning. New Zealand Veterinary Journal 57: 77-83.

Gibson, T.J., Johnson, C.B., Murrell, J.C., Mitchinson, S.L., Stafford, K.J. and Mellor, D.J., 2009c. Amelioration of electroencephalographic responses to slaughter by non-penetrative captive-bolt stunning after ventral-neck incision in halothane-anaesthetised calves. New Zealand Veterinary Journal 57: 96-101.

Gibson, T.J., Mason, C.W., Spence, J.Y., Barker, H. and Gregory, N.G., 2015a. Factors affecting penetrating captive bolt gun performance. Journal of Applied Animal Welfare Science 18: 222-238.

Gibson, T.J., Oliveira, S.E.O., Dalla Costa, F.A. and Gregory, N.G., 2019. Electroencephalographic assessment of pneumatically powered penetrating and non-penetrating captive-bolt stunning of bulls. Meat Science 151: 54-59.

Gibson, T.J., Ridler, A.L., Lamb, C.R., Williams, A., Giles, S. and Gregory, N.G., 2012. Preliminary evaluation of the effectiveness of captive-bolt guns as a killing method without exsanguination for horned and unhorned sheep. Animal Welfare 21: 35-42.

Goldhawk, C., Crowe, T., González, L.A., Janzen, E., Kastelic, J., Pajor, E. and Schwartzkopf-Genswein, K., 2014a. Comparison of eight logger layouts for monitoring animal-level temperature and humidity during commercial feeder cattle transport. Journal of Animal Science 92: 4161-4171.

Goldhawk, C., Crowe, T., Janzen, E., González, L.A., Kastelic, J., Pajor, E. and Schwartzkopf-Genswein, K.S., 2014b. Trailer microclimate during commercial transportation of feeder cattle and relationship to indicators of cattle welfare. Journal of Animal Science 92: 5155-5165.

González, L.A., Schwartzkopf-Genswein, K.S., Bryan, M., Silasi, R. and Brown, F., 2012a. Factors affecting body weight loss during commercial long-haul transport of cattle in North America. Journal of Animal Science 90: 3630-3639.

González, L.A., Schwartzkopf-Genswein, K.S., Bryan, M., Silasi, R. and Brown, F., 2012b. Relationships between transport conditions and welfare outcomes during commercial long-haul transport of cattle in North America. Journal of Animal Science 90: 3640-3651.

González, L.A., Schwartzkopf-Genswein, K.S., Bryan, M., Silasi, R. and Brown, F., 2012c. Space allowance during commercial long-distance transport of cattle in North America. Journal of Animal Science 90: 3618-3629.

Gottardo, F., Ricci, R., Fregolent, G., Ravarotto, L. and Cozzi, G., 2003. Welfare and meat quality of beef cattle housed on two types of floor with the same space allowance. Italian Journal of Animal Science 2: 243-253.

Grandin, T. and Deesing, M.D., 2008. Humane livestock handling. In: Grandin, T. and Deesing, M.D. (eds.) Understanding livestock behaviour and building facilities for healthier animals. Storey Publishing, North Adams, MS, USA, 228 pp.

Grandin, T. and Regenstein, J.M., 1994. Religious slaughter and animal welfare: a discussion for meat scientists. Meat Focus International 3: 115-123.

Grandin, T. and Shivley, Ch., 2015. How farm animals react and perceive stressful situations such as handling, restraint, and transport. Animals 5: 1233-1251.

Grandin, T., 1980. Mechanical, electrical and anesthetic stunning methods for livestock. International Journal for the Study of Animal Problems 1: 242-263.

Grandin, T., 1990. Humanitarian aspects of shehitah in the United States. Judaism 39: 436-446.

Grandin, T., 1992. Observations of cattle restraint devices for stunning and slaughtering. Animal Welfare 1: 85-90.

Grandin, T., 1993. Behavioural agitation during handling of cattle is persistent over time. Applied Animal Behaviour Science 36: 1-9.

Grandin, T., 1994. Euthanasia and slaughter of livestock. Journal of the American Veterinary Medical Association 204: 1354-1360.

Grandin, T., 1997a. Assessment of stress during handling and transport. Journal of Animal Science 75: 249-257.

Grandin, T., 1997b. The design and construction of facilities for handling cattle. Livestock Production Science 49: 103-119.

Grandin, T., 1998a. Reducing handling stress improves both productivity and welfare. The Professional Animal Scientist 4: 1-10.

Grandin, T., 1998b. Handling methods and facilities to reduce stress on cattle. Veterinary Clinics of North America Food Animal Practice 14: 325-341.

Grandin, T., 1998c. The feasibility of using vocalization scoring as an indicator of poor welfare during cattle slaughter. Applied Animal Behaviour Science 56: 121-128.

Grandin, T., 2001. Perspectives on transportation issues: the importance of having physically fit cattle and pigs. Journal of Animal Science 79: E201-E207.

Grandin, T., 2002. Return-to-sensibility problems after penetrating captive bolt stunning of cattle in commercial beef slaughter plants. Journal of the American Veterinary Medical Association 221: 1258-1261.

Grandin, T., 2007. Livestock handling and transport. CAB International, Wallingford, UK., 384 pp.

Grandin, T., 2009. Evaluation of methods of restraint for holding (fixation) of cattle, calves, and sheep for kosher and halal slaughter. Available at: https://www.grandin.com/ritual/evaluation.restraint.methods.kosher.halal.html.

Grandin, T., 2010. Auditing animal welfare at slaughter plants. Meat Science 86: 56-65.

Grandin, T., 2012. Auditing animal welfare and making practical improvements in beef-, pork- and sheep-slaughter plants. Animal Welfare 21: 29-34.

Grandin, T., 2013. Making slaughterhouses more humane for cattle, pigs, and sheep. Annual Review of Animal Bioscience 1: 491-512.

Grandin, T., 2014a. Behavioural principles of handling cattle and other grazing animals under extensive conditions. In: Grandin, T. (ed.) Livestock handling and transport. CABI, Wallingford, UK, pp. 39-64.

Grandin, T., 2014b. Improving welfare and reducing stress on animals in slaughter plants. In: Grandin, T. (ed.) Livestock handling and transport. CABI, Wallingford, UK, pp. 421-450.

Grandin, T., 2016. Evaluation of the welfare of cattle housed in outdoor feedlot pens. Veterinary and Animal Science 1-2: 23-28.

Grandin, T., 2017. Recommended animal handling guidelines and audit guide: a systematic approach to animal welfare. North American Meat Institute, Washington, DC, 131 pp. Available at: http://animalhandling.org/producers/guidelinesaudits.

Grandin, T., 2020. Livestock handling at the abattoir: effects on welfare and meat quality. Meat and Muscle Biology 4: 1-11.

Gregory, N.G. and Shaw, F., 2000. Penetrating captive bolt stunning and exsanguination of cattle in abattoirs. Journal of Applied Animal Welfare Science 3: 215-230.

Gregory, N.G. and Wotton, S.B., 1984. Time to loss of brain responsiveness following exsanguination in calves. Research in Veterinary Science 37: 141-143.

Gregory, N.G., 2005. Recent concerns about stunning and slaughter. Meat Science 70: 481-491.

Gregory, N.G., 2007. Stunning and slaughter animal welfare and meat production. In: Gregory, N.G. (ed.) Animal welfare and meat production. CABI Publishing, Wallingford, UK, pp. 191-212.

Gregory, N.G., 2008. Animal welfare at markets and during transport and slaughter. Meat Science 80: 2-11.

Gregory, N.G., Benson, T. and Mason C.W., 2009a. Cattle handling and welfare standards in livestock markets in the UK. Journal of Agricultural Science 147: 345-354.

Gregory, N.G., Fielding, H.R., von Wenzlawowicz, M. and von Holleben, K., 2010. Time to collapse following slaughter without stunning in cattle. Meat Science 85: 66-69.

Gregory, N.G., Schuster, P., Mirabito, L., Kolesar, R. and Mc Manus, T., 2012a. Arrested blood flow during false aneurysm formation in the carotid arteries of cattle slaughtered with and without stunning. Meat Science 90: 368-372.

Gregory, N.G., Shaw, F.D., Whitford, J.C. and Patterson-Kane, J.C., 2006. Prevalence of ballooning of the severed carotid arteries at slaughter in cattle, calves and sheep. Meat Science 74: 655-657.

Gregory, N.G., von Wenzlawowicz, M. and von Holleben, K., 2009b. Blood in the respiratory tract during slaughter with and without stunning in cattle. Meat Science 82: 13-16.

Gregory, N.G., von Wenzlawowicz, M., Alam, R.M., Anil, H.M., Sildere, T.Y. and Silva-Fletcher, A., 2008. False aneurysms in carotid arteries of cattle and water buffalo during shechita and halal slaughter. Meat Science 79: 285-288.

Gregory, N.G., von Wenzlawowicz, M., Holleben, K., Fielding, H.R., Gibson, T.J., Mirabito, L. and Kolesar, R., 2012b. Complications during shechita and halal slaughter without stunning in cattle. Animal Welfare 21: 81-86.

Grigor, P.N., Cockram, S.M., Steele, W.B., Le Sueur, C.J. and Forsyth, R.E., 2001. Effects of space allowance during transport and duration of midjourney lairage period on the physiological, behavioural and immunological responses of young calves during and after transport. Animal Science 73: 341-360.

Grist, A. and Wotton, S.B., 2021. A case of multiple stun attempts in a bovine due to chronic disease process causing cranial abnormalities. Animals 11: 116. https://doi.org/10.3390/ani11010116

Grist, A., Bock, R., Knowles, T.G. and Wotton, S.B., 2020. Further examination of the performance of blank cartridges used in captive bolt devices for the pre-slaughter stunning of animals. Animals 10: 2146. https://doi.org/10.3390/ani10112146

Grist, A., Lines, J.A., Bock, R., Knowles, T.G. and Wotton, S.B., 2019. An examination of the performance of blank cartridges used in captive bolt devices for the pre-slaughter stunning and euthanasia of animals. Animals 9: 55. https://doi.org/10.3390/ani9080552

Groβ, R., 1979. Elektroencephalographische und elektrokardiographische verlaufsuntersuchungen nach bolzenschuβbetäubung und nach töten durch entbluten in der form des rituellen schlachtens. Durch die Tierärztliche Hochschule, Hannover, Germany.

Haskell, M.J., Simm, G. and Turner, S.P., 2014. Genetic selection for temperament traits in dairy and beef cattle. Frontiers in Genetics 5: 1-18.

Hay, K.E., Barnes, T.S., Morton, J.M., Clements, A.C. and Mahoney, T.J., 2014. Risk factors for bovine respiratory disease in Australian feedlot cattle: use of a causal diagram-informed approach to estimates effects of animal mixing and movements before feedlot entry. Preventive Veterinary Medicine 117: 160-169.

Health of Animals Regulations, 2020. Regulations amending the health of animals regulations.

Hemsworth, P.H. and Coleman, G.J., 2011. Human-livestock interactions: the stockperson and the productivity of intensively farmed animals. CABI International, Wallingford, UK, 194 pp.

Hemsworth, P.H., 2003. Human-animal interactions in livestock production. Applied Animal Behaviour Science 81: 185-198.

Hemsworth, P.H., Coleman, G.J., Barnett, J.L., Borg, S. and Dowling S., 2002. The effects of cognitive behavioural intervention on the attitude and behaviour of stockpersons and the behaviour and productivity of commercial dairy cows. Journal of Animal Science 80: 68-78.

Herrán, L., Romero, M. and Herrán, L., 2017. Interacción humano-animal y prácticas de manejo bovino en subastas colombianas. Revista de Investigación Veterinaria del Perú 28: 571-585.

Heuston, C., 2017. Epidemiology of compromised and unfit cattle arriving at auction markets and abattoirs in Alberta. MSc. Thesis, University of Saskatchewan, Saskatoon, Canada.

Hogan, J.P., Petherick, J.C. and Phillips, C.J.C., 2007. The physiological and metabolic impacts on sheep and cattle of feed and water deprivation before and during transport. Nutrition Research Reviews 20: 17-28.

Hollerman, J.J., Fackler, M.L., Coldwell, D.M. and Benmenachem, Y., 1990. Gunshot wounds: 1. Bullets, ballistics, and mechanisms of injury. American Journal of Roentgenology 155: 685-690.

Hoppe, S., Brandt, H.R., Konig, S., Erhardt, G. and Gauly, M., 2010. Temperament traits of beef calves measured under field conditions and their relationships to performance. Journal of Animal Science 88: 1982-1989.

Hultgren, J., 2022. Avoiding live-animal transport to slaughter: mobile abattoirs. In: Faucitano, L. (ed.) Preslaughter handling and slaughter of meat animals. Wageningen Academic Publishers, Wageningen, the Netherlands, pp. 390-434.

Immonen, K., Kauffman, R.G., Schaefer, D.M. and Puolanne, E., 2000. Glycogen concentrations in bovine Longissimus dorsi muscle. Meat Science 54: 163-167.

Ishiwata, T., Uetake, K., Abe, N., Eguchi, Y. and Tanaka, T., 2006. Effects of an environmental enrichment using a drum can on behavioural, physiological and productive characteristics in fattening beef cattle. Animal Science Journal 77: 352-362.

Janloo, S.M., Dolezal, H.G., Gardner, B.A., Owens, F.N., Peterson, J. and Moldenhauer, M., 1998. Impact of withholding feed on performance and carcass measurements of feedlot steers. Animal Science Research Report, Department of Animal Science, OSU, pp.109-113.

Jarvis, A.M., Harrington, D.W.J. and Cockram, M.S., 1996a. Effect of source and lairage on some behavioural and biochemical measurements of feed restriction and dehydration in cattle at a slaughterhouse. Applied Animal Behaviour Science 50: 83-94.

Jarvis, A.M., Messer, C.D.A. and Cockram, M.S., 1996b. Handling, bruising and dehydration of cattle at the time of slaughter. Animal Welfare 5: 259-270.

Jarvis, A.M., Selkirk, L. and Cockram, M.S., 1995. The influence of source, sex class and preslaughter handling on the bruising of cattle at two slaughterhouses. Livestock Production Science 43: 215-224.

Jones, B., 2011. The slaughter of Australian cattle in Indonesia: an observational study. RSPCA Australia, 47 pp.

Jones, S.D.M., Schaefer, A.L., Robertson, W.M. and Vincent, B.C., 1990. The effects of withholding feed and water on carcass shrinkage and meat quality in beef cattle. Meat Science 28: 131-139.

Jones, S.D.M., Schaefer, A.L., Tong, A.K.W. and Vincent, B.C., 1988. The effects of fasting and transportation on beef cattle. 2. Body changes, carcass composition and meat quality. Livestock Production Science 20: 25-35.

King, D.A., Schuehle Pfeiffer, C.E., Randel, R.D., Welsh, T.H. Jr., Oliphint, R.A., Baird, B.E., Curley, K.O. Jr., Vann, R.C., Hale, D.S. and Savell, J.W., 2006. Influence of animal temperament and stress responsiveness on the carcass quality and beef tenderness of feedlot cattle. Meat Science 74: 546-556.

Kline, H.C., Edwards-Callaway, L. N., and Grandin, T., 2019. Field observation: pen stocking capacities for overnight lairage of finished steers and heifers at a commercial slaughter facility. Applied Animal Science 35: 130-133.

Kline, H.C., Wagner, D.R., Edwards-Callaway, L.N., Alexander, L.R. and Grandin, T., 2019. Effect of captive bolt gun length on brain trauma and post-stunning hind limb activity in finished cattle *Bos taurus*. Meat Science 155: 69-73.

Knowles, T.G., 1995. A review of post transport mortality among younger calves. Veterinary Record 137: 406-407.

Knowles, T.G., 1999. A review of road transport of cattle. Veterinary Record 144: 197-201.

Knowles, T.G., Warriss, P. and Vogel, K.D., 2014. Stress physiology of animals during transport. In: Grandin, T. (ed.) Livestock handling and transport, CABI, Wallingford, UK, p. 399-420.

Knowles, T.G., Warriss, P.D., Brown, S.N., and Edwards, J.D., 1999. The effect of transporting cattle by road for up to 31 hours. Veterinary Record 145: 575-582.

Knowles, T.G., Warriss, P.D., Brown, S.N., Edwards, J.D., Watkins, P.E., and Philips, A.J., 1997. Effects on calves less than one month old of feeding or not feeding them during road transport of up to 24 hours. Veterinary Record 140: 116-124.

Lambooij, E., Van der Werf, J.T.N., Reimert, H.G.M. and Hindle, V.A., 2012. Restraining and neck cutting or stunning and neck cutting of veal calves. Meat Science 91: 22-28.

Lay, D.C., Friend, T.H., Bowers, C.L., Grissom, K.K. and Jenkins, O.C., 1992a. Behavioural and physiological effects of freeze and hot iron branding on crossbred cattle. Journal of Animal Science 70: 330-336.

Lay, D.C., Friend, T.H., Bowers, C.L., Grissom, K.K. and Jenkins, O.C., 1992b. A comparative physiological and behavioural study of freeze and hot iron branding using dairy cows. Journal of Animal Science 70: 1121-1125.

Le Neindre, P., Boivin, X. and Boissy, A., 1996. Handling of extensively kept animals. Applied Animal Behaviour Science 49: 73-81.

Leon, A.F., Sanchez, J.A. and Romero, M.H., 2020. Association between attitude and empathy with the quality of human-livestock interactions. Animals 10: 1304. https://doi.org/10.3390/ani10081304

Levinger, I.M., 1961. Untersuchungen zum schlächtproblem. PhD. Thesis, University of Zürich, Zürich, Switzerland.

Levinger, I.M., 1995. Shechita in the light of the year 2000, critical review of the scientific aspects of methods of slaughter and shechita. Maskil L'David, Jerusalem, Israel, 204 pp.

Limon, G., Guitian, J. and Gregory, N.G., 2010. An evaluation of the humaneness of puntilla in cattle. Meat Science 84: 352-355.

Limon, G., Guitian, J. and Gregory, N.G., 2012. A review of the humaneness of puntilla as a slaughter method. Animal Welfare 21: 3-8.

Losada-Espinosa, N., Villarroel, M., María, G.A. and Miranda-de la Lama, G.C., 2018. Pre-slaughter cattle welfare indicators for use in commercial abattoirs with voluntary monitoring systems: a systematic review. Meat Science 138: 34-48.

Mach, N., Bach, A., Velarde, A. and Devant, M., 2008. Association between animal, transportation, slaughterhouse practices, and meat pH in beef. Meat Science 78: 232-238.

Machado, M., Guimarães, S.R.M., Medeiros, V.A.P., Giannoccaro, S.M.A., César, M.P., Mendez de Córdova, F. and Tony, R.A., 2013. Bone and brain lesions in horses following euthanasia with fire gun. Brazilian Journal of Veterinary Pathology 6: 102-105.

María, G.A., Villarroel, M., Chacon, G. and Gebresenbet, G., 2004. Scoring system for evaluating the stress to cattle of commercial loading and unloading. Veterinary Record 154: 818-821.

Marti, S., Wilde, R.E., Moya, D., Heuston, C.E.M., Brown, F. and Schwartzkopf-Genswein, K.S., 2017. Effect of rest stop duration during long-distance transport on welfare indicators in recently weaned beef calves. Journal of Animal Science 95: 636-644

Martin, M.S., Kline, H.C., Wagner, D.R., Alexander, L.R., Edwards-Callaway, L.N. and Grandin, T., 2018. Evaluation of different captive bolt lengths and breed influence upon post-stun hind limb and forelimb activity in fed cattle at a commercial slaughter facility. Meat Science 143: 159-164.

McNally, P.W. and Warriss, P.D., 1996. Recent bruising in cattle at abattoirs. Veterinary Record 138: 126-128.

Meléndez, D.M., Marti, S., Haley, D.B., Schwinghamer, T.D. and Schwartzkopf-Genswein, K.S., 2020. Effect of transport and rest stop duration on the welfare of conditioned cattle transported by road. PLOS ONE 15: e0228492. https://doi.org/10.1371/journal.pone.0228492

Meléndez, D.M., Marti, S., Haley, D.B., Schwinghamer, T.D. and Schwartzkopf-Genswein, K. S., 2021. Effect of conditioning, source and rest on welfare indicators of beef cattle transported by road. PLOS ONE 16: e0244854. https://doi.org/10.1371/journal.pone.0244854

Minka, N.S. and Ayo, J.O., 2007. Effects of loading behaviour and road transport stress on traumatic injuries in cattle transported by road during the hot-dry season. Livestock Science 107: 91-95

Mirabito, L., Stadig L., Terlouw, C., Bourguet, C., Marzin, V., Ducreux, B., Bergeaud-Blackler, F., Dalmau, A., Rodriguez, P., Pallisera Lloveras, J., Baltussen, W., Gerritzen, M., De Winter, M., Gibson, T., Cenci-Goga, B. and Novelli, S., 2015. Restraining systems for bovine animals slaughter without stunning, welfare and socio-economic implications (BOREST). Report for the European Commission, 279 pp.

Miranda-de la Lama, G., Leyva, I., Barrera-Serrano, A., Pérez-Linares, C., Sánchez-López, E., María, G. and Figueroa-Saavedra, F., 2012. Assessment of cattle welfare at a commercial slaughter plant in the northwest of México. Tropical Animal Health and Production 44: 497-504.

Miranda-de la Lama, G.C., Villarroel, M. and María, G.A., 2014. Livestock transport from the perspective of the pre-slaughter logistic chain: a review. Meat Science 98: 9-20.

Morris, C.A., Cullen, N.G., Kilgour, R. and Bremner, K.J., 1994. Some genetic factors affecting temperament in *Bos taurus* cattle. New Zealand Journal of Agricultural Research 37: 167-175.

Muñoz, D., Strappini, A. and Gallo, C., 2012. Animal welfare indicators to detect problems in the cattle stunning box. Archivos de Medicina Veterinaria 44: 297-302.

Nakyinsige, K., Che Man, Y.B., Aghwan, Z.A., Zulkifli, I., Goh, Y.M., Abu Bakar, F., Al-Kahtani, H.A. and Sazili, A.Q., 2013. Stunning and animal welfare from Islamic and scientific perspectives. Meat Science 95: 352-361.

Nangeroni, L.L. and Kennett, P.D., 1964. Electroencephalographic studies of the effects of shechita slaughter on cortical function in ruminants. Report of the New York State Veterinary College at Cornell University for the Year 1963-64. Cornell University, New York, NY, USA, pp. 38-39.

Neves, J.E.G. Paranhos da Costa, M.J.R., Roça, R.O., Faucitano, L. and Gregory, N.G., 2016. A note comparing the welfare of Zebu cattle following three stunning-slaughter methods. Meat Science 117: 41-43.

Newhook, J.C. and Blackmore, D.K., 1982. Electroencephalographic studies of stunning and slaughter of sheep and calves – 2. The onset of permanent insensibility in calves during slaughter. Meat Science 6: 295-300.

Nicholson, J.D.W., Nicholson, K.L., Frenzel, L.L., Maddock, R.J., Delmore, R.J., Lawrence, T.E., Henning, W.R., Pringle, T.D., Johnson, D.D., Paschal, J.C., Gill, R.J., Cleere, J., Carpenter, B.B., Machen, R.V., Banta, J.P., Hale, D.S., Griffin, D.B. and Savell, J.W., 2013. Survey of transport procedures, management practices, and health assessment related to quality, quantity, and value for value for market beef and dairy cows and bulls. Journal of Animal Science 91: 5029-5036.

OIE (World Organization for Animal Health), 2019. Terrestrial animal health code. Available at: https://www.oie.int/en/animal-welfare/animal-welfare-at-a-glance/.

Oliveira, S.E.O., Dalla Costa, F.A., Gibson, T.J., Dalla Costa, O.A. and Gregory, N.G., 2018b. Evaluation of brain damage resulting from penetrating and non-penetrating stunning in Nellore cattle using pneumatically powered captive bolt guns. Meat Science 145: 347-351.

Oliveira, S.E.O., Gregory, N.G., Dalla Costa, F.A., Gibson, T.J. and Paranhos da Costa, M.J.R., 2017. Efficiency of low versus high airline pressure in stunning cattle with a pneumatically powered penetrating captive bolt gun. Meat Science 130: 64-68.

Oliveira, S.E.O., Gregory, N.G., Dalla Costa, F.A., Gibson, T.J., Dalla Costa, O.A. and Paranhos da Costa, M.J.R., 2018a. Effectiveness of pneumatically powered penetrating and non-penetrating captive bolts in stunning cattle. Meat Science 140: 9-13.

Olson, C.A., Carstens, G.E., Herring, A.D., Hale, D.S., Kayser, W.C. and Miller, R.K., 2019. Effects of temperament at feedlot arrival and breed type on growth efficiency, feeding behaviour, and carcass value in finishing heifers. Journal of Animal Science 97: 1828-1839.

Önenç, A. and Kaya, A., 2004. The effects of electrical stunning and percussive captive bolt stunning on meat quality of cattle processed by Turkish slaughter procedures. Meat Science 66: 809-815.

Paranhos Da Costa, M.J.R., Huertas, S.M., Gallo, C. and Dalla Costa, O.A., 2012. Strategies to promote farm animal welfare in Latin America and their effects on carcass and meat quality traits. Meat Science 92: 221-226.

Park, R.M., Foster, M. and Daigle, C.L., 2020. A scoping review: the impact of housing systems and environmental features on beef cattle welfare. Animals 10: 565. https://doi.org/10.3390/ani10040565

Petherick, J.C. and Phillips, C., 2009. Space allowances for confined livestock and their determination from allometric principles. Applied Animal Behaviour Science 117: 1-12.

Petherick, J.C., 2005. Animal welfare issues associated with extensive livestock production: the northern Australian beef cattle industry. Applied Animal Behaviour Science 92: 211-234.

Petherick, J.C., Holroyd, R.G., Doogan, V.J. and Venus, B.K., 2002. Productivity, carcass and meat quality of lot-fed *Bos indicus* cross steers grouped according to temperament. Australian Journal of Experimental Agriculture 42: 389-398.

Pettiford, S.G., Ferguson, D.M., Lea, J.M., Lee, C., Paull, D.R., Reed, M.T., Hinch, G.N. and Fisher, A.D., 2008. Effect of loading practices and 6-hour road transport on the physiological responses of yearling cattle. Australian Journal of Experimental Agriculture 48: 1028-1033.

Phillips, C., Wojciechowska, J., Meng, J. and Cross, N., 2009. Perceptions of the importance of different welfare issues in livestock production. Animal 3: 1152-1166.

Piovezan, U., Gonçalves Cyrillo, J.N.S. and Paranhos da Costa, M.J.R., 2013. Breed and selection line differences in the temperament of beef cattle. Acta Scientiarum 35: 207-212.

Probst, J.K., Hillmann, E., Leiber, F., Kreuzer, M. and Spengler, N.A., 2013. Influence of gentle touching applied few weeks before slaughter on avoidance distance and slaughter stress in finishing cattle. Applied Animal Behaviour Science 144: 14-21.

Raj, M.A.B., Moss, B.W., McCaughey, J., McLauchlan, W., Kilpatrick, D.J. and McGaughey, S.J., 1991. Behavioural response to mixing of entire bulls, vasectomised bulls and steers. Applied Animal Behaviour Science 31: 157-168.

Rezac, D.J., Thomson, D.U., Seimens, M.G., Prouty, F.L., Reinhard, C.D. and Bartle, S.J., 2014. A survey of gross pathologic conditions in cull cows at slaughter in the Great Lakes region of the United States. Journal of Dairy Science 97: 4227-4235.

Robins, A., Pleiter, H., Latter, M. and Phillips, C.J., 2014. The efficacy of pulsed ultrahigh current for the stunning of cattle prior to slaughter. Meat Science 96: 1201-1209.

Romero, M.H., Gutiérrez, C. and Sánchez, J.A., 2012. Evaluation of bruises as an animal welfare indicator during preslaughter of beef cattle. Revista Colombiana de Ciencias Pecuarias 25: 267-275.

Romero, M.H., Uribe-Velásquez, L.F., Sánchez, J.A. and Miranda-de la Lama, G.C., 2013. Risk factors influencing bruising and high muscle pH in Colombian cattle carcasses due to transport and pre-slaughter operations. Meat Science 95: 256-263.

Rosen, S.D., 2004. Opinion piece: physiological insights into Shechita. Veterinary Record 154: 759-765.

Ross, M., Widowski, T.M. and Haley, D.B., 2016. The effects of feeding space on the behavioural responses of cattle during rest periods offered as part of long-distance transportation. Animal Welfare 25: 217-225.

Salvin, H.E., Lees, A.M., Café, L.M., Colditz, I.G. and Lee, C., 2020. Welfare of beef cattle in Australian feedlots: a review of the risks and measures. Animal Production Science 60: 1569-1590.

Sánchez-Hidalgo, M., Bravo, V. and Gallo, C., 2020. Behaviour and health indicators to assess cull cow's welfare in livestock markets. Frontiers in Veterinary Science 7: 471. https://doi.org/10.3389/fvets.2020.00471

Sánchez-Hidalgo, M., Rosenfeld, C. and Gallo, C., 2019. Associations between pre-slaughter and post-slaughter indicators of animal welfare in cull cows. Animals 9: 642. https://doi.org/10.3390/ani9090642

Sazili, A. Q., Norbaiyah, B., Zulkifli, I., Goh, Y. M., Lotfi, M. and Small, A.H., 2013. Quality assessment of longissimus and semitendinosus muscles from beef cattle subjected to non-penetrative and penetrative percussive stunning methods. Asian-Australasian Journal of Animal Sciences 26: 723-731.

Schwartzkopf-Genswein, K. and Grandin, T., 2019. Cattle transport in North America. In Grandin, T. (ed.) Livestock handling and transport. CABI, Wallingford, UK, pp. 153-183.

Schwartzkopf-Genswein, K.S., Faucitano, L., Dadgar, S., Shand, P., González, L.A. and Crowe, T.G., 2012. Road transport of cattle, swine and poultry in North America and its impact on animal welfare, carcass and meat quality: a review. Meat Science 92: 227-243.

Schwenk, B.K., Lechner, I., Ross, S.G., Gascho, D., Kneubuehl, B.P., Glardon, M. and Stoffel, M.H., 2016. Magnetic resonance imaging and computer tomography of brain lesions in water buffaloes and cattle stunned with handguns or captive bolts. Meat Science 113: 35-40.

Shragge, J.E. and Price, M.A., 2004. Religious slaughter. In: Jensen, W., Devine, C. and Dikeman, M. (eds.) Encyclopedia of meat sciences. Elsevier Ltd., Oxford, UK, pp. 1162-1167.

Šímová, V., Večerek, V., Passantino, A. and Voslářová, E., 2016. Pre-transport factors affecting the welfare of cattle during road transport for slaughter – a review. Acta Veterinari Brno 85: 303-318.

Stojkov, J., von Keyserlingk, M.A.G., Duffield, T. and Fraser, D., 2020. Fitness for transport of cull dairy cows at livestock markets. Journal of Dairy Science 103: 2650-2661.

Strappini, A.C., Frankena, K., Metz, J.H.M., Gallo, C. and Kemp, B., 2012. Characteristics of bruises in carcasses of cows sourced frosm farms or from livestock markets. Animal 6: 502-509.

Strappini, A.C., Frankena, K., Metz, J.H.M., Gallo, C.B. and Kemp, B., 2010. Prevalence and risk factors for bruises in Chilean bovine carcasses. Meat Science 86: 859-864.

Strappini, A.C., Metz, J.H.M., Gallo, C., Frankena, K., Vargas, R., De Freslon, I. and Kemp, B., 2013. Bruises in culled cows: when, where and how are they inflicted? Animal 7: 485-491.

Swanson, J.C. and Morrow-Tesch, J., 2001.Cattle transport: historical, research, and future perspectives. Journal of Animal Science 79: E102-E109.

Tadich, N., Gallo, C., Bustamante, H., Schwerter, M. and Van Schaik, G., 2005. Effects of transport and lairage time on some blood constituents of Friesian cross steers in Chile. Livestock Production Science 93: 223-233.

Tadich, N., Gallo, C., Echeverría, R. and Van Schaik, G., 2003. Efecto del ayuno durante dos tiempos de confinamiento y de transporte terrestre sobre algunas variables sanguíneas indicadoras de estrés en novillos. Archivos de Medicina Veterinaria 35: 171-185.

Tagawa, M., Okano, S., Sako, T., Orima, H. and Steffey, E.P., 1994. Effect of change in body position on cardiopulmonary function and plasma-cortisol in cattle. Journal of Veterinary Medical Science 56: 131-134.

Tarrant, P.V., Kenny, F.J. and Harrington, D., 1988. The effect of stocking density during 4-hour transport to slaughter on behaviour, blood constituents and carcass bruising in Friesian steers. Meat Science 24: 209-222.

Tarrant, P.V., Kenny, F.J., Harrington, D. and Murphy, M., 1992. Long distance transportation of steers to slaughter: Effect of stocking density on physiology, behaviour and carcass quality. Livestock Production Science 30: 223-238.

Teke, B., Akdag, F., Ekiz, B. and Ugurlu, M., 2014. Effects of different lairage times after long distance transportation on carcass and meat quality characteristics of Hungarian Simmental bulls. Meat Science 96: 224-229.

Terlouw, C. and Bourguet, C., 2022. Quantifying animal welfare preslaughter using behavioural, physiological and carcass and meat quality measures. In: Faucitano, L. (ed.) Preslaughter handling and slaughter of meat animals. Wageningen Academic Publishers, Wageningen, the Netherlands, pp. 12-61.

Terlouw, E.M.C. and Gibson, T.J., 2016. Physiology and behaviour of food animals. In: Velarde, A. and Raj, M. (eds.) Animal welfare at slaughter. 5M Publishing Ltd., Sheffield, UK, p. 16-32.

Terlouw, E.M.C., Arnould, C., Auperin, B., Berri, C., Le Bihan-Duval, E., Deiss, V., Lefèvre, F., Lensink, B. J. and Mounier, L., 2008. Pre-slaughter conditions, animal stress and welfare: current status and possible future research. Animal 2: 1501-1517.

Thomson, D.U., Wileman, B.W., Rezac, D., Miesner, M.D., Johnson-Neitman, J.L. and Biller, D.S., 2013. Computed tomographic evaluation to determine efficacy of euthanasia of yearling feedlot cattle by use of various firearm-ammunition combinations. American Journal of Veterinary Research 74: 1385-1391.

Titto, E.A.L., Titto, C.G., Gatto, E.G. Silva, N.C.M., Mourão, G.B., Nogueira Filho, J.C.M. and Pereira, A.M.F., 2010. Reactivity of Nellore steers in two feedlot housing systems and its relationship with plasmatic cortisol. Livestock Science 129: 146-150.

Tucker, C.B., Coetzee, J.F., Stookey, J., Thomson, D.R., Grandin, T., and Schwartzkopf-Genswein, K.S., 2015. Beef cattle welfare in the U.S.A.: identification of priorities for future research. Animal Health Research Reviews 16: 107-124.

Turner, S.P., Navajas, E.A., Hyslop, J.J., Ross, D.W., Richardson, R.I., Prieto, N., Bell, M., Jack, M.C. and Roehe, R., 2011. Associations between response to handling and growth and meat quality in frequently handled Bos Taurus beef cattle. Journal of Animal Science 89: 4239-4248.

USDA, 1994. 28 hour law of 1873 Citation: 49 USC 80502. Available at: https://www.nal.usda.gov/awic/twenty-eight-hour-law.

Valkova, L., Vecerek, V., Voslarova, E., Kaluza, M. and Takacova, D., 2021. The welfare of cattle, sheep, goats and pigs from the perspective of traumatic injuries detected at slaughterhouse postmortem inspection. Animals 11, 1406. https://doi.org/10.3390/ani11051406

Van Engen, N.K and Coetzee, J.F., 2018. Effects of transportation on cattle health and production: a review. Animal Health Research Reviews 19: 142-154.

Vanhonacker, F., Verbeke, W., Van Poucke, E. and Tuyttens, F.A.M., 2008. Do citizens and farmers interpret the concept of farm animal welfare differently? Livestock Science 116: 126-136.

Večerek, V., Vošlářová, E., Kameník, J., Machovcová, Z., Válková, L., Volfová, M. and Konvalinová, J., 2021. The effect of slaughtering skills on the welfare of cattle during stunning with a captive bolt. Acta Veterinaria Brno 90: 109-116.

Velarde, A. and Cenci-Goga, B., 2010. Religious slaughter: evaluation of current practices. Dialrel project report, 4 pp.

Velarde, A. and Dalmau, A., 2012. Animal welfare assessment at slaughter in Europe: moving from inputs to outputs. Meat Science 92: 244-251.

Velarde, A., Rodríguez, P., Dalmau, A., Fuentes, C., Llonch, P., von Holleben, K.V., Anil, M.H., Lambooij, J.B., Pleiter, H., Yesildere, T. and Cenci-Goga, B.T., 2014. Religious slaughter: evaluation of current practices in selected countries. Meat Science 96: 278-287.

Vogel, K.D., Claus, J.R., Grandin, T., Oetzel, G.R. and Schaefer, D.M., 2011. Effect of water and feed withdrawal and health status on blood and serum components, body weight loss, and meat and carcass characteristics of Holstein slaughter cows. Journal of Animal Science 89: 538-548.

Voisinet, B.D., Grandin, T., Connor, S.F.O., Tatum, J.D. and Deesing, M.J., 1997b. *Bos indicus* cross feedlot cattle with excitable temperaments have tougher meat and a higher incidence of borderline dark cutters. Meat Science 46: 367-377.

Voisinet, B.D., Grandin, T., Tatum, J.D., Connor, S.F.O. and Struthers, J.J., 1997a. Feedlot cattle with calm temperaments have higher average daily gains than cattle with excitable temperaments. Journal of Animal Science 75: 892-896.

Von Holleben, K., von Wenzlawowicz, M., Gregory, N., Anil, H., Velarde, A., Rodríguez, P., Cenci-Goga, B.T., Catanese, B. and Lambooij, E., 2010. Dialrel project – report on good and adverse practices – animal welfare concerns in relation to slaughter practices from the viewpoint of veterinary sciences, pp. 25-26.

Von Wenzlawowicz, M., von Holleben, K. and Eser, E., 2012. Identifying reasons for stun failures in slaughterhouses for cattle and pigs: a field study. Animal Welfare 21: 51-60.

Wagner, A.E., Muir, W.W. and Grospitch, B.J., 1990. Cardiopulmonary effects of position in conscious cattle. American Journal of Veterinary Research 51: 7-10.

Wagner, D.R., Kline, H.C., Martin, M.S., Alexander, L.R., Grandin, T. and Edwards-Callaway, L.N., 2019. The effects of bolt length on penetration hole characteristics, brain damage and specified-risk material dispersal in finished cattle stunned with a penetrating captive bolt stunner. Meat Science 155: 109-114.

Waiblinger, S., Boivin, X., Pedersen, V., Tosi M-V., Janczak, A.M., Visser, E.K. and Jones, R.B., 2006. Assessing the human-animal relationship in farmed species: a critical review. Applied Animal Behaviour Science 101: 185-242.

Warner, R.D., Ferguson, D.M., Cottrell, J.J. and Knee, B.W., 2007. Acute stress induced by the preslaughter use of electric prodders causes tougher beef meat. Australian Journal of Experimental Agriculture 47: 782-788.

Warriss, P.D., 2000. The effects of live animal handling on carcass and meat quality. In: Warriss, P.D. (ed.) Meat science: an introductory text. CABI, Wallingford, UK, pp. 131-155.

Warriss, P.D., Brown, S.N., Knowles, T.G., Kestin, S.C. and Edwards, J.E., 1995. Effects of cattle transport by road for up to 15 h. Veterinary Record 136: 19-323.

Warriss, P.D., Kestin, S.C., Brown, S.N. and Wilkins, L.J., 1984. The time required for recovery from mixing stress in young bulls and the prevention of dark cutting beef. Meat Science 10: 53-68.

Wathes, C.M., 2012. Cattle inversion for religious slaughter. Farm Animal Welfare Committee, London, UK, 7 pp.

Weary, D., Huzzey, J.M. and Keyserlingk, M., 2009. Using behaviour to predict and identify ill health in animals. Journal of Animal Science 87: 770-777.

Weeks, C.A., 2008. A review of welfare in cattle, sheep, and pig lairages, with emphasis on stocking rates, ventilation and noise. Animal Welfare 17: 275-284.

Weeks, C.A., McNally, P.W. and Warriss, P.D., 2002. Influence of the design of facilities at auction markets and animal handling procedures on bruising in cattle. Veterinary Record 150: 743-748.

Wigham, E.E., Butterworth, A. and Wotton, S., 2018. Assessing cattle welfare at slaughter – why is it important and what challenges are faced? Meat Science 145: 171-177.

Woods, J. and Grandin, T., 2008. Fatigue: a major cause of commercial livestock truck accidents. Veterinaria Italiana 44: 259-262.

Zavy, M.T., Juniewicz, P.E., Phillips, W.A. and von Tungeln, D.L., 1992. Effect of initial restraint, weaning and transport stress on baseline and ACTH stimulated cortisol responses in beef calves of different genotypes. Australian Journal of Veterinary Research 53: 551-557.

Zulkifli, I., Goh, Y.M., Norbaiyah, B., Sazili, A.Q., Lotfi, M., Soleimani, A.F. and Small, A.H., 2014. Changes in blood parameters and electroencephalogram of cattle as affected by different stunning and slaughter methods in cattle. Animal Production Science 54: 187-193.

Photo: Bert Lambooij, WageningenUR Livestock Research, Wageningen, The Netherlands

Chapter 3
Fish

M. Villarroel[1] *and E. Lambooij*[2]

[1]*Departmento de Producción Animal, ETSIAAB, Universidad Politécnica de Madrid, Avenida Puerta de Hierro 2, 28040 Madrid, Spain;* [2]*Wageningen UR Livestock Research, P.O. Box 338, 6700 AH Wageningen, the Netherlands (retired); morris.villarroel@upm.es*

Abstract

Despite being a relatively new field of research, many studies on fish welfare in aquaculture have considered the possible negative effects of preslaughter handling. In this Chapter we review the potential negative effects of feed withdrawal, crowding, emersion (removal of the fish from water), transport and different types of stunning and killing on fish welfare. As ectotherms fish require a rather prolonged fasting period before slaughter to empty their digestive tract, with some positive and negative effects depending on the duration. Recent findings are reviewed in terms of maximum fasting durations and the effects of water temperature. Crowding is normally performed using nets and is known to be stressful, especially when prolonged. Recent literature is reviewed in terms of its effects on physiological welfare indicators and the reduction of repeated or cumulative stressors. Emersion is often done by hand or using a crane and is possibly one of the more acute stressors during pre-slaughter handling. The literature is reviewed in terms of its negative effects and duration limits. Transport is not always necessary but can be of long duration and expensive. Loading and unloading procedures appear to be the most stressful while water quality changes become important for longer distances and higher densities. Lastly, regarding stunning and killing, many fish either die by asphyxia in air or in an ice slurry, with negative effects on welfare, but other mechanical methods and electronarcosis are discussed. Studies on the effects of dry stunning are considered as well as in-line in-water stunning, where fish are pumped through a long-electrified tube connected between the tank and the processing line. Other methods used to make fish unconscious are also discussed, such as gas stunning and the use of clove oil.

Keywords: fish, handling, preslaughter, slaughter, welfare

3.1 Introduction

In 2018, 54.3 million tonnes of fish were produced by aquaculture (FAO, 2020). Approximately 86% of those fish farms use fresh-water, and most are located in Asia (FAO, 2020). Fresh-water fish farming uses a wide diversity of species and systems, ranging from earthen ponds and raceways to above-ground tanks and net-pens (also called cages). The remaining 14% of fish farms use salt-water, also based on a wide variety of species but distributed around the globe, mostly using net-pens and coastal lagoons. Most of the fish farmed in developing nations are grown in earthen ponds (e.g. carps *Cyprinus* spp.) and are not fed directly, but subsist on natural production of algae, often in polyculture with other species. On the other hand, fish species that are more commonly consumed in industrialised countries (e.g. salmonids), are grown in relatively costly infrastructure (raceways and large high-density polyethylene floating cages) and fed formulated feeds several times daily. Overall, in both extensive and intensive aquaculture, the number of individual fish used in fish farming is estimated to be close to 100 billion and has increased by approximately 5% a year in the past decade (FAO, 2020) and, along with it, increasing problems related with fish health and welfare.

3.2 Advances in the science of fish welfare

According to the scientific literature, based on the Web of Science (Main collection), more than 2,126 articles have been published on fish welfare since 1980 (search terms 'fish' and 'welfare', year range 1980-2020, publication type 'article' and 'review'). Not all of those publications deal with the welfare of fish used on farms, some are also related to laboratory fish, like zebra fish (*Danio rerio*), but only 24 articles deal directly with fish welfare during the preslaughter period, which is the target of the current chapter. Most of those publications were published after the year 2005, making the whole field a relatively new endeavour. Nonetheless, in the past 15 years there have been several reviews of fish welfare in general, mostly focused on intensive production. All of those publications underline that preslaughter handling of fish is a critical point in the production chain.

One of the first important reviews of aspects related with the welfare of fish in aquaculture was by Conte (2004), who underlines that stress should be minimised in production centres, mostly since fish have very specific requirements related to their environment (i.e. water quality), outside of which they simply cannot survive or their production becomes unprofitable. Compared with terrestrial animals, fish are more demanding in terms of growing conditions and they can be stressed more readily, something which producers know quite well and try to minimise. Conte (2004) explains some special characteristics of aquaculture that make it complicated to assess welfare, including the great diversity of species used, the great variety of production systems (fresh-water, marine-water, intensive and extensive) and abundance or lack of on-farm monitoring systems (e.g. water quality, oxygen concentration).

In their review, Huntingford *et al.* (2006) summarise the arguments in favour of the hypothesis that fish can feel pain and can suffer. Fish are complex animals with a sophisticated behavioural repertoire and have the ability of suffer, although that ability can be different in degree and type compared to humans or other mammals (Braithwaite *et al.*, 2013). Although Huntingford *et al.* (2006) begin from a more philosophical viewpoint, for example debating why we should take care of fish, they underline four relevant problems or domains to carry out more research: the effects of the lack of food and water, the importance of monitoring the environment and installations, the effects of restricting movement and behaviour and how to avoid mental and physical suffering, all of which may occur during pre-slaughter handling and slaughter. In another pioneering review, Ashley (2007) describes the different handling procedures that can comprise fish welfare, including animal density and aggression, providing a seven-page table which summarises the main welfare problems, including pre-slaughter handling.

As these aforementioned reviews were being published, advances had also been made regarding fish sentience (Sneddon, 2003; Terlouw and Bourguet, 2022 – Chapter 1 of this book), so that beginning in 2008-2009 several large projects began to help study, certify and develop legislation about fish welfare, including the publication of fish welfare standards in the Aquatic Code of the OIE (for the updated version see OIE, 2019). In the same year, Branson (2008) published a highly cited book on fish welfare as the proceedings of a meeting held by the Fish Veterinary Society in November 2004 in Edinburgh (Scotland). Shortly afterwards, the European Food Safety Authority (EFSA, 2009a-g) published a general expert opinion on fish welfare and a series of species specific reports on stunning and slaughter. This was followed by a five-year EU COST project on fish welfare which led to another book some years later (Van de Vis *et al.*, 2013). More books, reviews, articles and handbooks have been published since then on the general topic of fish welfare, and in what follows we will attempt to summarise this growing amount of knowledge, specifically focusing on pre-slaughter handling stages: feed withdrawal, crowding, removal from water, transport (when required) and stunning and slaughter (Figure 3.1).

3.3 Feed withdrawal

At the end of the grow-out period, which can normally last for 1 to 2 years, depending on the species and market size (250 g to 3 kg), feeding is stopped well before slaughter. This begins a period of fasting, also called feed withdrawal, that can last a few days or up to several weeks, depending on the production system and species.

Feed is withdrawn from the fish before slaughter first and foremost to empty the gut. Eliminating feed and faeces from the gastrointestinal tract reduces the risk of contaminating the flesh during evisceration and increases shelf-life, since undigested feed will decompose quickly (Wall, 2001). Fasting also prevents regurgitation of feed during further handling (e.g. crowding or stunning), which contaminates the water and can be harmful for the fish (Cardia and Lovatelli, 2015). In addition, once the gut has been emptied and no more food is available, fish will decrease their basal metabolic rate, which reduces the need for oxygen and the excretion of waste products into the surrounding water (mostly ammonia and CO_2; Lines and Spence, 2012). That has a positive side-effect of improving water quality around the fish, which is better for welfare, especially if fish are going to be crowded or transported in small recipients after fasting, before moving to the point of slaughter. However, the rate of gut clearance (i.e. time taken to evacuate all feed and faeces from the tract) can vary according to the fish species and environmental factors, such as water temperature (see below). Fasting is also thought to have positive effects on flesh quality by reducing fish flesh fat content as well as gaping (Robb, 2008), but the results are contradictory depending on the species and study. Both short (a few days) and long fasting times (several weeks) can have negative effects on fillet shelf-life (Huidobro and Tejada, 2004) and flesh quality (Poli *et al.*, 2005), respectively.

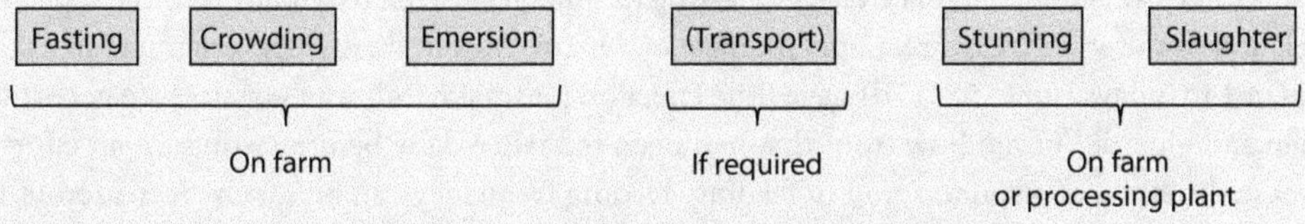

Figure 3.1. The phases of fish handling pre-slaughter (M. Villarroel, Universidad Politécnica de Madrid, Spain).

Compared to mammals, fish can spend more time without food since they are ectotherms. Their body temperature is regulated by external sources and their energy demand is four to five times lower than mammals (Van de Pol *et al.*, 2017). In the wild, many fish species are able to go for long periods without food, often for several weeks or months during migration or reproduction (Kiessling *et al.*, 1990). Apart from reducing their basal metabolic rate, fish can also decrease the activity of their organs related to exogenous nutrition, effectively reducing the size of their gastrointestinal tract (Waagbø *et al.*, 2017).

3.3.1 Effect of preslaughter fasting on fish welfare

Based on fish physiological characteristics and needs, the preslaughter fasting period requires a substantial amount of time and needs to be well-planned in advance. Concerns about the maximum permissible duration of preslaughter fasting have been raised for different fresh water and marine fish (Barcellos *et al.*, 2010; Mørkøre *et al.*, 2008). In rainbow trout, the Farm Animal Welfare Council (1996) recommends a 48 h maximum limit on fasting, arguing that the welfare of farmed fish that have been fed regularly will be affected negatively by a sudden cessation of feeding. Compassion in World Farming (2009) concludes that a 72 h limit is appropriate. The European Food Safety Authority (EFSA, 2009a) and, more recently, the Royal Society for the Protection of Animals (RSPCA, 2018) suggest limits based on degree days (°C d), 50 °C days and 54 °C days, respectively. Degree days represent the total sum of degrees over the days included (e.g. if the water temperature is 10 °C on Monday, and 11 °C on Tuesday, the total degree days over the two days is 21 °C d). In two recent handbooks on the welfare of Atlantic salmon (*Salmo salar*; Noble *et al.*, 2018) and rainbow trout (Noble *et al.*, 2020), the authors conclude that there is insufficient scientific knowledge about the effects of fasting on welfare to impose strict limits, but that it is important to take into account temperature (i.e. degree days).

Although fish can adapt their metabolism to withstand long periods without food, they present motivational mechanisms for feeding when nutritional reserves are low (Metcalfe and Thorpe, 1992). For example, farmed rainbow trout can fast for weeks with no apparent negative effects on stress physiology, behaviour or flesh quality (Jentoft *et al.*, 2005; Pottinger *et al.*, 2003). However, as reviewed by Huntingford *et al.* (2012), underfeeding and feed deprivation have negative effects on welfare in general and tend to increase aggression among fish. In non-salmonids, such as tilapia (*Oreochromis niloticus*) or surubim (*Pseudoplatystoma* spp.), feed withdrawal has also been found to increase aggression and decrease energy reserves and body condition so that individuals have a decreased ability to respond to possible stressors (Fantini *et al.*, 2020).

What is less clear is the immediate effect of fasting fish that have been habitually feeding, especially in relation to water temperature. López-Luna *et al.* (2013) showed that typical commercial fasting times of 1 to 3 days (up to 58 °C d) have little effect on plasmatic indicators of stress (e.g. cortisol, lactate and glucose) in rainbow trout that had been fed twice daily before fasting. In an effort to decrease the stress of a sudden stop in feeding, feeding frequency can be slowly decreased in the weeks before the fasting period. Bermejo-Poza *et al.* (2015) found that rainbow trout that had been fed once every two days (skip-a-day feeding) for a month before slaughter adapted better to a two day fast than trout fed daily or once every four days, based on muscle glycogen reserves and their hepato-somatic index. In a more recent study, Bermejo-Poza *et al.* (2017) found that approximately 22 °C days of fasting are required to empty the gut of rainbow trout, while liveweight decreased significantly after only 42 °C d. Based on these studies, and in agreement with EFSA (2009a), it appears that fasting durations of less than 50 °C d do not impose important stress levels on rainbow trout, and are sufficient to empty the gut, which suggests that fasting times above 50-55 °C d should be avoided. More studies are required in other species, including tropical fish, to find out whether this recommendation can be generalised.

3.4 Crowding

At the end of the fasting period, fish density (crowding) needs to be increased in one area of the water volume, in order to facilitate removing fish from the water, either using different types of nets or suction pumps. Without this procedure, too much effort would be required to catch all fish in a reasonable amount of time, with the risk of making the catching procedure even more stressful.

The crowding period can be quite brief (10 min; Robb, 2008) or last for several hours (2-4 h; Erikson *et al.*, 2016; Robb, 2008), depending on the size of the water volume and the amount of fish to be removed. Based on the evidence that crowding increases the concentration of toxic elements in the water (e.g. ammonia; Lines and Spence, 2012), fish physical exertion, exposes individuals to bright light at the surface, decreases oxygen uptake and can cause physical injury (Erikson *et al.*, 2016; Robb, 2008), most producers agree that crowding duration should be as short as possible.

One important factor is that, depending on the size of the cage, operators are never certain about how many fish will be caught in the seine or sweep net, so that some fish may be crowded and caught immediately, while others will just have to wait in the net until they are removed later or eventually released. Cardia and Lovatelli (2015) provide informative visuals of different ways to crowd fish in net-pens using seine nets to help avoid crowding/catching too few or too many fish.

Crowding should be also done calmly (Ashley, 2007; Robb, 2008). Where possible, crowding can begin by slowing draining water from the home tank or raceway, over a period of several hours, until it is easier to handle the fish. Where this is impossible, such as for undrainable ponds, a large seine net is slowly dragged through the water (Figure 3.2A,B; Pillay and Kutty, 2005). For net-pens, the depth of the pen can be reduced gradually by raising the bottom net or using a seine net (Brown *et al.*, 2010), gradually moving fish closer to the surface (Figure 3.2C,D).

3.4.1 Effect of crowding on fish welfare

Several studies have shown that crowding depletes energy stores, increases blood lactate, cortisol and haematocrit levels, and disturbs osmoregulation in fish (Cooke and Suski, 2005; Merkin *et al.*, 2010). In one of the first studies on the effects of crowding on fish welfare, using sea bass and sea bream, Bagni *et al.* (2007) compared no crowding (i.e. fish taken individually from the cage using a dip net) versus fish crowded for 3 h in a small tank (at a density above 70 kg/m^3) and reported an earlier and faster *rigor mortis* onset, an indirect measure of preslaughter stress, in the flesh of crowded fish.

Figure 3.2. Different types of crowding depending on the production system: (A) sweeping an earthen pond with a seine net fo catla (*Labeo catla*); (B) sweeping a trout pond; (C) manually emptying a net-pen using a dip-net for sea-bream (*Sparus aurata*); (D emptying a net-pen using dry brailing for sea-bass (*Dicentrarchus labrax*) (M. Villarroel, Universidad Politécnica de Madrid, Spain

In salmon, Robb (2008) found that two consecutive sweeps to crowd/catch were less stressful, based on plasma cortisol levels, than one sweep, where the last fish were removed after 4 h. Also in salmon, Roth *et al.* (2012) took flesh samples at the beginning and end of a short (10 min) and long (3 h) crowding period and report that the latter fish presented an earlier onset of *rigor mortis* and a more significant drop in muscle pH compared to the former. The initial crowding density used in that study was 200 kg/m^3, which is quite high compared to normal grow-out densities (ranging from 20 to 40 kg/m^3 (Conte, 2004)), and represents an example of the high densities reached during crowding in the industry. Erikson *et al.* (2016) confirm that crowding either increased or maintained high levels of plasma stress indicators (e.g. cortisol, lactate) in salmon, although its effects varied depending on prior handling procedures. In another salmonid, Brijs *et al.* (2018) implanted trout with heart rate bio-loggers and found that crowding increased heart rate frequency by approximately 8 beats/min from normal values of approximately 50 beats/min (in fasted fish), with a concomitant rise in cortisol levels. Moreover, the stress from crowding in addition to brailing, fasting and transport had a cumulative effect, which maintained an higher heart rate (approximately 10 beats/min) for at least 12 h post-stress, emphasising the importance of minimising repeated or sequential stressors. Also in trout, Bermejo-Poza *et al.* (2019) found that the combined effect of fasting for 135 °C days and a prolonged crowding duration (20 min), significantly increased plasma cortisol, glucose and lactate levels compared to control groups with a crowding time of 30 s.

Studies in other species tend to confirm the results found in salmonids. Fantini *et al.* (2020) report that, when allowed to rest after crowding, surubim (*Pseudoplatystoma* spp.) can recover normal levels of oxygen saturation in blood. In cod (*Gadus morhua*), Brown *et al.* (2010) compared fish that had been rapidly line-caught and those that had been crowded for approximately 20 min and found that the latter had 7-10 times higher blood cortisol levels than the former. In carp, fasting for 14 days makes them more susceptible to crowding stress as opposed to shorter fasting times (Hoseini *et al.*, 2019).

3.5 Emersion

After crowding, fish must be removed from the water in order to be transported or handled for stunning and slaughter. One exception is in-line stunning where fish 'swim' through a tube and are stunned in the water, but few companies have adopted this practice, despite its possible benefits (Section 3.7.5.3).

Normally fish are removed from their 'home' tank or net-pen using a net. Traditionally this is done by hand, but on an industrial level hand-netting, even in a highly crowded net, is not usually cost effective. Instead, an elongated sock-shaped net is used, often called a brail net or landing net, which is attached to a retractable crane to help drag it through the water and onto the boat (Figure 3.2D). The upper part of the net is strung onto a circle of heavy metal tubing which maintains the shape of the net and can sink into the water. The bottom of the net can be open or closed using a rope. This type of net can also be enclosed in a tarpaulin or metal casing to keep a certain amount of water in the net whilst being raised by the crane. Keeping the fish immersed in water in the net is thought to decrease stress levels since more oxygen is available and fish on the bottom layer are not crushed by those above. The latter conditions underline one of the key aspects of netting with regards to welfare, to try and ensure that not too many fish are caught in the net at once.

3.5.1 Effect of emersion on fish welfare

Removing fish from water is believed to impose a great amount of stress, so much so that a brief emersion (10-30 s) is used as a reliable stress test for fish in laboratory trials (Ibarra-Zatarain *et al.*, 2016). Fish densities in the net are at least as high as in crowding, with the added negative effects of forced contact, possible compression on all sides, extreme struggling and lack of oxygen. Both dry and wet brailing can cause physical damage (e.g. scale loss and rupture of air bladder) and abrasion from contact with other fish and the net (Sneddon *et al.*, 2016).

Based on its serious negative effects on fish welfare, emersion time should be kept to a minimum (Lines and Spence, 2014), probably less than 5 min for the majority of fish species, but there is no general recommendation for a minimum time in aquaculture. According to a few studies on this topic, in turbot, Mugnier *et al.* (1998) found that air exposure of less than 4 min had little effect on plasma cortisol levels. In carp, Van de Vis *et al.* (2017) report that fish will struggle during air exposure for about 1-10 min, after which they become exhausted. In sturgeon (*Acipenser shrenckii*), Lu *et al.* (2005) found prolonged damage to the central nervous system after 30 min air exposure.

3.6 Transport

After crowding and emersion, fish are taken to the stunning/slaughter area. For many species, since it is both difficult and expensive to transport live fish, stunning and slaughter occur on the farm. If transport is required, however, it can be quite brief (less than 1 h) or last many hours, e.g. from 8 h (Gatica *et al.*, 2008) to over 1 day (Iversen *et al.*, 2005), depending on the species and production system. Transport before slaughter can be by boat (well-boats) or road and very rarely by air, or, in the case of net-pens, the whole pen can be towed to a processing area, effectively towing the whole 'farm'.

In some species (i.e. sea bream and sea bass) no transport is used after emersion, since fish are immediately placed into containers with ice water on boats beside the net-pens. In other species, like salmon, preslaughter transport logistics can be more complex. In northern Europe live salmon are often transported in well-boats, which carry the fish from the net-pen or fish farm to the slaughter site on land (Erikson *et al.*, 2016). The well-boats can be open-hold, where water from outside runs through the boat, but tend to be closed-hold to avoid any possible spread of diseases. When closed, oxygen diffusers are added and if transport duration is long, small recirculation units are used to decrease CO_2 and avoid ammonia accumulation. Often well-boats have underwater cameras to observe the conditions of the fish in real-time. After arrival to the dock near the processing unit, fish are unloaded and placed into holding cages or transferred directly to the processing line to be stunned and slaughtered. If the processing site has a high through-put, fish are preferably placed in holding cages which allows the well-boat to leave for another load (Erikson *et al.*, 2016).

Although in Europe fish are included in the main transport regulation for live animals (EU Regulation 1/2005 (EC, 2005)), the regulations are less enforceable compared to other species, since there are no specific requirements for fish, or maximum limits on transport duration (Van de Vis *et al.*, 2017). Since fish production covers a great variety of species, with their own specific requirements and cut-off values regarding compatible water quality parameters (e.g. oxygen requirements), it is also understandable that a general transport code has not yet been developed, although producers are obliged to avoid pain and suffering, where possible.

3.6.1 Effect of transport on fish welfare

If and when fish need to be moved from their home tank into a well-boat or onto a road vehicle, both loading and unloading appear to be the most stressful stages, in addition to changes in water temperature and water quality during the journey (Gatica *et al.* 2008; Humborstad *et al.*, 2020; Santurtun *et al.*, 2018). The live transport of fish requires well equipped well-boats or vehicles and, as such, can be quite costly in terms of equipment to maintain water quality and personnel needed to move and observe the fish (Southgate, 2008). In their review of fish welfare during transport in the European Union, Van de Vis *et al.* (2017) found that it largely follows OIE standards (OIE, 2019) in terms of fish welfare, especially for Atlantic salmon, rainbow trout, European sea bass and gilthead sea bream, and to a lesser degree for common carp.

On the other hand, other species are transported live without water on the way to market or before applying ice (e.g. eel, carp, tilapia), sometimes referred to as dry transport (Lines and Spence, 2012), where fish welfare is most likely compromised since they lack both water and oxygen, and are exposed to rapid temperature changes, vehicle vibrations and the physical shocks when dropped or crushed by other fish.

In general, there is no legislation that restricts stocking density during transport, so that producers need to account for how transport duration may affect water quality, most critically related to oxygen levels, for which air or pure oxygen can be diffused into the water (Terlouw *et al.*, 2008). Depending on the duration of the journey, the accumulation of metabolic waste products (i.e. CO_2 and NH_3) in the surrounding water will be the next important issue, but that will depend on fish density. According to Santurtun *et al.* (2018), journey duration in a well-boat should not exceed 2.5 h at a fish density of 70 kg/m^3 and less than 20 min at 170 kg/m^3. After transport, it is also beneficial for the fish to have a recovery or lairage period, as performed in many terrestrial species, before stunning/slaughter.

In summary, before stunning and slaughter, fasting, crowding, emersion and transport can all have important negative effects on the welfare of the fish if not performed quickly or correctly (Hultmann *et al.*, 2016). Based on key aspects to be considered for fish welfare outlined in Huntingford *et al.* (2006), Table 3.1 summarises how these aforementioned pre-slaughter handling procedures can affect fish welfare.

3.7 Pre-slaughter stunning and killing with effects on unconsciousness and insensibility

3.7.1 Consciousness in fish

Animals are stunned to induce a state of unconsciousness and insensibility of sufficient duration to ensure that they do not recover before death. Death may be caused by the stunning method itself or by further handling, which we will refer to as slaughter. Based on studies in a wide range of species, there is evidence for different types of consciousness in both livestock and fish (Terlouw and Bourguet, 2022 – Chapter 1 of this book). The forebrain system is developed in vertebrate terrestrial animals, whereas only the basal subcortical system is well developed in fish (Le Neindre *et al.*, 2017). In mammals the amygdala and hippocampus are crucial cerebral structures for awareness. The anatomical counterparts in fish are slowly becoming identified in the telencephalic medial pallial regions. Studies have provided evidence for such higher brain functions in fish (Broglio *et al.*, 2005; Nilsson *et al.*, 2008a,b; Rodriguez *et al.*, 2006). The balance of evidence indicates that some fish species have the capacity to experience pain. However, research and developments in the area of cognition and brain imaging techniques should be carried out in fish to further our knowledge and understanding of pain perception (Le Neindre *et al.*, 2017).

Table 3.1. A summary of the possible impacts of different pre-slaughter handling procedures on fish welfare in terms of confinement, density, new environment and water quality.

Process	Are fish confined?	Are fish at high density?	Are fish in a new environment?	Is water quality an issue?
Fasting	not normally	no	no	no, can even improve
Crowding	yes	yes	no	yes, can be fatal
Emersion	yes	yes	yes	yes, can be fatal
Transport	not normally	yes	yes	yes, can be fatal

The main function of subcortical brain regions, with a variety of brainstem structures, is to coordinate and control both the cephalic segment, including eye movements, and the remaining parts of the body. The periaqueductal grey structure is of fundamental importance for the emotional-motor integration as it coordinates a large variety of emotion-related behaviours, such as defence, aggression, reproduction, voice, and pain, from fish to human beings (Merker, 2013; Panksepp and Northoff, 2009; Parvizi and Damasio, 2001). When confronted with a stressor, fish produce adrenalin and noradrenalin and the stress-steroid cortisol (Schreck and Tort, 2016), as do all vertebrates. These endocrine responses support fight/flight responses and counteract adverse stressor effects: stress requires a resetting of energy flows (Wendelaar Bonga, 1997). Fish show several adaptive behavioural and physiological responses that have evolved to cope with stressors. Many of these responses are homologous with those of other vertebrates (EFSA, 2009a).

In fish, consciousness and sensibility can be measured by behavioural and clinical indicators, EEG (electro encephalogram) including VER's (visual evoked responses), VEP's (visual evoked potentials) and SER's (somatosensory evoked responses). Behavioural indicators are self initiated behaviour (swimming and equilibrium), responses to stimuli, such as handling (i.e. catching), prick (painful prick on the lip with a sharp needle and stimulation with 6 V on the lip), and clinical reflexes, such as eye roll (vestibulo-ocular reflex) and breathing (opercular movement) (Kestin *et al.*, 2002). The methods may be applied to assess the effects of stunning procedures on the welfare of fish, however, they are indicators. Similar to humans and other animal species, an EEG is also necessary in fish to assess the state of (un)consciousness, (in)sensibility and brain disorder. The electrical activity recorded on the EEG can be classified into δ (<4 Hz), τ (4 to 7 Hz), α (8 to 13 Hz), β (13 to 32 Hz) and γ (32 to 45 Hz) frequency bands. In α, β and γ rhythms, the animal is thought to be conscious (Schomer and Lopes da Silva, 2012). Power spectral analysis is a method for the analysis of EEG signals. To analyse the EEG signals power spectral parameters can be used to quantify the effects on the brain and depth of anaesthesia and there several variations. VER's, VEP's and SER's are used to measure changes in electrical potential in the brain as a response on light flesh or needle skin scratches.

To measure EEG signals electrodes are implanted via drilling in the skull and fixed with dental cement or inserted subcutaneously (Lambooij *et al.*, 2002a,b). A new development is a non-invasive method. A silicone suction cup was created and fitted with silver chloride electrodes. The electrodes were connected with and amplifier via shielded wires. To be water tied a peristaltic pump is used (Bowman *et al.*, 2019).

3.7.2 Criteria and restrictions

Different criteria are important to choose a stunning method. In general, the method is influenced by the fish species, its size, place of stunning, apparatuses used for restraining, degree of contamination and whether it will be consumed by humans. These include physical and electrical euthanasia methods (gas mixtures, lethal substances and novel/experimental methods). These methods can be evaluated in relation to the following criteria:

- Animal welfare. The Treaty of Amsterdam (EFSA, 2004) sets rules in a 'protocol on the protection and welfare of animals'. It recognises the sentience of animals and requires that

European Institutions pay full regard to welfare requirements of animals in the formulation and implementation of Community regulations. All animals, including fish, should be protected from anthropogenic sources of excitement, pain or suffering during transport, lairage, restraint, stunning, slaughter or killing.

- Effect. The effect should be realisation of the process, which does not say anything about the process. The effectiveness of a treatment is varying in terms of the effort required to carry it out and how successful the operator is at reaching the defined goal.
- Efficiency. Efficiency in this case is the extent to which resources are used to reach the specific goal. A process is efficient when fewer resources are necessary in relation to a standard. When translated to stunning and killing methods this involves concepts such as time, materials and cost.
- Consideration of the operator. This criterion includes the active number of personnel, the physical and emotional burden for the operator or controller and possible harmful effects on the operator. Accuracy is essential to obtain a correct stunning and killing process for the fish. The operator needs to be educated and know what is happening with the animal. The method should also be operator-friendly.
- Risks and restrictions. The apparatus needs to be temper proof, adapted to the animal and not susceptible to interference. The risk of microbial contamination of fish during stunning can be related to contamination of water, increased handling, invasive stunning and exsanguination (EFSA, 2009f).

3.7.3 Physical methods

3.7.3.1 Head percussion

Percussive stunning of slaughter animals by means of a blow on the head with a special instrument is the oldest stunning method. Cerebral concussion is generally agreed to be a traumatically induced derangement of the nervous system, resulting in an instantaneous diminution or loss of consciousness without gross anatomical changes in the brain (EFSA, 2009b). To use carp as an example, they are hand netted from the lairage tank and placed into a container without water or with a limited volume of water. Fish are stunned one by one with blows to the skull using a wooden or plastic club, named 'priest' (Figure 3.3). In practice, several blows are needed to obtain an acceptable result (Lambooij *et al.*, 2007). The procedure is followed by evisceration (Białowąs *et al.*, 2004). Silver carp (*Hypophthalmichthys molitrix*) were stunned one by one with one or two blows on the head with a wooden club. Meat quality was compared with immersion in ice slurry and gill cutting. The last treatment result in a faster ATP depletion, and a higher deterioration rate in texture properties (72 h post-mortem) compared to percussion and immersion in ice slurry. Fillets of silver carp stunned with both last methods show a better quality in terms of K value and cooking loss (Zhang *et al.*, 2017). In silver carp, stunning stress leads to deterioration in muscle texture and may impair the water-holding capacity of fillets (Zhang *et al.*, 2019). Manual percussive stunning with a priest of African sharptooth catfish results in immediately and irreversibility insensibility, although 36% of them regained signs of consciousness. It is suggested that it is difficult to administer an accurate blow of sufficient force on a live struggling catfish (Brijs *et al.*, 2021).

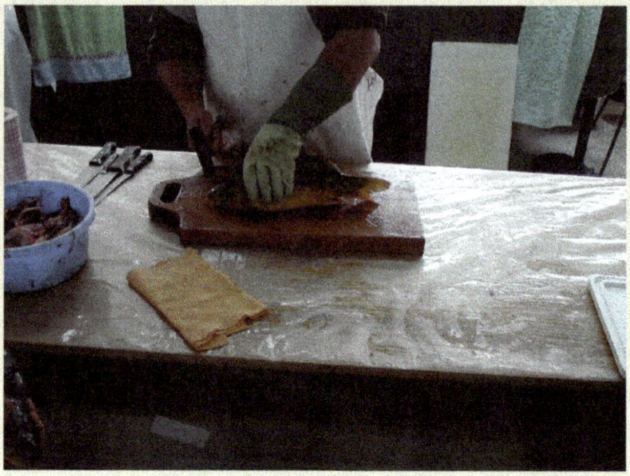

Figure 3.3. Blow on the head of a common carp (E. Lambooij, Wageningen University, the Netherlands).

Another option is to use a captive bolt gun, a rounded nylon cylinder of 60 g that is fired mid-dorsally with an air pressure 7.5 bars on the skull, slightly rostral to the eyes. This method can either be done automatically on fish swimming into the system, or semi-automatically on fish either stunned or sedated prior to manual loading into the machine. For automated systems the challenge is not only that all fish must swim in the correct direction, but also in the correct position and be aligned correctly. The stunning efficiency also depends on the distribution of size within the fish population (EFSA, 2009g). However, in some studies the use of this system failed to render all carp unconscious before slaughter (Lambooij *et al.*, 2007). The method is also used for Atlantic salmon, trout and yellowtail kingfish. When the force is set higher than 8 bars, salmon become unconscious immediately, however, with negative consequences for the carcass, such as broken jaws and burst eyes (Lambooij *et al.*, 2010). For yellowtail kingfish this force might be even insufficient to induce unconsciousness and not recommended. An alternative is electrical stunning (Llonch *et al.*, 2012). This implies that just a hitting region above or around the brain may be ineffective, unless the force is so great that the skull is destroyed, including eyes and jaws. From a commercial viewpoint this is even more disturbing, since percussive machines are used over a broad range of sizes, affecting stunning efficiency (Robb and Roth, 2003).

3.7.3.2 Brain penetration

The penetrating captive bolt in combination with exsanguination or insertion of a pithing rod into the cavity is an effective and efficient method used on cattle, pigs, sheep and goats (see species-related chapters in this book: Cockram and Velarde, 2022; Faucitano and Raj, 2022; Gallo *et al.*, 2022; Lambooij and Algers, 2016; Miranda-de la Lama *et al.*, 2022). A captive needle pistol was adapted to eel and African catfish with respect to the length and shape of the needle, by which compressed air was injected in three directions radial and in one direction (caudally to the spinal cord), into the brain using 3 bars during 1.5 s. It is likely that the compressed air administered through the captive needle pistol results in damage of the brain by laceration of higher brain regions, leading to unconsciousness and insensibility. The flow in the caudal direction causes laceration of the upper spinal cord, which may prevent post-stun convulsions (Lambooij *et al.*, 2002b, 2003). Severe convulsions, previously observed in some eels and African catfish, might be related to a sub-optimal positioning of the pistol, and consequently less damage to the upper spinal cord. At lower body temperatures, however, muscle activity is significantly reduced, making it possible to gut or bleed. Therefore, chilling fish between stunning and slaughter is recommended (Brijs *et al.*, 2021; Lambooij *et al.*, 2002b, 2003).

Large tuna is killed one-by-one by a diver with an underwater shot to the head using a power-head device which resembles a short gun barrel. When the bullet hits the head, it expands on impact and so damages the head and brain. The same ammunition is used in shot guns for shooting from above the surface (EFSA, 2009e).

3.7.3.3 Spiking

A neck-cut is performed to immobilise the animal and facilitate evisceration. The neck-cut consists of cutting the spinal cord just behind the head, without disruption of the soft tissue and flow of blood to the brain (EFSA, 2009d,e; Van de Vis *et al.*, 2003). Spiking, coring or 'ike jime' are the terms used to describe this killing method (Lines and Spence, 2014). For effective use the core or spike is rapidly inserted into the brain via the pineal window followed by pithing (EFSA, 2009d,e) by passing a wire down the spinal column to immobilise fish (Lines and Spence, 2014). This killing method is effective for all fish species, including small bluefin tuna, except for eels that based on laboratory studies, may survive after the neck-cut (EFSA, 2009d,e; Flight and Verheijen, 1993).

3.7.3.4 Decapitation

The head is separated from the body by a sharp cut which may result in bleeding leading to death through anoxia. However, decapitation without prior stunning is not considered an ideal killing method for any species of animals because the brain continues to function for an appreciable time due to some recoil in the arteries that will stem blood flow resulting in delaying induction of unconsciousness and death (EFSA, 2009d; Lines and Spence, 2014). Eels may lose reactions 30 min after decapitation (Verheijen and Flight, 1997) and it takes 13 min before brain function ceased completely (Van de Vis *et al.*, 2003).

3.7.3.5 Live chilling

Live chilling means a transfer from water at ambient temperature into in a relatively small tank or bin filled with ice flakes, liquid ice or ice water slurry (EFSA, 2009c,e,f; Figure 3.4). Here fish are firstly cooled rapidly and then deprived of oxygen (Robb and Kestin, 2002), resulting in fish immobilisation due to exposure to ice water and death by asphyxiation on ice (Robb and Kestin, 2002).

It is likely that eels, which are transferred from water at 18 °C to ice water are stressed, as specific behaviour (i.e. vigorous attempts to escape) and an irregular heart rate were observed (Lambooij

Figure 3.4. Live chilling of seabass in ice slurry (E. Lambooij, Wageningen University, the Netherlands).

et al., 2002d). Brain activity and responses to pain stimuli disappeared at a body temperature of approximately 8 °C, which occurred after approximately 12 min (Lambooij *et al.*, 2002d). In African catfish, low brain activity and loss of responses to pain stimuli were observed after approximately 15 min at a body temperature of approximately 13 °C (Lambooij *et al.*, 2006a). Seabream transferred to an ice slurry mixture at 0.5 °C from water at 23 °C had an average time to loss of brain function of 5 min (Van de Vis *et al.*, 2003) and common carp (Van den Burg *et al.*, 2005) showed that exposure to a sudden temperature drop is stressful and not recommended. Immersion in ice water may not induce unconsciousness, although the brain activity decreased to a lower level in turbot (Lambooij *et al.*, 2015) as this fish is known to be extremely tolerant of thermal insult and, in commercial practice, are often chilled, transported live on ice and sold live (Roth *et al.*, 2009).

In terms of animal welfare, the impact of this process on the welfare of fish is generally considered to have a detrimental effect (Van de Vis *et al.*, 2003). Dropping 15 °C in temperature is a questionable practice which involves the release of primary and secondary stress responses (i.e. increased levels of cortisol, lactate and osmolality in blood plasma, as well as forcing the muscles to contract in response to severe pain; Robb and Kestin, 2002). Based on this evidence, the Farm Animal Welfare Council (1996) suggested that this killing method should be prohibited for live trout.

Because of its greater resilience to low temperatures (<8 °C), turbot did not undergo hypoxia, as shown by the lack of variation in muscle pH and pO_2 levels, and flesh quality remained unaffected (Roth *et al.*, 2009). However, in this fish and sole a negative effect on the osmoregulatory capacity was observed (Foss *et al.*, 2019). The internal cooling rate was faster in turbot when starting at a high environmental temperature in summer (approximately 19 °C) and was comparable for both summer and winter temperatures after 5 min in the ice water. A higher or lower initial environmental temperature seems to have little effect on the physiological processes seen in the brain or on heart activity of turbot (Lambooij *et al.*, 2015).

Flesh quality of grass carp (*Ctenopharyngodon idellus*) fillets stored at 4 °C and stunned in ice slurry showed lower total volatile basic nitrogen (TVB-N), electrical conductivity, total viable counts total biogenic amines, and higher sensory scores compared to percussion stunning. Stunning in ice slurry could improve fillets shelf life (Qin *et al.*, 2016).

3.7.3.6 Saltwater bath

This method is generally only used for eels, which are difficult to kill. This practice consists in placing live eels in a dry tank to which NaCl or a combination of NaCl and aqueous Na_2CO_3 is added, which results in denaturation of the mucus proteins and clotting (EFSA, 2009d). Another option is to add 25% ammonia solution. After approximately 4 min eels are paralysed and the container is then filled with water. The eels are left in the container for approximately 20 min (EFSA, 2009d). The upper layer of skin may be damaged and the eyes become opaque or white during the exposure. To remove the clotted slime and salt, eels are later transferred into a special mixing and rotating machine (tumbler construction or mixer principle) and are washed with water for about 10 min. The whole process is called 'desliming' and takes approximately 30 min.

Eels make extremely vigorous attempts to escape from salt or ammonia (Kuhlmann and Münkner, 1996) and take a long time to lose consciousness. When the water is added to the tumbler, active movements of the eels can sometimes still be observed and likely finally stop due to exhaustion. Exposing eels to salt or ammonia is now considered a bad practice in Germany and has been prohibited since 1999 (Van de Vis *et al.*, 2003).

3.7.4 Asphyxia

Many farmed fish species are killed by asphyxiation in air or in ice flakes. Afterwards, some species, including pangasius, tilapia, trout, seabass, seabream, eel and many flat fish species, are bled, decapitated or even eviscerated, even while conscious (Lines and Spence, 2014). Asphyxia is considered one of the most stressful killing methods, resulting in high intensity rigor after 6 h (Erikson, 1997). Asphyxia in air, which is usually achieved by netting or pumping fish and placing them in free draining bins or boxes, severely affects fish welfare as it leads to a much more prolonged pre-mortem activity in comparison with asphyxia in chilled water (Skjervoldt *et al.*, 1999, 2001). Seabream and seabass are killed by this method simply by removing them from water and leaving them to die in air. In most cases, violent attempts to escape are made and maximal stress responses are initiated (Robb and Kestin, 2002). The time necessary to kill seabass by asphyxia in air range from about 70 min to up to 128 min (Acerete *et al.*, 2009; Poli *et al.*, 2005). Flesh from fish killed by asphyxia in air had higher lactate and lower pH levels, and an earlier rigor onset compared with spiking and percussion (Poli *et al.*, 2005). The analysis of TVB-N, thiobarbituric acid reactive substances (TBARS), pH, trimethylamine and sensory evaluation showed that asphyxia in ice-water could maintain better quality of Chilean jack mackerel (*Trachurus murphyi*) during refrigerated storage in an industrial setting (Lyu *et al.*, 2015). Atlantic cod and haddock are wild-caught in commercial fisheries using nets, transferred from sea to deck, dumped and stored until evisceration. These fish can survive landing and storing for at least 2 h with brain and heart activity maintained after landing on-board and on-board storage, although they lose the capacity to respond behaviourally (Lambooij *et al.*, 2012).

3.7.5 Electronarcosis

Electronarcosis has been used to stun fish species before killing. Effective electrical stunning can be achieved by induction of a general epileptiform seizure in the brain measured by an EEG and expressed by tonic and clonic muscle convulsions and ultimately strong depression of activity (Lambooij, 2014). The specifications depend on the species, waveform, field strength of the current applied, and water conductivity.

There are two methods applicable for use in practice: stunning in water or after withdrawal from water (dry stunning).

3.7.5.1 Head-only dry stunning

Electrical stunning of fish species by head-only stunning of each individual animal is very difficult to perform because of the size and mobility. After electrical stunning the fish should remain unconscious and insensible until death. This can be achieved when electrical stunning is combined as soon as possible with decapitation, bleeding or chilling in ice (EFSA, 2009b,d,e).

In head-only stunning for eels, electricity is applied to the head by a pair of stunning tongs in air. A general epileptiform seizure was observed with 255 V, 545 mA for 1 s. The general epileptiform insult lasted for at least 0.5 min (Lambooij *et al.*, 2002c). In various head-only stunning studies, the minimum current has been determined for a number of fish species: 500 mA for rainbow trout (Kestin *et al.*, 1995), 600 mA in eel (Lambooij *et al.*, 2002d), 629 mA in African catfish (Lambooij *et al.*, 2004), and 240 mA in carp (Lambooij *et al.*, 2007).

3.7.5.2 Water bath

Typically, a tank with electrodes attached to opposite sides or bottom and lid is filled with water and fish before an electric current is applied. A current is passed between the electrodes, using the water and the fish within as a conductor. Provided certain parameters are met, fish are immediately stunned. After the current is turned off, fish are removed from water. The different methods are summarised in Table 3.2.

The conductivity of water in the stunning tank can influence the field strength in fish. At low water conductivities (e.g. 20 μS/cm) the electrical field in fish may be only half the nominal field, while at higher conductivities (e.g. 1000 μS/cm) the electric field in fish may be 50% stronger. In addition, a fish stunned side by side or top to bottom requires a 35% stronger electric field (Lines and Kestin, 2004).

able 3.2. Electrical currents necessary to stun fish in a tank when plate electrodes are placed head-to-tail or opposite sides or op-to-bottom.[1]

Species	Electrodes	Stunning parameters	Killing parameters	Reference
el (*Anguilla Anguilla*)	individual, head to tail	0.64 A/dm^2; 50 Hz; 500 µS/cm	may recover by chilling	Lambooij *et al.*, 2002a,b
African catfish (*Clarias gariepinus*)	individual, top to bottom	1.5 A/dm^2; 50 Hz; 5 s; 500 µS/cm	decapitation and bleeding or chilling in ice slurry	Brijs *et al.*, 2020; Lambooij *et al.*, 2006b
arp (*Cyprinus carpio*)	individual, head to tail	0.14 A/dm^2; 50 Hz; 200 µS/cm	follow up chilling in flaked ice	Lambooij *et al.*, 2007
Nile tilapia (*Oreochromis niloticus*)	individual, top to bottom	1.0 A/dm^2; 50 Hz; 700 µS/cm; 5 s	follow up chilling in ice water	Lambooij *et al.*, 2008a
eabass (*Dicentrarchus labrax*)	individual, head to tail	3 A/dm^2; 133 Hz; 43% duty cycle; 50 µS/cm; 5 s	follow up chilling in ice water	Lambooij *et al.*, 2008b
	side to side	4.05 A/dm^2		
	group	1.5 Vrms/cm		
Atlantic cod (*Gadus morhua*)	individual, side to side	0.93 V/cm; 3.6 A /dm^2; 170 Hz; 33% duty cycle; 5 s	follow up chilling in ice water	Digre *et al.*, 2010
Pacu (*Piaractus mesopotamicus*)	group	205 V; 50 Hz; duty cycle 70%; 1.3 A; 45 s	none	Rucinquea *et al.*, 2018
obia (*Rachycentron canadum*)	group	150 V, (AC), 60 Hz, AC, 7.3 A, 3 min		Vargas Baldi *et al.*, 2018
African sharptooth catfish (*Clarias gariepinus*)	individual, side to side	1.69 A/dm^2; 15.98 V/cm r.m.s; 10 s	follow up chilling in ice water	Brijs *et al.*, 2021

A/dm^2 = amperage per square dm electrode; A rms = amperage root mean square; V/cm = volt per cm water; Vrms = volt root mean square.

Seabass can be stunned effectively by electricity in water bath and killed by chilling in ice water (Lambooij *et al.*, 2008b). The relatively high initial pH values of the flesh on the day of slaughter are similar to those recorded in eels (Morzel and Van de Vis, 2003). After 2 days of storage, the pH values increased and decreased later on. The same figure with a peak in the pH value at day 2, followed by a slow increase, was also seen in carp (Bialowas *et al.*, 2004). After 3 days of storage, the pH was consistently and significantly higher, in comparison with the traditional salt bath method to deslime conscious eel (Morzel and Van de Vis, 2003).

The exposure to electricity may cause a very high blood pressure increase lasting a few seconds which combined with skeletal muscle contractions may result in haemorrhagic damage and blood spotting in fish flesh (Sandblom *et al.*, 2012). Based on behavioural indicators pacu is considered unconscious after group wise electrical stunning. Each fish was controlled after stunning. Rigor mortis was reached after 6 h and no abnormalities were found in the meat (Rucinquea *et al.*, 2018). Also, cobia were considered unconscious after electrical stunning using clinical indicators and an adequate stunning method compared to CO_2 narcosis and hypothermia for animal welfare. The values of pH, TVB-N and TBARS measured in the fillets were similar among the different stunning methods (Vargas Baldi *et al.*, 2018). When tilapia were stunned with 203 V, 3.3 A during 15 s using 400, 800 or 1,200 Hz, white muscle catalase, glutathione reductase and lactate dehydrogenase were higher in stunned fish with 400 and 800 Hz compared to stunned with 1,200 Hz or eugenol. It is suggested that higher frequencies and eugenol cause lower stress responses in tilapia (Venturini *et al.*, 2018).

3.7.5.3 In-line in-water stunning

Semi-automatic continuous throughput electrical stunning devices have also been developed as electric stunning tunnels or pipes. This device achieves stunning by pumping fish through a long-electrified tube connected between the tank and the processing line. Fish are exposed to an electric field during their transfer through the pipe and arrive unconscious or dead into the harvest container. The effectiveness of the stunning depends on the strength of the electric field applied, and the density and distribution of fish in the tank. For portion size trout, field strength between 3 and 6 V/cm had to be applied for 30 to 60 s in order to achieve permanent insensibility (Lines and Kestin, 2004). Fresh water conductivity may vary on a broad scale (20 to 1000 μS/cm) which influences the strength electric field and consequently the efficacy of the stun. This can be calculated by the formula: P=0.001 EC, where P: power in watts per litre of water, E: electric field strength in V/cm and C: water conductivity in μS/cm. In consequence, at higher water conductivity, insensibility can be achieved at lower field strength (Lines and Kestin, 2004). If fish (e.g. trout) are stunned and brain function is lost, they may enter a stage of mild tonic and clonic spasms that can last approximately 20-50 s (Kestin *et al.*, 1995; Robb and Roth, 2003). However, the duration of spasms depends on water temperature: the higher the temperature, the shorter the spasms last. If trout are not killed by the process, recovery of consciousness may happen within a few minutes depending on the stunning parameters. Higher field strength, increased water conductivity and longer electrical application times may be associated with longer periods of unconsciousness and a larger proportion of fish killed by the process. Higher frequencies (up to 2,000 Hz) are associated with shorter periods of unconsciousness and lower mortality (Roth *et al.*, 2002).

Because effective stunning mainly depends on electric field strength, numerous designs and parameters for electrical stunning exist in terms of duration, voltage, ac/dc and frequencies. Some commercial machines are available and many farms have developed and build their own system in order to obtain good fillet quality. Because of safety reasons, several systems are operated at low voltage, i.e. 24 or 50 V.

3.7.5.4 Dry stunning

For dry stunning fish need to be removed from the water by netting or pumping. Fish are placed in a device which consists of a conveyor as a negative electrode and above are suspended rows of steel flaps (Figure 3.5), which are the positive electrodes. The size of the device and number of rows can be adapted to the fish species, speed and number of animals (SeaSide A/S, Stranda, Norway).

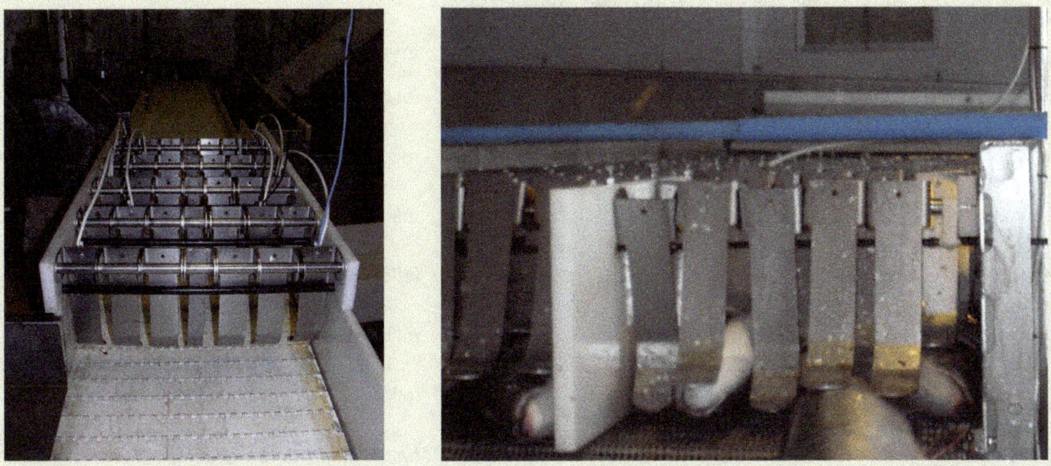

'igure 3.5. Dry stunning of fish (E. Lambooij, Wageningen University, the Netherlands).

The most common difficulty with dry stunning is to ensure that the fish are not exposed to pre-stun shocks caused, for example, by entering the machine tail first or spasms of fish, resulting in fish losing contact with the electrodes (EFSA, 2009g). When fish enter the device head first or side-first they can be stunned immediately without any problem. When they enter tail-first the device can be adapted, such that the current starts when the whole fish is positioned under the rows of flaps. For eel, the device is adapted in such a way that the whole conveyor is filled with eel first and the current is started together with a shower applied afterwards. Showering is used to prevent burning the slime layer and skin. In all cases fish should be killed by chilling, bleeding or evisceration. When common sole and pike-perch are dry-stunned for 5 s they may recover during 10 min chilling (Llonch *et al.*, 2012). In head or tail-first position the common sole and turbot can be stunned correctly, when stunned for 20 s and chilled in ice water (Daskalova *et al*, 2016; Table 3.3). A slower post-mortem pH decline and redder and darker colour is recorded in turbot flesh after dry stunning compared to sole (Daskalova and Pavlov, 2015). Furthermore, a more rapid rigor mortis onset due to increased ATP depletion has been also observed in the flesh of dry-stunned cod (Digre *et al.*, 2010). However, no blood spots and discolouration in cod fillets or spinal fractures were observed (Erikson *et al.*, 2012). A higher ultimate pH, possibly due to pre-stunning exercise depleting muscle energy levels, was observed 8 days post-mortem in cod stunned in air compared to electrical stunning (Erikson *et al.*, 2012). Also Atlantic mackerel can be stunned for 5 s head to body using dry stunning followed by chilling in shallow freshwater/ice slurry. However, unconsciousness was indicated by clinical assessment (i.e. reflex and stimuli responses; Anders *et al.*, 2019).

Table 3.3. Electrical currents necessary to stun fish in dry stunning device.[1]

Species	Parameters	Position	Killing	Reference
African cat fish (*Clarias gariepinus*)	0.91 A_{rms}; 150 ac/dc; 100 Hz; 5 s	head-body	follow up decapitation and chilling in ice slurry	Brijs *et al.*, 2021; Sattari *et al.*, 2010
Claresse® (*Heteroclarias* sp.)	0.57 A_{rms}; 150 V ac/dc; 100 Hz; 5 s	head-body	follow up decapitation and chilling in ice slurry	Sattari *et al.*, 2010
Atlantic Salmon (*Salmo solar*)	0.67 A_{rms}; 107 V ac/dc; 100 Hz; 5 s	head-body	follow up exsanguination	Lambooij *et al.*, 2010
Atlantic cod (*Gadus morhua*)	41V; 0.2 A; dc 100 Hz, 15 s	head-body	follow up chilling	Digre *et al.*, 2010; Erikson *et al.*, 2012
Yellowtail kingfish (*Seriola lalandi*)	0.54 A_{rms}; 124 V dc and 11 V_{rms} ac; 100 Hz; 5 s	head-body	follow up for 10 min chilling	Llonch *et al.*, 2012
Common sole (*Solea solea*)	0.65 A_{rms}; 98 V dc and 8.4 V_{rms} ac; 100 Hz); 5 s	head-body	follow up chilling	Llonch *et al.*, 2012
Pike-perch (*Stizostedion lucioperca*)	0.75 A_{rms}; 144 V dc and 13 V_{rms} ac; 100 Hz; 5 s	head-body	follow up for 10 min chilling	Llonch *et al.*, 2012
Wild cod (*Gadus morhua*)	0.34 A_{rms}; 52 V ac/dc; 100 Hz; >3 s	head-body	follow up throat cut and bleeding out	Lambooij *et al.*, 2012
Haddock (*Melanogrammus aeglefinus*)	0.36 A_{rms}; 52 V ac/dc; 100 Hz; >3 s	head-body	follow up throat cut and bleeding out	Lambooij *et al.*, 2012
Turbot (*Scophthalmus maximus*)	1.27A_{rms} for 1 sec. and 0.65 A_{rms} for 19 sec.	head-body	follow up chilling	Daskalova *et al.*, 2016
Common sole (*Solea solea*)	1.20 A_{rms} for 1 sec. and 0.36 A_{rms} for 19 sec.	tail-body	follow up chilling in ice water	Daskalova *et al.*, 2016
Turbot (*Scophthalmus maximus*)	3.88 A_{rms} for 1 sec., and 1.44 A_{rms} for 19 s	tail-body	follow up chilling in ice water	Daskalova *et al.*, 2016
Atlantic mackerel (*Scomber scombrus*)	0.68 A_{rms}; 110 V ac/dc; 5 s	head-body	follow up chilling in shallow freshwater/ice slurry	Anders *et al.*, 2019

[1] A/dm^2 = amperage per square dm electrode; A rms = amperage root mean square; V/cm = volt per cm water; Vrms = volt root mean square.

3.7.6 Gases

3.7.6.1 Carbon dioxide

Use of carbon dioxide (CO_2) results in rapid unconsciousness, being an analgesic and anaesthetic gas. Furthermore, this gas is inexpensive, non-flammable, non-explosive, heavier than air, and poses minimal hazard to personnel when used with properly designed equipment. As CO_2 is highly soluble in water, in order to be sedated fish are placed in water saturated with this gas. However, exposure to high concentrations of CO_2 is stressful for animals due to irritation of the mucous membranes and eyes (EFSA, 2009f,g). Experiments show that measurement of VER's does not correspond with visual indicators. Especially, in colder environmental conditions consciousness my be induced in a later stage when only visual indicators are used. It is suggested that immersion of fish for 12 min in water with a high concentration (saturated) of CO_2 result in a slow induction of unconsciousness and insensibility and to avoidable stress (Bowman *et al.*, 2020).

Under commercial slaughter conditions for Atlantic salmon, CO_2 is administered to a tank filled with seawater. As it becomes saturated with CO_2 the water pH falls to about 5.5-6.0 (Erikson, 2008). Fish are pumped into the water and are left there until struggling stops (after 2-4 min of exposure; Erikson, 2008; Erikson *et al.*, 2006; Robb *et al.*, 2000b). Subsequently, fish are removed and bled. These aversive reactions cause injury and scale loss (Robb *et al.*, 2002; Roth *et al.*, 2002) and also gill haemorrhage, loss of mucus, high metabolic activity and stress (Roth *et al.*, 2002). Because fish become immobile before losing consciousness (Robb *et al.*, 2000a), there is a risk that fish could be exsanguinated or gutted whilst still being conscious.

3.7.6.2 Carbon monoxide

Carbon monoxide (CO) induces a rapid state of unconsciousness and results in death for warm-blooded animals. At concentrations between 2 and 6% its inhalation makes animals unconscious within a few min. This gas is inflammable at room concentrations above 12%. To protect the operator the area needs to be well ventilated. The principle behind the use of CO is that it binds to the haem-iron proteins preventing neuroglobin, myoglobin and haemoglobin from binding oxygen, and an animal will lose consciousness and die of anoxia (EFSA, 2004).

Preliminary studies have shown that Atlantic salmon exposed to seawater saturated with CO causes no flight reaction or any reaction associated with stress or discomfort. After 35 min fish might be unconscious (as shown by no swimming activity, problems with operculum ventilation and total loss of equilibrium; Secci *et al.*, 2018). In another experiment (Bjørlykke *et al.*, 2011), a tank filled with sea water was flushed with 100% CO at the bottom of the tank for 10 min at 2-3 bar. Initially salmon in the water showed normal behaviour, which became slower until they showed loss of reactivity after 7-8 min (onset of anaesthesia), although some minutes later some convulsions appeared. The fillets were redder, rigor mortis was more intense and pH decrease rate was higher than the electrical stunned controls. The same behaviour and quality aspects were observed in tench (*Tinca tinca*) which proved the efficiency of CO as a good agent to stun fish (Secci *et al.*, 2018). Fillets of rainbow trout (*Oncorhynchus mykiss)* were stored cooled down for 3 days and 30 days frozen after CO, electrical or asphyxia stunning. Fillets of trout stunned with CO were after short cooling intermediate regarding drip loss, pH and colour. After long storage these fillets were low in yellowness and the panel observed the lowest odour intensity and the highest juiciness scores (Concollato *et al.*, 2019).

3.7.6.3 Nitrogen

Nitrogen gas (until an oxygen level of <1 mg/l) was experimentally used instead of CO_2 resulting in effective stunning with no strong aversive reactions in rainbow trout. In the fillets of fish stunned with nitrogen the ATP levels and pH values were higher compared to those stunned with asphyxia (Wills *et al.*, 2006). In another experiment with Atlantic salmon aversive reactions were observed. After induction they show strong tail flaps and gasping and appeared not sedated or anaesthetised. Nitrogen is therefore not recommended (Erikson, 2011).

3.7.7 Chemical methods

3.7.7.1 Clove oil and Aqui-S™

The main components in clove oil are eugenol (4-allyl-2-methoxyphenol) and iso-eugenol (4-propenyl-2-methoxyphenol), which comprise 90-95% of clove oil by weight. Clove oil has been used in fish at concentrations of 25-100 mg/l in the water, depending on the species and degree of anaesthesia needed. The positive features reported for clove oil as anaesthetic led to the development of a new anaesthetic in New Zealand, called Aqui-S™ (AQUI-S Ltd., Lower Hutt, New Zealand), whose active compound is isoeugenol (4-propenyl-2-methoxyphenol). Exposure to this agent result in gradual loss of equilibrium, swimming and finally in unconsciousness in Atlantic cod and African sharptooth catfish (Brijs *et al.*, 2021; Erikson *et al.*, 2012). When farmed gilthead seabream (*Sparus aurata* L) is immersed in clove essential oil encapsulated in β-cyclodextri embedded in ice crystals of slurry ice is considered to reduce stress for stunning at slaughtering (López-Cánovas *et al.*, 2019). In the European Union and Norway, Aqui-S™ is not allowed for stunning/slaughtering of fish due to the cost of overcoming the legislative requirements, etc. for food fish (Van de Vis *et al.*, 2014). Carp (*Cyprinus carpio* L) immersed in a solution with 125 ppm clove oil is sufficient to induce unconsciousness. Afterwards carp could survive a 90 min freezing time at a temperature of -13 °C. This procedure is of interest for selling live fish (Suwandi *et al.*, 2020).

3.8 Conclusions

- Prolonged fasting periods can have negative consequences regarding fish welfare and probably need not be longer than 50-60 °C days.
- Crowding should be performed calmly but in the shortest possible time frame as to avoid negative effects on fish welfare by either direct physical harm or secondary effects of increased fish density.
- Emersion is one of the most aversive handling methods, where fish are preferably kept out of water for less than 5 min.
- Regarding fish transport, loading and unloading procedures seem to be the most stressful while water quality parameters must be kept within reasonable limits for longer distance transport, in conjunction with fish density.
- The balance of evidence indicates that some fish species have the capacity to experience pain and stress. The level of consciousness and sensibility is still under debate.
- Physical methods are used for stunning and killing, however several methods are not effective and efficient. For many fish species asphyxia in air or in ice water is practised. The method is stressful and it takes long before unconsciousness and insensibility is induced. As the salt bath method they are not recommended for animal welfare reasons.
- Electronarcosis is developed for fish species and recommended for use in practice when applied in a correct manner. The in-line in-water device achieves stunning by pumping fish through a long-electrified tube connected between the tank and the processing line. For dry stunning fish is removed from the water and placed in a device which consists of a conveyor as a negative electrode and above are suspended rows of steel flaps, which are the positive electrodes.

- Use of CO_2 results in rapid unconsciousness, being an analgesic and anaesthetic gas. Exposure to high concentrations of CO_2 is stressful for animals due to irritation of the mucous membranes and eyes and for animal welfare reasons not recommended. CO induces a rapid state of unconsciousness and results in death without aversive responses in experiments. The responses to nitrogen were different in different species.
- Exposure to clove oil results in gradual loss of equilibrium, swimming and finally in unconsciousness. The agent is not allowed for stunning and slaughtering fish in the EU and Norway.

References

Acerete, L., Reig, L., Alvarez, D., Floz, R. and Tort, L., 2009. Comparison of two stunning/slaughtering methods on stress response and quality indicators of European sea bass (*Dicentrarchus labrax*). Aquaculture 287: 139-144.

Anders, N., Roth, B., Grimsbø, E. and Breen, M. 2019. Assessing the effectiveness of an electrical stunning and chilling protocol for the slaughter of Atlantic mackerel (*Scomber scombrus*). PLoS ONE 14: e0222122. https://doi.org/10.1371/journal.pone.0222122

Ashley, P.J., 2007. Fish welfare: current issues in aquaculture. Applied Animal Behaviour Science 104: 199-235.

Bagni, M., Civitareale, C., Priori, A., Ballerini, A., Finoia, M., Brambilla, G. and Marino, G., 2007. Pre-slaughter crowding stress and killing procedures affecting quality and welfare in sea bass (*Dicentrarchus labrax*) and sea bream (*Sparus aurata*). Aquaculture 263: 52-60.

Barcellos, L.J.G., Marqueze, A., Trapp, M., Quevedo, R.M. and Ferreira, D., 2010. The effects of fasting on cortisol, blood glucose and liver and muscle glycogen in adult jundiá *Rhamdia quelen*. Aquaculture 300: 231-236.

Bermejo-Poza, R., De la Fuente, J., Pérez, C., Lauzurica, S., González, E., Diaz, M.T. and Villarroel, M., 2015. The effect of intermittent feeding on the pre-slaughter fasting response in rainbow trout. Aquaculture 443: 24-30.

Bermejo-Poza, R., De la Fuente, J., Pérez, C., González de Chavarri, E.G. Diaz, M.T., Torrent, F. and Villarroel, M., 2017. Determination of optimal degree days of fasting before slaughter in rainbow trout (*Oncorhynchus mykiss*). Aquaculture 473: 272-277.

Bermejo-Poza, R., Fernández-Muela, M., De la Fuente, J., Pérez, C., De Chavarri, E.G. Díaz M.T., Torrent, F. and Villarroel, M., 2019. Physio-metabolic response of rainbow trout during prolonged food deprivation before slaughter. Fish Physiology and Biochemistry 45: 253-265.

Bialowas, H., Pilarcyzyk, M., Van de Vis, H., Lambooij, B. and Veldman, M., 2004. The influence of stunning method on pH value of the carp flesh. Zootechnika 501: 25-30.

Bjørlykke, G.A., Roth, B., Sørheim, O., Kvamme, B.O. and Slinde, E., 2011. The effects of carbon monoxide on Atlantic salmon (*Salmo salar* L.). Food Chemistry 127: 1706-1711.

Bowman, J., Hjelmstedt, P. and Gräns, A., 2019. Non-invasive recording of brain function in rainbow trout: evaluations of the effects of MS-222 anaesthesia induction. Aquaculture Research 50: 3420-3428.

Bowman, J., Van Nulan, N., Per Hjelmstedt, P., Berg, C. and Gräns, A., 2020. Evaluation of the reliability of indicators of consciousness during CO2 stunning of rainbow trout and the effects of temperature. Aquaculture Research 51: 5194-5202.

Braithwaite, V.A., Huntingford, F. and Van den Bos, R., 2013. Variation in emotion and cognition among fishes. Journal of Agricultural and Environmental Ethics 26: 7-23.

Branson, E.J., 2008. Fish welfare. John Wiley & Sons, London, UK, 299 pp.

Brijs, J., Sandblom, E., Axelsson, M., Sundell, K., Sundh, H., Huyben, D., Broström, R., Kiessling, A., Berg, C. and Gräns, A., 2018. The final countdown: continuous physiological welfare evaluation of farmed fish during common aquaculture practices before and during harvest. Aquaculture 495: 903-911.

Brijs, J., Sundell, E., Hjelmstedt, P., Berg, C., Senčić, I., Sandblom, E., Axelsson, M., Lines, J. Bouwsema, J., Ellis, M., Saxer, A. and Grans, A., 2021. Humane slaughter of African sharptooth catfish (*Clarias gariepinus*): effects of various stunning methods on brain function. Aquaculture 531: 735887. https://doi.org/10.1016/j.aquaculture.2020.735887

Broglio, C., Gomez., A., Duran, E., Ocana, F.M., Jiminez-Moya, F., Rodriguez, F. and Salas, C., 2005. Hallmarks of a common forebrain vertebrate plan: specialized pallial areas for spatial, temporal and emotional memory in actinopterygian fish. Brain Research Bulletin 66: 277-281.

Brown, J.A., Watson, J., Bourhill, A. and Wall, T., 2010. Physiological welfare of commercially reared cod and effects of crowding for harvesting. Aquaculture 298: 315-324.

Cardia, F. and Lovatelli, A., 2015. Aquaculture operations in floating HDPE cages: a field handbook. Fisheries and Aquaculture Technical Paper no. 593. FAO, Rome, Italy, 176 pp.

Cockram, M.S. and Velarde, A., 2022. Sheep. In: Faucitano, L. (ed.) Preslaughter handling and slaughter of meat animals. Wageningen Academic Publishers, Wageningen, the Netherlands, pp. 266-310.

Compassion in World Farming, 2009. The welfare of farmed fish. Available at: https://www.ciwf.org.uk/media/3818654/farmed-fish-briefing.pdf

Concollato, A., Dalle Zotte, A., Vargas, S.C., Cullere, M., Secci, G. and Parisi, G., 2019. Effects of three different stunning/slaughtering methods on physical, chemical, and sensory changes in rainbow trout (*Oncorhynchus mykiss*). Journal Science Food Agriculture 99: 613-619.

Conte, F.S., 2004. Stress and the welfare of cultured fish. Applied Animal Behaviour Science 86: 205-223.

Cooke, S.J. and Suski, C.D., 2005. Do we need species-specific guidelines for catch-and-release recreational angling to effectively conserve diverse fishery resources? Biodiversity and Conservation 14: 1195-1209.

Daskalova, A.H. and Pavlov, A.I., 2015. Effect of two stunning methods on post mortem muscle pH and meat quality of common carp (*Cyprinus carpio* L.). Bulgarian Journal Veterinary Medicine 18: 83-90.

Daskalova, A.H., Bracke, M.B.M., Van de Vis, J.W., Roth, B., Reimert, H.G.M., Burggraaf, B. and Lambooij, E., 2016. Effectiveness of tail-first dry electrical stunning, followed by immersion in ice water as a slaughter (killing) procedure for turbot (*Scophthalmus maximus*) and common sole (*Solea solea*). Aquaculture 455: 22-31.

Digre, H., Erikson, U., Misimi, E., Lambooij, B. and Van de Vis, H., 2010. Electrical stunning of farmed cod *Gadus morhua* L.: a comparison of an industrial and experimental method. Aquaculture Research 41: 1190-1202.

European Food Safety Authority (EFSA), 2004. Welfare aspects of animal stunning and killing methods. EFSA Scientific Report AHAW/04-027. EFSA, Parma, Italy, 241 pp.

European Food Safety Authority (EFSA), 2009a. General approach to fish welfare and to the concept of sentience in fish. EFSA Journal 954: 1-27.

European Food Safety Authority (EFSA), 2009b. Species-specific welfare aspects of the main systems of stunning and killing of farmed carp. EFSA Journal 1013: 1-37.

European Food Safety Authority (EFSA), 2009c. Species-specific welfare aspects of the main systems of stunning and killing of farmed fish: rainbow trout. EFSA Journal 1013: 1-55.

European Food Safety Authority (EFSA), 2009d. Species-specific welfare aspects of the main systems of stunning and killing of farmed eels (*Anguilla anguilla*). EFSA Journal 1014: 1-42.

European Food Safety Authority (EFSA), 2009e. Species-specific welfare aspects of the main systems of stunning and killing of farmed tuna. EFSA Journal 1072: 1-53.

European Food Safety Authority (EFSA), 2009f. Food safety considerations concerning the species-specific welfare aspects of the main systems of stunning and killing of farmed fish. EFSA Journal 1190: 1-16.

European Food Safety Authority (EFSA), 2009g. Species-specific welfare aspects of the main systems of stunning and killing of farmed Atlantic salmon. EFSA Journal 2012: 1-77.

Erikson, U., 1997. Muscle quality of Atlantic salmon (*Salmo salar*) as affected by handling stress. PhD-thesis, Norwegian University of Science and Technology, Trondheim, Norway.

Erikson, U., 2008. Live chilling and carbon dioxide sedation at slaughter of farmed Atlantic salmon: a description of a number of commercial case studies. Journal of Applied Aquaculture 20: 38-61.

Erikson, U., 2011. Assessment of different stunning methods and recovery of farmed Atlantic salmon (*Salmo salar*): isoeugenol, nitrogen and three levels of carbon dioxide. Animal Welfare 20: 365-375.

Erikson, U., Hultmann, L and Steen, E.J., 2006. Live chilling of Atlantic salmon (*Salmo salar*) combined with mild carbon dioxide anaesthesia. I. Establishing a method for large-scale processing of farmed fish. Aquaculture 252: 183-198.

Erikson, U., Lambooij, B., Digre, H., Reimert, H.G.M., Bondø. and Van de Vis, H., 2012. Conditions for instant electrical stunning of farmed Atlantic cod after de-watering, maintenance of unconsciousness, effects of stress, and fillet quality – a comparison with Aqui-S™. Aquaculture 324-325: 135-144.

Erikson, U., Gansel, L., Frank, K., Svendsen, E. and Digre, H., 2016. Crowding of Atlantic salmon in net-pen before slaughter. Aquaculture 465: 395-400.

European Commission (EC), 2005. Council Regulation (EC) No 1/2005 on the protection of animals during transport and related operations and amending Directives 64/432/EEC and 93/119/EC and Regulation (EC) No 1255/97. No 1/2005 of 22 December 2004. Official Journal of the European Union L 3: 1-43. Available at: http://data.europa.eu/eli/reg/2005/1/oj

Fantini, L.E., Rodrigues, R.A., Honorato, C.A., Dos Reis Goes, E.S., Julien Ferraz, A.L., De Lara, J.A.F., Hanson, T. and De Campos, C.M., 2020. Resting time before slaughter restores homeostasis, increases rigor mortis time and fillet quality of surubim *Pseudoplatystoma* spp. PLoS ONE 15: e0233636. https://doi.org/10.1371/journal.pone.0233636

Farm Animal Welfare Council (FAWC), 1996. Report on the welfare of farmed fish. FAWC Report 2765. Available at: https://www.gov.uk/government/publications/fawc-report-on-the-welfare-of-farmed-fish

Faucitano, L. and Raj, M., 2022. Pigs. In: Faucitano, L. (ed.) Preslaughter handling and slaughter of meat animals. Wageningen Academic Publishers, Wageningen, the Netherlands, pp. 178-230.

Flight, W.G.F. and Verheijen, F.J., 1993. The 'neck-cut' (spinal transection): not a humane way to slaughter eel (*Anguilla anguilla* L.). Aquaculture and Fisheries Management 24: 523-528.

Food and Agriculture Organization of the United Nations (FAO), 2020. The state of world fisheries and aquaculture 2020. FAO, Rome, Italy.

Foss, A., Kjartan, A., Imsland, D., Van de Vis, H., Abbink, W., Lambooij, B. and Roth, B., 2019. Physiological response of temperature shocks in turbot and sole. Journal of Applied Aquaculture 31: 34-47.

Gallo, C., Schwartzkopf-Genswein, K. and Gibson, T., 2022. Cattle. In: Faucitano, L. (ed.) Preslaughter handling and slaughter of meat animals. Wageningen Academic Publishers, Wageningen, the Netherlands, pp. 62-116.

Gatica, M.C., Monti, G., Gallo, C., Knowles, T.G. and Warriss, P.D., 2008. Effects of well-boat transportation on the muscle pH and onset of rigor mortis in Atlantic salmon. Veterinary Record 163: 111-116.

Hoseini, S.M., Abtahi, B. and Yousefi, M., 2019. Effects of fasting on metabolic and immunological responses of common carp (*Cyprinus carpio*) to a further acute stress. Aquaculture Research 50: 1177-1185.

Huidobro, A. and Tejada, M., 2004. Gilthead seabream (*Sparus aurata*): suitability for freezing and commercial alternatives. Journal of the Science of Food and Agriculture 84: 1405-1413.

Hultmann, L., Tobiassen, T., Aas-Hansen, Ø., Phu, T.M. and Rustad, T., 2016. Muscle quality and proteolytic enzymes of farmed Atlantic cod (*Gadus morhua*) during storage: effects of pre-slaughter handling and increased storage temperature. Journal of Aquatic Food Product Technology 25: 540-554.

Humborstad, O.B., Noble, C., Sæther, B.S., Midling, K.Ø. and Breen, M., 2020. Fish welfare in capture-based aquaculture (CBA). In: Kristiansen, T.S., Fernö, A., Pavlidis, M.A. and Van de Vis, H. (eds.) The welfare of fish. Springer Nature, Cham, Switzerland, pp. 439-462.

Huntingford, F.A., Adams, C., Braithwaite, V.A., Kadri, S., Pottinger, T.G., Sandøe, P. and Turnbull, J.F., 2006. Current issues in fish welfare. Journal of Fish Biology 68: 332-372.

Huntingford. F., Kadri, S. and Jobling, M., 2012. Aquaculture and behavior. Wiley-Blackwell, Chichester, UK, 340 pp.

Ibarra-Zatarain, Z., Fatsini, E., Rey, S., Chereguini, O., Martin, I., Rasines, I., Alcaraz, C. and Duncan, N., 2016. Characterization of stress coping style in Senegalese sole (*Solea senegalensis)* juveniles and breeders for aquaculture. Royal Society Open Science 3: 160495. https://doi.org/10.1098/rsos.160495

Iversen, M., Finstad, B., McKinley, R.S., Eliassen, R.A., Carlsen, K.T. and Evjen, T., 2005. Stress responses in Atlantic salmon (*Salmo salar* L.) smolts during commercial well boat transports, and effects on survival after transfer to sea. Aquaculture 243: 373-382.

Jentoft, S., Aastveit, A.H., Torjesen, P.A. and Andersen, Ø., 2005. Effects of stress on growth, cortisol and glucose levels in non-domesticated Eurasian perch (*Perca fluviatilis*) and domesticated rainbow trout (*Oncorhynchus mykiss*). Comparative Biochemistry and Physiology Part A: Molecular and Integrative Physiology 141: 353-358.

Kestin S.C, Wotton, S. and Adams, A., 1995. The effect of CO_2, concussion or electrical stunning of rainbow trout (*Oncorhynchus mykiss*) on fish welfare. In: European Aquaculture Society (ed.) Proceedings of Aquaculture Europe '95, Trondheim, Norway. Special Publication 23: 380.

Kestin, S.C., Van de Vis, J.W. and Robb, D.H.F. 2002. Protocol for assessing brain function in fish and the effectiveness of methods used to stun and kill them. Veterinary Record 150: 302-307.

Kiessling, A., Johansson, L. and Kiessling, K.H., 1990. Effects of starvation on rainbow trout muscle: I. Histochemistry, metabolism and composition of white and red muscle in mature and immature fish. Acta Agriculturae Scandinavica 40: 309-324.

Kuhlmann, H. and Münkner, W., 1996. Gutachterliche Stellungnahme zum tierschutzgerechten Betäuben/Töten von Aalen in größeren Mengen. Fischer and Teichwirt 47: 404-495.

Lambooij, E., 2014. Electrical stunning. In: Devine, C. and Dikeman, M. (eds.) Encyclopedia of meat sciences. Elsevier Ltd., Oxford, UK, pp. 407-412.

Lambooij, B. and Algers, B., 2016. Mechanical stunning and killing methods. In: Velarde, A. and Raj, M. (eds.) Animal welfare at slaughter. 5M Publishing Ltd., Sheffield, UK, pp. 111-132.

Lambooij, E., Van de Vis, J.W., Kuhlmann, H., Münkner, W., Oehlenschläger, J., Kloosterboer, R.J. and Pieterse, C., 2002a. A feasible method for humane slaughter of eel (*Anguilla anguilla* L.): electrical stunning in fresh water prior to gutting. Aquaculture Research 33: 643-652.

Lambooij, E., Van de Vis, J.W., Kloosterboer, R.J. and Pieterse, C., 2002b. Evaluation of captive needle stunning of farmed eel (*Anguilla anguilla*, L.): suitability for humane slaughter. Aquaculture 212: 141-148.

Lambooij, E., Van de Vis, J.W., Kloosterboer, R.J. and Pieterse, C., 2002c. Evaluation of head-only and head-to-tail electrical stunning of farmed eels (*Anguilla anguilla*, L.) for the development of a humane slaughter method. Aquaculture Research 33: 323-331.

Lambooij, E., Van de Vis, J.W., Kloosterboer, R.J. and Pieterse, C., 2002d. Welfare aspects of live chilling and freezing of farmed eel (*Anguilla anguilla*, L.): neurological and behavioural assessment. Aquaculture 210: 159-169.

Lambooij, E., Kloosterboer, R.J., Pieterse, C., Gerritzen, M.A. and Van de Vis, J.W., 2003. Stunning of farmed African catfish (*Clarias gariepinus*) using a captive needle pistol: assessment of welfare aspects. Aquaculture Research 34: 1353-1358.

Lambooij, E., Kloosterboer, R.J., Gerritzen, M.A. and Van de Vis, J.W., 2004. Head-only electrical stunning and bleeding of African catfish (*Clarias gariepinus*): assessment of loss of consciousness. Animal Welfare 13: 71-76.

Lambooij, E., Kloosterboer, R.J., Gerritzen, M.A. and Van de Vis, J.W., 2006a. Assessment of electrical stunning in fresh water of African catfish (*Clarias gariepinus*) and chilling in ice water for loss of consciousness and sensibility. Aquaculture 254: 388-395.

Lambooij, B., Kloosterboer, K., Gerritzen, M.A., André, G., Veldman, M. and Van de Vis, H., 2006b. Electrical stunning followed by decapitation or chilling of African catfish (*Clarias gariepinus*): assessment of behavioural and neural parameters and product quality. Aquaculture Research 37: 61-70.

Lambooij, E., Pilarczyk, M., Bialowas, H., Van den Boogaart, J.G.M. and Van de Vis, J.W., 2007. Electrical and percussive stunning of the common carp (*Cyprinus carpio* L.): neurological and behavioural assessment. Agricultural Engineering 37: 171-179.

Lambooij, E. Gerritzen, M.A., Reimert, H., Burggraaf, D. and Van de Vis, J.W., 2008a. A humane protocol for electro-stunning and killing of Nile tilapia in fresh water. Aquaculture 275: 88-95.

Lambooij, E. Gerritzen, M.A., Reimert, H., Burggraaf, D. Andre., G. and Van de Vis, J.W., 2008b. Evaluation of electrical stunning of sea bass (*Dicentrarchus labrax*) in sea water and killing by chilling: welfare aspects, product quality and possibilities for implementation. Aquaculture Research 39: 50-58.

Lambooij, E., Grimsbø, E., Van de Vis, J.W., Reimert, H.G.M., Nortvedt, R. and Roth, B., 2010. Percussion and electrical stunning of Atlantic salmon (*Salmo salar*) after dewatering and subsequent effect on brain and heart activities. Aquaculture 300: 107-112.

Lambooij, B., Digre, H., Reimert, H.G.M., Aursand, I.G., Grimsmo, L. and Van de Vis, H., 2012. Effects of on-board storage and electrical stunning of wild cod (*Gadus morhua*) and haddock (*Melanogrammus aeglefinus*) on brain and heart activity. Fisheries Research 127-128: 1-8.

Lambooij, B., Bracke, M., Reimert, H., Foss, A, Imsland, A. and Van de Vis, H., 2015. Electrophysiological and behavioural responses of turbot (*Scophthalmus maximus*) cooled in ice water. Physiology & Behavior 149: 23-28.

Le Neindre, P., Bernard, E., Boissy, A., Boivin, X., Calandreau, L., Delon, N., Deputte, B., Desmoulin-Canselier, S., Dunier, M., Faivre, N., Giurfa, M., Guichet, J.L., Lansade, L., Larrère, R., Mormède, P., Prunet, P., Schaal, B., Servière, J. and Terlouw, C., 2017. Animal consciousness. EFSA Supporting publication 2017 EN-1196. Available at: https://doi.org/10.2903/sp.efsa.2017.EN-1196

Lines, J. and Kestin, S., 2004. Electrical stunning of fish: the relationship between the electric field strength and water conductivity. Aquaculture 241: 219-234.

Lines, J.A. and Spence, J., 2012. Safeguarding the welfare of farmed fish at harvest. Fish Physiology and Biochemistry 38: 153-162.

Lines, J.A. and Spence, J., 2014. Humane harvesting and slaughter of farmed fish. Revue Scientifique et Technique Office International des Epizooties 33: 255-264.

Llonch, P., Lambooij, E., Reimert, H.G.M., and Van de Vis, J.W., 2012. Assessing effectiveness of electrical stunning and chilling in ice water of farmed yellowtail kingfish, common sole and pike-perch. Aquaculture 365: 143-149.

López-Cánovas, A.E., Cabas, I., Ros-Chumillas, M., Navarro-Segura, L., López-Gómez, A. and García-Ayala, A., 2019. Nanoencapsulated clove essential oil applied in low dose decreases stress in farmed gilthead seabream (*Sparus aurata L.*) during slaughter by hypothermia in ice slurry. Aquaculture 504: 437-445.

López-Luna, J., Vásquez, L., Torrent, F. and Villarroel, M., 2013. Short-term fasting and welfare prior to slaughter in rainbow trout, *Oncorhynchus mykiss*. Aquaculture 400: 142-147.

Lu, G., Mak, Y.T., Wai, S.M., Kwong, W.H., Fang, M., James, A., Randall, D. and Yew, D.T., 2005. Hypoxia-induced differential apoptosis in the central nervous system of the sturgeon (*Acipenser shrenckii*). Microscopy Research and Technique 68: 258-263.

Lyu, F., Huang, R.-I., Liu, L., Zhou, X. and Ding, Y.-T., 2015. Effect of slaughter methods on the quality of Chilean jack mackerel (*Trachurus murphyi*) during refrigerated storage. Journal Food Science Technology 52: 1742-1747.

Merker, B., 2013. The efference cascade, consciousness, and its self: naturalizing the first-person pivot of action control. Frontiers in Psychology 4: 501. https://doi.org/10.3389/fpsyg.2013.00501

Merkin, G.V., Roth, B., Gjerstad, C., Dahl-Paulsen, E. and Nortvedt, R., 2010. Effect of pre-slaughter procedures on stress responses and some quality parameters in sea-farmed rainbow trout (*Oncorhynchus mykiss*). Aquaculture 309: 231-235.

Metcalfe, N.B. and Thorpe, J.E., 1992. Anorexia and defended energy levels in over-wintering juvenile salmon. Journal of Animal Ecology 61: 175-181.

Miranda-de la Lama, G.C., Mamani-Linares, W. and Estévez-Moreno, L.X., 2022. Species destined for non-traditional meat production: 2. Goats and South American domestic camelids. In: Faucitano, L. (ed.) Preslaughter handling and slaughter of meat animals. Wageningen Academic Publishers, Wageningen, the Netherlands, pp. 348-389.

Mørkøre, T., Tahirovic, V. and Einen, O., 2008. Impact of starvation and handling stress on rigor development and quality of Atlantic salmon (*Salmon salar* L). Aquaculture 277: 231-238.

Morzel, M. and Van de Vis, H., 2003. Effect of the slaughter method on the quality of raw and smoked eels (*Aguilla anguilla* L). Aquaculture Research 34: 1-11.

Mugnier, C., Fostier, A., Guezou, S., Gaignon, J.L. and Quemener, L., 1998. Effect of some repetitive factors on turbot stress response. Aquaculture International 6: 33-45.

Nilsson, J., Kristiansen, T.S., Fosseidengen, J.E., Fernö, A. and Van den Bos, R., 2008a. Learning in cod (*Gadus morhua*): long trace interval retention. Animal Cognition 11: 215-222.

Nilsson, J., Kristiansen, T.S., Fosseidengen, J.E., Fernö, A. and Van den Bos, R., 2008b. Signal and goal tracking in Atlantic cod (*Gadus morhua*). Animal Cognition 11: 651-659.

Noble, C., Gismervik, K., Iversen, M.H., Kolarevic, J., Nilsson, J., Stien, L.H. and Turnbull, J.F., 2018. Welfare indicators for farmed Atlantic salmon: tools for assessing fish welfare. Nofima, Tromsø, Norway.

Noble, C., Gismervik, K., Iversen, M.H., Kolarevic, J., Nilsson, J., Stien, L.H. and Turnbull, J.F., 2020. Welfare indicators for farmed rainbow trout: tools for assessing fish welfare. Nofima, Tromsø, Norway.

World Organisation for Animal Health (OIE), 2019. Welfare of farmed fish. Aquatic animal health code. Section 7, pp. 133-144. Available at: https://tinyurl.com/9t9rtay2

Panksepp, J. and Northoff, G., 2009. The trans-species core SELF: the emergence of active cultural and neuro-ecological agents through self-related processing within subcortical-cortical midline networks. Consciousness and Cognition 18: 193-215.

Parvizi, J. and Damasio, A., 2001. Consciousness and the brainstem. Cognition 79: 135-160.

Pillay, T.V.R. and Kutty, M.N., 2005. Harvesting and post-harvest technology. In: Pillay, T.V.R. and Kutty, M.N. (eds.) Aquaculture, principles and practices. Blackwell Publishing, Oxford, UK, pp. 260-273.

Poli, B.M., Parisi, G., Scappini, F. and Zampacavallo, G., 2005. Fish welfare and quality as affected by pre-slaughter and slaughter management. Aquaculture International 13: 29-49.

Pottinger, T.G., Rand-Weaver, M. and Sumpter, J.P., 2003. Overwinter fasting and re-feeding in rainbow trout: plasma growth hormone and cortisol levels in relation to energy mobilisation. Comparative Biochemistry and Physiology Part B: Biochemistry and Molecular Biology 136: 403-417.

Qin, N., Li, D., Hong, H., Zhang, Y., Zhu, B. and Luo, Y., 2016. Effects of different stunning methods on the flesh quality of grass carp (*Ctenopharyngodon idellus*) fillets stored at 4 °C. Food Chemistry 201: 131-138.

Robb, D.H.F., 2008. Welfare of fish at harvest. In: Branson, E.J. (ed.) Fish welfare. Blackwell Publishing Ltd., Oxford, UK, pp. 217-242.

Robb, D.H.F. and Kestin, S.C., 2002. Methods used to kill fish: field observations and literature reviewed. Animal Welfare 11: 269-282.

Robb, D.H.F. and Roth B., 2003. Brain activity of Atlantic salmon (*Salmo salar*) following electrical stunning using various field strengths and pulse durations. Aquaculture 216: 363-369.

Robb, D.H.F., Kestin, S.C. and Warriss, P.D., 2000a. Muscle activity at slaughter: I. Changes in flesh colour and gaping in rainbow trout. Aquaculture 182: 261-269.

Robb, D.H.F., Wotton S.B., McKinstry J.L., Sørensen N.K. and Kestin S.C., 2000b. Commercial slaughter methods used on Atlantic salmon: determination of the onset of brain failure by electroencephalography. Veterinary Record 147: 298-303.

Robb, D.H.F., O'Callaghan, M., Lines, J.A. and Kestin, S.C., 2002. Electrical stunning of rainbow trout (*Oncorhynchus mykiss*): factors that affect stun duration. Aquaculture 205: 359-371.

Rodriguez, F., Broglio, C., Duran, E., Gomez, A. and Salas, C., 2006. Neural mechanisms of learning in teleost fish. In: Brown, C., Laland, K. and Krause, J. (eds.) Fish cognition and behaviour. Blackwell Publishing, Oxford, UK, pp. 242-277.

Roth, B., Veland, J.O., Moeller, D. Imsland, A. and Slinde, E., 2002. The effect of stunning methods on rigor mortis and texture properties of Atlantic salmon (*Salmo salar*). Journal of Food Science 67: 1462-1466.

Roth, B., Imsland, A.K. and Foss, A., 2009. Live chilling of turbot and subsequent effect on behaviour, muscle stiffness, muscle quality, blood gases and chemistry. Animal Welfare 18: 33-41.

Roth, B., Grimsbø, E., Slinde, E., Foss, A., Stien, L.H. and Nortvedt, R., 2012. Crowding, pumping and stunning of Atlantic salmon, the subsequent effect on pH and rigor mortis. Aquaculture 326: 178-180.

Royal Society for the Prevention of Cruelty to Animals (RSPCA), 2018. RSPCA welfare standards for farmed rainbow trout. RSPCA, West Sussex, UK.

Rucinquea, D.S., Watanabeb, A.L. and Maiolino Molentoa, C.F., 2018. Electrical stunning in pacu (*Piaractus mesopotamicus*) using direct current waveform. Aquaculture 497: 42-48.

Sandblom, E., Djordjevic, B., Sundh, H., Seth, H., Sundell, K., Lines, J.A. and Kiessling, A., 2012. Effects of electric field exposure on blood pressure, cardio ventilatory activity and the physiological stress response in Arctic char, *Salvelinus alpinus* L. Aquaculture 344-349: 135-140.

Santurtun, E., Broom, D.M. and Phillips, C.J.C., 2018. A review of factors affecting the welfare of Atlantic salmon (*Salmo salar*). Animal Welfare 27: 193-204.

Sattari, A., Lambooij, E., Sharifi, H., Abbink, W., Reimert, H. and Van de Vis, J.W., 2010. Industrial dry electro-stunning followed by chilling and decapitation as a slaughter method in Claresse (*Heteroclarias* sp.) and African catfish (*Clarias gariepinus*). Aquaculture 302: 100-105.

Schreck, C.B. and Tort, L., 2016. The concept of stress in fish. In: Schreck, C.B., Tort, L. and Farrell, A.P. (eds.) Biology of stress in fish. Academic Press, Amsterdam, the Netherlands, pp. 2-33.

Schomer, D.L. and Lopes da Silva, F.H., 2012. Niedermeyer's electroencephalography – basic principles, clinical applications and related fields. Wolters Kluwer, Philadelphia, PA, USA, 1296 pp.

Secci, G., Parisi, G., Meneguz, M., Iaconisi, V., Cornale, P, Macchi, E., Gasco, L. and Gai, F., 2018. Effects of a carbon monoxide stunning method on rigor mortis development, fillet quality and oxidative stability of tench (*Tinca tinca*). Aquaculture 493: 233-239.

Skjervold, P.O., Fjaera, S.O. and Østby, P.B., 1999. Rigor in Atlantic salmon as affected by crowding stress prior to chilling before slaughter. Aquaculture 175: 93-101.

Skjervold, P.O., Fjæra, S.V., Østby, P.B. and Einen, O., 2001. Live-chilling and crowding stress before slaughter of Atlantic salmon (*Salmo salar*). Aquaculture 192: 265-280.

Sneddon, L.U., 2003. The evidence for pain in fish: the use of morphine as an analgesic. Applied Animal Behaviour Science 83: 153-162.

Sneddon, L.U., Wolfenden, D.C.C. and Thomson, J.S., 2016. Stress management and welfare In: Schreck, C.B., Tort, L. and Farrell, A.P. (eds.) Biology of stress in fish. Academic Press, Amsterdam, the Netherlands, pp. 463-539.

Southgate, P.J., 2008. Welfare of fish during transport. In: Branson, E. (ed.) Fish welfare. John Wiley & Sons, London, UK, pp. 185-194.

Suwandi, R., Nugraha, S., Fathihatunnisa, R., Jacoeb, A.M. and Suptijah, P., 2020. Effect of stunning and freezing on carp (*Cyprinus carpio* L) survival rate. IOP Conference Series: Earth Environmental Science 414: 012025. https://doi.org/10.1088/1755-1315/414/1/012025

Terlouw, C. and Bourguet, C., 2022. Quantifying animal welfare preslaughter using behavioural, physiological and carcass and meat quality measures. In: Faucitano, L. (ed.) Preslaughter handling and slaughter of meat animals. Wageningen Academic Publishers, Wageningen, the Netherlands, pp. 12-61.

Terlouw, E.M.C., Arnould, C., Auperin, B., Berri, C., Le Bihan-Duval, E., Deiss, V., Lefèvre, F., Lensink, B.J. and Mounier, L., 2008. Pre-slaughter conditions, animal stress and welfare: current status and possible future research. Animal 2: 1501-1517.

Van de Pol, I., Flik, G. and Gorissen, M., 2017. Comparative physiology of energy metabolism: fishing for endocrine signals in the early vertebrate pool. Frontiers in Endocrinology 8: 36. https://doi.org/10.3389/fendo.2017.00036

Van de Vis, H., Kestin S., Robb, D., Oehlenschlaeger, J., Lambooij, B., Muenkner, W., Kuhlmann, H., Kloosterboer, K., Tejada, M., Huidobro, A., Ottera, H., Roth, B., Soerensen, N.K., Akse, L., Byrne, H. and Nesvadba, P., 2003. Is humane slaughter of fish possible for industry? Aquaculture Research 34: 211-220.

Van de Vis, J.W., Kiessling, A., Flik, G. and Mackenzie, S., 2013. Welfare of farmed fish in present and future production systems. Springer, London, UK, 302 pp.

Van de Vis, H., Abbink, W., Lambooij, B. and Bracke, M., 2014. Stunning and killing of farmed fish: how to put it into practice? In: Devine, C. and Dikeman, M. (eds.) Encyclopedia of meat sciences. Elsevier Ltd., Oxford, UK, pp. 421-426.

Van de Vis, H., Bergevoet, R., Stokkers, R., Schrijver, R., Dewar, D., Van de Braak, K. and Witkamp, S., 2017. Welfare of farmed fish: common practices during transport and at slaughter. Final report SANTE/2016/G2/009. European Commission, Directorate-General for Health and Food Safety, Brussels, Belgium, 186 pp.

Van den Burg, E.H., Peeters, R.R., Verhoye, M., Meek, J., Flik, G. and Van der Linden, A., 2005. Brain responses to ambient temperature fluctuations in fish: reduction of blood volume and initiation of a whole-body stress response. Journal of Neurophysiology 93: 2849-2855.

Vargas Baldi, S.C., Parisi, G., Bonelli, A., Balieiro, J.C.C., Lapa Guimarães, J. and Macedo Viegas, E.M., 2018. Effects of different stunning/slaughter methods on frozen fillets quality of cobia (*Rachycentron canadum*). Aquaculture 486: 107-113.

Venturini, F.P., Vargas Baldi, S.C., Parisi, G., Duarte Costa, T., Rucinque, D.S., Pires Melo, M. and Macedo Viegas, E.M., 2018. Effects of different stunning methods on blood markers and enzymatic activity of stress responses of tilapia (*Oreochromis niloticus*). Italian Journal of Animal Science 17: 1094-1098.

Verheijen, F.J. and Flight, W., 1997. Decapitation and brining: experimental tests show that after these commercial methods for slaughtering eel *Anguilla anguilla* (L.), death is not instantaneous. Aquaculture Research 28: 361-366.

Waagbø, R., Jørgensen, S.M., Timmerhaus, G., Breck, O. and Olsvik, P.A., 2017. Short-term starvation at low temperature prior to harvest does not impact the health and acute stress response of adult Atlantic salmon. PeerJ 5: e3273. https://doi.org/10.7717/peerj.3273

Wall, A.J., 2001. Ethical considerations in the handling and slaughter of farmed fish. In: Kestin, S.C. and Warriss, P.D. (eds.) Farmed fish quality. Blackwell Science, Oxford, UK, pp. 108-115.

Wendelaar Bonga, S.E., 1997. The stress response in fish. Physiological Reviews 77: 591-625.

Wills, C.C., Zampacavallo, G., Poli, B.M., Proctor, B.M.M. and Henehan, G.T.M., 2006. Nitrogen stunning of rainbow trout. International Journal of Food Science and Technology 41: 395-398.

Zhang, L., Li, Q., Lyu, J., Kong, C., Song, S. and Luo, Y., 2017. The impact of stunning methods on stress conditions and quality of silver carp (*Hypophthalmichthys molitrix*) fillets stored at 4 °C during 72 h post mortem. Food Chemistry 216: 130-137.

Zhang, L., Zhang, Y., Jia, S., Li, Y., Li, Q., Li, K., Hong, H. and Luo, Y., 2019. Stunning stress-induced textural softening in silver carp (*Hypophthalmichthys molitrix*) fillets and underlying mechanisms. Food Chemistry 295: 520-529.

Photo: Bert Driessen, Research, Control & Consult-Belgium, Paal, Belgium

Chapter 4
Horses

B. Driessen[1], D. Marlin[2] and J. Buyse[3]*

[1]Research, Control & Consult-Belgium bv, 3583 Paal, Belgium; [2]AnimalWeb Ltd., Tennyson House, Cambridge Business Park, Cambridge, CB4 0WZ, United Kingdom; [3]Laboratory of Livestock Physiology, Department of Biosystems, KU Leuven, 3001 Heverlee, Belgium; bert@rcc-belgium.be

Abstract

Horse slaughter and the consumption of horsemeat occur worldwide. It is remarkable that there is no evidence of a direct relationship between horse slaughter and horse meat consumption within different countries. For example, horses are slaughtered in Canada and South America, and their meat for human consumption is exported to Japan or Europe. In contrast to the transport of sport and recreational horses, commercial transport of horses, especially slaughter horses, has been studied to a lesser extent. However, it is clear that road transport of horses for slaughter is a complex operation which includes pre-transport handling, loading, transport itself, unloading and lairage. For welfare reasons, the transport must be carried out with the utmost care to induce as little stress to the horses as possible. In the drive to optimise animal welfare, meat quality often improves as well. The welfare of these horses must always be optimal no matter what the final destination of their journey is. Before horses are transported, individual fitness for travel should be determined. Although the practical guidelines on fitness for travel vary from country to country, the general recommendation is not to transport unfit horses. The infrastructure at the loading site and the expertise of the handlers determines the ease of loading and the level of stress experienced by the horses. Several stressors during the transport process influence both behaviour and welfare, for example the skilfulness of the driver, the loading density, duration of journey, vehicle design and internal environmental conditions. In addition, after arrival at the slaughterhouse, the handling during unloading, lairage conditions and stunning and slaughter can have a significant impact on stress and horse welfare.

Keywords: fitness for travel, loading, stunning, transport, unloading

4.1 Introduction

In certain cultures and geographic areas and for many millennia, horses have been and still are a source of red meat suitable for human consumption. Horse bones found at Solutré in France and cave paintings in Lascaux show that Ice Age hunters regularly ate horses (Peeters, 2020). In some European countries, horses were used as food source in periods of war, poverty and famine. Although there was not a previous tradition of horsemeat consumption, the high cost of food products during the French Revolution induced the people to consume horsemeat (Stanciu, 2015). Horse meat consumption became popular in Europe after World War II, due to low availability of beef meat resulting in increased slaughter of draught horses supplying more affordable meat (Padalino, 2015). Around 1960, prosperity following World War II increased and industrialisation and mechanisation made horses, all but obsolete for farm work and transportation. Ever since, the horsemeat industry has been slowly shrinking (Peeters, 2020).

According to FAOSTAT (http://www.fao.org/faostat/en/#data/QL), the number of horses slaughtered in the USA, Australia and Europe has been decreasing strongly since the last decade. But, even so, around five million horses are still slaughtered annually worldwide from an estimated global population of 59 million. China currently slaughters the most horses worldwide, around 1.47 million heads, while the USA owns the largest stock of horses, estimated at 10.7 million heads. After the prohibition of horse slaughter in the USA in 2007, the number of horses transported and slaughtered in Canada and Mexico has increased radically (Stull, 2012; Taylor and Sieverkropp, 2013). This made Mexico the third largest producer of horse meat in the world after China and Kazakhstan (http://www.fao.org/faostat/en/#data/QL). In 2019, the total horsemeat production was reported to be 752,000 tons (http://www.fao.org/faostat/en/#data/QL). It is clear that horse slaughter and consumption of horsemeat occurs worldwide. It is therefore remarkable that there is no direct relationship between horse slaughter and horsemeat consumption within individual countries. For example, horses are slaughtered in Canada and South America, but processed meat for human consumption is exported to Japan or Europe. In Central Asia, horse meat is part of the traditional diet, while some European or African regions consider horse meat a delicacy (Garifulovich *et al.*, 2013).

Horse meat is known to contain a high protein and mineral content, and is especially high in iron (Banu, 2009), and low contents of intramuscular fat (Palenik *et al.*, 1980) and cholesterol (Robelin *et al.*, 1984). It is also characterised by a very dark red colour and yellow fat (Stanislawczyk and Znamirowska, 2005) and presents a slightly sweet taste which is caused by the high content of carbohydrates, and in particular, intramuscular glycogen (Rødbotten *et al.*, 2004; Stanislawczyk and Znamirowska, 2005).

Similarly to other livestock species, for welfare reasons and an optimal horse meat quality, horses should undergo as little stress as possible to prevent from bruises and/or injuries (CATGP, 2018; Marlin *et al.*, 2011; Pawshe *et al.*, 2016). Preslaughter stress is caused by suboptimal environmental temperature and relative humidity, gas concentrations (ammonia, carbon monoxide and dioxide), inappropriate infrastructures (e.g. poor lighting, sharp protrusions and too narrow corridors), social interactions (e.g. fighting for the determination of dominance rank in unfamiliar groups), and transport and handling activities, such as loading, unloading and the transport itself, and weighing (Nivelle *et al.*, 2020).

In the past few decades, the circumstances in which horses are transported has resulted in public concern and discussion, more specifically the transport of slaughter horses (Marlin *et al.*, 2011; Nivelle *et al.*, 2020; Vermeulen *et al.*, 2019). While transport conditions of sport and recreational horses and their responses to transport stress are well studied (Leadon *et al.*, 2008; Marlin *et al.*, 2011; Nivelle *et al.*, 2020), a limited number of studies deals with horses destined for slaughter (Collins *et al.*, 2000; Iacono *et al.*, 2007a; Marlin *et al.*, 2011; Nivelle *et al.*, 2020; Werner and Gallo, 2008). In addition, most studies about loose travelling horses deal with healthy horses that are used to being loaded and transported (Friend *et al.*, 1998; Friend, 2000).

There are two types of horses intended for slaughter: on one hand, extensively kept in feedlots (in USA and South America) naïve horses (Nivelle *et al.*, 2020; Roy *et al.*, 2015a) and, on the other hand, cull animals from the work and recreational groups (Houpt and Wickens, 2014). These latter horses are sent for slaughter when they have reached the end of their productive use as recreational or sport horse (Lenz, 2009) and are commonly culled due to muscle-skeletal and chronic disorders (Ireland *et al.*, 2012; Jonsson *et al.*, 2013). Most of these horses are collected by dealers from multiple sources or from auctions and markets to accumulate sufficient horses to form a load to be shipped to a slaughter plant.

Both groups are generally loaded onto the truck in loose groups (Houpt and Wickens, 2014) and are transported by road or, although to a much lesser extent, by rail (Marlin *et al.*, 2011), by sea (Cavallone *et al.*, 2002; Stull, 2001) or by air (Harper, 2017). Therefore, this chapter focuses on road transport of slaughter horses.

4.2 Pre-transport handling

4.2.1 Fitness for travel

The preparation for the transport procedure starts with the selection of fit animals in order to optimise welfare during and after transport as horses in good physical condition can better cope with transport stress (Vermeulen *et al.*, 2019). Except for transport to a veterinary centre, unfit horses should not be loaded onto a vehicle as they hardly keep their balance on a moving vehicle and require more physical effort, resulting in an earlier onset of fatigue (Vermeulen *et al.*, 2019). In addition, weak, lame or debilitated animals are difficult to handle in a low stress manner (Grandin, 2014a). Many countries have described the fitness for travel requirements in legal regulations (Vermeulen *et al.*, 2019). Although they differ from country to country, according to all legal regulations, it is evident that sick, injured, weak, disabled, or fatigued horses are not to be loaded and transported (Australian Animal Welfare Standards and Guidelines, 2021; European Regulation 1/2005; http://www.oie.int/international-standard-setting/terrestrial-code/accessonline/; TQA, 2011). In addition, horses that cannot stand steadily or cannot walk without pain are not allowed to be transported. Low body condition can also be a feature on which to declare fit or not for travel, especially when the body condition would result in poor welfare because of the expected climatic conditions during transport (http://www.oie.int/international-standard-setting/terrestrial-code/accessonline/; Vermeulen *et al.*, 2019). Miranda-de La Lama *et al.* (2021) suggest nasal discharge, lameness and ocular discharge as key indicators to define fitness for transport and to integrate these criteria in a monitoring tool to measure welfare of slaughter horses. A checklist of nine animal parameters to determine the fitness for travel is described by CATGP (2018). Each regulation also defines that each horse must be inspected by an animal handler, a veterinarian or another person in charge. The person in charge not only must assess the fitness for travel of each equid, but must also take into account other circumstances, such as the weather and the transport distance, in which the horse will be transported.

However, the requirements to declare the animal as fit for transport are sometimes too vague and debatable. For example, the terms 'weak' or 'injured' are not defined and it is not clear what depth of a wound would be considered to render a horse unfit for transport. Because of these concerns,

Vermeulen *et al.* (2019) launched a call for a more detailed description around fitness for travel requirements to exclude subjective interpretations.

4.2.2 Hydration and feed withdrawal

Horses must be sufficiently hydrated before departure, irrespective of whether they come from a farm, auction, market or assembly centre. According to Messori *et al.* (2016), water provision before transport is even more preferable than during transport considering the issues in watering horses in a moving truck (see later in this chapter).

While sport and recreational horses are often allowed to eat hay during transport, slaughter horses usually are not fed during transport. To our knowledge, it is not known how long or whether it is necessary to fast slaughter horses before loading and during transport. The application of fasting reduces the risk of soiling and resulting meat contamination (Padalino, 2015). However, it is questioned if fasting before transport is a necessary practice as it may significantly compromise gastrointestinal function and lead to an increased risk of gastric squamous ulceration and colic (Padalino *et al.*, 2020). These findings suggest that horses should have access to forage until the time of loading to decrease the contact between the squamous mucosa and gastric secretions or small intestinal reflux (Padalino *et al.*, 2020).

4.2.3 Regrouping

Regrouping horses before loading is not an uncommon practice to form groups adapted to the size of the vehicle. Regrouping can be done just before loading or a few days before loading at the farm or collection centre. However, mixing groups disturbs the social cohesion between the horses in the original groups (Van Dierendonck and Spruijt, 2012) and may result in fighting (i.e. biting and kicking) to establish a social hierarchy and bruises on the carcass (Houpt and Wickens, 2014), especially in loose transported groups. During the journey, aggression between horses in mixed groups may lead to sudden head movements and difficulty in balancing resulting in a higher risk of the head striking against the interior of the vehicle (Roy *et al.*, 2015b). Furthermore, regrouping can interfere with the loading of the horses because stressed animals within the groups are more difficult to handle. Regrouping usually looks at the negatives of transporting animals in groups, but there is very little evidence showing that transporting equines together is the correct practice. For example, unhandled youngsters display anxiety when not transported in a group and for this reason their individual transport is not recommended (Knowles *et al.*, 2010).

As social animals, the more nervous and agitated horses can affect their neighbours in the vehicle. According to Saslow (2002) and McGreevy (2012) anxiety and arousal can be socially transmitted to herd mates. This should be taken into account when mixing horses with different temperaments and travel experiences (Broom and Johnson, 1993; Fazio *et al.*, 2013). The temperamental horses should be isolated and should be kept separate in the transport vehicle (European Regulation 1/2005; http://www.oie.int/international-standard-setting/terrestrial-code/accessonline/).

4.2.4 Loading

Loading onto a vehicle is the most stressful procedure for both naïve and for experienced horses (Casamassima *et al.*, 2008; Friend *et al.*, 1998) as they are taken out of their familiar surroundings and are confronted with alleys, pens, trailers, limited visual and physical space and interactions with the handlers, all causing a stress response, with effects lasting up to slaughter (Terlouw *et al.*, 2008). This stress response is showed by the greater average heart rate recorded at loading than during transport. The increase in heart rate is caused on one hand by the physical exertion in climbing the ramp, but predominantly by the emotional fear experienced during loading (Waran, 1993).

Loading horses in groups may be easier than loading individual horses as they are herd animals and usually maintain visual contact copying one another and keeping social cohesion (Houpt and Houpt, 1998; Mal *et al.*, 1991). Anyway, when moving horses either in a group or singly stalled, the nature of horses must be taken into account. Horses are driven by a very strong flight response (reactive running) showing blood adrenaline levels around 10 times higher than in humans in stressful situations (Fraser, 2010; Snow *et al.*, 1992). Horses are frightened by sudden and unfamiliar sounds and visual impressions to which they can react by stopping, retreating, stepping or jumping aside, moving backwards, or bolting (Ekesbo and Gunnarsson, 2018).

This response is even greater in slaughter horses that, differently from sport or recreational horses, are either not or barely familiar with transport. Regular loading or transport can be considered as training or conditioning and would facilitate easier and more rapid loading (Fraser, 2010). According to Grandin (1989), there is a correlation between the animal's previous experience with handling and its reaction to handling in the future.

4.2.4.1 Loading facility design

A good loading dock design, featuring a loading ramp, prevents distress, excitement and injury and decreases the time to load the horses (CATGP, 2018; European Regulation 1/2005; Horse Meat Federation, 2016). Ideally, the loading dock/ramp should be level with the floor of the vehicle deck. However, in practice, it is not common to have such a loading facility. There are two ways to overcome the difference in height between the ground and the vehicle deck. Loading can be done by a ramp built into the vehicle or by a stand-alone ramp at the loading place. Loading ramps can be made from wood, concrete, steel or even an earthen raised bank can be used.

The slope of the ramp should not exceed 36.4% (or 20°), but good practice demands a maximum slope of no more than 10° (17.3%; CATGP, 2018). A ramp with a steep angle will induce an elevated heart rate and cause stress (Tateo *et al.*, 2012). If the ramp is steeper than 10°, it should be equipped with stair steps or foot battens reducing the risk of slipping (European Regulation 1/2005; Grandin, 2015a; Horse Meat Federation, 2016). The stair steps should rise a maximum of 10 cm and have a length of 30 cm. In each step two grooves with a minimum of 2.5 cm depth should be made. Wooden or steel cleats can also be used, although placing cleats too far apart can cause slipping, whilst placing cleats too close together will result in hooves sliding over the top of the cleats. Cleats should have a dimension of 5 cm x 5 cm (Grandin, 2015a), 25 mm high and spaced at 20-35cm (SCARM, 2002).

According to Grandin (2015a) a ramp design is inadequate and must be improved if more than 1% of horses fall at loading. To prevent horses from slipping mud on the ramp should be removed and in case of concrete ramps sand or gravel can be used (Nivelle *et al.*, 2020).

Reflections in standing water at the loading place can make horses hesitant and make it more difficult to load (Grandin, 2015b). Sometimes when horses step on a metal ramp, they can be frightened by the noise and flee. In this case, a rubber mat or some sand can be used as it buffers the hollow sound of the ramp when a horse steps on it (Nivelle *et al.*, 2020; Weeks *et al.*, 2012).

To prevent an attempt to escape or falling off, side walls should be placed at each side of the ramp (Fraser, 2010). Side walls limits the distraction of the horses by activities in the near environment and also prevent escape attempts due to the blocked vision of the horses. As an indirect result the ease of loading will increase (as showed by less baulking and less reluctant to move forward) and the loading duration will be shorter (CATGP, 2018; European Regulation 1/2005; Horse Meat Federation, 2016; Nivelle *et al.*, 2020). Nivelle *et al.* (2020) registered an average loading duration of 30 s per horse with a standard deviation of 15 s which indicates a considerable variation in the speed of loading in practical conditions.

The vehicle entry is also an issue. There are two types of entries, i.e. the trapdoor (Figure 4.1) and the broad tail gate. The trapdoors allows the horses to enter the vehicle one after the other, while when using a broad tail gate, two or three horses can be loaded side by side. The height of the

Figure 4.1. South American vehicle for horse transportation featuring a trapdoor entry (B. Driessen, Research, Control & Consult, Belgium).

trapdoors requires consideration because too low trapdoors result in head bumping. The risk of head-bumping is greater in stressed horses that keep their head high (Nivelle *et al.*, 2020). However, closing trapdoors is easy to do, but raising a tail gate can cause a dangerous situation. The latching mechanism should be easy to close to avoid injuries of the handlers when the horses are bumping into the door and the tail gate would open (Houpt and Wickens, 2014). Inexperienced horses can be difficult to load because the vehicle gives them a feeling of enclosure. In that case an open-top type of vehicle can encourage a horse to enter because it creates the illusion of openness or a potential exit (Fraser, 2010).

Horses are reluctant to enter darkened passageways (Ekesbo and Gunnarsson, 2018; Houpt and Wickens, 2014), although they have the capability for both nocturnal and diurnal vision (Ekesbo and Gunnarsson, 2018). Lighting the interior of the vehicle can help loading as, similarly to other livestock species, horses have a tendency to move from a darker to a brighter area (Grandin, 1989), provided that the light does not shine directly in their eyes. Neveux *et al.* (2017) found that if the LED light inside the trailer was stronger than outside, the horses needed significantly less time to enter the trailer. Cross *et al.* (2008) noticed that horses loading from a lit arena to a dark trailer sniffed the ground more, showing increased exploration of their environment. According to the findings of Neveux *et al.* (2017), on non-cloudy days a light intensity above 4,500 lx would be needed to load the horses into the trailer. There are no specific recommendations for lighting to inspect horses. However, similarly to pigs (Faucitano and Lambooij, 2019), a minimum of 40 lx is recommended to inspect horses during the loading process.

4.2.4.2 Handler's attitude

The quality of the handling during loading and unloading is influenced by the experience, attitude, education and training of the handlers (Nivelle *et al.*, 2020). Excessive force during loading only increases fear in horses and may lead to falls and injuries. A calmer approach and communication between handlers to prevent conflicting signals to the horses are recommended (Nivelle *et al.*, 2020). For example, one handler should not stand too close to the passageway at the same time as another handler is driving horses to that direction (Nivelle *et al.*, 2020). According to Grandin (2015b), handlers with expertise in low-stress handling can control the movements of animals by simply adjusting their body position.

To minimise stress to the animals, to avoid their strong reactions and to prevent dangerous work situations during handling, handlers must have knowledge of the concept of flight distance and the point of balance. The horse will start to move away when the handler enters the flight zone. When the person stands still, the horse will turn and face the person. When the handler stands in front of the point of balance at the shoulder, the horse will go backward. To move the horse forward, the handler should enter the flight zone behind the point of balance (Grandin, 2014b).

Handlers should not scream, whistle, clap their hands, flap their arms or make sudden movements in order to keep the horses calm as possible (Grandin, 1999; Horse Meat Federation, 2016; http://www.oie.int/international-standard-setting/terrestrial-code/accessonline/). Flags can be used as moving tools (Figure 4.2), provided they are used calmly as waving flags can cause stress (Houpt and

Figure 4.2. Moving horses inside a trailer using a flag (photo: B. Driessen, Research, Control & Consult, Belgium).

Wickens, 2014). Whips, electric prods, sticks and dogs are not recommended because they induce fear and anxiety to the horses (Dai *et al.*, 2021). More specifically, the use of electric prods, besides not being necessary, is not recommended as it causes agitation and results in dangerous animals (Grandin, 1999; Nivelle *et al.*, 2020).

4.3 Transport

Transport or being in a moving vehicle is a stressful event for horses as showed by the increased activity and use of energy compared to horses that are not transported (Doherty *et al.*, 1997). The welfare of horses during transport depends on their behaviour, environmental conditions, vehicle design, and travel and vehicle driving conditions.

4.3.1 Behavioural response

The normal standing position adopted by horses during transport is a potential risk if horses cannot maintain their balance as they may get injured as a result of falling or hitting the internal structures of the vehicle (Houpt and Wickens, 2014). When horses fall during transport they may be unable to stand again either when being transported in a loose group or in stalls. They thus have to make a physical effort to continually adjust the posture to resist the transport vehicle movements and react to noises and visual stressors, all resulting in anxiety and physical stress as showed by their increased heart rate during transport (Doherty *et al.*, 1997).

4.3.2 Microclimate

The microclimate within the vehicle, in terms of temperature, relative humidity (RH) and solar radiant load and wind speed, has an impact on the behaviour and welfare of horses during transport. For horses, the thermoneutral zone ranges from a minimum of 5 °C to a maximum of 25 °C (Morgan, 1998). When the environmental temperature exceeds these limits of the thermoneutral zone, horses have to invest energy to keep their body temperature constant. When exposed to cold stress conditions (<5 °C), horses exhibit shivering or huddling (Grandin, 2015a), while in warmer conditions (>25 °C), they regulate their temperature initially by increased skin vasodilation (convective heat loss) and then through sweating and increased respiratory rate (CATGP, 2018; Luz *et al.*, 2015). The limits of the thermoneutral zone are not static, but can be influenced by animal characteristics (e.g. condition, winter or summer coat, being acclimated) or operational procedures, i.e. stocking density or rugging (Cuddeford, 2003; Hoopes, 2018; Roy and Cockram, 2015). The only practical way to assess cold or heat stress in horses during transport is by direct observation by trained personnel.

Relative humidity (RH) also plays a crucial role in the horses' ability to thermoregulate. Sweating, as heat dissipation, at higher ambient temperatures will be more efficient at lower RH. There is no temperature humidity index (THI) framework specifically for horses. However, THI must be interpreted carefully, since this parameter does not take solar load and wind speed into account (Gaughan *et al.*, 2008). Moreover, the lack of a reference framework with limit values for heat stress in horses complicates interpretation (Nivelle *et al.*, 2020).

During transport in an open trailer or in closed (roofed) passively- or actively-ventilated vehicles the climate has an impact on the well-being of the horses. In open trailers, commonly used in South America, air displacement enhances heat dissipation through forced convection and accelerate evaporation of sweat (Nivelle *et al.*, 2020). In hot weather conditions, the stationary time of the loaded trailer should be kept to a minimum, because heat stroke can be caused by extreme temperature increases in a passively-ventilated, open or roofed trailer (Houpt and Wickens, 2014). During rest stops in such conditions it is advised to park the van or trailer in shaded areas with all windows (in vans) and ramps open during rest stops (Waran *et al.*, 2007). Research about the ventilation in horse trailers is limited. However, Purswell *et al.* (2006) concluded that in under-ventilated trailers the exposure to ammonia, nitric oxide and carbon monoxide in the trailer can compromise the horses' respiratory clearance mechanism. Long-term exposure to toxic gases, i.e. ammonia and carbon monoxide, in high concentration can cause irritation of the respiratory tract and eyes in horses during transportation. Ammonia, resulting from the degradation of urine by bacteria, is the most important gas to consider. The concentration of ammonia can be reduced by the use of absorbent floor bedding (CATGP, 2018), like sawdust or straw, although this bedding type may worsen respiratory health due to associated dust, pollen, moulds and endotoxin (Padalino, 2017; Stull, 1999). However, no recommendations on the permissible levels of toxic gases are available for horses during transport (Stull, 1999).

In winter conditions one side of a passively-ventilated roofed vehicle can be covered with tarpaulin. However, this protection can negatively influence the ventilation in the vehicle and affect the thermal comfort of horses as showed by increased sweating (Messori *et al.*, 2016). For this reason, the use of tarpaulin is not recommended in warmer climatic conditions (Messori *et al.*, 2016). Houpt and Wickens (2014) advises in very cold weather conditions to use blankets on horses located in open trailers.

4.3.3 Vehicle design

Nowadays, horses are transported in single deck trailers or lorries with or without a ramp. Roy *et al.* (2015c) described the loading of horses in Iceland where a hydraulic tailgate lift was used. Initially, the caged platform was hydraulically lowered to the ground level to load the horses. After moving the horses onto this platform, the platform was lifted until the level of the floor of the vehicle was reached. Horse transport vehicles vary in style from a single horsebox to large stock trailers loading 20 to 40 horses (Nivelle *et al.*, 2020). Nowadays, double deckers used for the transport of other species are not used to transport horses because they lack sufficient height and the internal ramps are difficult for the horses to negotiate (Houpt and Wickens, 2014). Stull (1999) compared horse transports in a potbelly trailer with horse transports in a straight-deck trailer (Figure 4.3A,B). Horses transported in potbelly trailers showed approximately 3.5 times more injuries than in a straight-deck trailers. Because the head and face were the most prevalent area for injury, the width of the door opening may be a factor in the greater injury rate in the horses transported in potbelly trailers (Stull, 1999). Whereas no effect of vehicle type (foot- vs gooseneck trailer) on the prevalence of bruises was reported by Miranda-de la Lama *et al.* (2021).

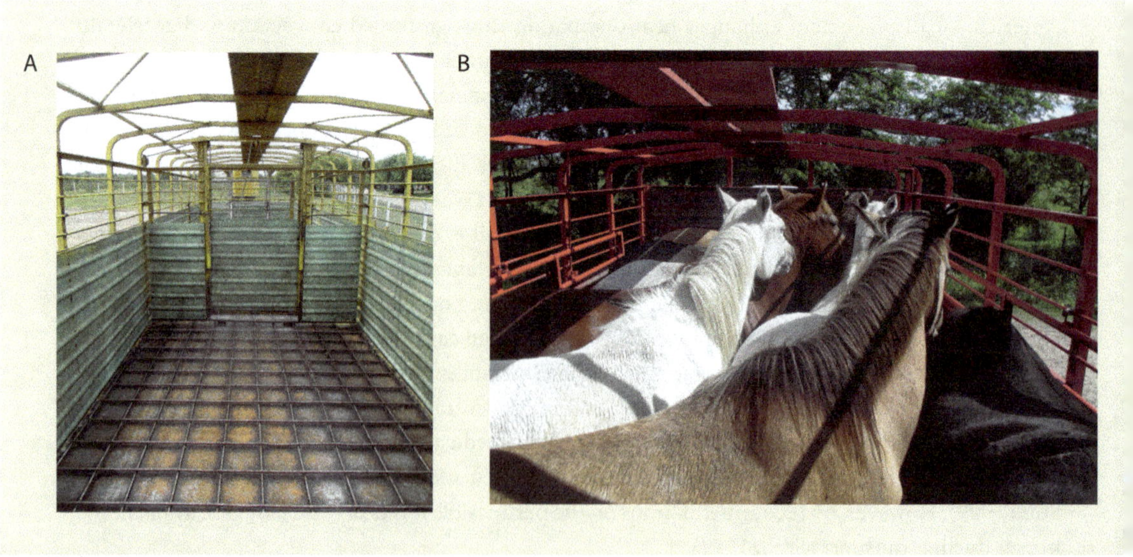

Figure 4.3. A typical South American straight-deck trailer to transport horses, (A) unloaded and (B) loaded (B. Driessen, Research Control & Consult, Belgium).

4.3.3.1 Flooring

The floor of all transport vehicles must be designed to protect horses against slipping as this occurrence is a source of anxiety and stress for them (Grandin, 2015a). To improve their stability, in some vehicles a wire mesh is laid on the floor with a 30×30 cm square pattern (Nivelle *et al.*, 2020). The rods should not lay criss-cross on top of each other, but the mat of rods must lie flat to prevent gaps from injuring the hooves (Grandin, 2015a). In general, aluminium and metal floor can frighten horses due to the noise produced by the hooves stepping on the floor and the lack of grip. For this reason, a rubber mat or a thick layer of litter is preferable as it buffers the hollow sound of the flooring when a horse steps on it (Nivelle *et al.*, 2020; Weeks *et al.*, 2012). According to CATGP (2018), a layer thickness of 1 cm per 100 km is advised for wood shavings. According to European Regulation 1/2005, a layer of bedding material is only required for journeys that exceed 8 h. The bedding material is used to absorb spilled water, faeces and urine produced by the animals during the journey (CATGP, 2018).

4.3.3.2 Vehicle height

The vehicle should be high enough to provide sufficient head room to ensure adequate ventilation and physical comfort while in the standing position (Fraser, 2010). However, the existing recommendations are conflicting, ranging from at least 20 cm of free head space when standing (AATA, 2005) to a minimum internal height of at least 75 cm higher than the height of the withers of the highest animal (European Regulation 1/2005).

4.3.4 Transport conditions

4.3.4.1 Stocking density

Different authorities and guidelines have specified minimum and maximum loading density for horse transport. For example, minimal space allowance for an adult horse during a long journey is 1.75 m^2 in Europe (European Regulation 1/2005), while it is only 1.2 m^2 in Australia (Australian Animal Welfare Standards and Guidelines, 2021).

It has been shown that at high stocking densities (1.28 m^2/horse) the risk of imbalance, falls, fighting and mounting and injuries in loose transported horses is higher (Collins *et al.*, 2000; Iacono *et al.*, 2007b; Knowles *et al.*, 2010) and heat dissipation is reduced during transport. However, it is difficult to make a comparison between transports just based on the stocking densities. Because adult horses can vary in weight, height and length, and thus vary in dimensions and contours, stocking density requirements in terms of m^2 per animal are not reliable. Two vehicles may have the same stocking density in m^2/horse, but in practice the density in kg/m^2 may vary greatly due to differences in body sizes and shape. On the other hand, loading a number of horses based on the estimated the average weight of the group can also lead to estimation errors since this does not take the variation in animal weight into account (Nivelle *et al.*, 2020). Furthermore, the individual horse weight is often not known at the moment of loading.

During transport, horses need more space than just the contours of their body because of their physics. Like all equidae, horses have a high centre of gravity and, in addition, 60% of their weight is carried on the forelegs (Cregier, 1982). To cope with accelerations and decelerations in the moving, horses position their feet outside the normal position under the body to adopt a wider base and adjust their head and neck position in the direction in which they are facing to keep their balance (Broom, 2008; Stull, 1997). To brace and adjust this position, horses require more floor room (Marlin *et al.*, 2011). Besides the floor surface, other factors affecting their ability to balance are vehicle design, the type of suspension, the experience of the driver and the type of road (Houpt and Wickens, 2014; Waran *et al.*, 2007). During transport, horses endeavour not to contact other animals or the sidewalls of the vehicle (Clark *et al.*, 1993; Waran, *et al.*, 1996; Waran *et al.*, 2007). According to Racklyeft and Love (1990), the horse needs space to lower its head, at least to shoulder height, because failure to be able to do so has a negative impact on the respiratory function during journeys of more than a few hours. Hence, horses during transport require more space than when standing still.

4.3.4.2 Compartmentation

Compartmentation (Figure 4.4) is useful and necessary for several reasons. Horses of significantly different sizes or ages and sexually mature mares and stallions, must be handled and transported separately, unless the animals have been reared together (European Regulation EC 1/2005). The use

Figure 4.4. Compartmentation of a group of loose housed horses (B. Driessen, Research, Control & Consult, Belgium).

of separate compartments improves the travel management and reduces the risk of injuries due to kicking and pawing, especially in temperamental and dominant horses, such as stallions (Iacono *et al.*, 2007b; Weeks *et al.*, 2012). Stallions, in fact, are more aggressive and difficult to handle in comparison with mares and geldings (Roy *et al.*, 2015a). However, within a load of slaughter horses, the proportion of stallions is usually small because of their small proportion in the total population and in commercial practical conditions no more than two stallions are usually included in a load. The majority of the transported horses are mares (50 to 58%) and geldings (40 to 47%) (Roy and Cockram, 2015; Stull, 1999). According to Stull (1999), mainly adult horses (average age of 11.4±0.4 years, ranging from 1 to 30 years) are transported to slaughter.

To our knowledge, no scientific information about the optimal dimensions of a compartment or about the partition design are available. Partitions should be solid, start from the floor level encompassing the whole width of the vehicle and should be designed in such a way to protect and isolate physically individuals and groups, but not reducing the ventilation in the trailer and not increasing the risk of overheating.

4.3.4.3 Restraint during transport

Tied and untied animals should not be transported in the same compartment (European Regulation 1/2005). The ropes or other attachments used must be in good condition, resistant and long enough to allow horses to lie down (AATA, 2005). Short ropes can cause entanglement and loss of balance and subsequent panic (Tasker, 1990) or, in the worst case, death due to choking (Broom, 2014). Furthermore, when the rope is too short, horses have to maintain their heads in an elevated position, which increases the accumulation of purulent airway secretions and the risk of running normal pharyngeal flora into the lower respiratory tract (Raidal *et al.*, 1995, 1996; Raycklyeft and Love, 1990) leading to an increased risk of 'shipping fever'.

4.3.4.4 Water provision

During transport, specifically long hauls (defined by the European Regulation 1/2005 as a haul of more than 8 h (EC, 2005)), horses can be provided with water on the vehicle. Transport in hot weather conditions and without water for more than 24 h will induce severe dehydration, even in healthy horses (Friend, 2000). Even in cold weather conditions, horses must have access to drinking water. However, watering horses during transport is questionable. Horses often start drinking between 20 and 60 min after water is offered (Iacono *et al.*, 2007a), but their water intake is low (Iacono *et al.*, 2007a; Mars *et al.*, 1992). Some horses do not drink at all because of the fear of the new, unfamiliar water source (Weeks *et al.*, 2012), taste of the water or transport stress. Horses often may be reluctant to drink when inside a moving vehicle, since to do so they need to adopt a vulnerable position, such as risking banging into the drinkers with their mouth and lips (McGreevy, 2004). Furthermore, providing water in moving trucks may result in slippery floor due to water spillage and thus increasing the risk of poor stability, slips and injuries. Thus, water provision during stops is preferable (Messori *et al.*, 2016).

4.3.4.5 Vehicle driving style

When the transport vehicle is driven smoothly, it is easier for the horses to keep their balance and to avoid contact with the vehicle interior and or others. When the driver negotiates corners and bends at high speed, or drives with too high accelerations or to violent and sudden braking, horses are subjected to substantial lateral and or cranio-caudal movements. This poor driving style may also lead to toppling, sliding and the requirement for excessive corrective muscular action, resulting in bruising, muscular fatigue (as showed by increased muscle enzymes released in the blood), fear and injuries to the animals (Randall *et al.*, 1995; Tateo *et al.*, 2012). In addition, poor condition of the brakes can cause improper braking and loss of balance which can cause the horse to be propelled forward or backwards inside the trailer with consequent injuries to the front of the head (Dai *et al.*, 2021).

4.3.4.6 Journey duration

Medica *et al.* (2019) noticed that the sympatho-adrenal system response (based on blood adrenaline and noradrenaline concentration) of stallions transported in in a trailer featuring individual compartments was greater after short than medium-longer period (100 vs 200-300 km) of transportation which may indicate an adaptation process during transport to alleviate the effects of the perceived transportation stress. In addition, they noticed also a great variability of the sympatho-adrenergic response in different individuals, which may reflect individual variations in response to transport-induced stress (Medica *et al.*, 2019).

There is a wide variation in the length of the legally allowed journey duration. In the current European legislation, long journeys are defined as those exceeding 8 h (European Regulation EC 1/2005). However, this legislation defines that horses may be transported for a maximum period of 24 h (European Regulation EC 1/2005) if they are watered and fed every 8 h. After this journey time, animals must be unloaded, fed and watered and be rested for at least 24 h. Transporting unbroken horses on long journeys is not allowed (European Regulation EC 1/2005). In Canada, the maximum transport time is 28 h without water, feed and rest (CFIA, 2021). Thereafter the horses must have a rest period of at least 8 consecutive hours before resuming their journey. In the US, horses must be offloaded after a maximum duration in confinement of 28 h and be provided rest, water, and feed for at least 6 h (USDA, 1994). Argentinian legislation allows for a transport duration up to 36 h without feeding, watering or rest (SNSCA, 1999). In Uruguay, no maximum transport duration is mentioned (Uruguay Government, 2009).

Journey duration affects horse welfare because of some additive factors, such as restricted space and comfort, fasting and environmental conditions (Broom, 2014). For example, horses that have been fasted for a long period (i.e. during a long journey) are more vulnerable to cold temperatures (Roy and Cockram, 2015). Miranda-de la Lama *et al.* (2021) reported increased prevalence of bruising in horses transported for more than 12 h, independently of the country of origin, lairage duration, vehicle type, commercial category or sex. For this reason, these authors concluded that long journeys should be planned with frequent and long stops, during which horses should be fed and watered. If animals have adequate rest, food and water at resting points, the adverse effects of long journeys are reduced (Broom, 2014). According to European Regulation (EC 1/2005), horses must be given water during long journeys every 8 h without unloading. In addition, stops would also allow vehicle cleaning to prevent horses from slipping during the journey (Padalino, 2015).

4.4 Handling at the slaughter plant

4.4.1 Unloading

After arrival at the slaughter plant, the horses should be unloaded as quickly as possible, especially from passively-ventilated vehicles during hot weather, because the temperature inside these vehicles will increase rapidly whilst stationary (Grandin, 2014c).

The unloading procedure can be hindered by horses' unwillingness to go downhill and their response to an unfamiliar environment characterised by specific sounds and smells, interactions with other animals, etc. (Grandin, 2014c; Messori *et al.*, 2016). As the result of previous experiences at loading or during transport, horses can exhibit reluctance to exit the vehicle or leave it at excessive speed (Siniscalchi *et al.*, 2014). While exiting the vehicle, horses should be allowed additional time to inspect the new environment (e.g. the ramp) and to walk off quietly and calmly, so that the risk of injury is reduced (Messori *et al.*, 2016). To this end, the handler(s), including the driver, should understand the behavioural signals given by the animals during unloading and guiding and adapt their actions accordingly. They also need to assess the effect of their actions on the animals, not only on the horse on which he is focused, but also the horses in the surrounding area which will also pick up the signals and may be stressed by these (Nivelle *et al.*, 2020). Handling aids, such as flags or driving paddles should also be used properly. They should only be used as an extension of the handler's body and never to hit an animal (https://www.hsa.org.uk/; Nivelle *et al.*, 2020). Furthermore, to ease handling out of the vehicle, the exit gate should be as large as possible and the door must be open at its maximum width (SCARM, 2002). Grandin (1980) recommends a width of 2.5-3.0 m at the unloading facility. This provides a wide, clear path for animals to move off the vehicle without crowding and tripping (Grandin, 2008). The minimum door/gate width is not defined, but based on an average-size horse (158 cm tall, approximately 90 cm large and 550 kg body weight; Waran *et al.*, 2007), a walkway width of about 1 m would permit the passage of one horse at a time (CECM, 1987).

At unloading, horses should be observed by the personnel at the slaughter plant to identify unfit, ill or injured animals. If injured horses are detected in the vehicle, sound stock should be unloaded first. In the case of severe injuries, emergency killing and bleeding must be carried out in the vehicle or on the unloading platform. Less severely injured horses should be slaughtered immediately without keeping them for a period in lairage. Holding injured animals with obvious signs of stress or of pain arriving at night or on weekends until the slaughter floor is operating is unacceptable (SCARM, 2002).

When manoeuvred to pull it over to the unloading dock, the vehicle should be correctly aligned with the edge and walls of the dock to avoid gaps (SCARM, 2002). A ramp is needed to fill a gap between the vehicle deck and the unloading dock. Because of the significant correlation (r=0.39) between the steepness of the ramp (steeper than an angle of 20° or 36.4% to the horizontal for horses which is the maximum level specified in European Regulation 1/2005) and the prevalence of falling, the ramp floor should be covered by a rubber to reduce slipping (Messori *et al.*, 2016).

4.4.2 Path to the lairage pens

To avoid stops in the flow of horses to the lairage pens, alleys design should not feature corners. Wide alleys from 3 to 3.5 m as used in cattle facilities can be used to drive the horses from the unloading dock to the pens (Grandin, 2014d). Horses are social animals and experience stress when separated, therefore it is recommended to move them in small groups to help keep them calm (Grandin, 1999). Also the floor should be made of solid, non-slip concrete and be uniform in texture or colour. Solid walls are recommended to calm down nervous horses as they impede visual contacts with other horses (Grandin, 2014c). Rubber or coatings can be used to avoid metal on metal contacts and to dampen loud noises of metal gates, locks and doors.

4.4.3 Lairage time and holding pens design

Recommendations about the lairage time for horses are rare and optimum lairage times, in terms of processing requirements, have not been studied for most horses (FAWC, 2003). Under commercial conditions, after arrival at the slaughter plant, horses are either slaughtered immediately, after a few hours of lairage, after an overnight lairage or are kept some days (Roy *et al.*, 2015a). A minimum lairage time of 6 h for horses is recommended in Chile (Werner and Gallo, 2008). However, Werner and Gallo (2008) state that, in terms of animal welfare, the slaughtering of horses soon after the arrival to the slaughterhouse is good practice, because increasing lairage time results in increasing stress (e.g. noise, fear, thirst, hunger, etc.) and deterioration in the quality of the horse meat. Meat quality losses can occur as a result of excessive stress, bruises or injuries (CATGP, 2018; Nivelle *et al.*, 2020; Pawshe *et al.*, 2016). For the same reasons, regrouping unfamiliar horses should be avoided because of the risk of fighting (Cavallone *et al.*, 2002).

During lairage, horses must have access to drinking water and, should they be kept in lairage for a day or more, they must be fed with forage (European Regulation 1/2005; http://www.oie.int/international-standard-setting/terrestrial-code/accessonline/).

Pens should be long and narrow to allow efficient animal movement, with horses entering through one end of the pen and leaving through the other, and reduce fighting (Grandin, 1990).

Under adverse weather conditions (solar radiation or heavy rainfall), the presence of shelter in the holding pens is recommended to provide shade (e.g. trees) and prevent the floor from becoming soggy/muddy in case the holding pens are not or partially roofed. To prevent muddy, slippery floors, the floor should be permeable to drain water surplus quickly. To ensure adequate drainage, slightly slope the surface of the pen. Laying porous material, such as gravel, crushed stone or sand, on a geotextile in muddy pens, allows drainage of water away from the surface. The geotextile ensures that the layer of porous material is kept in place and lasts longer (Living on the Land, 2021).

4.4.4 The path to the stun box

In the passageway leading to the stun box, horses, which are herd animals, have to follow each other one by one in line, which can induce fear and agitation (Grandin, 1999). Fearful horses can be guided to the stunner using a halter or a head rope. According to Micera *et al.* (2012), applying mentholated ointment on the horse nostrils 45 min before stunning can decrease its olfactory perception of danger at slaughter and reduces the stress response as showed by the lower concentrations of epinephrine and norepinephrine in blood.

In some slaughterhouses, horses are showered (Figure 4.5) before entering the stun box in order to reduce the amount of dust in their coats. However, this practice makes the passageway more slippery and produces stress and panic, i.e. kicking or flight reaction, in horses as they are not familiar to showers (Grandin, 2014c; Vermeulen *et al.*, 2019). A shower that automatically starts up with low pressure and produces a very fine water spray is recommended to reduce stress-related behaviour (B. Driessen, unpublished data).

In lairage, the most important requirement is to reduce disruptive noises. High-pitched noises caused by metal on metal or metal on concrete are the main ones to be avoided as they excite the animals making them more difficult to handle and their stunning less effective (Bulens *et al.*, 2013; Grandin, 2015c). The reduction of these types of noises can be achieved by applying damping rubber

igure 4.5. In some slaughterhouses, horses are showered before slaughter, but this practice can induce stress (B. Driessen, esearch, Control & Consult, Belgium).

to the walls and floors of loading docks, latches, weighing scales, etc. In addition, the lairage area can be separated from the stun box by using sandwich panels, so that disturbing noises coming from the slaughtering area that would cause stress to the horses in the stun box are reduced (Cockram, 2020; Grandin, 2020a).

Horses are flight animals and may only enter the stun box voluntarily if:

- There is no solid wall is in front of them (http://www.oie.int/international-standard-setting/terrestrial-code/accessonline/).
- There are solid sight panels to prevent external distractions (http://www.oie.int/international-standard-setting/terrestrial-code/accessonline/).
- The floor is non-slip and stable (Houpt and Wickens, 2014).
- Lamps light up a dark stun box. LED lighting is recommended as it reduces shadows, but should not be directed in the horse's eyes (Grandin, 2020a).
- There are no light spots causing light reflections on stainless steel (Grandin 2020a).

4.5 Stunning and slaughter

During the stunning phase, the welfare of the horses is determined by the design of the stun box, the calmness of the horses, the type of stunner used, the skilfulness of the operators and the bleeding technique. Regularly training of personnel is necessary to maintain the experiences and understanding about handling (based on behavioural principles, such as flight zone and point of balance), restraining, stunning techniques, signs of (un)consciousness, hoisting, bleeding, and equipment care and maintenance (European Regulation 1/2005; https://www.hsa.org.uk/).

4.5.1 Stun box

The function of the stun box is to restrict the movements of the horse sparing avoidable pain, fear or agitation in order to facilitate effective stunning and killing and to ensure the safety of operators and the proper application of stunning (European Regulation 1/2005). However, restraining is likely to create distress to the animals and should therefore be applied for as short a period as possible. To this end, business operators are required to ensure that animals are not placed in the box until the person in charge of stunning or bleeding is ready to perform these procedures as quickly as possible. However, a specific maximum period between restraining and stunning is not determined in the European Regulation 1/2005.

Stun boxes should be designed to allow horses enter willingly with minimal stress. The appropriate width for stun boxes and chutes tends to be overestimated (AVMA, 2016). The box should not be too wide to prevent an horse from turning around. The European Regulation 1/2005 states that the restraint facilities must be adjustable to the horse size, but does not provide any recommendation for stun box maximum dimensions (EC, 2005). Whereas, AVMA (2016) recommends an appropriate width of 81 cm for a horse stun box. The box must also be easily and safely accessible to handle an animal that is lying down or has fallen (SCARM, 2002).

Clamping the horse's head by a mechanised device is an active restraint form and should not be used (AVMA, 2016). Restraining or trying to restrain a horse's head results in more fear and panic, and due their long neck head restraint cannot be done safely (EC, 2017; Harvey, 2020). Passive devices, such as a tray to prevent the horse from putting its head down, which restrain head movement without clamping the head can be used (AVMA, 2016).

4.5.2 Stunning

Unlike most meat species, there is limited information on the stunning of horses (Gibson *et al.*, 2015). To induce immediate insensibility in horses a penetrating captive bolt gun powered by gunpowder (Western Europe) or compressed air (South America; Figure 4.6) or a long rifle (Canada) can be used (Algers and Atkinson, 2014; Gibson *et al.*, 2015). The length of the bolt and the power of the weapon determine the depth of penetration into the skull (Algers and Atkinson, 2014). Pneumatic/compressed bolt guns have a higher power and bolt velocity than a gun powered captive bolt (Algers and Atkinson, 2014). The use of a non-penetrating captive bolt and the electronarcosis techniques have been also reported by Cáraves and Gallo (2007) in Chilean horse slaughterhouses, but without good results. Failures in gun powered or compressed air captive bolt stunning have been explained by the use of damp cartridges (Grandin, 2015c) and too low air pressure (Bulens *et al.*, 2013), respectively. When using a rifle, ricochet of the bullet is possible and therefore, the operators must use care in positioning of themselves and others when the procedure is performed (see species-related chapters in this book: Gallo *et al.*, 2022; Hultgren, 2022; Needham and Hoffman, 2022; Shearer and Ramirez, 2013).

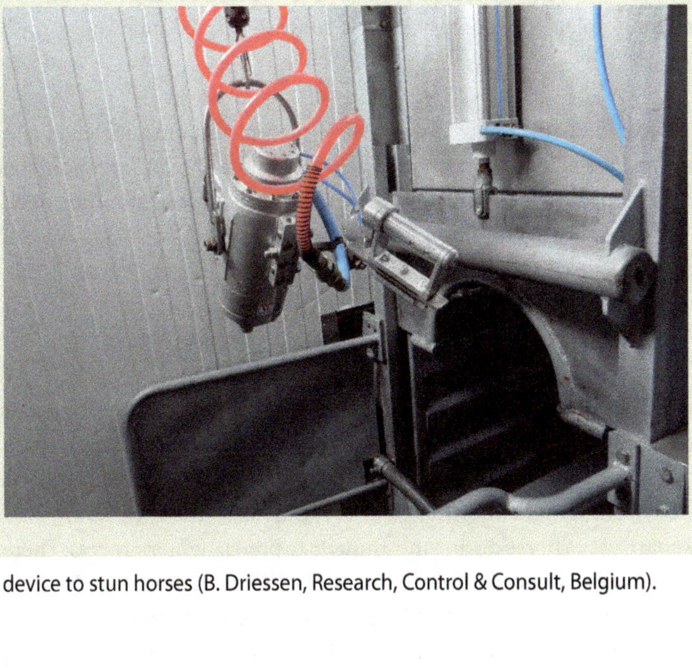

Figure 4.6. A pneumatic device to stun horses (B. Driessen, Research, Control & Consult, Belgium).

For penetrating captive bolt stunning, the device is held against the animal's forehead so that the bolt can penetrate the skull and brain. Because the brain of the horse is located high on the forehead, the shot must be placed slightly above the intersection of an X drawn on the forehead, with each line beginning at the inside corner of the eye and progressing to the opposite ear (Woods and Shearer, 2015). When properly applied, the shot results in immediate unconsciousness and breathing stop due to trauma to the cerebral hemisphere and the brainstem (Grandin, 2020b), while if the location is not precise, horses may regain consciousness during exsanguination (Werner and Gallo, 2008). However, there have been some questions about whether or not a captive bolt actually kills or just stuns an animal as most horses continue kicking immediately after being killed. An untrained person might confuse these spasms with movements and wrongly assumes that the shot animal would not be stunned (Bulens *et al.*, 2013). The stunning operative must be aware that stressed or agitated horses may lunge forward or rear up when shot (Woods and Shearer, 2015). Similarly to cattle (Finnie, 1995; Gregory *et al.*, 2007), signs of effective captive bolt penetration in horses are immediate collapse and a several seconds period of tetanic spasm, followed by slow hind limb movements of increasing frequency, absence of corneal reflex and open eyes.

4.5.3 Slaughter

After successful stunning, the animal falls into the roll-out area, is hoisted up and exsanguinated immediately to prevent the horse from regaining consciousness (Algers and Atkinson, 2014). The acceptable maximum stun-to-stick interval is 60 s (https://www.hsa.org.uk/). Exsanguination should be performed using a pointed and very sharp knife with a rigid blade (Woods and Shearer, 2015) that severs the jugular vein, carotid artery, and windpipe. In addition, a chest stick can be performed to optimise bleeding. When properly performed, blood should flow freely (Woods and Shearer, 2015). Great care must be practiced while performing exsanguination because horses are still capable of making violent, involuntary, muscle contractions that may injure the handler (Woods and Shearer, 2015).

Animals must be observed during the stun-to-stick interval and during bleeding to detect signs of ineffective stunning. Presence of the corneal reflex, recovery of breathing, spontaneous blinking of the eye lids, vocalisation and the righting reflex are signs of ineffective stunning or slaughter (Bulens *et al.*, 2013). Subsequent port-mortem operations, such as carcass dressing or scalding, shall only be performed once the absence of signs of life of the animal has been verified (Bulens *et al.*, 2013).

4.6 Conclusions

Similarly to other livestock species, the welfare of horses is influenced by a complex range of parameters just before transport (e.g. fitness for transport), during transport to the slaughterhouse, during holding in lairage and in the final hours of the slaughter process. Infrastructures, animal-related factors, environmental conditions, but certainly also the interaction with the handlers, determine the welfare of the horse during the immediate pre-slaughter process. There are similarities with the transport of recreational and sport horses, but the main difference is that slaughter horses are often transported in loose groups at a higher stocking density and in less well adapted vehicles. From region to region, there are also differences in preslaughter handling of slaughter horses in practice. These differences are based on years of habits, regulations and specific regional needs. To obtain more knowledge and to provide more scientific evidence-based recommendations, more research is needed.

References

Animal Transportation Association (AATA), 2005. AATA manual for the transportation of live animals by road. T.C. Harris, Redhill, UK, 94 pp.

American Veterinary Medical Association (AVMA), 2016. The AVMA guidelines for the humane slaughter of animals. AVMA, Schaumburg, IL, USA, 64 pp.

Algers, B. and Atkinson, S., 2014. Mechanical stunning. In: Devine, C. and Dikeman, M. (eds.) Encyclopedia of meat sciences. Elsevier, Oxford, UK, pp. 413-417.

Australian Animal Welfare Standards and Guidelines, 2021. Land transport of livestock. Available at: http://www.animalwelfarestandards.net.au/land-transport/

Banu, C., 2009. Food industry treatise. ASAB Publishing House, Bucharest, Romania, pp. 113-119.

Broom, D.M. and Johnson, K.G., 1993. Stress and animal welfare. Chapman and Hall, London, UK, 211 pp.

Broom, D.M., 2008. The welfare of livestock during road transport. In: Appleby, M., Cussen, V., Garcés, L., Lambert, L. and Turner J. (eds.) Long distance transport and the welfare of farm animals. CAB International, Wallingford, UK, pp. 157-181.

Broom, D.M., 2014. Welfare of transported animals: factors influencing welfare and welfare assessment. In: Grandin, T. (ed.) Livestock handling and transport. CAB International, Wallingford, UK, pp. 23-38.

Bulens, A., Van Beirendonck, S., Van Thielen, J. and Driessen, B., 2013. Functionarissen voor het dierenwelzijn: handboek. KU Leuven|Thomas More Kempen, Geel, Belgium, 144 pp.

Canadian Food Inspection Agency (CFIA), 2021. Health of animals regulations. Part XII: transport of animals-regulatory amendment. Interpretive guidance for regulated parties. Available at: https://tinyurl.com/2385wnt8

Cáraves, M. and Gallo, C., 2007. Characterization and evaluation of the stunning systems used for horses in Chile. Archivos de Medicina Veterinaria 39: 105-113.

Casamassima, D., Palazzo, M., Presutti, T. and Cinone, M., 2008. Parametri fisiologici in cavallo di razza araba sottoposti a prove di salita sul trailer utilizzando due tipi di doma. Ippologia 3: 13-19.

Cavallone, E., Di Giancamillo, M., Secchiero, B., Belloli, A., Pravettoni, D. and Rimoldi, E.M., 2002. Variations of serum cortisol in Argentine horses subjected to ship transport and adaptation stress. Journal of Equine Veterinary Science 22: 541-545.

Clark, D.K., Friend, T.H. and Dellmeier, G., 1993. The effect of orientation during trailer transport on heart rate, cortisol and balance in horses. Applied Animal Behaviour Science 38: 179-189.

Cockram, M., 2020. Condition of animals on arrival at the abattoir and their management during lairage. In: Grandin, T. and Cockram, M. (eds.) The slaughter of farmed animals. Practical ways of enhancing animal welfare. CAB International, Wallingford, UK, pp. 49-77.

Consortium of the Animal Transport Guides Project (CATGP), 2018. Guide to good practices for the transport of horses destined for slaughter. Available at: http://animaltransportguides.eu/wp-content/uploads/2016/05/EN-Guides-Horses-final.pdf

Council of Europe Committee of Ministers (CECM), 1987. Recommendation No. R (87) 17 of the Committee of ministers to member states on the transport of horses. Available at: https://tinyurl.com/zvasvuav

Collins, M.N., Friend, T.H., Jousan, F.D. and Chen, S.C., 2000. Effects of density on displacement, falls, injuries, and orientation during horse transportation. Applied Animal Behaviour Science 67: 169-179.

Cregier, S.E., 1982. Reducing equine hauling stress: a review. Journal of Equine Veterinary Science 2: 186-198.

Cross, N., Versnel, C., Cawdell-Smith, J. and Phillips, C., 2008. Effects of lighting conditions on the welfare of horses being loaded for transportation. Journal of Veterinary Behavior: Clinical Applications and Research 3: 20-24.

Cuddeford, D., 2003. Equine nutrition, Crowood Press Ltd, Marlborough, UK, 160 pp.

Dai, F., Zappaterra M., Minero, M., Bocchini, F., Riley, C.B. and Padalino, B., 2021. Equine transport-related problem behaviors and injuries: a survey of Italian horse industry members. Animals 11: 223. https://doi.org/10.3390/ani11010223

Doherty, O., Booth, M., Waran, N., Salthouse C. and Cuddeford, D., 1997. Study of the heart rate and energy expenditure of ponies during transport. Veterinary Record 6: 589-592.

Ekesbo, I. and Gunnarsson, S., 2018. Farm animal behaviour: characteristics for assessment of health and welfare. CAB International, Wallingford, UK, 342 pp.

European Commission (EC), 2005. European Regulation (EC) No 1/2005 of 22 December 2004 on the protection of animals during transport and related operations and amending. Official Journal of the European Union L 3, 5.1.2005: 1-44.

European Commission (EC), 2017. Preparation of best practices on the protection of animals at the time of killing. Final Report. Publications Office of the European Union, Luxembourg, Luxembourg, 397 pp.

Farm Animal Welfare Council (FAWC), 2003. Report on the welfare of farmed animals at slaughter or killing. Part 1: red meat animals. Defra Publications, London, UK, 72 pp.

Faucitano, L. and Lambooij, B., 2019. Transport of pigs. In: Grandin, T. (ed.) Livestock handling and transport. CAB International, Wallingford, UK, pp. 307-327.

Fazio, E., Medica, P., Cravana, C. and Ferlazzo, A., 2013. Cortisol response to road transport stress in calm and nervous stallions. Journal of Veterinary Behavior 8: 231-237.

Finnie, J.W., 1995. Neuropathologic changes produced by non-penetrating captive bolt percussive captive bolt stunning of cattle. New Zealand Veterinary Journal 43: 183-185.

Fraser, A.F., 2010. The behaviour and welfare of the horse. CAB International, Oxfordshire, UK, 255 pp.

Friend, T.H., 2000. Dehydration, stress, and water consumption of horses during long-distance commercial transport. Journal of Animal Science 78: 2568-2580.

Friend, T.H., Martin, M.T., Householder, D.D. and Bushong, D.M., 1998. Stress responses of horses during a long period of transport in a commercial truck. Journal of the American Veterinary Medical Association 212: 838-844.

Gallo, C., Schwartzkopf-Genswein, K. and Gibson, T., 2022. Cattle. In: Faucitano, L. (ed.) Preslaughter handling and slaughter of meat animals. Wageningen Academic Publishers, Wageningen, the Netherlands, pp. 62-116.

Garifulovich, K.S., Kaspievna, M.S., Satyvaldievich, A.R. and Kuralbaevna, T.G., 2013. About the status of meat industry in Kazakhstan and in the World, Middle-East. Journal of Scientific Research 17: 434-439.

Gaughan, J.B., Mader, T.L., Holt, S.M. and Lisle, A., 2008. A new heat load index for feedlot cattle. Journal of Animal Science 86: 226-234.

Gibson, T.J., Bedford, E.M., Chancellor, N.M. and Limon, G., 2015. Pathophysiology of free-bullet slaughter of horses and ponies. Meat Science 108: 120-124.

Grandin, T., 1980. Livestock behavior as related to handling facilities design. International Journal for the Study of Animal Problems 1: 33-52.

Grandin, T., 1989. Behavioural principles of livestock handling. Professional Animal Scientist 5: 1-11.

Grandin, T., 1990. Design of loading facilities and holding pens. Applied Animal Behaviour Science 28: 187-201.

Grandin, T., 1999. Safe handling of large animals (cattle and horses). Occupational Medicine 14: 195-212.

Grandin, T., 2008. Engineering and design of holding yards, loading ramps and handling facilities for land and sea transport of livestock. Veterinaria Italiana 44: 235-245.

Grandin, T., 2014a. A whole systems approach to assessing animal welfare during handling and restraint. In: Grandin, T. (ed.) Livestock handling and transport. CAB International, Wallingford, UK, pp. 1-13.

Grandin, T., 2014b. Behavioural principles of handling cattle and other grazing animals under extensive conditions. In: Grandin, T. (ed.) Livestock handling and transport. CAB International, Wallingford, UK, pp. 39-64.

Grandin, T., 2014c. Improving welfare and reducing stress on animals at slaughter plants. In: Grandin, T. (ed.) Livestock handling and transport. CAB International, Wallingford, UK, pp. 421-450.

Grandin, T., 2014d. Behaviour principles of handling beef cattle and the design of corrals, lairages, races and loading ramps. In: Grandin, T. (ed.) Livestock handling and transport. CAB International, Wallingford, UK, pp. 80-109.

Grandin, T., 2015a. Welfare during transport of livestock and poultry. In: Grandin, T. (ed.) Improving animal welfare: a practical approach. CAB International, Wallingford, UK, pp. 222-246.

Grandin, T., 2015b. Improving livestock, poultry, and fish welfare in slaughter plants with auditing programs. In: Grandin, T. (ed.) Improving animal welfare: a practical approach. CAB International, Wallingford, UK, pp. 164-193.

Grandin, T., 2015c. How to improve livestock handling and reduce stress. In: Grandin, T. (ed.) Improving animal welfare: a practical approach. CAB International, Oxfordshire, UK, pp. 69-95.

Grandin, T., 2020a. Behavioural principles of stockmanship and abattoir design. In: Grandin, T. and Cockram, M. (eds.) The slaughter of farmed animals. Practical ways of enhancing animal welfare. CAB International, Wallingford, UK, pp. 90-110.

Grandin, T., 2020b. Recommended captive bolt stunning techniques for cattle. Available at: https://www.grandin.com/humane/cap.bolt.tips.html

Gregory, N.G., Lee, C.J. and Widdicombe, J.P., 2007. Depth of concussion in cattle shot by penetrating captive bolt. Meat Science 77: 499-503.

Harper, M., 2017. Ethical question of the month – December 2017. Canadian Veterinary Journal 58: 1253.

Harvey, A., 2020. End-of-life options and the problems with horse slaughter. Available at: https://horsesandpeople.com.au/end-of-life-options-for-horses-and-the-problems-with-slaughter/

Hoopes, K.H., 2018. Caring for horses in cold weather. Utah State University, Logan, UT, USA, 4 pp.

Horse Meat Federation, 2016. Manual for the animal welfare of horses during transport and slaughtering. Available at: https://tinyurl.com/2ue22bf2

Houpt, K.A. and Houpt, T.R., 1988. Social and illumination preferences of mares. Journal of Animal Science 66: 2159-2164.

Houpt, K.A. and Wickens, C.L., 2014. Handling and transport of horses. In: Grandin, T. (ed.) Livestock handling and transport. CAB International, Wallingford, UK, pp. 315-341.

Hultgren, J., 2022. Avoiding live-animal transport to slaughter: mobile abattoirs. In: Faucitano, L. (ed.) Preslaughter handling and slaughter of meat animals. Wageningen Academic Publishers, Wageningen, the Netherlands, pp. 390-434.

Iacono, C.M., Friend, T.H., Johnson, R.D., Krawczel, P.D. and Archer, G.S., 2007a. A preliminary study on the utilization of an on board watering system by horses during commercial transport. Applied Animal Behaviour Science 105: 227-231.

Iacono, C.M., Friend, T., Keen, H., Martin, T. and Krawczel, P., 2007b. Effects of density and water availability on the behaviour, physiology, and weight loss of slaughter horses during transport. Journal of Equine Veterinary Science 27: 355-361.

Ireland, J., Clegg, P., McGowan, C., McKane, S., Chandler, K. and Pinchbeck, G., 2012. Comparison of owner-reported health problems with veterinary assessment of geriatric horses in the United Kingdom. Equine Veterinary Journal 44: 94-100.

Jonsson, L., Roepstorff, L., Egenvall, A., Nasholm, A., Dalin, G. and Philipsson, J., 2013. Prevalence of clinical findings at examinations of young Swedish warmblood riding horses. Acta Veterinaria Scandinavica 55: 34.

Knowles, T.G., Brown, S.N., Pope, S.J., Nicol, C.J., Warriss, P.D. and Weeks, C.A., 2010. The response of untamed (unbroken) ponies to conditions of road transport. Animal Welfare 19: 1-15.

Leadon, D., Waran, N., Herholz, C. and Klay, M., 2008. Veterinary management of horse transport. Veterinaria Italiana 44: 149-163.

Lenz, T.R., 2009. The unwanted horse in the United States: an overview of the issue. Journal of Equine Veterinary Science 29: 253-258.

Living on the Land, 2021. Reduce mud & keep water clean. Sacrifice areas. Available at: https://s3.wp.wsu.edu/uploads/sites/2079/2014/02/sacrifice-areas.pdf

Luz, C.S.M., Fonseca, W.J.L., Vogado, G.M.S., Fonseca, W.L., Oliveira, M.R.A., Sousa, G.G.T., Farias, L.A. and Sousa Júnior, S.C., 2015. Adaptative thermal traits in farm animals. Journal of Animal Behaviour and Biometeorology 4: 6-11.

Mal, M.E., Friend, T.H., Lay, D.C., Vogelsang, S.G. and Jenkins O.C., 1991. Behavioural responses of mares to short-term confinement and social isolation. Applied Animal Behaviour Science 31: 13-24.

Marlin, D., Kettlewell, P., Parkin, T., Kennedy, M., Broom, D. and Wood, J., 2011. Welfare and health of horses transported for slaughter within the European Union Part 1: methodology and descriptive data. Equine Veterinary Journal 43: 78-87.

Mars, L.A., Kiesling, H.E., Ross, T.T., Armstrong, J.B. and Murray, L., 1992. Water acceptance and intake in horses under shipping stress. Journal of Equine Veterinary Science 12: 17-20.

McGreevy, P., 2004. Equine behavior. Journal of Equine Veterinary Science 24: 397-398.

McGreevy, P., 2012. Equine behavior – a guide for veterinarians and equine scientists. W.B. Saunders, London, UK, 378 pp.

Medica, P., Bruschetta, G., Cravana, C., Ferlazzo, A. and Fazio, E., 2019. Effect of transportation on the sympatho-adrenal system responses in horses. Research in Veterinary Science 125: 401-404.

Messori, S., Visser, E.K., Buonanno, M., Ferrari, P., Barnard, S., Borciani, M. and Ferri, N., 2016. A tool for the evaluation of slaughter horse welfare during unloading. Animal Welfare 25: 101-113.

Micera, E., Moramarco, A.M. and Zarrilli, A., 2012. Reduction of the olfactory cognitive ability in horses during preslaughter: stress-related hormones evaluation. Meat Science 90: 272-275.

Miranda-de la Lama, G.C., González-Castro, C.A., Gutiérrez-Piña, F.J., Villarroel, M., Maria, G.A. and Estévez-Moreno, L.A., 2021. Horse welfare at slaughter: a novel approach to analyse bruised carcasses based on severity, damage patterns and their association with pre-slaughter risk factor. Meat Science 172: 108341. https://doi.org/10.1016/j.meatsci.2020.108341

Morgan, K., 1998. Thermoneutral zone and critical temperatures of horses. Journal of Thermal Biology 23: 59-61.

Needham, T. and Hoffman, L.C., 2022. Species destined for non-traditional meat production: 1. African game species, cervids, ostriches, crocodiles and kangaroos. In: Faucitano, L. (ed.) Preslaughter handling and slaughter of meat animals. Wageningen Academic Publishers, Wageningen, the Netherlands, pp. 312-347.

Neveux, C. Melac, E., Ferard, M. and Pousset, N., 2017. LED lighting eases and secures the loading of horses and appeases them during confinement in a stationary horse trailer. Proceedings of the 13th International Equitation Science Conference, Wagga, Australia.

Nivelle, B., Vermeulen, L., Van Beirendonk, S., Van Thielen, J. and Driessen, B., 2020. Horse transport to three South American horse slaughterhouses: a descriptive study. Animals 10: 602. https://doi.org/10.3390/ani10040602

Padalino, B., 2015. Effects of the different transport phases on equine health status, behavior, and welfare: a review. Journal of Veterinary Behavior 10: 272-282.

Padalino, B., 2017. Transportation of horses and the implications for health and welfare. PhD-thesis, University of Sydney, Sydney, Australia.

Padalino, B., Davis, G.L. and Raidal, S.L., 2020. Effects of transportation on gastric pH and gastric ulceration in mares. Journal of Veterinary Internal Medicine 34: 922-932.

Palenik, S., Blechova, H. and Palanska, O., 1980. Chemical composition and quality of the meat of cold and warm blooded foals. Zirocisna Vyroba 25: 269-278.

Pawshe, M.D., Badhe, S.R., Khedkar, C.D, Pawshe, R.D. and Pundkar, A.Y., 2016. Horse meat. In: Caballero, B., Finglas, P. and Toldrá, F. (eds.) Encyclopedia of food and health. Elsevier, Amsterdam, the Netherlands, pp. 353-356.

Peeters, H., 2020. Horse. The Eriskay Connection, Breda, the Netherlands, 208 pp.

Purswell, J.L., Gates, R.S., Lawrence, L.M., Jacob, J.D., Stombaugh, T.S. and Coleman, R.J., 2006. Air exchange rate in a horse trailer during road transport. Transactions of the ASABE 49: 193-201.

Raidal S.L., Love, D.N. and Bailey, D.G., 1995. Inflammation and increased numbers of bacteria in the lower respiratory tract of horses within 6 to 12 h of confinement with the head elevated. Australian Veterinary Journal 72: 45-50.

Raidal, S.L., Love, D.N. and Bailey, D.G., 1996. Effects of posture and accumulated airway secretions on tracheal mucociliary transport in the horse. Australian Veterinary Journal 72: 45-49.

Randall, J.M., Duggan, J.A. and Alami, M.A., 1995. Influence of motion and vibration on animals. Fleischwirtschaft 75: 158-160.

Raycklyeft, D.J. and Love, D.N., 1990. Influence of head posture on the respiratory tract of healthy horses. Australian Veterinary Journal 67: 402-405.

Robelin, J., Boccard, R., Martin-Rosset, W., Jussiaux, M. and Trillaud-Geyl, A., 1984. Caracteristiques des carcasses et qualités de la viande de cheval. In: Jarrige, R. and Martin-Rosset, W. (eds.) Le cheval. INRA, Paris, France, pp. 601-610.

Rødbotten, M., Kubberød, E., Lea, P. and Ueland, Ø., 2004. A sensory map of the meat universe. Sensory profile of meat from 15 species. Meat Science 68: 137-144.

Roy, R.C. and Cockram, M.S., 2015. Patterns and durations of journeys by horses transported from the USA to Canada for slaughter. Canadian Veterinary Journal 56: 581-586.

Roy, R.C., Cockram, M.S. and Dohoo, I.R., 2015a. Welfare of horses transported to slaughter in Canada: assessment of welfare and journey risk factors affecting welfare. Canadian Journal Animal Science 95: 509-522.

Roy, R.C., Cockram, M.S., Dohoo, I.R. and Riley, C.B., 2015b. Injuries in horses transported to slaughter in Canada. Canadian Journal of Animal Science 95: 523-531.

Roy, R.C., Cockram, M.S., Dohoo, I.R. and Ragnarsson, S., 2015c. Transport of horses for slaughter in Iceland. Animal Welfare 24: 485-495.

Saslow, C.A., 2002. Understanding the perceptual world of horses. Applied Animal Behaviour Science 78: 209-224.

Servicio Nacional de Sanidad y Calidad Agroalimentaria (SNSCA), 1999. Resolución No. 97/1999. Créase el Registro nacional de medios de transporte de animales. características técnicas. habilitación. condiciones para el embarque y transporte. lavado y desinfección. Disposiciones Generales. Available at: https://tinyurl.com/udf7v3bb

Shearer, J.K. and Ramirez, A., 2013. Procedures for humane euthanasia. Iowa State University, Ames, IA, USA, 13 pp.

Siniscalchi, M., Padalino, B., Lusito, R. and Quaranta, A., 2014. Is the left forelimb preference indicative of a stressful situation in horses? Behavior Processes 107: 61-67.

Snow, D.H., Harris, R.C., MacDonald, I.A., Forster, C.D and Marlin, D.J., 1992. Effects of high-intensity exercise on plasma catecholamines in the thoroughbred horse. Equine Veterinary Journal 24: 462-467.

Stanciu, S., 2015. Horse meat consumption – between scandal and reality. Procedia Economics and Finance 23: 697-703.

Standing Committee on Agriculture and Resource Management (SCARM), 2002. Livestock at slaughtering establishments. CSIRO Publishing, Collingwood, Australia, 34 pp.

Stanislawczyk, R. and Znamirowska, A., 2005. Changes in physicochemical properties of horsemeat during frozen storage. Acta Scientiarum Polonorum, Technologia Alimentaria 4: 89-96.

Stull, C.L., 1997. Physiology, balance, and management of horses during transportation. Proceedings of the Horse Breeders and Owner Conference. Red Deer, AB, Canada.

Stull, C.L., 1999. Responses of horses to trailer design, duration, and floor area during commercial transportation to slaughter. Journal of Animal Science 77: 2925-2933.

Stull, C.L., 2001. Evolution of the proposed federal slaughter horse transport regulations. Journal of Animal Science 79: E12-E15.

Stull, C.L., 2012. The journey to slaughter for North American horses. Animal Frontiers 2: 68-71.

Tasker, W.J., 1990. Transport problems. Equine Behaviour 25: 19-20.

Tateo, A., Padalino, B., Boccaccio, M., Maggiolino, A. and Centoducati, P., 2012. Transport stress in horses: effects of two different distances. Journal of Veterinary Behavior: Clinical Applications and Research 7: 33-42.

Taylor, M. and Sieverkropp, E., 2013. The impacts of US horse slaughter plant closures on a western regional horse market. Journal of Agricultural and Resource Economics 38: 48-63.

Terlouw, E.M.C., Arnould, C., Auperin, B., Berri, C., Le Bihan-Duval, E., Deiss, V., Lefèvre, F., Lensink, B.J. and Mounier, L., 2008. Pre-slaughter conditions, animal stress and welfare: current status and possible future research. Animal 2: 1501-1517.

Transport Quality Assurance (TQA), 2011. Commercial transportation of equines to slaughter: final rule. US Federal Register 76: 173. Available at: https://tinyurl.com/u255eyhz

United States Department of Agriculture (USDA), 1994. Transportation of animals. USDA, Washington, DC, USA. Available at: https://www.nal.usda.gov/awic/twenty-eight-hour-law

Uruguay Government, 2009. Ley Uruguaya No. 18.471. Tenencia responsable de animales. Uruguay Government, Monteviedeo, Uruguay. Available at: https://legislativo.parlamento.gub.uy/temporales/leytemp2855352.htm

Van Dierendonck, M.C. and Spruijt, B.M., 2012. Coping in groups of domestic horses – review from a social and neurobiological perspective. Applied Animal Behaviour Science 138: 194-202.

Vermeulen, L., Van Beirendonck, S., Van Thielen, J. and Driessen, B., 2019. A review: today's practices about the fitness for travel on land of horses toward the slaughterhouse. Journal of Veterinary Behavior: Clinical Applications and Research 29: 102-107.

Waran, N.K., 1993. The behaviour of horses during and after transport by road. Equine Veterinary Education 5: 129-132.

Waran, N.K., Robertson, V., Cuddleford, D., Kokoszko, A. and Marlin, D.J., 1996. Effects if transporting horses facing either forwards or backwards on their behaviour and heart rate. Veterinary Record 139: 7-11.

Waran, N., Leadon, D. and Friend, T., 2007. The effects of transportation on the welfare of horses, In: Waran, N. (ed.). The welfare of horses. Springer, Dordrecht, the Netherlands, pp. 125-150.

Weeks, C.A., McGreevy, P. and Waran, N.K., 2012. Welfare issues related to transport and handling of both trained and unhandled horses and ponies. Equine Veterinary Education 24: 423-430.

Werner, M. and Gallo, C., 2008. Effects of transport, lairage and stunning on the concentrations of some blood constituents in horses destined for slaughter. Livestock Science 115: 94-98.

Woods, J. and Shearer, J.K., 2015. Recommended on-farm euthanasia practices. In: Grandin, T. (ed.) Improving animal welfare: a practical approach. CAB International, Wallingford, UK, pp. 194-221.

Photo: Eleonora Nannoni, Università di Bologna, Bologna, Italy

Chapter 5
Pigs

L. Faucitano[1] and M. Raj[2]*

[1]Agriculture and Agri-Food Canada, Sherbrooke Research and Development Centre, 2000 College Street, Sherbrooke, Quebec, J1M 0C8, Canada; [2]Department of Clinical Veterinary Science, University of Bristol, Langford, BS40 5DU, United Kingdom (retired); luigi.faucitano@agr.gc.ca

Abstract

This chapter deals with the following key factors that, both individually and additively, are relevant for animal welfare and pork meat quality: (1) the farm of origin, in terms of housing conditions, characterisation of the pig herd (gender, genetics and slaughter weight), alleys and loading dock design, animal fitness for transport and preparation of pigs for transport, i.e. feed withdrawal; (2) transport, i.e. handling, equipment and design of facilities at loading and unloading, season, travel conditions (travel duration, loading density and driver) and truck design (loading/unloading system and internal ambient control); (3) lairage conditions, i.e. rest time, handling and climate control; and (4) stunning and slaughter, including restraining of animals, application of stunning methods (i.e. electrical and gas stunning) and exsanguination (i.e. stun to stick interval and bleeding technique). Research provided the evidence that withdrawing feed prior to transport helps reduce transport losses and carcass condemnations, short journeys may be more stressful than longer journeys and the use of truck designs featuring tail-gate lifts or hydraulic decks can help reduce the workload of handlers and improve the welfare of pigs during transport. Both short and long lairages can result in increased proportions of bruised carcasses and poor meat quality. However, these effects are exacerbated by the poor control of ambient and housing conditions, mixing of unfamiliar pigs and rough handling. Research is needed to identify stunning systems alternative to the use of CO_2 gas and electrical stunning that are controversial on animal welfare grounds and not entirely satisfactory on products quality grounds, respectively. The objective of this review is to provide science-based evidence that poor management of practices during these preslaughter periods and the lack of training programs for the correct animal handling can result in lower animal welfare and in significant losses for the pork chain economy.

Keywords: animal welfare, handling, meat quality, pigs, slaughter

5.1 Introduction

At each stage of the preslaughter period, starting from the time they leave their home pen and ending with their slaughter, pigs are exposed to different stressors, including on-farm feed withdrawal, loading and transport, mixing, interactions with humans and poor design of facilities (e.g. ramps, alleys and docks). These factors, both individually and/or additively, can contribute to animal losses and non-ambulatory (because of fatigue or injury) animals on arrival at the abattoir or at slaughter (Faucitano, 2018; Goumon and Faucitano, 2017; Ritter *et al.*, 2009), physiological condition, carcass downgrading due to skin lesions and weight losses, and meat quality defects due to abnormal post-mortem muscle acidification (Dalla Costa *et al.*, 2007; Faucitano, 2001, 2018; Rioja-Lang *et al.*, 2019). In an observational study involving more than 300 loads shipped from 60 farms to three different slaughter plants, Dalla Costa *et al.* (2019b) reported a 3% increase in the incidence of non-ambulatory and injured pigs between loading at the farm and at unloading at the plant. Preslaughter stress may also lead to a deterioration of the physiological condition of pigs until slaughter with detrimental effects on meat quality. A number of studies showed increased heart rates (Correa *et al.*, 2010), body temperature (Rocha *et al.*, 2019) and blood lactate concentrations (Edwards *et al.*, 2010, 2011; Rocha *et al.*, 2015) after each handling event (loading, unloading and moving to slaughter) from farm to slaughter, all contributing to poor pork quality production (Faucitano, 2018). In a study involving 19 shipments to slaughter, Dalla Costa *et al.* (2007) showed that the proportion of pigs showing bruises on the body doubled from the farm to slaughter (39 to 83%). Aaslying *et al.* (2013) following shipments from 6 commercial farms also reported an increasing incidence of clear deep and/or long (>3 cm) lesions, mostly due to fighting, from the delivery facility to the abattoir.

This chapter will overview the factors influencing the response of pigs to preslaughter stress and the handling and management practices to apply on them at the farm prior to transport, during transportation, in lairage and at stunning/slaughter at the abattoir, with the objective to limit their impact on animal welfare and pork carcass and meat quality.

5.2 The farm of origin

The farm of origin was the major source of variation of animal losses during transport (25 vs 16% for transporter and packer) reported in an epidemiological study run in Canada (Dewey *et al.*, 2009; Sunstrum *et al.*, 2006). Fitzgerald *et al.* (2009) reported a 0.93% difference in dead-on-arrival (DOA) and non-ambulatory pigs between the poorest and best performing farms, defined in terms of pigs' growth rate and health control. The effect of the farm also largely explains the difference in the ease of handling between batches of pigs observed at the slaughter plant (Grandin and Vogel, 2011) and in the incidence of bruised carcasses (Dalla Costa *et al.*, 2007) and pork meat quality defects (Rocha *et al.*, 2016).

The responsibility of the producer for the occurrence of economic losses from farm to slaughter is in terms of choice of the housing conditions and barn design, and genetics and gender of the herd. Furthermore, at the time of dispatching the animals to slaughter, the producer is responsible for the selection of animals by fitness for transport and weight (in group splitting market situations), the application of pre-transport feed withdrawal and the handling of pigs from the home pen to the truck gate.

5.2.1 Housing conditions

During the grow-to-finish period, pigs are normally kept under intensive, with little environmental enrichment housing conditions, resulting in pigs showing a high degree of reactivity to novel stimuli, less developed social behaviour or increased fearfulness (Faucitano and Goumon, 2018). In contrast, pigs raised in an enriched environment, i.e. lower stocking density, bedding and experience with human handling, appear easier to handle, more resilient to preslaughter stress (based on lower blood lactate levels at slaughter) and less aggressive in mixed group conditions than those raised under intensive housing conditions (Faucitano and Goumon, 2018; Rocha *et al.*, 2016). Recently, Faucitano *et al.* (2020) reported a greater ease of handling, i.e. less round-turns, back-up and stops, and less handler interventions at loading, all resulting in shorter loading time in pigs fattened in enriched conditions (straw bedding and 1.3 m^2/pig) compared to conventionally-housed pigs (slatted floor and 0.79 m^2/pig).

5.2.2 Genetics

Genetics contributes to pigs' perception and adaptation to their environment and to explain their susceptibility to stress (Rydhmer and Lundeheim, 2008). It has been cautioned that over-selection for a single trait, such as the high efficiency of lean growth applied over the last 50 years (Clutter and Brascamp, 1998), may lead to undesirable effects on behaviours and physical capacities of the animals through correlated effects on behaviours, metabolism, and health (Rydhmer and Canario, 2014). When compared to genotypes selected for more fat deposition, pigs from genetic lines selected for lean meat have been reported to be more fearful, more aggressive, more reluctant to move (Rydhmer and Lundeheim, 2008), and, finally, more prone to die or become non-ambulatory prior to slaughter (Dalla Costa *et al.*, 2019b). Genetic-dependent changes in individual conformation can also explain the variation in skin lesions (Driessen *et al.*, 2020b), with more conformed pigs having a more sensitive skin that can be more easily damaged (Candek-Potokar and Skrlep, 2012; Knecht and Duzinski, 2016).

A specific gene, such as the Halothane gene (HAL-1843 gene) causing malignant hyperthermia syndrome, also called Porcine Stress Syndrome (PSS), which is characterised by intense muscular contractions leading to a generalised rigidity of the muscles, rapid increase in body temperature and metabolic acidosis in the tissues, all resulting in increased PSE (pale, soft, exudative) and blood-splashed pork meat, has been shown to be directly related to pigs' response to handling and transportation stress (Fàbrega *et al.*, 2002; Gispert *et al.*, 2000), and stunning (Velarde *et al.*, 2001). Since the 1990s, pigs can now be identified as normal (Hal^N), heterozygous (Hal^{Nn}), or homozygous (Hal^n) for the PSS mutation using a DNA-based test (De Vries *et al.*, 2000). This PSS mutation occurs at a high frequency in certain breeds, such as Piétrain (Gispert *et al.*, 2007). Piétrain terminal sires are still used in the selection programs in order to get leaner and more efficient crossbreds (Weschenfelder *et al.*, 2012). However, Weschenfelder *et al.* (2013a) found a greater level of fatigue (based on greater blood creatine kinase, CK, concentration) at slaughter and more exudative hams and loins from Piétrain Hal^{Nn} crossbred pigs loaded through steep ramps (22 to 41° angle) and transported to slaughter over a short time (45 min) compared with Piétrain Hal^N crossbreds.

Ritter *et al.* (2008b) failed to find evidence of the presence of the HAL-1843 mutation in fatigued and dead pigs tested on arrival at the slaughter plant, which suggests that mutations of other genes (e.g. dystrophin gene; Nonneman *et al.*, 2012) could account for the stress symptoms still found in HAL^{NN} pigs and contribute to explain these transport losses. The mutations of this gene may also explain the greater sensitivity to the halothane gas test in HAL-1843 gene-free pigs and their two-fold higher expression of the non-ambulatory condition after long transport (>1000 km) and simulated aggressive handling at a slaughter plant compared to HAL-1843 gene-free pigs with low sensitivity to the halothane gas test (9.0 vs 19%; Allison *et al.*, 2005, 2006).

5.2.3 Gender

The effects of the most common genders in swine production, i.e. gilts and barrows or castrated males, on the response to handling and transport are unclear, ranging from no difference between these genders in the physiological response as assessed by blood acid-base parameters or rectal temperature to handling using paddles or electric prods (Hamilton *et al.*, 2004) to a greater rise in rectal temperature up to 2 h after handling in barrows than in gilts (Ritter *et al.*, 2005) or from no gender effect to greater losses (DOA and fatigued pigs) after handling and transport in barrows than in gilts loads (Carr *et al.*, 2005; Sutherland *et al.*, 2006). There is also a clear evidence that the proportion of bruised carcasses is higher in barrows than gilts (Driessen *et al.*, 2020b; Mota-Rojas *et al.*, 2006; Warriss and Brown, 1985), which would indicate a gender effect on fighting behaviours in mixed group situation.

Entire males are traditionally raised in some countries (Spain, UK, Ireland and Australia) and will be increasingly produced in a number of other EU countries (Squires *et al.*, 2020) and in Brazil (MAPA, 2020) in response to the welfare concern due to the pain and stress associated with the surgical castration procedure (Von Borrell *et al.*, 2009). The use of entire male pigs is raising interest because they favour the productivity and sustainability of swine production thanks to a greater growth rate, feed conversion and carcass lean yield, and lower environmental impact due to reduced energy costs and manure production (Gispert *et al.*, 2010; Squires *et al.*, 2020). However, besides the need to monitor and control entire male carcasses for boar taint, an off-odour and off-flavour in cooked pork meat, higher levels of aggression (biting and mounting) and greater level of skin lesions and fighting-type bruises have been reported for this pig gender after mixing with unfamiliar conspecifics compared to castrated males and gilts (Teixeira and Boyle, 2014; Warriss and Brown, 1985). Aggressive behaviours and skin lesions can be limited in boars during preslaughter handling and transport by keeping the group intact (no mixing; Rydhmer *et al.*, 2013) or by mixing them with gilts instead of with other entire males (Van Staaveren *et al.*, 2015). However, the expression of aggressive behaviours and skin lesion score may also vary within a group of boars according to their testicular development which, depends, among others, on the social rank of the individual within the group (Giersing *et al.*, 2000). An inverse relationship between testicular function and carcass lesion scores due to fighting in the truck during the wait before unloading within mixed groups of boars has been reported by Wesoly *et al.* (2015), which confirms the evidence that high ranking boars present lower skin lesions compared with low ranked ones that are mounted more frequently (Rydhmer *et al.*, 2006).

Immunocastration through the injection of a GnRF analogue a few weeks before slaughter proved to be a valid alternative to the painful practice of surgical castration, while keeping the growth performance of boars until a few weeks before slaughter and reducing aggressions and sexual behaviours during the late finishing phase at the farm (Fàbrega *et al.*, 2010) and during the preslaughter period (Guay *et al.*, 2013; Rocha *et al.*, 2013). When compared to surgical castrates, immunocastrates vocalise less during handling and loading (Guay *et al.*, 2013) and have similar fighting behaviours when mixed with unfamiliar pigs of the same gender in lairage (Rocha *et al.*, 2013).

5.2.4 Slaughter weight

Market weight has been increasing worldwide over the past decades. For example, average market weight in Canada increased from 105 kg in 1989 to the current weight of 130 kg (Correa, 2011), while in US it increased by 5 kg from 2004 to 2013 (Wu *et al.*, 2017). It is presently forecasted that market weights will get heavier in the years to come based on the weight increase of 0.5 kg/year (NASS, 2014). This increase in market weight is driven by the dilution of fixed production cost through the reduction of the total number of pigs required to produce a given quantity of pork and the improvement of genetic selection for leaner-type pigs (Wu *et al.*, 2017), resulting in better carcass yields and meat to bone ratio (Ellis and Bertol, 2001). However, it has been reported that heavier pigs are less easy to handle than lighter pigs (Bertol *et al.*, 2011; Rocha *et al.*, 2016) and need more handler interventions (i.e. more electric prodding) to move forward. Because of their larger size, heavier pigs are also more likely to get stuck in narrow barn hallways, chutes, or ramps during loading or unloading (Zurbrigg *et al.*, 2017a) which may result in increased risk of transport losses and/or bruised carcasses (Correa, 2011; Passafaro *et al.*, 2019). To reduce the physical effort of heavier pigs (150 kg) and ease their handling through loading/unloading ramps, a maximum ramp slope of 15° angle is recommended (Goumon and Faucitano, 2017). When compared to light pigs, heavier pigs are also particularly vulnerable to heat stress resulting from overcrowding or poorly ventilated environments (Renaudeau *et al.*, 2011) as they generate more body heat (+2% per 5 kg body weight increase) and are less able to dissipate it (Brown-Brandt *et al.*, 2004). Based on significant, although rather weak, correlation (r=0.35) between body weight and conformation and fighting behaviour (Olesen *et al.*, 1996), heaviest and largest pigs in a mixed group fight more.

5.2.5 Animal health and fitness for transport

According to legislation (EC Council Regulation, No 1/2005 (EC, 2005)), guidelines (www.animaltransportguides.eu) and codes of practice (Australian Government, 2012; NFACC, 2014), it is essential that all animals are checked before loading to determine their fitness for transportation. Unfit pigs, i.e. unable to move or keep balance, showing respiratory distress, prolapsed, profuse and continuous bleeding, and in late pregnancy, or having just given birth should not be loaded for transport as these poor health conditions make them weak and with less chance to cope with extended feed deprivation (Zurbrigg *et al.*, 2017a), to move away from aggressions, and more likely to lose balance due to sudden stops or accelerations or changes of direction of the transport vehicle (www.animaltransportguides.eu). This will eventually result in 1.3 greater risk for a pig in this health or physical condition to die during transit (Dalla Costa *et al.*, 2019b) or to become non-ambulatory on arrival at the slaughter plant (Zurbrigg *et al.*, 2017a). Based on data collected in a survey of more than 300 loads (38,000 pigs) transported to three Brazilian slaughter plants, Dalla Costa *et al.* (2019b) reported that of the pigs identified as non-ambulatory after transport (0.38%), half of them were already in this condition at the farm before loading and calculated a 1.32 greater risk of dying during transport in pigs non-ambulatory or presenting hernias or arthritis at the farm before loading.

The relatively small heart size (0.3% of the body weight; Niewold *et al.*, 2000) with bradycardia combined with the rapid growth of modern pigs, heart abnormalities and acute stress during transport can result in tachycardia and death due to acute heart failure (Faucitano and Lambooij, 2019; Zurbrigg *et al.*, 2017a). A pig affected by cardiac lesions, such as hypertrophy, may likely be incapable to respond to the increased cardiac workload required during sorting, loading and transport to slaughter and be more susceptible to dying or becoming non-ambulatory during transport (Zurbrigg *et al.*, 2017a,b, 2018).

Loading pigs with respiratory problems, i.e. pneumonia based on post-mortem lung inspection, has been associated to increased number of panting pigs at unloading at the slaughter plant (Rocha *et al.*, 2016). However, based on the findings from a number of studies reviewed by Zurbrigg *et al.* (2017a), respiratory health does not appear to be a primary cause for transport losses. More recently, Driessen *et al.* (2020a) related the proportion of pigs coughing in the finishing room due to inadequate ventilation to a faster post-mortem acidification rate (lower initial pH) in the loin muscle and greater carcass temperature.

The annual culling rate for sows can be up to 50% in Europe (Herskin *et al.*, 2020a) and more than 50% in North America (Blair and Lowe, 2019), the latter rate corresponding to over 3 million sows shipped to slaughter every year (Blair and Lowe, 2019). Sows are usually culled due to poor body condition, resulting from the multiple parturitions and lactations, lameness or failure to rebreed (Grandin, 2016). Due to these physical conditions, sows may be weakened, suffer from pain and systemic disease, and hardly walk and cope with preslaughter challenges, i.e. being handled on and off the truck and keeping balance and extreme thermal conditions during transport (Cockram, 2019, 2020). Hence, producers must pay particular attention when selecting fit culled sows for transportation as handling and transport may deteriorate pre-existing clinical signs, such as vulva and udder ulcers, skin lesions, wounds and torn hoofs (Thodberg *et al.*, 2019), resulting in a greater risk of arriving fatigued, lame or even dead at the slaughter plant compared with market weight pigs (Peterson *et al.*, 2017).

5.2.6 Feed withdrawal

Feed withdrawal is a recommended practice for on-farm preparation of pigs before slaughter (NFACC, 2014; SCAHAW, 2002) in order to prevent transport losses, hyperthermia and travel sickness, to reduce carcass contamination due to lower risk of gut contents spillage during carcass processing and to improve pork quality (Faucitano, 2018).

Death in unfasted pigs has been either associated to the pressure of the full stomach on the vena cava, which reduces its diameter and decreases the blood circulatory efficiency, in terms of venous return (Warriss, 1998), or to cardiac insufficiency due to impaired blood flow to the gastrointestinal system during the digestion and absorption of food process after eating caused by the additive effect of the exertion of sorting and loading on pigs with full stomach (Zurbrigg *et al.*, 2017a). A 12 h fasting before slaughter can lead to a loss of 1.3 kg of gastrointestinal fluid, which may be sufficient to help pigs cope with the stress of handling and transportation (Warriss, 1998). Indeed, there is evidence that a 12-18 h fasting time can help reduce transport mortality compared to <12 or >18 h fasting intervals prior to shipping (Faucitano and Goumon, 2018).

A total interval (from last feed to slaughter) of 24 h has been proposed as an acceptable compromise between carcass yield, food safety and pork meat quality (Faucitano *et al.*, 2010). However, these advantages of fasting prior to slaughter period may be in conflict with some animal welfare objectives, particularly in relation to the hydration status of the animal, hunger-related behaviours and ease of handling, especially when long fasting times are applied. In pigs, in absence of water provision, dehydration becomes apparent at 6 h fasting time (Warriss, 1982) and, based on drinking behaviour and water intake, may result in significant thirst starting from 18 h fasting time (Brown *et al.*, 1999; Goumon *et al.*, 2013c). The increased drinking behaviour has been associated to a behavioural response to hunger and the pigs' need to fill their stomach and feel satiated (Saucier *et al.*, 2007). A pre- and post-fasting feeding behaviour study showed that pigs are already hungry after 12 h feed deprivation (Beattie *et al.*, 2002). Other hunger-related behaviours are increased physical activity and fighting rate (Brown *et al.*, 1999; Dalla Costa *et al.*, 2016), both exacerbating hunger through a decrease in the hepatic and muscle glycogen content (Faucitano, 2018), and reluctance to move during handling (Acevedo-Giraldo *et al.*, 2020; Dalla Costa *et al.*, 2016). When compared with unfasted pigs, fasted groups of pigs (18 h prior to loading) appear to be more difficult to handle at loading, as they go backward, round-turn and vocalise more (Dalla Costa *et al.*, 2016). More recently, Acevedo-Giraldo *et al.* (2020) also reported more handler interventions (vocalisation, hits, pushes and slaps) to drive fasted pigs (8 h) at loading and unloading at the slaughter plant compared with unfasted ones.

Overall, research findings showed that when it is not confounded by other preslaughter events (e.g. mixing, long transportation and lairage, etc.), feed withdrawal time alone has little or no effect on body weight losses, fighting behaviours and pork meat quality (Dalla Costa *et al.*, 2016; Faucitano *et al.*, 2006).

5.2.7 Loading

Loading pigs onto the truck is considered the most critical stage of the transport period as showed by the increased heart rate (reaching up to 160 heart beats; Correa *et al.*, 2013) and salivary cortisol and blood lactate concentrations compared to the basal values (Anderson, 2010; Bradshaw *et al.*, 1996a), with effects lasting until slaughter and eventually influencing pork meat quality (Correa *et al.*, 2010; Edwards *et al.*, 2010, 2011). The stress associated with loading procedure results from a combination of factors, such as group splitting in the finishing pen, the design of the loading facility, the size of the group and the handling system. These factors also have an impact on loading time. Longer loading times (up to 3 h) due to poor alley, ramp or dock design and large group size may result in greater risk of skin lesions (Guàrdia *et al.*, 2009) and leg fractures, resulting in carcass condemnation (Dalla Costa *et al.*, 2019b). Shortening loading is of advantage for the producer and trucker who aim to save time, but should not be to the detriment of animal welfare and product quality. Speeding up the loading process though rough handling, i.e. fast pace and electric prodding, may increase the risk of fatigued pigs on arrival at the slaughter plant (Correa *et al.*, 2010) and result in more skin lesions and PSE pork meat production (Correa *et al.*, 2010; Guàrdia *et al.*, 2004). The loading process is also physically demanding for handlers and truckers as showed by their heart rate increasing by more than 40 beats/min (Goumon *et al.*, 2013a). Repeated loading at multiple farms during the same day or handling large loads may cause fatigue-related rough interventions on pigs, resulting in increased animal losses during transport and carcass lesions (Dalla Costa *et al.*, 2019b; Ritter *et al.*, 2006).

5.2.7.1 Group splitting

The most stressful challenge a pig must cope with before transport to slaughter is being sorted out and driven out the finishing pen as it involves a close interaction with handlers and a separation from the familiar group (Goumon and Faucitano, 2017). Group splitting at the time of loading is applied to handle small groups of pigs (Lewis and McGlone, 2007) through the alleys up to the truck gate in both all-out and split-marketing strategies. All-out marketing system may ease and speed-up the loading process as each pen is emptied in one shot, but has the disadvantage to promote aggressiveness between pen mates near the end of the finishing stage due to increasing reduction of space for the heavier pigs (Conte *et al.*, 2012) and increase the risk of animal losses during transport due to uncontrolled provision of adequate space in the truck for pigs of different weight (Ritter *et al.*, 2007). In contrast, the split-marketing strategy, which is common in some European countries (i.e. France and Spain) and North America, implies the removal of the heaviest 25 to 50% of pigs from a pen to ship them to slaughter 1-2 weeks earlier than the other pen-mates (Goumon and Faucitano, 2017). Its advantage is the reduction of production costs due to shorter finishing phase and the opportunity to ship loads of more uniform market weight to the abattoir in respect for the physical needs for space in the truck. Keeping pigs in groups of uniform weight would help reduce fighting in mixed conditions and skin lesions as well (Driessen *et al.*, 2020b). Furthermore, the removal of the heaviest, dominant pigs from the pen increases feed efficiency and growth rate of pigs remaining in the pen due to greater floor and feeder space (Faucitano and Goumon, 2018). However, this practice may also alter the social environment in the pen leading remaining pigs to fight in order to re-stablish the social hierarchy.

5.2.7.2 Walking the distance to the truck gate

At loading, the distance over which pigs are moved to reach the loading area can vary from a few meters up to more than 100 m (Ritter *et al.*, 2008a). Walking a longer distance (46 to 91 vs 15 to 30 m) to reach the loading area showed to result in signs of fatigue, such as increased frequency of open-mouth breathing, skin discoloration, heart rate and greater salivary cortisol and blood lactate concentrations (Edwards *et al.*, 2011; Ritter *et al.*, 2007, 2008a; Rocha *et al.*, 2019). The physiological condition of pigs after having been handled for longer distance can be aggravated by rough handling as recently found by Rocha *et al.* (2019) who in a simulated loading study found increased infrared body temperature, heart rate and salivary cortisol levels in pigs imposed to walk a longer distance (90 vs 30 m) at a faster pace and through more frequent handler's interventions. Furthermore, Dalla Costa *et al.* (2019b) reported a three-fold greater risk of carcass condemnations due to leg factures in pigs moved a longer distance to reach the loading ramp positioned at the opposite end vs in the middle of the barn.

The use of shipping rooms or pens usually located near the barn exit and which pigs are moved to at least 2-4 h before loading helps minimise the distance to walk shortening handling time by 30 min (for a load of 100 pigs), while reducing pigs' heart rate and DOAs (-25%; Chevillon, 2001).

Furthermore, this practice is recommended for biosecurity and feed withdrawal control (Faucitano and Goumon, 2018). However, the use of shipping pens also implies the need to mix pigs originating from different finishing pens (Goumon and Faucitano, 2017) and for this reason they are one of the major sources of fighting-type lesions on the carcass (Brandt and Aaslying, 2015). To limit fighting or other agonistic behaviours during the pre-loading wait, it has been recommended to keep pigs in small shipping pens featuring divisions/barriers to create hiding/protection areas (McGlone and Curtis, 1985) and adjust the stocking density according to the wait time (from 0.45 m^2/pig for <0.5 h wait to 0.65 m^2/pig for periods longer than 3 h; SCAHAW, 2002). Furthermore, to ensure thermal comfort and avoid further sorting and mixing, the pen should be ventilated and of a size similar to that of the truck compartment (Goumon and Faucitano, 2017).

5.2.7.3 Group size

The correct choice of the size of the batch of pigs to drive forward at loading is paramount to guarantee a smooth flow of pigs through the barn alleys and ramps, without stressing them, and reduce the workload for handlers, When compared to larger groups (≥7 pigs), pigs moved in small groups (4 to 6 pigs) present lower increase in heart rate, are easier to handle (less round-turns) and faster to load (Lewis and McGlone, 2007), and are more fit for transport as showed by the reduced proportion of dead (0.19 vs 0.56%) or non-ambulatory pigs (0.36 vs 0.70%) on arrival at the slaughter plant (Berry *et al.*, 2009). Moving small groups has been specifically recommended to ease handling through narrow alleys (up to 1.2 m; Faucitano and Goumon, 2018). However, research showed that compared to groups of 12 or 20 pigs, moving a group of 4 or 8 animals is preferable to ease handling, based on the number of turn-backs, handling scores, squeals and handlers interventions, regardless of the alley width (2.4 or 0.9 m; Kavanagh, 2010).

5.2.7.4 Handling tools

Plastic rattle paddles and boards, electric prods and flags are common tools for moving pigs at loading under commercial conditions. However, these tools do not have the same efficiency and the same effects on pigs during handling (Faucitano and Goumon, 2018). Although, electric prods are still largely used to accelerate the loading procedure and reduce the workload of handling crews through alleys and ramps (Correa, 2011; Faucitano, 2001), regulations and codes of practice recommend to limit, if not avoid, their use at all preslaughter stages (EC, 1993; NFACC, 2014; SCAHAW, 2002). The use of this handling tool makes handling harder (McGlone *et al.*, 2004) as pigs back-up, round turn, slip, fall and jam more (Faucitano and Goumon, 2018) and increase heart rates and blood lactate concentrations (Correa *et al.*, 2010; Edwards *et al.*, 2010), all resulting in a greater incidence of fatigued pigs (Correa *et al.*, 2010) and PSE and blood-splashed pork (Correa *et al.*, 2010; Van de Perre *et al.*, 2010). The implementation of on-farm animal welfare program including, among others, the removal of electric prods at loading, contributed to reduce the proportion of transport losses from 0.41 to 0.08% and of condemned carcasses from 0.13 to 0.03% over a 3 years period (Correa, 2011). However, handlers can be still allowed to use electric prods as a last resort when a fit animal refuses to move forward (Grandin, 2002), but cannot abuse them. To limit a negative impact on pigs' physiological condition as assessed by the measure of rectal temperature and blood lactate concentration, the number and duration of electric prod hits on a pigs should not be more than two and last less than 1 s each, respectively (EC, 1993; Ritter *et al.*, 2008c). Alternatively, plastic boards, rattle paddles and flags appear efficient alternative devices for moving pigs along the alley and the loading ramp (McGlone *et al.*, 2004), with positive effects lasting until slaughter, in terms of lower exsanguination blood lactate, carcass lesion scores and blood splashed pork meat compared with the electric prod (Correa *et al.*, 2010).

5.2.7.5 Alleys and ramp design, and loading equipment

To ensure a smooth flow of pigs from the home pen to the barn exit and reduce loading time, alleys with 45° angles should be preferred to those featuring 90° corners as sharp angles are perceived as a dead-end by pigs resulting in stops, round-turns and jamming (Faucitano and Goumon, 2018). Goumon *et al.* (2013c) used behaviour observations and heart rate monitoring to show that pigs prefer negotiating with 30° corners than with 60 and 90° angles.

Climbing and descending a ramp are the main factors contributing to the highest peaks in heart rate of pigs at loading. This physiological response along with longer loading time and greater frequency of turn backs and baulking is exacerbated as the ramp angle increases from 0 to 45° (Faucitano and Goumon, 2018). Dalla Costa *et al.* (2019b) compared ramp slopes of <20° vs >20° and reported a 4-fold greater risk for DOA when steeper ramps were used. It has been also showed that handling pigs through steeper ramps also requires a physical effort (approx. 21 heart beats increase from 0 to 26° slope) for the handler (Goumon *et al.*, 2013b). According to current guidelines, fixed or adjustable ramp slopes should not exceed an angle of 20° or 25°, respectively (National Pork Board, 2016; EFSA Panel on Animal Health and Welfare, 2011). However, as market pigs have become heavier and difficult to move (Bertol *et al.*, 2011; Rocha *et al.*, 2016), recent recommendations point to a reduction of maximum ramp slope to 15° (Goumon and Faucitano, 2017).

Dalla Costa *et al.* (2019b) reported that pig falls at loading (up to 37%) were responsible for increased number of carcass condemnations due to leg fractures and bruises. Slips, falls and vocalisation on the loading ramp can be minimised and loading time reduced by covering the floor with feed, sand or dry wood shavings dry wood shavings when compared with no bedding in winter (Garcia and McGlone, 2015). Avoidance of steps higher than 15 cm can also make pigs less reluctant to move forward and decrease handler's interventions (SCAHAW, 2002; Weschenfelder *et al.*, 2012, 2013a).

Given the difficult handling through ramps, alternative systems, such as the loading gantry and modularised box, have been developed and tested for their efficiency and effects on pig welfare at loading (Berry *et al.*, 2012; Teixeira *et al.*, 2020). The loading gantry consists of an aluminium covered chute with slopes ranging from 7 to 18° feeding the bottom and the upper deck of the trailer, respectively. In a study using a standard loading ramp (19° slope) as a control, Berry *et al.* (2012) reported an improvement in the loading process when using the loading gantry, in terms of reduction of electric prod use and pigs' overlaps, slips, falls, baulking and vocalisations, all resulting in lower transport losses (0.6 vs 1%). The idea of using modular system to limit the stress of handling during the loading procedure dates from the mid-2000, when Brown *et al.* (2005) studied the response of pigs being loaded from level ground to the truck deck using containers positioned near the fattening pen and lifted up with a forklift. This method actually reduced loading time and decreased heart rate and body temperature in pigs during the loading process, but still caused a certain degree of stress in pigs as showed by the increased salivary cortisol level when compared to pigs loaded with an hydraulic tail-lift and a 18° angle tail board ramp. This result has been interpreted as a pigs' response to the novelty of the module environment with effects lasting up to 30 min after loading (Brown *et al.*, 2005). The commercial use of this system remains questionable as recently confirmed by Teixeira *et al.* (2020). When compared to a conventional loading procedure by a ramp, the use of a TRANSUS modularised box lifted by a crane from the ground level resulted in longer loading time (+3 min) with no effect on behaviours and physiological response or pork meat quality parameters.

5.3 Transport

Although most pigs already had a previous transport experience during their transfer as piglets from the nursery to the grow-to-finish facilities, based on the proportion of pigs that are are found dead on the truck or that are non-ambulatory on arrival at the abattoir (Ritter *et al.*, 2009, 2020), transportation to slaughter is considered to be one of the most stressful events in pigs life (Faucitano and Goumon, 2018). However, only when the degree of transport stress, as assessed by animal welfare indicators, is severe it can influence the conversion of muscle to meat and result in pork meat quality defects (Table 5.1).

Over the years, transport research has been above all addressed to meet animal welfare, regulatory and pork industry economic concerns related to the loss of market weight in transit or carcass condemnations and maintain public trust towards meat production practices (Faucitano and Lambooij, 2019). Clear examples of these efforts come from the four major pork exporting regions of the world, i.e. US, Canada, Europe and Brazil, in response to transport losses statistics published in governmental and research reports (Table 5.2) and related effects on the respective pork chain

Table 5.1. Summary of results of several studies (2010-2021) on the effect of transport on measures of animal welfare and carcass and meat quality (adapted from Schwartzkopf-Genswein *et al.*, 2012).[1]

Study	Welfare measures[2]	Carcass/meat quality
Correa *et al.* (2010)	***	**
Weschenfelder *et al.* (2012)	*	NS
Correa *et al.* (2013)	**	*
Weschenfelder *et al.* (2013a)	**	*
Goumon *et al.* (2013c)/Scheeren *et al.* (2014)	**	*
Correa *et al.* (2014)	***	**
Newman *et al.* (2014)	***	*
Arduini *et al.* (2017)	*	NS
Moak (2021)	**	*
Urrea *et al.* (2021)	*	NS

[1] *P<0.05; **P<0.01; ***P<0.001; NS = non-significant.
[2] Includes animal losses (dead and non-ambulatory pigs) and physiological parameters.

Table 5.2. Distribution of reported dead-on-arrival (DOA) and non-ambulatory (NA) pigs on arrival at the slaughter plant by world location.

Location	DOA, %	NA, %	References
USA	0.19-0.26	0.27-0.63	FSIS (2011-2016), Passafaro *et al.* (2019), Peterson *et al.* (2017), Ritter *et al.* (2009, 2020), Sutherland *et al.* (2009)
Europe	0.03-0.50	0.36	Averós *et al.* (2008), Christensen *et al.* (1994), Nannoni *et al.* (2016), Vitali *et al.* (2014), Warriss (1996)
Canada	0.07-0.17	0.27	CFIA (2021), Dewey *et al.* (2009), Sunstrum *et al.* (2006)
Brazil	0.3-1.2	0.79-1.5	Correia (2021), Dalla Costa *et al.* (2019b), Diesel (2016)

economies. For example, transport losses have been reported to cost the US and Brazilian pork industries $89 and $11 million per year, respectively (Diesel, 2016; Ritter *et al.*, 2020).

The responsibility for animal losses during transport may be either equally shared among the producer, trucker and abattoir (in the case of integrated production systems) or shouldered alone by the trucker. The application of personnel training sessions and rewards programs and the enforcement of legislation through the imposition of monetary penalties are paramount to improve handling and reduce animal losses (Dalla Costa *et al.*, 2019b; Faucitano, 2018; Rocha *et al.*, 2016).

Transportation to slaughter, encompassing the period between the entrance of pigs into the truck and their exit at the slaughter plant, is a multi-component phenomenon, including multiple factors that, *per se* or additively, have an impact on the response of pigs to stress with effects lasting up to slaughter (Faucitano and Lambooij, 2019; Passafaro *et al.*, 2019). These factors are journey conditions (season, travel duration, space allowance and driver skills), truck design, animal location (deck and compartment) in the truck, and handling at the reception and unloading at the slaughter plant.

5.3.1 Journey conditions

5.3.1.1 Season

Pigs may experience extreme fluctuations in the ambient conditions during their travel to slaughter as showed by the variation in ambient temperatures recorded in recent Canadian transport trials, ranging from -28.8 to 1.9 °C and from 9.1 to 40.1 °C in winter and summer, respectively (Rioja-Lang *et al.*, 2019). These temperature ranges often fall below or exceed the recommended thermal comfort zone for pigs during transport (15 to 30 °C for 111 to 160 kg pigs; Bracke *et al.*, 2020). Exceeding the maximum threshold of 30 °C with relative humidity (RH) levels >88% pigs' evaporative cooling mechanisms through behavioural adaptations, i.e. reduced activity, accelerated breathing and contact with cool and wet surfaces, are significantly compromised, and may result in death (Bracke *et al.*, 2020; Rioja-Lang *et al.*, 2019). For this reason, the European Commission has established 35 °C as the maximum allowed temperature for the long-distance transport of pigs (EC, 2005).

Greater in-transit losses have been reported in summer than in winter (Faucitano and Goumon, 2018; Nannoni *et al.*, 2016), with the highest death rate being recorded in July and August (Haley *et al.*, 2008a; Vitali *et al.*, 2014). Haley *et al.* (2008b) reported the highest DOA rate (0.40%) at an ambient temperature peak of 34 °C. In general, the risk of having DOA and non-ambulatory pigs on arrival at the plant is greater at temperatures higher than 20 and starting from 17 °C, respectively (Haley *et al.*, 2008a, 2010; Kephart *et al.*, 2010; Sutherland *et al.*, 2009; Warriss and Brown, 1994). A recent retrospective observational study using ambient data collected from 2010 to 2015 in the area of US federal slaughter plants reported that the risk of market-weight pigs losses during transport increased by 1.4 times between 29 and 33 °C compared to a temperature range of 12-26 °C (Peterson *et al.*, 2017). In the former temperature range, the risk of death during transport was even two-fold greater for cull sows (Peterson *et al.*, 2017). Increased transport losses have been also recorded at low ambient temperatures, with even greater rates of DOA and non-ambulatory pigs on arrival at the plant being reported in winter than in summer (Fitzgerald *et al.*, 2009; Guàrdia *et al.*, 1996; Sutherland *et al.*, 2009). For example, highest peaks of transit losses have been observed in the late fall and early winter months (Ellis and Ritter, 2006), with rates of non-ambulatory pigs increasing at ≤5 °C (Sutherland *et al.*, 2009). In the aforementioned retrospective transport data analysis, Peterson *et al.* (2017) found that the risk of death was approximately one time greater in hauls occurring at 4-10 °C than at 12-26 °C for both market weight pigs and cull sows.

Transport losses under warm and cold ambient conditions can be associated to a lower ease of handling at loading and unloading (Torrey *et al.*, 2013a,b), reduced resting in the truck (Goumon *et al.*, 2013c), and greater body temperature, heart rate and blood CK, and lactate concentrations at slaughter (Conte *et al.*, 2015; Correa *et al.*, 2014; Goumon *et al.*, 2013c; Sommavilla *et al.*, 2017; Xiong *et al.*, 2015). These factors can also explain the greater lesion scores and number of frostbites on the carcass of pigs transported in winter (Correa *et al.*, 2013; Goumon *et al.*, 2013c; Scheeren *et al.*, 2014).

5.3.1.2 Travel time

Both long and short journeys can affect the welfare of pigs and pork meat quality during transport (Rioja-Lang *et al.*, 2019). Transports longer than 3 h and up to 24 h have been shown to result in increased risk of animal and body weight losses (Dalla Costa *et al.*, 2019b; Sutherland *et al.*, 2009; Warriss *et al.*, 1990), higher blood glucose, cortisol, lactate and haematocrit levels, indicators of fatigue and dehydration, at slaughter (Faucitano and Lambooij, 2019; Newman *et al.*, 2014), and DFD (dark, firm, dry) pork meat (Mota-Rojas *et al.*, 2006). A recent study monitoring the North American cull sow marketing network reported that sows could travel straight to slaughter or through assembly points between 24 to 120 h (Blair and Lowe, 2019). In a Danish survey of cull sows loads shipped to slaughter, long transport duration (up to 8 h) was identified as one of the most important causing factors for the deterioration of clinical signs on arrival at the abattoir (Thodberg *et al.*, 2019). However, these results are more likely due to the effects of other concurrent factors, such as time on fasting, season of the year, type of vehicle, within truck ambient conditions and loading density than to transport time *per se* (Faucitano and Goumon, 2018; Thodberg *et al.*, 2019). For example, in unfasted pigs the risk of DOA increases for journeys up to 8 h (Averós *et al.*, 2008), while in summer increased risk for DOA has been reported in hauls longer than 2 h (Vitali *et al.*,

2014). The effects of long travel duration (18 vs 12 and 6 h) on the incidence of DFD pork meat is also aggravated in transports occurring in winter in the top and bottom compartments of the truck (Scheeren *et al.*, 2014).

Lower ease of handling in the lairage area, greater blood concentrations of cortisol and lactate at slaughter, and increased proportion of PSE loins are also reported in pigs experiencing shorter journeys (15 min to 2 h; Faucitano and Lambooij, 2019). A possible reason for these effects is the lack of time for pigs to recover from the stress of loading, especially onto poorly designed trucks featuring internal ramps, upon arrival at the abattoir (Weschenfelder *et al.*, 2013a).

Internationally, allowed maximum transport times range from 8 to 28 h (Faucitano and Goumon, 2018). Passed this time, it is recommended or mandated to unload pigs at a feed, water and rest station (Government of Canada, 2020; www.controlpost.eu), where they are kept from 5 to 24 h, depending on the region of the world (Faucitano and Goumon, 2018), to allow them recover from the effects of dehydration, hunger and general fatigue. However, this practice implies the additive stress of unloading and loading, and mixing pigs in a novel environment as showed by the greater heart rate recorded in pigs loaded on and off the truck compared with those kept in the truck (Chevillon *et al.*, 2003).

5.3.1.3 Floor space in the truck

The choice of truck floor space provided to pigs during transport is the result of a compromise between economic pressure to maximise profit from a single journey by loading as many animals as possible, and the comfort of animals during the journey (Rioja-Lang *et al.*, 2019). However, the importance of loading density should not be underestimated as it can account for the largest portion of variation in total transport losses compared to other transport variables, such as driver, handling crew, ambient conditions, loading duration and wait time before unloading (Fitzgerald *et al.*, 2009).

Internationally, loading density or space allowance, expressed as kg/m^2 and $m^2/100$ kg, respectively, are regulated by legislation, codes of practice or guidelines based on climatic conditions and pig market weight (Faucitano and Lambooij, 2019). The loading density of 235 kg/m^2 or space allowance of 0.425 $m^2/100$ kg are the maximum and minimum, respectively, legal requirements, in the EU (EC 95/29/EC) to ensure that pigs can stand and lie down to rest in their natural position without being disturbed and pushed to continually change their position during transport (Lambooij, 2014). The respect of these requirements have been associated with more resting and less drinking behaviour during transport and lairage, respectively, lower body temperature, heart activity, exsanguination blood CK levels, skin lesions, rectal prolapses, eventually resulting in decreased proportion of DOA and non-ambulatory pigs at unloading at the plant, and DFD pork meat (Faucitano and Goumon, 2018; Faucitano and Lambooij, 2019; Urrea *et al.*, 2021). However, results arising from recent studies using planimetric measures of the floor surface covered by pigs in different positions during transport showed that a pig of modern genetics, being larger and heavier, may need a minimal greater space (approx. 0.49 m^2) to entirely lie down in fully lateral recumbent position on the truck floor, which suggests a revision of the EU legislation (Arndt *et al.*, 2019).

In some guidelines, recommended load size is simply determined by the number of animals per m^2 according to the truck width and length, and climate conditions (from <14 to 32 °C; https://www.ontariopork.on.ca/Resources/Transportation/Calculator). Fitzgerald *et al.* (2009) showed that when the number of pigs is reduced from 3 to 2/m^2 the risk of animal losses is decreased from 1 to 0.2%. Whereas the risk of carcass condemnation due to leg fractures can increase by 4.4 times for each extra pig allocated per m^2 in the truck (Dalla Costa *et al.*, 2019b).

Besides providing more floor space to lie down, greater space allowance (>0.425 m^2/pig) also gives pigs a greater freedom to move around in the compartment during transport causing muscle fatigue due to the attempt of pigs to maintain their balance and cope with the unexpected and sudden movements of the truck, or fighting (Barton-Gade and Christensen, 1998). The fatigue-related muscle glycogen depletion was the cause of the 11% greater risk of producing DFD pork meat in pigs transported at space allowance increasing from 0.37 to 0.50 m^2/100 kg (Guàrdia *et al.*, 2005).

The application of loading densities should be adjusted according to ambient conditions, body weight and transport time. Animal losses are lower when more truck floor space (0.54 vs 0.46 and 0.39 m^2/100 kg) is provided in summer (Ritter, 2007). For this reason, it is recommended to provide 15 to 25% more space during hot/humid ambient conditions (CARC, 2001; EC, 2019). However, an increased risk for skin lesions (+ 28%) has been reported when more floor space (0.50 vs 0.25 m^2/100 kg) is provided in summer transports (Guàrdia *et al.*, 2009). In winter, an almost 10% greater risk of carcass lesions has been found when pigs are transported with a lower space allowance (0.25 m^2/100 kg) as due to lack of space they overlap and fight to seek for space to lie down and huddle with the compartment mates to keep their body temperature (Guàrdia *et al.*, 2009).

The application of the EU loading density during transport of heavy pigs (160 kg) appears to result in overcrowding (0.7 pig/m^2) leading to a greater risk of DOA (Nannoni *et al.*, 2016). Because of their different physical and thermal needs (Renaudeau *et al.*, 2011), for the transport of heavier pigs (from 114 kg to 182 kg) it is recommended to increase the minimum floor space from 0.40 to 0.61 m^2/pig and from 0.46 to 0.65 m^2/pig in winter and summer, respectively (Grandin, 2017).

The effects of space allowance on the welfare of pigs during transport and meat quality also depend on travel time and range from a reduced incidence of non-ambulatory pigs on arrival at the plant after short transport (<1 h) compared with longer journeys (3 h) at greater floor space (0.40-0.49 to 0.52 m^2/100 kg; Pilcher *et al.*, 2011) to a lower risk of PSE pork meat production after short journeys (1 vs 3 h) at reduced floor space (0.25 vs 0.50 m^2/100 kg; Guàrdia *et al.*, 2004). These latter results may be explained by the fact that in the initial travel phase (first 30 min mostly on unpaved road) pigs do not lie down, but stand to better cope with the fear and instability caused by the frequent truck accelerations/decelerations and vibrations (Chevillon, 2001) until they settle down (up to 2 h later; Barton-Gade and Christensen, 1998; Lambooij *et al.*, 1985). Providing pigs with less space in hauls of less than 3 h would help standing/sitting pigs keep their balance by holding each other while the vehicle negotiates bends or poor road surfaces (Barton-Gade and Christensen, 1998) and would reduce the effects of physical stress on muscle metabolism (Guàrdia *et al.*, 2004).

5.3.1.4 Driver type and driving conditions

In a transport survey involving more than 100 loads (approx. 5 million market pigs) shipped to Canadian and US processors, the truck driver accounted for 8% of the 0.2% total animal losses recorded in a 3 years transport survey (Haley *et al.*, 2008a). Significant associations between drivers and in-transit losses have been confirmed in later studies (Fitzgerald *et al.*, 2009; Passafaro *et al.*, 2019). The effect of driver has ben explained by the difference in handling ability during the loading/unloading process and truck driving skills. The poor handling skills explained the different proportion of pigs reluctant to move, slipping and turning-back between two drivers while loading and loading-off the same trailer (Rocha *et al.*, 2016). Passafaro *et al.* (2019) reported a 0.89 lower risk of total transport losses when the truck is driven by the producer compared to contracted truckers and explained this result by the greater carefulness of the former to move his own pigs compared to employees.

The effect of driving conditions has been associated with the fact that pigs do not lie down on rough journeys (Bradshaw *et al.*, 1996b; Randall *et al.*, 1995) and wild driving (Peeters *et al.*, 2008), likely because of the level of vibrations, noise and sudden truck braking, accelerations and turning, resulting in pigs' postural instability and muscular effort to keep the balance (Peeters *et al.*, 2008). Furthermore, increased animal losses have been associated to truck instability during transport on poor roads due to excessive wind and precipitations (Passafaro *et al.*, 2019). These transport conditions can also result in greater proportion of bruised carcasses (Driessen *et al.*, 2020b) and incidence of PSE pork meat (Hoffman and Fisher, 2010; Zou *et al.*, 2020), and higher stress level (i.e. increased blood CK concentration at slaughter; Zou *et al.*, 2020). This evidence supports the need of training programs for livestock transporters (i.e. TQA; National Pork Board, 2016) as mandated by the EU legislation (EC Council Directive 1/2005 (EC, 2005)).

5.3.2 Truck design

The design of trucks used for pig transportation to the abattoir varies from single-deck trucks to three-deck trailers, i.e. 'pot-belly' (PB) or straight/flat-deck trailer models. These latter trailer types are frequently used as they can accommodate a larger number of animals (more than 200) on a single journey. The variation in design between truck types implies the use of ramps or hydraulic devices for loading and unloading, mechanical or passive ventilation and airflow patterns, water misting/sprinkling, deck height and drinkers and feeders. Consequently, truck design features that mostly influence the comfort of pigs during transport include the loading/unloading system, the deck/compartment position and the internal ambient control (Faucitano and Lambooij, 2019).

5.3.2.1 Loading/unloading systems

Fixed decks combined with internal ramps within the truck make the procedures of loading and unloading more difficult, resulting in greater risk of animal losses (Barton-Gade *et al.*, 2007; Correa *et al.*, 2013) and PSE and DFD pork (Correa *et al.*, 2013, 2014; Guàrdia *et al.*, 2004). Comparative studies, including the PB trailer, featuring three fixed decks and two to five, often steep (up to 32° slope), ramps and 180° turns, dead-ends and steps, and other trailer models featuring full or semi-hydraulic decks, associated the use of the PB trailer with greater use of electric prods (up to 32%) and longer loading time (Ritter *et al.*, 2008a; Torrey *et al.*, 2013a,b; Weschenfelder *et al.*, 2013a). These observations can help explain the greater percent of dead and fatigued pigs on arrival at the abattoir, and blood cortisol, CK and lactate concentration at slaughter reported in a number of transport trials (Faucitano and Goumon, 2018; Rioja-Lang *et al.*, 2019). However, even the use of hydraulically moving decks has been reported to result in some handling issues on and off the truck. More pigs' overlaps and round-turns, and handler interventions have been reported in hydraulically lift deck trailer models by Weschenfelder *et al.* (2012, 2013a). In this study, the lower ease of handling was explained by the pigs' response to presence of a step near the trailer gate resulting from the piling up of the semi-moving middle and top decks on the bottom deck floor. The increase in slaps and vocalisations in this trailer type at unloading was also interpreted as a response of pigs to the challenge of leaving the trailer (Pedersen-MacNab *et al.*, 2019).

In European trucks featuring fixed decks, ramps are replaced by a tailgate lift where groups of pigs are hydraulically lifted up to the targeted deck with no or very limited human handling. Although, when compared to sloped ramps (16 to 18° angle), the use of the tail-gate lift does not result in any variation in heart rate and body temperature at loading and blood cortisol level at slaughter in pigs (Brown *et al.*, 2005; Nanni Costa *et al.*, 1999, 2002), it appears to reduce the risk of PSE pork meat (Guàrdia *et al.*, 2004).

5.3.2.2 Deck and compartment position

Within truck deck and compartment positions have an impact on pig welfare during transport and pork meat quality (Bench *et al.*, 2008). In European transport trials, a higher incidence of PSE or DFD pork meat along with greater body temperature, blood cortisol levels and dehydration state at slaughter, and carcass lesion scores have been found in pigs transported on the bottom deck of the truck (Driessen *et al.*, 2020b; Lambooij and Engel, 1991; Lambooij *et al.*, 1985). Whereas, greater blood norepinephrine concentrations have been found in heavy pigs transported on the upper deck compared to the bottom deck of a truck featuring internal ramps (Bozzo *et al.*, 2020).

Greater body weight losses, percent of DOA, body temperatures and blood cortisol and CK concentrations at slaughter have been reported in pigs transported in the front and top front compartments compared with rear ones in European and Brazilian trucks (Christensen and Barton-Gade, 1999; Dalla Costa *et al.*, 2007; Warriss *et al.*, 2006). In North American transport studies, increased core and skin temperatures after loading and during transport, higher heart rate after loading and longer unloading time have been reported in pigs located in the top front and rear, and bottom rear compartments (Conte *et al.*, 2015; Goumon *et al.*, 2013c; Torrey *et al.*, 2013b; Xiong *et al.*, 2015). Whereas, a greater proportion of DFD pork meat has been found in the *Longissimus*, *Semimembranosus* and *Adductor* muscles of pigs transported in the middle front compartment of this trailer type (Correa *et al.*, 2014).

In these studies, the effects of deck and/or compartment position were likely caused by the combined effect of handling and heat stress during the loading and unloading process through internal ramps, poor ambient conditions due to insufficient airflow or the high rate of vibrations imposing pigs to a do a physical effort to keep their balance while in standing position or the loading order (Faucitano and Goumon, 2018; Faucitano and Lambooij, 2019).

Brown *et al.* (2012) hypothesised that the different floor vibration rate could explain the lower proportion of lying pigs in the front compartments compared with the rear ones in a PB trailer. However, recently Morris *et al.* (2021) reported vibration issues in both top and bottom compartments of this trailer type, with greater vibration rate being found in the bottom rear compartment.

Other factors, such as truck type, ambient conditions and transport duration, may exacerbate the effect of animal position in the truck on its response to transport stress. In Canadian transport trials, the greater increase in gastro-intestinal tract temperature was found in pigs transported in the top front compartment of a PB trailer in summer compared to winter (Conte *et al.*, 2015). Whereas, Arduini *et al.* (2017) found the lowest body surface temperature in heavy pigs located on the upper deck of a flat-hydraulic deck trailer under warmer and more humid ambient conditions. Furthermore, a greater proportion of pale pork meat has been reported in pigs transported in the top and bottom decks compared with the middle deck of a PB trailer in summer than in winter (Correa *et al.*, 2013). Similarly, pigs transported on the bottom deck of a PB trailer for longer time (18 vs 6 and 12 h) in summer (vs winter) presented higher blood cortisol values at slaughter (Sommavilla *et al.*, 2017), while under the same transport conditions the risk for DFD pork meat production was greater for pigs transported in the middle front compartment in winter (Scheeren *et al.*, 2014). Within the same trailer type, compared with the bottom deck, the proportion of exudative pork was higher in pigs transported on the top deck in the fall, while it was lower in winter (Newman *et al.*, 2014).

5.3.2.3 Within truck climate control

The quality of the within truck microclimate conditions varies considerably through the journey (from -16 to 38 °C and from 43 to 100% RH; Brown *et al.*, 2011; Miranda de-la Lama *et al.*, 2021; Moak, 2021; Sommavilla *et al.*, 2017) and its effects on pigs' thermal comfort during transport depend of the external ambient conditions and truck design. Each 1 °C and 1% increase in the external ambient temperature and relative humidity can increase internal truck temperature by 0.99 and 0.11 °C, respectively, resulting in greater risk of animal losses at each temperature degree increase (Dewey *et al.*, 2009). This internal temperature increase is more likely in passively ventilated vehicles, such as the PB trailer, where the internal temperature can rise up to 3-4 °C in 5 min at a rate of approximately 1 °C/min, especially in stationary situations (Lambooij, 1988; Xiong *et al.*, 2015). Canadian transport trials showed that in a stationary PB trailer overall temperature can increase above 30 °C, with the bottom front compartments being up to 10 °C warmer than the external ambient temperature (Faucitano and Goumon, 2018). After a 30 min wait before unloading the middle and bottom front compartments can be 6 to 10 °C warmer than the external temperature (Fox *et al.*, 2014; Weschenfelder *et al.*, 2012). In these internal thermal conditions, the risk of DOA can increase by 2.2 times at every 30 min increase in wait time (Haley *et al.*, 2008a,b) resulting in approximately 0.3% DOA after 4 h stop time at >20 °C ambient temperatures (Sutherland *et al.*, 2009).

However, internal microclimate can also change while the vehicle is moving depending on the airflow rate, especially in passively ventilated vehicles. The greater temperature recorded in moving PB trailer compared to flat-deck trailer (Ritter *et al.*, 2008a; Weschenfelder *et al.*, 2012) may be due to the punch-type pattern openings on the sides of the PB trailer reducing the air flow through the compartments compared with the slatted openings of the flat-deck trailer (Weschenfelder *et al.*, 2012). Reduced within truck air circulation in passively ventilated trailers has been also associated to low outside wind speed (<4.5 m/s) during transport, resulting in high temperature-related increased risk of total animal losses and skin damages (Driessen *et al.*, 2020b; Passafaro *et al.*, 2019). Insufficient head room prevents pigs from receiving adequate ventilation and keeping a normal posture while standing, resulting in transport losses (Christensen and Barton-Gade, 1999; Nielsen, 1982). According to a EU recommendation (SCAHAW, 2002), a 30 cm minimum head room between the deck ceiling and the greatest point on the pig (middle of the back) is needed to ensure adequate airflow, temperature and humidity regulation and removal of noxious gases in vehicles with natural ventilation.

Under this poor thermal conditions, pigs kept in a stationary trailer should be cooled-off by water sprinkling/misting or fan-assisted ventilation or, even better, the two systems combined (Brown *et al.*, 2011). The application of 5 min water sprinkling at the departure from the farm after loading and 5 min before unloading at the slaughter plant at ambient temperature above 20 °C showed to improve the thermal conditions of pigs on their arrival at the plant, resulting in less fatigued pigs at slaughter and less exudative pork meat, particularly in those transported in the middle front and rear compartments (Fox *et al.*, 2014; Nannoni *et al.*, 2014). However, in these studies RH also increased (up to +7.5%) in the sprinkled trailer, which may result in an inefficient evaporative cooling (Fox *et al.*, 2014). In situations of excessive humidity in the truck, such after the application of water sprinkling/misting, in order to improve internal trailer microclimate conditions and pigs' thermal state by increasing evaporative cooling, mechanical ventilation should be used (Christensen and Barton-Gade, 1999; Passafaro *et al.*, 2019). Recently, the use of fan banks applying forced ventilation for 0.5 h and water misting for 10 min on pigs located in a stationary trailer prior to unloading reduced the internal trailer THI and improved pigs' thermal conditions, in terms of lower body heat loss (Pereira *et al.*, 2018).

Lower temperatures have been recorded in the upper compartments of PB trailers, which can be explained by the increased thermal radiation cooling these compartments due to the presence of more exposed surfaces and lack of roof insulation (Brown *et al.*, 2011; Weschenfelder *et al.*, 2012). These ambient conditions can be especially detrimental to pigs' thermal comfort in winter transports. A simple Styrofoam insulation to the ceiling of the upper deck helped increase the internal temperature by 10 °C during hauls occurring at -20 °C (Gonyou and Brown, 2012). Other strategies to mitigate the impact of winter conditions on the vehicle internal environment and maintain thermo-neutral zone of pigs are the use of boarding and bedding. As recommended by the TQA guidelines (National Pork Board, 2016), at temperatures between -12 and 9.4 °C trucks must be fitted with 90% boarding to keep the pig heat in the vehicle and reduce transport losses (Faucitano and Goumon, 2018). When these recommendations are not met, i.e. low boarding level (0-30%) at temperatures below 5 °C, the proportion of DOA increases (McGlone *et al.*, 2014a).

If pigs are provided with a well-bedded floor, the lower critical temperatures during transport can be reduced by 3 to 5 °C (Randall, 1993). In winter transports, insufficient bedding (less than 6 bales of bedding; McGlone *et al.*, 2014b) may result in pigs less prone to lie down in order to avoid the contact with the truck cold aluminium floor (Goumon *et al.*, 2013c) and greater number of carcass bruises, including frostbites (Goumon *et al.*, 2013c; Scheeren *et al.*, 2014).

5.3.3 Arrival and unloading

The timeliness of arrival of the truck is one of core criteria scored by animal welfare auditors at the slaughter plant (Grandin, 2017) for pork meat quality certification (i.e. Certified Humane Raised and Handled®; Humane Farm Animal Care, 2018). To meet this criterion, pigs should be unloaded within 0.5 h of their arrival at the slaughterhouse and unloading should be completed within an hour (Grandin, 2017). However, under commercial conditions, wait times can vary from 4 min up to 4 h (Driessen *et al.*, 2020b; Faucitano and Pedernera, 2017). Longer waits can be due to logistical reasons, i.e. lack of multiple unloading docks allowing more than one truck to unload simultaneously, or poor coordination of truck arrivals based on lairage capacity and slaughter speed (Faucitano and Pedernera, 2017). The risk of DOA can increase by almost 2 times at each 30 min increase in the wait time prior to unloading (Haley *et al.*, 2008a,b), reaching up 0.28% death rate after a maximum of 4 h wait at temperatures above 20 °C (Sutherland *et al.*, 2009). Thodberg *et al.* (2019) reported that increasing wait time prior to unloading from 0 to 75 min was one of the most important causing factors for the deterioration of clinical signs in cull sows on arrival at the abattoir.

The greater risk of pigs found dead or non-ambulatory at unloading in relation to wait time in the slaughter plant yard has been explained by climate changes inside the stationary vehicle and heat stress (Faucitano and Pedernera, 2017). Weschenfelder *et al.* (2012, 2013a) showed that, even at mild external ambient temperatures (8 °C), after only 10 min wait before unloading internal temperatures in the top rear and bottom front compartments of a PB trailer were 6 °C higher than the external temperature. These thermal conditions in the truck may result in increased number of panting pigs (Van de Perre *et al.*, 2010), while their effects on pork meat quality are unclear (Driessen *et al.*, 2020b; Van de Perre *et al.*, 2010).

Exiting the truck is a source of intense physical stress for pigs as showed by the increased heart rate (Correa *et al.*, 2010, 2013) and blood lactate concentration (Edwards *et al.*, 2011; Rocha *et al.*, 2015). Recommendations regarding handling and facility design (i.e. ramp, dock and gaps) are similar to those provided for the loading process. To empty the truck smoothly and faster, pigs should be unloaded by compartment and in small batches using rattle paddles (Faucitano and Geverink, 2008). The use of electrical prods is banned at many slaughter plants as it increases the number of pigs mounting, slipping and round-turning (Rabaste *et al.*, 2007). These behaviours may have an impact on pork meat quality. Slips, caused by poor handling or slippery floor (Torrey *et al.*, 2013b), at unloading have been significantly correlated to increased pork meat exudation (r=0.63; Rocha *et al.*, 2016), while a frequency of slips, falls and overlaps greater than 6% in this phase has been associated to increased risk for PSE pork meat (Sardi *et al.*, 2020). Furthermore, the vocalisation of pigs generating a noise level of up to 78 dB in the unloading area was identified as the major factor (compared to proportion of panting pigs at unloading, use of electric prods and season) responsible for the faster loin muscle pH drop rate, resulting in greater proportion of PSE pork meat (Van de Perre *et al.*, 2010).

5.4 Lairage

The abattoir manager is responsible for the optimisation of lairage conditions (layout, ambient control and handling systems) with the objective to allow pigs recover from the stress of transport and handling and ensure optimal, consistent and uniform carcass and meat quality (Faucitano, 2018). According to Bucher and Scheibl (2014), the quality of this recovery phase can have an influence on stunning effectiveness through the effects on pigs' cardiovascular systems. High-stress lairages (based on poor design and excessive sound level) can prevent pigs from resting properly as showed by greater blood lactate and CK concentration in pigs at slaughter and exudative pork meat compared to low-stress ones (Warriss *et al.*, 1994) and offset all efforts and investments made by the pork chain to improve animal performance and welfare, and carcass and meat quality.

The recovery rate of pigs in lairage and its related effects on the physiological condition of pigs at slaughter and carcass and meat quality depends on the time spent in lairage (or on rest), the quality of handling systems, design of critical areas (pens, alleys and chutes) and environmental control.

5.4.1 Rest time

Under commercial conditions, reported rest times vary from <1 to 24 h depending on the loads delivery schedule and lairage capacity (Geverink *et al.*, 1996; Gispert *et al.*, 2000; Sardi *et al.*, 2020), with the latter having to be larger enough to offer at least 1.1 times the slaughter capacity per hour (Holmes *et al.*, 2020). However, there is evidence that the recovery of pigs' physiological conditions (i.e. return of cortisol to the basal concentrations in blood at slaughter) after transport and handling is completed with a resting period of 2-3 h (Faucitano, 2010). Unless required by poor ambient conditions (i.e. >30 °C or >10 ppm ammonia levels; Fraqueza *et al.*, 1998; Weeks, 2008), rest periods of one hour or less are not recommended. Based on their latency to settle and lie down in the pen (30-60 min; Geverink *et al.*, 1996; Moss, 1978), unrested pigs would go to slaughter while still under the effects of transport and handling stress. Goumon *et al.* (2013b) showed that exposing pigs to a physical exercise after a short rest (35 vs 75 or 150 min) makes them more susceptible to stress as showed by the increased heart and respiration rate, and skin temperature. This physiological response either combined with greater muscle lactate level or decreased residual glycogen content in the muscle, both indicators of fatigue, at slaughter may result in greater incidence of PSE or DFD pork meat in pigs kept in lairage for insufficient time (Driessen *et al.*, 2020b; Faucitano, 2018; Hambrecht *et al.*, 2005). However, while longer lairage time (overnight to 22 h) helps reduce the risk of PSE pork meat, it increases the risk of dead-in-pen (DIP, at >12 h rest; Nannoni *et al.*, 2016) and pork meat with DFD characteristics (Dalla Costa *et al.*, 2016; Guàrdia *et al.*, 2005) as a result of reduced muscle glycogen content and overall glycolytic potential at slaughter (Rocha *et al.*, 2015). This muscle metabolic condition at slaughter is caused by the combined effect of fasting duration and fighting rate (Dalla Costa *et al.*, 2016). A positive association between skin lesions and carcass condemnations with lairage time (2 min to 22 h) has been reported in a number of studies (Dalla Costa *et al.*, 2019b; Driessen *et al.*, 2020b; Faucitano, 2010). More specifically, Guàrdia *et al.* (2009)

reported an almost two-fold higher risk of skin lesions due to fighting in pigs kept in lairage for 15 h compared to 3 h (18 vs 10%).

5.4.2 Rest conditions: mixing, space allowance and climate control

The increased level of aggression over rest time and its effects on skin lesions and pork meat quality are aggravated by mixing unfamiliar groups of pigs (Driessen *et al.*, 2020c; Warriss, 2003), which is a common practice in lairage considering the difference in size between the truck compartment and lairage pen. Rabaste *et al.* (2007) showed that a 10 times increased fighting rate when pigs are mixed in larger groups (30 pigs) in the pen compared to small groups (10 pigs).

An average space allowance of 0.5 m^2/100 kg or 1-2 pigs/m^2 is recommended to allow pigs to rest and drink (Faucitano, 2010). However, under commercial conditions, space allowances can vary from 0.28 to 0.73 m^2/pig, depending on the country and lairage time (Dalmau *et al.*, 2016; Gispert *et al.*, 2000). Greater space allowance (0.64-0.85 vs 0.42-0.46 m^2/pig or 1.2 vs 3.1 pigs/m^2) has been associated with greater fighting rate in mixed groups of pigs (Geverink *et al.*, 1996; Moss, 1978). Based on the evidence that most fighting occurs within the first hour of lairage (Geverink *et al.*, 1996; Moss, 1978), stocking densities of 0.42 m^2/pig have been proposed for short lairage (<3 h) to restrict fighting between mixed pigs, while more space (0.66 m^2/pig) should be provided for lairages longer than 3 h to allow pigs to lie down and move around (Weeks, 2008). However, results from a survey of lairage conditions at three Italian slaughter plants showed that the risk of DIP was the lowest when pigs were provided with more space (0.64 vs 0.46 m^2/100 kg) in the lairage pen (Vitali *et al.*, 2014).

The positive effects of space allowance on DIP reported by Vitali *et al.* (2014) were favoured by a better control of ambient conditions, in terms of greater ventilation rate through larger windows on the roof and side walls, low light brightness and efficient cooling systems, during lairage. To ensure the thermal comfort of pigs air temperature should be kept between 15 and 18 °C and combined with relative humidity between 59 and 65% (Honkavaara, 1989). At colder or warmer temperature conditions, pigs can either suffer from cold or heat stress, as showed by increased shivering and huddling (Fraqueza *et al.*, 1998; Knowles *et al.*, 1998), and panting (Holmes *et al.*, 2020), respectively. Cold and heat stress, and related behavioural response have been associated with a greater incidence of DFD pork meat, due to muscle energy depletion aimed at maintaining a constant body temperature (Knowles *et al.*, 1998) and PSE pork meat (Lammens *et al.*, 2007; Van de Perre *et al.*, 2010). The heat stress condition, which is particularly detrimental to the welfare of sows on the day of slaughter (Holmes *et al.*, 2020), is especially exacerbated at warmer (>30 °C) and more humid (RH>80%) ambient conditions (Santos *et al.*, 1997). This physiological condition may explain the two-fold greater DIP percent reported at temperatures above 32 °C compared with 16 °C (Faucitano and Velarde, 2021).

Water spraying pigs in the lairage pen (Figure 5.1) helps reduce the heat-related respiration rate (Huynh *et al.*, 2006), the number of DIP (Vitali *et al.*, 2014) and the incidence of PSE pork through a 2 °C drop in muscle mass temperature (Long and Tarrant, 1990). However, at temperatures below 5 °C, showering is not recommended as it causes animal shivering and may lead to DFD pork due to muscle energy depletion to maintain a constant body temperature (Knowles *et al.*, 1998). To get the

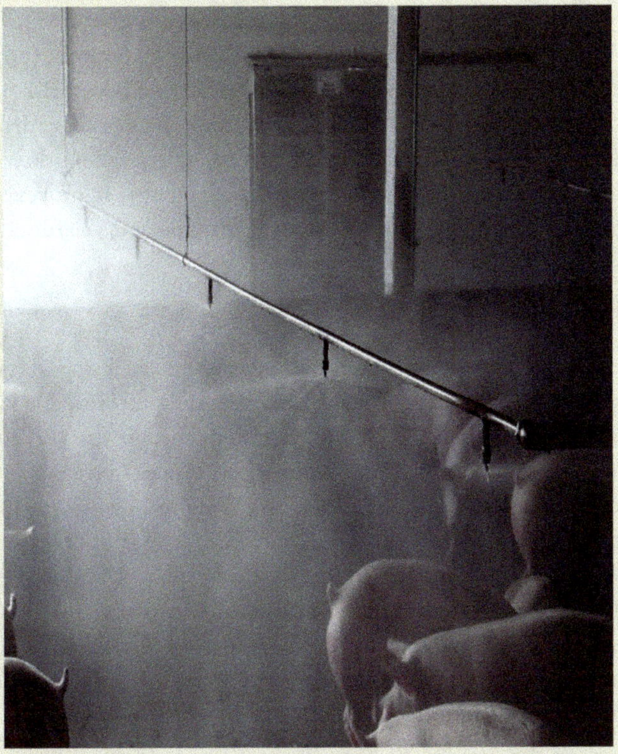

Figure 5.1. Water misting in the lairage pen (L. Faucitano, AAFC, Canada).

best results in terms of pigs' thermoregulation, current recommendations are to apply water misting during the first 5-10 min after arrival in the pen only (Holmes *et al.*, 2020). Extended water misting may increase pig activities, including fighting, at detriment of rest time (Knowles *et al.*, 1998), create puddles on the floor reducing space for resting comfortably (Holmes *et al.*, 2020) and may increase air humidity above 90% (Weeks *et al.*, 2008). To improve pigs' thermal conditions by evaporative cooling as well as limiting the rise of air temperature to less than 3 °C and of concentrations of ammonia levels to less than 10 ppm (Weeks, 2008), the lairage area must be properly ventilated (135 m^2/h; Brent, 1986; Weeks, 2008).

In a study comparing the noise levels of cattle, sheep and pig lairage areas, Weeks *et al.* (2009) found that pig lairages were the loudest ones (110 vs 97 and 91 dB, respectively). The highest peak (120 dB) is usually recorded in the peri-mortem area (Faucitano, 2010; Iulietto *et al.*, 2018). Unexpected or intermittent loud noises produce a fear response in pigs, as showed by the number of pigs huddling in the pen looking for protection or escaping from the source of noise (Geverink *et al.*, 1998), increased heart rate, and greater blood lactate, CK and cortisol levels at slaughter (Faucitano, 2010), all resulting in increased production of PSE pork meat (Faucitano and Velarde, 2021). Vermeulen *et al.* (2015) associated noise level lower than 85 dB in the peri-mortem to a reduced risk for PSE pork meat. To avoid stops and round turns while moving pigs forward, lairage alleys must be sufficiently and consistently lighted to avoid shadows and contrasts in colour on the floor (Faucitano and Velarde, 2021). Research has shown that driving pigs from the unloading dock to the lairage pen under low light brightness (40 lux) can result in a lowest risk of DIP (Vitali *et al.*, 2014), while greater light intensity (>160-215 lux) would make pigs easier to drive forward into the stunning chute reducing electric prodding (Grandin, 2010). Observations done at some European and Canadian slaughter plants showed that the use of green lighting improves ease of moving pigs through the alleys as it reduces shadows on the floor (Faucitano and Velarde, 2021).

5.4.3 Moving pigs in lairage and to the stunner

A greater proportion of pigs slipping, jamming, backing-up and vocalising has been reported when moving them through the lairage alleys, from unloading to the stun area, due to the additive effect of high slaughter speed, poorly designed lairage layout and group size (Faucitano, 2018), These behavioural responses have been associated with greater heart rate (Correa *et al.*, 2010, 2013), blood lactate and CK levels at slaughter (Edwards *et al.*, 2010; Hambrecht *et al.*, 2005; Rocha *et al.*, 2015) and skin lesions scores (Rabaste *et al.*, 2007), and more exudative pork meat. Rocha *et al.* (2016) reported a significant positive correlation between the frequency of slips and electric prod use in the alley feeding the stunning chute and pork meat exudation (r=0.74 and r=0.69, respectively). Recently, it has been demonstrated that when physical effort and psychological stress prior to stunning are kept at minimal levels, the impact of the stunning system (electrical vs gas stunning) on pork meat quality is minor (Terlouw *et al.*, 2021a).

The arrangement of the pens and the alleys as well the presence of angles or dead-ends combined with group size are critical factors influencing the flow of pigs to the lairage pen and to the stun area. The herringbone arrangement of pens, where pens are positioned on an angle of about 45° to the entry and exit passages, helps easy movement in and out of the pen (Faucitano and Geverink, 2008). Alleys should be large enough for pigs to walk next to and behind each other in groups and should be level or have a slope angle lower than 3° (Holmes *et al.*, 2020).

In the peri-mortem area, the design of the alley feeding the stunning chute and the chute itself is also very important. Compared with the funnel-shape entrance, the use of the batch flow system or curved raceway allows a smoother flow of pigs into the single/double electric stun raceway (Figure 5.2A,B), provided it is not overfilled with pigs (<75% of the capacity; Grandin, 2017), as showed by the reduced number of pigs backing-up/stopping and electric prod interventions compared with the straight single file entrance (Edwards *et al.*, 2010). The traditional entrance of pigs in single line into a CO_2 gas stunning system can also result in handling issues due to the stop-start forward movement of pigs along the chute to allow the first pigs in line to enter the cradle (Faucitano, 2010). A significant improvement is represented by the group-wise stunning system, where groups of 15 pigs are moved forward using automatic push gates and loaded in groups of 5 into the cradle (Christensen and Barton-Gade, 1997). Lower proportion of PSE and blood-splashed pork meat, and bruised carcasses have been reported in pigs handled through this system (Christensen and Barton-Gade, 1997; Franck *et al.*, 2003).

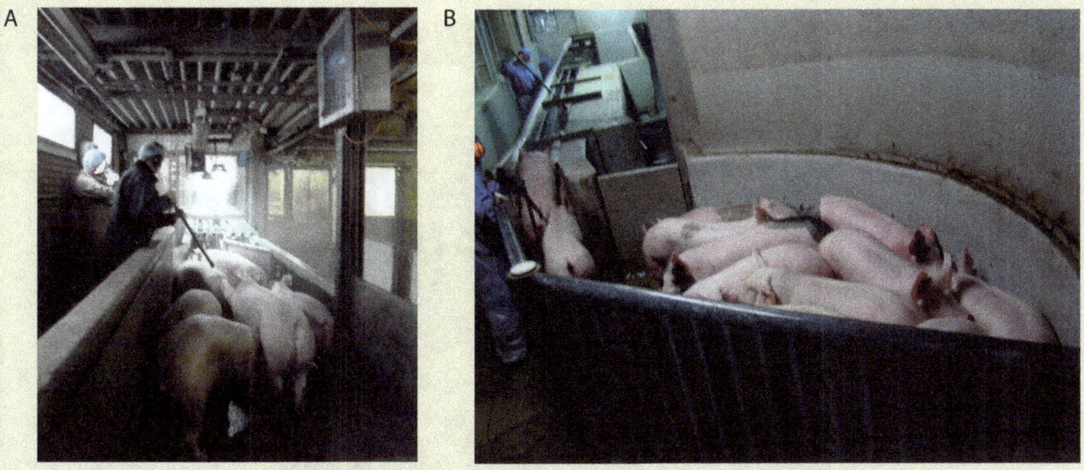

igure 5.2. (A) Funnel-shape and (B) curved race entrance into the stunning chute (Photo courtesy (A) Eyes on Animals, the etherlands; (B) L. Faucitano, AAFC, Canada).

The stun chute should be short (<3 m) with no bends (Faucitano and Geverink, 2008). The use of a too long and large (>1 pig wide) raceways combined with the absence of hold-down bars (Figure 5.3) has been associated with increased heart rate and blood lactate and cortisol levels at slaughter, and higher skin damage scores resulting from overlapping between pigs lining-up in the chute and greater electric prod use (Faucitano, 2010).

The ease of handling and moving of pigs through the lairage alleys increases as the group size decreases from 10 to 5 pigs (Dalla Costa *et al.*, 2019a). Moving too large groups of pigs (up to 45 pigs per batch) may result in a greater incidence of PSE and blood-splashed pork meat due to increased need for electric prod use (Barton-Gade *et al.*, 1992). The difficulty of moving batches of large size is exacerbated by the presence of corners and dead-ends in the handling area. Recently, Herskin *et al.* (2020b) reported an additive effect of alley design and group size during sow handling at the slaughter plant, with more stops and 180° turns being observed when larger groups of sows (35-95 sows) had to negotiate a 90° corner compared with smaller groups (5-6 sows).

5.5 Stunning and slaughter

Slaughter, including the restraining of the animal, the stunning application and the exsanguination, for human consumption is the last stage of pig production.

Consumers are increasingly concerned about what they eat and aware that animals are sentient beings and demand high welfare standards throughout the production chain, especially at the time of killing. Humane slaughter regulations and guidelines therefore require that animals are rendered immediately unconscious and insensible by a stunning method prior to bleeding and they must

Figure 5.3. Too long stunning chutes with no hold-down bars slow down the flow of pigs to the stunner and are a source of stres and skin lesions in pigs (L. Faucitano, AAFC, Canada).

remain so until death occurs through exsanguination (blood loss). Stunning can be a process that causes immediate loss of consciousness and sensibility (reversible stunning methods) or induces immediate death (irreversible stunning methods). As a general guide to good practice at slaughter, pigs should not be restrained if the operator is not ready to stun them and they should not be stunned if the operator is not ready to bleed them, i.e. without any delay between the processes. The European Food Safety Authority (EFSA, 2020) has recently published a scientific opinion on welfare of pigs at slaughter, which involved identification of hazards, welfare consequences (pain, fear and relevant animal-based measures) due to the hazard, its origin (staff, equipment or facility) and suggested preventive and corrective measures at each of the slaughter processes, including restraining, stunning and bleeding, among others.

The main stunning methods used in the slaughter of pigs are electrical and exposure to gas mixtures.

5.5.1 Electrical stunning

Electrical stunning may be applied head-only or head-to-body. In both cases, to facilitate the correct application of electrical stunning pigs must be immobilised using a restrainer matching the size of the animal in order to facilitate the correct placement of the electrodes, and a continuous flow of current.

5.5.1.1 Restraint systems

In electrical stunning, pigs can be either restrained in a V-type or conveyor belt system, where pigs are lifted up from the ground level and either conveyed by the flanks or astride a shaped belt, respectively, to the stunner. Comparative studies showed that pigs restrained in the V-type restrainer presented greater heart rate (+60 heart beats/min) and proportion of PSE and blood-splashed pork meat compared with those handled with the conveyor belt (Faucitano, 2010). As a general guide to good practice at slaughter, pigs should not be restrained if the operator is not ready to stun them as it may cause pain, fear or distress in pigs as showed by increased slips or falls, struggle, escape attempts and vocalisation (Grandin, 2016; Velarde and Dalmau, 2012). Weschenfelder *et al.* (2013b) and Rocha *et al.* (2015) reported increased skin temperature (up to 38 °C) and blood lactate concentration (+2 mM compared to rest levels) in pigs stuck in the restrainer for a few seconds.

5.5.1.2 Head-only electrical stunning

Head-only electrical stunning involves application of a current of sufficient magnitude through the brain to induce generalised epileptiform activity in the electroencephalogram (EEG; Figure 5.4), which is indicative of unconsciousness and insensibility.

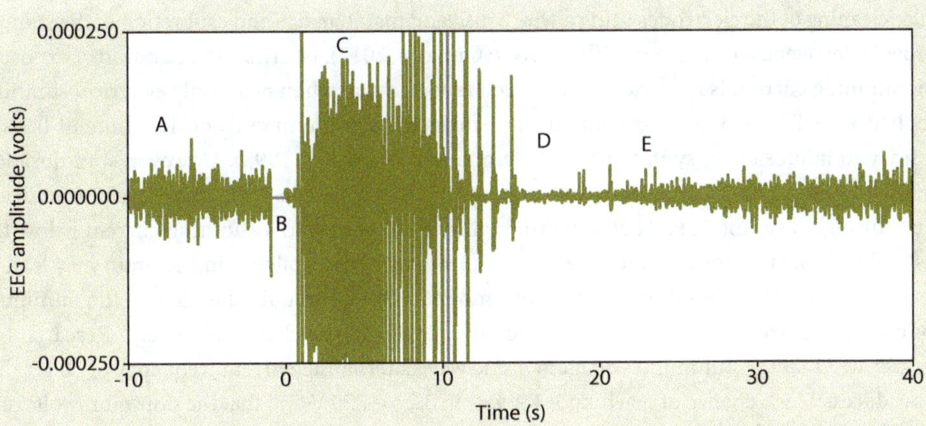

Figure 5.4. Example of head-only electrical stunning induced changes in the electroencephalogram (EEG); A = pre-stunning activity; B = EEG not recorded during stunning; C = epileptiform activity; D = quiescent EEG; and E = recovery of normal activity resembling pre-stunning, indicative of consciousness. Pigs are unconscious during epileptic and quiescent phases in the EEG.

The effectiveness of head-only electrical stunning depends upon several factors. Firstly, the position of stunning electrodes (tongs), which should be ideally placed on either side of the head, between the eyes and base of the ears, such that they span the brain. Under commercial slaughterhouse conditions, involving manual stunning in low throughput or mechanical stunning in high throughput, it is not always possible to apply the electrodes in this recommended position. In addition, the stunning electrodes placed in the ideal position may slip and slide due to collapse of the animal as soon as stunning current is applied. Research has shown that stunning electrodes needs to be placed trans-cranially, which can be bitemporal, dorsoventral or diagonal placement of stunning electrodes on the head, such that they span the brain (Anil and McKinstry, 1998). Secondly, the amount of voltage/current required to inducing epileptiform activity in the brain would vary according to the tissue electrical resistance/impedance in the pathway between the electrodes. For example, under the ideal bitemporal tong position, a current of 0.5 A delivered using a 50 Hz sine wave alternating current (AC) would be sufficient, whereas, when the tongs are placed just behind the ears a current of 1.3 A would be necessary to stun pigs (Anil, 1991; Hoenderken, 1978). Under commercial conditions, given the difficulty to place the electrodes in the ideal position, a minimum of 1.3 A is recommended in Europe when the electrodes are placed behind the ears. Wotton and O'Callaghan (2002) demonstrated that the impedance of a pig's head was predominantly a function of the stunning voltage and decreased non-linearly with increasing voltage used to stun the animal. It was suggested that the applied voltage should be high enough, a minimum of 240 V, to breakdown the impedance rapidly and produce an immediate stun. Thirdly, the electrical contact between the animal and stunning electrodes should be sufficient to facilitate current flow necessary to achieve immediate stunning. Poor electrical contact can occur due to the presence of dense hair at the electrode position or due to uneven bone surface in the skull of some animals, especially in mature pigs. It has been shown that the pressure applied during the application of stunning, i.e. how hard the stunning tongs were pressed against the skull to penetrate the fleece and dirt, is important to ensuring good electrical contact and it should be maintained during the entire duration of stunning. A minimum pressure of 25 kg has been suggested to be optimum (DEFRA, 2005). In addition, regular cleaning of the electrodes using a wire brush and maintenance and calibration of the stunner improves effectiveness of stunning (EFSA, 2004; Grandin, 2017). Fourthly, the duration of exposure to the stunning current is important. It has been reported that when head-only electrical stunning is applied with 1.3A using a constant current stunner, a minimum current flow time of 0.3 s is necessary to induce epilepsy in the brain (Berghaus and Troeger, 1998). However, stun duration should be at least 2 to 3 s to produce a duration of unconsciousness, which will last until death occurs through bleeding (EFSA, 2004). Prolonging the exposure to a stunning current below the threshold necessary to induce immediate onset of a generalised epilepsy in the brain may lead to painful electrical immobilisation. Finally, the amount of voltage required to deliver the minimum recommended current depends upon the frequency (Hz) of the electrical current. Research has demonstrated that the minimum voltage at a sine wave alternating current frequency of 50 Hz that will produce a 97.5% chance of reaching 1.3 A within 0.2 s is 250 V and that the minimum voltage at a sine wave alternating current frequency of 1,500 Hz that will produce a 97.5% chance of reaching 1.3 A within 0.2 s is 350 V. Evidently, the minimum voltage required to delivering 1.3 amps increases with increasing frequency of the stunning current (DEFRA, 2005).

Monitoring of the effectiveness of head-only electrical stunning has to be performed routinely to ensure good animal welfare and identify the causes of failure. Electrical stunning induced changes in the EEG can be correlated with the animal-based measures that could be used reliably (EFSA, 2013). In slaughterhouse conditions, effective head-only electrical stunning induced loss of consciousness can be recognised from the immediate collapse of the animal and tonic immobility during exposure to the stunning current. Immediately after exposure to the current, pigs show tonic seizure followed by clonic seizures, indicative of a generalised epilepsy in the brain. During the tonic seizure, pigs are in a state of tetanus and stretch out their fore- and hind-legs, breathing is absent, and the eyeballs are fixed or rotated into the socket. The tonic seizure is followed by the clonic seizures which manifests as kicking of legs, paddling or galloping movements (Berghaus and Troeger, 1998; EFSA, 2013). Pigs may show two episodes of clonic seizures and they may start breathing during the second episode, leading to recovery of consciousness.

Reflexes that would require brain control are abolished during generalised epilepsy. The animal stops breathing during tonic seizure and the clonic seizures may occur in two episodes. The palpebral (elicited by touching eyelashes or inner or outer canthus of the eye), corneal (elicited by touching the cornea) and pupillary (elicited by focusing bright light into the pupil) reflexes and response to external stimuli including pain (e.g. nose prick) are also abolished during the period of unconsciousness (Anil, 1991). However, presences of corneal reflex during seizures may not be a reliable indicator of consciousness (Vogel *et al.*, 2011). A persistent negative corneal reflex indicates a profound brain dysfunction (Gregory, 1998). Pigs start breathing during the second episode of clonic seizures, which is the beginning of recovery of consciousness. Pigs will show positive eye reflexes and start to vocalise soon after resumption of breathing, and therefore, any animal showing spontaneous breathing should be re-stunned or a back-up method should be applied immediately (Anil, 1991; Anil and McKinstry, 1998).

5.5.1.3 Head-to-body electrical stunning

Pigs can also be electrically stunned by inducing cardiac ventricular fibrillation (cardiac arrest) through a two current cycle application of the electric current across the head followed by a second current applied between the head and the chest, behind the position of the heart (automated head-to-chest stunning) or on the back (manual head-to-back stunning). This electrical stunning should always be performed using a 50 Hz sine wave alternating current (AC), as higher frequencies do not induce cardiac ventricular fibrillation, and a minimum current of 1.3 A should be still applied (EFSA, 2004). Vogel *et al.* (2011) investigated the efficacy of head-only stunning for 3 s immediately followed by application of the same stunning electrodes to the cardiac region of the animal for 3 s while lying in lateral recumbency. Based on the results it was concluded that the two-cycle method reduced the incidence of recovery of consciousness following stunning without significant effects on meat quality or slaughterhouse operation speed. Nodari *et al.* (2014) reported that head-to-body automatic stunning in a pig slaughterhouse involving application of on average 2.5 A across the head and 1.0 A between the head and chest electrodes resulted in the best animal welfare outcomes with none of the animals showing signs of consciousness.

Effective head-to-body electrical stunning is characterised by tonic seizure during exposure to the stunning method. After exposure to the current, pigs continue to show tonic seizure followed by clonic seizures, leading to flaccid carcass (loss of muscle tone) that can be recognised from drooping ears and limp legs. Breathing is absent and eyes are fixed or rotated in their sockets. Corneal and palpebral reflex are abolished and reaction to pain stimuli are absent.

However, during manual head to body stunning involving two electric current cycle method, the time interval between the two applications is critical. For instance, tonic-clonic seizures induced by the first head-only electrical stunning may impede the application of second current cycle across the chest to induce cardiac ventricular fibrillation. Therefore, additional care should be taken to apply the second cycle before the effectively stunned animals recover consciousness, which can be recognised from the resumption of spontaneous breathing.

5.5.1.4 Exposure to gas mixtures

Pigs may be exposed individually or in a small group to CO_2 at high concentrations, CO_2 mixed with inert gases, or inert gases alone. Published research so far involved exposure of pigs to argon (Ar) or nitrogen (N_2) as an alternative to high concentrations of CO_2. This method has two animal welfare advantages compared with electrical stunning: (1) animals can be exposed to gas mixtures in groups with the minimum amount of restraint and handling stress; and (2) irreversible stunning of pigs eliminates the chance of recovery of consciousness (EFSA, 2004; Velarde *et al.*, 2007). Two main systems are used commercially, the dip-lift system and the paternoster system. Dip-lift systems has a cage (Figure 5.5) into which a group of up to 6 pigs can be loaded and the cage is lowered into maximum concentrations of CO_2 at the bottom of the pit (EFSA, 2004). The paternoster system has

Figure 5.5. A CO_2 stunning system featuring a cradle or gondola (photo courtesy of A. Velarde, IRTA, Spain).

up to 7 cages or gondolas, each with a capacity to hold 2 to 6 pigs, depending upon the throughput, rotating through the CO_2 gradient in a 3-8 m deep pit, stopping at various intervals for loading of live pigs on one side and unloading unconscious pigs on the other side for sticking (EFSA, 2004). The number of pigs per group, the time taken to reach maximum CO_2 concentrations, and total exposure times are manipulated by the individual abattoirs according to their own discretion (Atkinson *et al.*, 2012). The space allowance in the cage shall be enough to allow the animals to lie down or to stand without being on top of each other or stacked even at maximum permitted throughput. Gondolas overloading is a criterion in animal welfare auditing protocols, based on the observation of 25 to 50 gondolas, depending on the slaughter plant size. Animal welfare scores range from excellent (no gondola is overloaded) to serious problem (one or more pigs forced by the handler to jump on top of the other pigs in the gondolas with an electric prod or by hitting, shoving or kicking). For example, for a gondola measuring 2.7×1.2 m wide, the target group size by body weight should be up to 8 pigs of 134 kg and less to 5 pigs heavier than 145 kg (Grandin, 2017).

Exposure to high concentrations of CO_2 in air

Carbon dioxide stunning of pigs is widely used in high throughput slaughterhouses, based on its positive effects on pork meat quality compared with electrical stunning (Velarde *et al.*, 2000b, 2001), although best results are reported at higher CO_2 gas concentration (90 vs 80%) and longer gas exposure time (100 vs 70 s; Nowak *et al.*, 2007). However, Webster (1994) argued that 'the best of the existing stunning systems based on CO_2 undoubtedly cause more distress at the point of stunning than the best of the high voltage electrical stunning systems'. Indeed, the use of this stunning system remains to be controversial from the animal welfare perspective as CO_2 in high concentration causes pain, fear and respiratory distress (EFSA, 2020) based on signs of aversion, such as retreat attempts, head-shaking, sneezing, breathlessness, freezing, escape attempts and gasping, shown by pigs during CO_2 exposure (Faucitano and Velarde, 2021). These behaviours have at least three important animal welfare reasons. Firstly, CO_2 being an acidic gas, is pungent to inhale in high concentrations (40% by volume or more in air). It has been reported that majority of pigs (88%) avoid an atmosphere containing high (>80% by volume) concentrations of CO_2 (Cantieni, 1976; Raj and Gregory, 1995). This aversion was found to be greater than the motivation to obtain a reward (apples) in the CO_2 atmosphere, even after 24 h fasting (Raj and Gregory, 1995). Velarde *et al.* (2007) reported that aversion was higher at 90 than at 70% CO_2, likely due to greater irritation of the nasal mucosa membranes and more severe hyperventilation. However, 90% CO_2 induced anaesthesia faster, as shown by the reduced latency to lose posture (Velarde *et al.*, 2007). Secondly, CO_2 being a potent respiratory stimulant, causes severe hyperventilation prior to loss of consciousness. Hyperventilation occurs during exercise as well as inhalation of carbon dioxide. Carbon dioxide-induced hyperventilation differs, however, in that carbon dioxide is more potent than exercise in inducing dyspnoea (breathlessness), and with carbon dioxide inhalation there is a greater feeling of being unable to breathe deeply enough (Gregory *et al.*, 1990). Exercise induces dyspnoea at a given concentration of CO_2 and it also results in gradual increases in the blood CO_2 concentration. In humans, a blood CO_2 tension increase of 5 mm Hg above normal level will stimulate respiration, whereas the blood oxygen (O_2) tension has to decrease by about 60 mm Hg from the normal level before hypoxia stimulates the respiratory centres in the brain. Owing to the respiratory distress, some pigs exposed to less than 70% CO_2 in air were found to show escape attempts (Raj and Gregory, 1996).

Although the time to onset of breathlessness depends upon the concentration of CO_2, the cumulative respiratory distress occurring in conscious pigs would appear to be very similar between 40 to 90% CO_2 in air (Raj and Gregory, 1996). Nevertheless, exposure of pigs to CO_2 leads to metabolic acidosis (reduction in blood pH; Mota-Rojas *et al.*, 2012) and, as a consequence, significant reduction in the pH of the cerebrospinal fluid bathing the brain, which, in turn, induces gradual loss of consciousness and sensibility through inhibition of the spontaneous brain activity (Rodriguez *et al.*, 2008). The time to loss of consciousness during exposure of pigs to 80 to 90% CO_2 in air is highly variable and it could be as long as 38 s (Hoenderken *et al.*, 1979). It has been reported that somatosensory evoked potentials in the brain was abolished on average 36 s during exposure to more than 80% carbon dioxide in air (Raj *et al.*, 1997). Rodriguez *et al.* (2008) recorded changes occurring in the middle latency auditory evoke potentials (MLAEP) in the brain of pigs during exposure to 90% CO_2 in air and reported that loss of consciousness began on average after 60 s of exposure to the gas and all the pigs exhibited aversive reactions prior to loss of consciousness.

Exposure to a mixture of carbon dioxide and inert gases

Research has shown that the majority (75%) of pigs did not show aversion to the presence of 30% CO_2 in air in the feeding chamber to obtain a reward of apples (Raj and Gregory, 1995). However, exposure of pigs to 30% by volume of CO_2 in Ar mixture induced moderate respiratory distress (assessed from the audio tape) prior to loss of consciousness (Raj and Gregory, 1996). Exposure of pigs to a mixture of less than 40% CO_2 by volume and inert gases, Ar or N_2, with less than 2% residual O_2 by volume leads to hypercapnic anoxia. It has been reported that the brain evoked potentials were abolished on average at 17 s during exposure of pigs to a mixture of 30% CO_2 in Ar with 2% residual O_2 (Raj *et al.*, 1997). While using a mixture of CO_2 and inert gases, the concentration of CO_2 as well as the residual O_2 are critical, and the duration of exposure required to stun pigs will vary depending upon the concentrations of these gases.

Gas mixtures of Ar or N_2 and up to 30% CO_2 have high stability and uniform concentrations in the pit (Dalmau *et al.*, 2010), and therefore, this mixture can be used in the existing gas stunning equipment. Llonch *et al.* (2012a, 2013) investigated the effects of exposure of pigs to mixtures of 70% N_2 and 30% CO_2, 80% N_2 and 20% CO_2, or 85% N_2 and 15% CO_2 and the results indicated that pigs show signs of aversion during exposure to all the mixtures compared to atmospheric air but the aversive response did not increase in consecutive sessions. Furthermore, the onset of unconsciousness based on the changes occurring in the EEG (index of unconsciousness, IoC) was reached after 45, 47, 43 s, respectively, of exposure. However, it resulted in increased incidence of PSE meat and ecchymosis in the musculature when compared with 90% CO_2 (Llonch *et al.*, 2012b). More recently, Atkinson *et al.* (2020) compared the welfare and meat quality effects of stunning pigs with 80 to 90% CO_2 or a mixture of 20% CO_2 and 80% N_2 in a commercial slaughterhouse. The results of this study showed that: (1) exposure to both the gases provoked aversion and discomfort, but these were less severe in N_2 mixture when compared with high CO_2; (2) the stun quality of the N_2 mixture was poorer (7.4% showed signs of poor stunning) than high CO_2; (3) the stunning quality of the N_2 mixture was improved when residual oxygen (O_2) concentration was maintained below 2%; and (4) meat quality was slightly poorer in N_2 stunning compared to high CO_2, with a higher percentage of carcasses showing PSE pork meat. More recently, the behavioural reactions and heart rates of pigs exposed to different gas mixtures, i.e. 80% CO_2/air for 90 s, 40% CO_2/30% O_2/air for 180 s and 70% N_2O/30% CO_2 for 90 s, were assessed by Terlouw *et al.* (2021b). In this study, all gas mixtures produced discomfort as, before losing consciousness, most pigs presented respiratory difficulties and increased heart rates.

Exposure to inert gases

The mechanism of induction of unconsciousness by anoxia is due to the inhibition of N-methyl-D-aspartate (NMDA) receptor channels in the brain, which is essential for maintaining neuronal arousal during conscious state (EFSA, 2004). Anoxia occurring as a result of exposure to Ar or N_2 induces unconsciousness by depriving the brain of O_2. Cerebral dysfunction occurs when the partial pressure of O_2 in cerebral venous blood falls below 19 mmHg.

Research has shown that pigs do not show any aversion to the presence of 90% Ar in air with 2% residual O_2 (Raj and Gregory, 1995). It has been reported that exposure to Ar induced minimal respiratory distress and pigs lost posture without any evidence of behavioural arousal or escape attempts (Raj and Gregory, 1996). It has been reported that the brain evoked potentials were abolished on average at 15 s during exposure of pigs to Ar with 2% residual O_2 (Raj *et al.*, 1997). It is worth noting that unlike exposure to high concentrations of CO_2, inert gases will expand from liquid to gaseous form as soon as they are released from the pressurised containers, using heat from the environment. Therefore, the exposure of the animals to low temperature is not a hazard.

Further investigation carried out under commercial conditions determined the feasibility of using anoxia to stun/kill pigs in a paternoster system (Raj, 1999). The results indicated that, pigs should be ideally exposed to anoxia for 3 min and stuck within 15 s or exposed for 5 min and stuck within 45 s of exiting the gas to prevent resumption of consciousness. The results also indicated that some pigs require a minimum of 7 min to die during exposure to 90% Ar in air.

There is lack of will to change from using high concentrations of CO_2 to inert gases. It is frequently argued that Ar is expensive without performing real cost benefit analysis, taking animal welfare into account. Although research into anoxia stunning of pigs involved Ar as an anoxic agent, N_2 can also be used either alone or in combination with a low concentration of argon. It is reassuring to animal welfare scientists to note that a mixture of 20 to 25% Ar in N_2 is being used to stun/kill broiler chickens and turkey, whilst they are in transport containers, in the UK.

Some people may argue that exposure to inert gas induced anoxia requires longer exposure times than high concentrations of CO_2 to kill pigs, which will lead to reduced throughput rate. A number of options have been proposed previously for commercial development (Raj, 2008):

- exposure for 90 s to anoxia induced anoxia followed by induction of cardiac arrest, whilst the pigs are within the chamber, using an electric current;

- exposure for 3 min to anoxia induced anoxia followed by induction of cardiac arrest or bleeding within 15 s of exiting the chamber;
- exposure for 5 min to anoxia followed by bleeding within 45 s; and
- exposure for 7 min followed by bleeding within 60 s.

Signs of unconsciousness that could be used for monitoring effectiveness of exposure to gas mixtures include fixed eyes, dilated pupil, absence of the palpebral, corneal and pupillary reflexes and absence of response to painful stimuli, such as nose prick or ear pinch (Raj, 1999; Rodriguez *et al.*, 2008).

Nitrogen filled high expansion foam

Owing to the fact N_2 is lighter than air, it would be difficult to contain this gas within the existing gas stunning chambers and maintain an anoxic atmosphere with less than 2% residual O_2 necessary to stun pigs effectively. However, N_2 filled foam could be used to fill a chamber and create anoxic atmosphere and the feasibility of using N_2 filled high expansion foam to kill poultry has been tested (McKeegan *et al.*, 2013; Raj *et al.*, 2008). More recently, Lindahl *et al.* (2020) evaluated the feasibility using N_2 filled high expansion foam for killing pigs and the results indicated that pigs did not show any strong aversive behaviours when exposed to foam, however, they avoided putting their heads and snouts into the foam, and the rate of escape attempts through the lid increased when foam levels became high. The authors nevertheless concluded that this technique may be a viable alternative to using high concentrations of CO_2 for stunning pigs.

Overall, there is an urgent need to find potential alternative for stunning pigs at slaughter. In this regard, Sindhoj *et al.* (2021) reviewed the scientific literature published between 2010 and 2020 to find recent developments or evaluations of alternative methods that could lead to the replacement of CO_2 for stunning pigs at slaughter. The concluding remarks from the authors were: from the perspective of pig welfare, viable alternatives should consider low stress pre-handling, group stunning without restraint, immediate onset of unconsciousness or absence of avoidable pain, distress and suffering until the loss of consciousness, consistency of stunning and duration of unconsciousness and insensibility until death. However, in order to be widely adopted by industry, alternative methods must also be considered viable in terms of operator safety, production efficiency (processing speed and capacity), integration into slaughter production lines, improved meat and carcass quality, food safety and profitability.

5.5.2 Slaughter

Bleeding of pigs without delay following stunning is an important step in the slaughter process intended to cause death in unconscious animals. The return to consciousness after stunning and bleeding efficiency depend on the stunning method that is used, the interval between stun and sticking and the bleeding procedure.

It should be emphasised that the duration of unconsciousness induced by head-only electrical stunning should last longer than the sum of time between the end of stunning and sticking and the time to onset of death due to blood loss following sticking. The bleed out time should be long enough to allow for death to occur in animals and death should be confirmed before carcass processing begins (e.g. entering the scalding tank). The appropriate stun-to-stick interval needs to be calculated under the prevailing electrical stunning parameters and slaughter situations. The minimum time to return of rhythmic breathing, which is the earliest sign of recovery of consciousness, following effective head-only electrical stunning of pigs is reported to be 37 s (Anil, 1991). Research has also shown that, when chest sticking is performed in pigs, the time to loss of evoked potentials in the brain, which is indicative of loss of brain responsiveness to external stimuli, is 22 s (Wotton and Gregory, 1986). By subtracting the time to loss of evoked potentials following an accurate sticking from the minimum duration of apparent unconsciousness produced by an accurate stun, the maximum acceptable stun-to-stick interval can be calculated as 15 s. The stun-to-stick interval is frequently prolonged in small slaughterhouses if one operator performs head-only electrical stunning on groups (e.g. of four or more) of free-standing pigs in a stunning pen and then performs hoisting and sticking of these pigs, not necessarily in the sequential order of stunning. Inevitably, the stun-to stick interval will be considerably longer for the last pig to be stuck in a group. Incomplete sectioning of the brachiocephalic trunk can be a problem, as studies have shown that up to 13% of pigs required more than one sticking attempt to bleed out properly (Anil *et al.*, 2000; Spencer and Veary, 2010).

Head-to-body (chest or back) application of an electric current eliminates the chances of recovery of consciousness as it induces cardiac arrest and thus the death of the animal. However, the animal has to be bled-out within 3 min in order to prevent blood clots in the large vessels and ensure a satisfactory bleeding (Gregory, 1998).

Pig stunning with a high concentration (>80% by volume in air) of CO_2 can be reversible or irreversible. The depth and duration of unconsciousness achieved with CO_2 depends upon the animal, the CO_2 concentration gradient in the pit, the speed at which animals are lowered towards the bottom of the pit, where the highest CO_2 concentration is achieved, and the duration of exposure to the maximum concentration (Raj and Gregory, 1996; Troeger and Woltersdorf, 1991). Sticking must start as soon as possible after pigs exiting the gas to prevent resumption of consciousness. Velarde *et al.* (2000a) showed a lower proportion of pigs showing signs of consciousness (sensibility to pain and righting reflex) when the duration of the carbon dioxide cycle was longer (103 vs 98 s) and the interval between discharge from the system to sticking was shorter (43 vs 58 s). Prolonged exposure may result in irreversible stunning and eliminate the chances of recovery of consciousness. Exposure of pigs to a minimum of 90% by volume of CO_2 in air for 3-5 min results in death of most of the pigs at the exit from the gas.

Raj (1999) reported that exposure of pigs to the CO_2/Ar mixture with 2% residual O_2 for 3 min resulted in satisfactory stunning, however, bleeding should commence within 15 s to avoid resumption of consciousness, as in head-only electrical stunning. A 5 min exposure to this gas mixture followed by bleeding within 45 s prevented carcass convulsions during bleeding. The exposure of pigs to CO_2/Ar mixture for 7 min resulted in death in all the pigs.

Monitoring the state of consciousness at the exit and during bleeding can be performed using relevant ABMs. In this sense, complete cessation of any respiratory activity, including gagging, and loss of muscle tone at the exit of the chamber has been reported (Raj, 1999). Other signs of unconsciousness include fixed eyes, dilated pupil, absence of the palpebral, corneal and pupillary reflexes and absence of response to painful stimuli such as nose prick or ear pinch (Raj, 1999; Rodriguez *et al.*, 2008).

Normally, pigs are bled with a chest stick aimed at severing the brachiocephalic trunk which gives rise to the carotid arteries that supply oxygenated blood to the brain. Chest sticking is performed by inserting a knife on the ventral aspect of the base of the neck, just in front of the sternum, towards the thoracic inlet. The blade of the knife should be long enough to reach the brachiocephalic trunk. The size of the sticking incision should be large enough (e.g. 10 cm) to allow profuse bleeding and rapid onset of death (Anil *et al.*, 2000).

5.6 Conclusions

The research findings overviewed in this chapter provide the evidence about the impact of the farm of origin, transport to slaughter and lairage conditions, and stunning and slaughter practices on the welfare and carcass and meat quality of pigs. They also highlight the importance of training programs for handlers and truckers through the whole marketing process and the usefulness of animal welfare audit approval. However, more has to be done. So far, most handling and transportation research focused on market weight pigs (up to 160 kg), but there is a considerable lack of information regarding heavier pigs, including cull sows and boars. Future research should also focus on truck design, i.e. air flow patterns and vibration rate, and insulation and cooling systems under different ambient conditions, management of rest stops during long distance transportation and strategies to limit fighting in mixed groups of pigs in lairage.

There is a clear evidence that the use of CO_2 for stunning pigs remains controversial on animal welfare grounds and the use of electrical stunning is not entirely satisfactory on carcass and meat quality grounds. Owing to this conflict between animal welfare and meat quality, further research and development are needed to implement a commercially feasible and economically viable method ensuring to eliminate pain, fear and respiratory distress during the induction to unconsciousness and meeting consumers' demand for higher welfare at slaughter.

References

Aaslyng, M., Brandt, P., Blaabjerg, L. and S. Støier, 2013. Assessment and incidence of skin damage in slaughter pigs. In: Proceedings of the 59th International Congress of Meat Science and Technology. 18-23 August, 2012. Izmir, Turkey. Available at: http://icomst-proceedings.helsinki.fi/papers/2013_04_13.pdf.

Acevedo-Giraldo, Sánchez, J.A. and Romero, M.H., 2020. Effects of feed withdrawal times prior to slaughter on some animal welfare indicators and meat quality traits in commercial pigs. Meat Science 167: 107993. https://doi.org/10.1016/j.meatsci.2019.10799

Allison, C.P., Johnson, R.C. and Doumit, M.E., 2005. The effects of halothane sensitivity on carcass composition and meat quality in HAL-1843-normal pigs. Journal of Animal Science 83: 671-678.

Allison, C.P., Marr, A.L., Berry, N.L., Anderson, D.B., Ivers, D.J., Richardson, L.F., Keffaber, K., Johnson, R.C. and Doumit, M.E., 2006. Effects of halothane sensitivity on mobility status and blood metabolites of HAL-1843-normal pigs after rigorous handling. Journal of Animal Science 84: 1015-1021.

Anderson, D.B., 2010. Relationship of blood lactate and meat quality in market hogs. In: Proceedings of the 63rd Reciprocal Meat Conference of the American Meat Science Association. June 20-23, 2010. Texas Tech University, Lubbock, TX, USA.

Anil, M.H. and McKinstry, J.L., 1998. Variations in electrical stunning tong placements and relative consequences in slaughter pigs. The Veterinary Journal 155: 85-90.

Anil, M.H., 1991. Studies on the return of physical reflexes in pigs following electrical stunning. Meat Science 30: 13-21.

Anil, M.H., Whittington, P.E. and McKinstry, J.L., 2000. The effect of the sticking method on the welfare of slaughter pigs. Meat Science 55: 315-319.

Arduini, A., Redaelli, V., Luzi, F., Dall'Olio, S., Pace, V. and Costa, L.N., 2017. Relationship between deck level, body surface temperature and carcass damages in Italian heavy pigs after short journeys at different unloading environmental conditions. Animals 7: 10. https://doi.org/10.3390/ani7020010

Arndt, H., Volkmann, N., Spindler, B., Hartung, J. and Kemper, N., 2019. Do pigs have adequate space in animal transportation vehicles? Planimetric measurement of the floor area covered by finishing pigs in various body positions. Frontiers in Veterinary Science 5: 330. https://doi.org/10.3389/fvets.2018.00330

Atkinson, S., Algers, B., Pallisera, J., Velarde, A. and Llonch, P., 2020. Animal welfare and meat quality assessment in gas stunning during commercial slaughter of pigs using hypercapnic-hypoxia (20% CO_2 2% O_2) compared to acute hypercapnia (90% CO_2 in air). Animals 10: 2440-2456.

Atkinson, S., Velarde, A., Llonch, P. and Algers, B., 2012. Assessing pig welfare at stunning in Swedish commercial abattoirs using CO2 group-stun methods. Animal Welfare 21: 487-495.

Australian Government, 2012. Australian animal welfare standards and guidelines – land transport of livestock. Animal Health Australia (AHA). Available at: https://tinyurl.com/fbnma3m6

Averós, X., Knowles, T.G., Brown, S.N., Warriss, P.D. and Gosálvez, L.F., 2008. Factors affecting the mortality of pigs being transported to slaughter. Veterinary Record 163: 386-390.

Barton-Gade, P. and Christensen, L., 1998. Effect of different loading densities during transport on welfare and meat quality in Danish slaughter pigs. Meat Science 48: 237-247.

Barton-Gade, P., Blaabjerg, L.O. and Christensen, L., 1992. New lairage system for slaughter pigs: effect on behaviour and quality characteristics. In: Proceedings of the 38th International Congress of Meat Science and Technology, August 1992. Clermont-Ferrand, France, pp. 161-164.

Barton-Gade, P., Christensen, L., Baltzer, M. and Petersen, V.J., 2007. Causes of pre-slaughter mortality in Danish slaughter pigs with special emphasis on transport. Animal Welfare 16: 459-470.

Beattie, V.E., Burrows, M.S., Moss, B.W. and Weatherup, R.N., 2002. The effect of food deprivation prior to slaughter on performance, behaviour and meat quality. Meat Science 62: 413-418.

Bench, C., Schaefer, A.L. and Faucitano, L., 2008. The welfare of pigs during transport. In: Faucitano, L. and Schaefer, A.L. (eds.) The welfare of pigs – from birth to slaughter. Wageningen Academic Publishers, Wageningen, the Netherlands, pp. 161-195.

Berghaus, A. and Troeger, K., 1998. Electrical stunning of pigs: minimum current flow time required to induce epilepsy at various frequencies. In: Proceedings of the 44th International Congress of Meat Science and Technology. August 30th – September 4th, 1998. Barcelona, Spain, pp. 1070-1081.

Berry, N.L., Johnson, A.K., Hill, J., Lonergan, S., Karriker, L.A. and Stalder, K.J., 2012. Loading gantry versus traditional chute for the finisher pig: effect on welfare at the time of loading and performance measures and transport losses at the harvest facility. Journal of Animal Science 90: 4028-4036.

Berry, N.L., Ritter, M.J., Brunton, E., Stremsterfer, W., Hoag, B., Wolfe, J., Fitzgerald, N., Porth, M., Delaney, D. and Weldon, T., 2009. Effects of moving market weights pigs in different group sizes during loading on stress responses and transportation losses at the packing plant. In: Proceedings of the Midwest American Society of Animal Science. Des Moines, IA, USA, pp. 5.

Bertol, T.M, Braña, D.V., Ellis, M., Ritter, M.J., Peterson, B.A., Mendoza, O.F. and McKeith, F.K., 2011. Effect of feed withdrawal and dietary energy source on muscle glycolytic potential and blood acid-base responses to handling in slaughter-weight pigs. Journal of Animal Science 89: 1561-1573.

Blair, B. and Lowe, J., 2019. Describing the cull sow market network in the US: a pilot project. Preventive Veterinary Medicine 162: 107-109.

Bozzo, G., Padalino, B., Bonerba, E., Barasso, R., Tufarelli, V., Zappaterra, M. and Ceci, E., 2020. Pilot study of the relationship between deck level and journey duration on plasma cortisol, epinephrine and norepinephrine levels in Italian heavy pigs. Animals 10: 1578. https://doi.org/10.3390/ani10091578

Bracke, M.B.M., Herskin, M.S., Marahrens, M., Gerritzen, M.A. and Spoolder, H.A.M., 2020. Review of climate control and space allowance during transport of pigs. European Reference Centre for Animal Welfare (EURCAW) – pigs. Available at: https://www.eurcaw.eu/en/eurcaw-pigs/output/scientific-output.htm

Bradshaw, R.H., Parrott, R.F., Forsling, M.L., Goode, J.A., Lloyd, D.M., Rodway, R.G. and Broom, D.M., 1996a. Stress and travel sickness in pigs: effects of road transport on plasma concentrations of cortisol, beta-endorphin and lysine vasopressin. Animal Science 63: 507-516.

Bradshaw, R.H., Parrott, R.F., Forsling, M.L., Goode, J.A., Lloyd, D.M., Rodway, R.G. and Broom, D.M., 1996b. Behavioral and hormonal responses of pigs during transport: effect of mixing and duration of journey. Animal Science 62: 547-554.

Brandt, P. and Aaslyng, M.D., 2015. Welfare measurements of finishing pigs on the day of slaughter: a review. Meat Science 103: 13-23.

Brent, G., 1986. Housing the pig. Farming Press Ltd, Ipswich, UK.

Brown, J.A., Samarakone, T.S., Crowe, T., Bergeron, R., Widowski, T., Correa, J.A., Faucitano, L., Torrey, S. and Gonyou, H.W., 2011. Temperature and humidity conditions in trucks transporting pigs in two seasons in eastern and western Canada. Transactions of the American Society of Agricultural and Biological Engineers 54: 1-8.

Brown, J.A., Scheeren, M.B., Faucitano, L., Bergeron, R., Widowski, T.M., Crowe, T. and Gonyou, H.W., 2012. Truck compartment and trip duration affect pork quality of pigs transported under Canadian conditions. In: Proceedings of the 38th Animal Transportation Association (AATA) Annual Meeting. March 18-21, 2012. Vancouver, BC, Canada.

Brown, S.N., Knowles, T.G., Edwards, J.E. and Warriss, P.D., 1999. Relationship between food deprivation before transport and aggression in pigs held in lairage before slaughter. Veterinary Record 145: 630-634.

Brown, S.N., Knowles, T.G., Wilkins, L.J., Chadd, S.A. and Warriss, P.D., 2005. The response of pigs being loaded or unloaded onto commercial animal transporters using three systems. The Veterinary Journal 170: 91-100.

Brown-Brandt, T.M., Nienaber, J.A., Xin, H. and Gates, R.S., 2004. A literature review of swine heat production. Transactions of the American Society of Agricultural and Biological Engineers 47: 259-270.

Bucher, M. and Scheibl, P., 2014. Animal welfare – stunning and bleeding. In: Ninios, T., Lundén, J., Korkeala, H. and Fredriksson-Ahoma, A. (eds.) Meat inspection and control in the slaughterhouse. John Wiley & Sons, Oxford, UK, pp. 47-71.

Canadian Agri-Food Research Council (CARC), 2001. Recommended code of practice for the care and handling of farm animals transportation. CARC, ON, Canada, 63 pp.

Canadian Food Inspection Agency (CFIA), 2021. Red meat condemnation report by species-hogs. CFIA, Ottawa, ON, Canada. Available at: https://tinyurl.com/59pycwzk

Candek-Potokar, M. and Skrlep, M., 2012. Factors in pig production that impact the quality of dry-cured ham: a review. Animal 6: 327-338.

Cantieni, J., 1976. Ein Bbeitrag zur CO2-betabung von schlachtschweinen. Schweizer Archiv für Tierheilkunde 119: 355-375.

Carr, S.N., Gooding, J.P., Rincker, P.J., Hamilton, D.N., Ellis, M., Killefer, J. and McKeith, F.K., 2005. A survey of pork quality of downer pigs. Journal of Muscle Foods 16: 298-305.

Chevillon, P., 2001. Pig welfare during pre-slaughter and stunning. In: Proceedings of the 1st International Virtual Conference on Meat Quality. November 2000. Concordia, Brazil, pp. 145-158.

Chevillon, P., Frotin, P., Rousseau, P. and Kerisit, R., 2003. Incidence sur le bien-être des porcs de la densité de chargement et du protocole d'alimentation lors d'un transport des porcs de 36 heures incluant 9 heures d'arrêt. Journées de la Recherche Porcine en France 35: 179-186.

Christensen, L. and Barton-Gade, P., 1997. New Danish developments in pig handling at abattoirs'. Fleischwirtschaft 77: 604-607.

Christensen, L. and Barton-Gade, P., 1999. Temperature profile in double-decker transporters and some consequences for pig welfare during transport. In: Russel, A.J.F. (eds.) Farm animal welfare – who writes the rules? British Society of Animal Science, White Colne, UK, pp. 125-128.

Christensen, L., Barton-Gade, P. and Blaabjerg, L.O, 1994. Investigation of transport conditions in participating countries in the EC project: PL 920262. Proceedings of 40th International Congress of Meat Science and Technology, 28 August-2 September 1994. The Hague, the Netherlands. Available at: icomst-proceedings.helsinki.fi/papers/1994_12_02.pdf.

Clutter, A.C. and Brascamp, E.W., 1998. Genetics of performance traits. In: Rothschild, M.F. and Ruvinsky, A. (eds.) The genetics of the pig. CAB International, Wallingford, UK, pp. 427-462.

Cockram, M.S., 2019. Fitness of animals for transport to slaughter. Canadian Veterinary Journal 60: 423-429.

Cockram, M.S., 2020. Welfare issues associated with the transport of cull sows to slaughter. Veterinary Record 186: 183-184.

Conte, S., Faucitano, L., Bergeron, R., Torrey, S., Gonyou, H.W., Crowe, T. and Widowski, T.M., 2015. Effects of season, truck type and location within truck on gastrointestinal tract temperature of market-weight pigs during transport. Journal of Animal Science 93: 5840-5848.

Conte, S., Lawlor, P.G., O'Connell, N. and Boyle, L.A., 2012. Effect of split marketing on the welfare, performance, and carcass traits of finishing pigs. Journal of Animal Science 90: 373-380.

Correa, J.A., 2011. Effect of farm handling and transport on physiological response, losses and meat quality of commercial pigs. Advances in Pork Production 22: 249-256.

Correa, J.A., Gonyou, H.W., Torrey, S., Widowski, T., Bergeron, R., Crowe, T.G., Laforest, J.P. and Faucitano, L., 2013. Welfare and carcass and meat quality of pigs being transported for 2 hours using two vehicle types during two seasons of the year. Canadian Journal of Animal Science 93: 43-55.

Correa, J.A., Gonyou, H.W., Torrey, S., Widowski, T., Bergeron, R., Crowe, T.G., Laforest, J.P. and Faucitano, L., 2014. Welfare of pigs being transported for long distance using a pot-belly trailer during winter and summer. Animals 4: 200-213.

Correa, J.A., Torrey, S., Devillers, N., Laforest, J.P., Gonyou, H.W. and Faucitano, L., 2010. Effects of different moving devices at loading on stress response and meat quality in pigs. Journal of Animal Science 88: 4086-4093.

Correia, E.R., 2021. Incidência e causas de ocorrência de DOA e NANI em suínos durante o manejo pré-abate. MSc-thesis, Department of Animal Science, Universidade Estadual de Londrina, Londrina, Brazil.

Dalla Costa, F.A., Devillers, N., Paranhos da Costa, M.J.R. and Faucitano, L., 2016. Effects of applying preslaughter feed withdrawal at the abattoir on behaviour, blood parameters and meat quality in pigs. Meat Science 119: 89-94.

Dalla Costa, O.A., Dalla Costa, F.A., Di Castro, I.C., Gregory, N.G., Di Campos, M.S., De Medeiros Leal, G.B. and De Castro Tavernari, F., 2019a. Ease of handling and physiological parameters of stress, carcasses, and pork quality of pigs handled in different group sizes. Animals 9: 798. https://doi.org/10.3390/ani9100798

Dalla Costa, O.A., Dalla Costa, F.A., Feddern, V., Dos Santos Lopes, L., Coldebella, A., Gregory, N.G. and Monteiro de Lima, G.J.M., 2019b. Risk factors associated with pig pre-slaughtering losses. Meat Science 155: 61-68.

Dalla Costa, O.A., Faucitano, L., Coldebella, A., Ludke, J.V., Peloso, J.V., Dalla Roza, D. and Paranhos da Costa, M.J.R., 2007. Effects of the season of the year, truck type and location on truck on skin bruises and meat quality in pigs. Livestock Science 107: 29-36.

Dalmau, A, Nande, A, Vieira-Pinto, M., Zamprogna, S., Di Martino, G., Ribas, J.C.R., Paranhos da Costa, M., Halinen Elemoe, K. and Velarde, A., 2016. Application of the Welfare Quality® protocol in pig slaughterhouses of five countries. Livestock Science 193: 78-87.

Dalmau, A., Rodríguez, P., Llonch, P. and Velarde, A., 2010. Stunning pigs with different gas mixtures: part 1: gas stability. Animal Welfare 19: 315-323.

De Vries, A.G., Faucitano, L., Sosnicki, A. and Plastow, G.S., 2000. The use of gene technology for optimal development of pork meat quality. Food Chemistry 69: 397-405.

Department for Environment, Food and Rural Affairs (DEFRA), 2005. Electrical stunning in pigs: evaluation of the voltages and frequencies required for effective stunning while maintaining satisfactory carcass quality. Evidence project final report MH0110. DEFRA, London, UK.

Dewey, C., Haley, C., Widowski, T., Poljak, Z. and Friendship, R., 2009. Factors associated with in-transit losses of fattening pigs. Animal Welfare 18: 355-361.

Diesel, T.A., 2016. Fatores de risco associados às perdas quantitativas e econômicas ocorridas no manejo pré-abate de suínos. PhD-thesis, Universidade Estadual Paulista, Jaboticabal, Brazil.

Driessen, B., Beirendonck, S.V. and Buyse, J., 2020a. Effects of housing, short distance transport and lairage on meat quality of finisher pigs. Animals 10: 788. https://doi.org/10.3390/ani10050788

Driessen, B., Beirendonck, S.V. and Buyse, J., 2020b. Effects of transport and lairage on the skin damage of pig carcasses. Animals 10: 575. https://doi.org/10.3390/ani10040575

Driessen, B., Beirendonck, S.V. and Buyse, J., 2020c. The impact of grouping on skin lesions and meat quality of pig carcasses. Animals 10: 544. https://doi.org/10.3390/ani10040544

Edwards, L.N, Engle, T.E., Grandin, T.A., Ritter, M.J., Sosnicki, A., Carlson, B.A. and Anderson, D.B., 2011. The effects of distance traveled during loading, lairage time prior to slaughter, and distance traveled to the stunning area on blood lactate concentration of pigs in a commercial packing plant. The Professional Animal Scientist 27: 485-491.

Edwards, L.N., Grandin, T.A., Engle, T.E., Porter, S.P., Ritter, M.J., Sosnicki, A. and Anderson, D.B., 2010. Use of exsanguination blood lactate to assess the quality of preslaughter pig handling. Meat Science 86: 384-390.

EFSA Panel on Animal Health and Welfare (AHAW), 2011. Scientific opinion concerning the welfare of animals during transport. EFSA Journal 9: 1966. https://doi.org/10.2903/j.efsa.2011.1966

Ellis, M. and Bertol, T.M., 2001. Effects of slaughter weight on pork and fat quality. Proceedings of the 2nd International Virtual Conference on Pork Quality. Concordia, Brazil, pp. 213-224.

Ellis, M., Ritter, M.J., 2006. Management practices and meat quality. Proceedings of the 59th American Meat Science Association (AMSA) Reciprocal Meat Conference. Champaign-Urbana, IL, USA, pp. 29-32.

European Commission (EC), 1993. Council Directive (93/119/CEE) of 22 December 1993 on the protection of animals at the time of slaughter or killing. Official Journal of the European Union L340: 21-34. Available at: http://data.europa.eu/eli/dir/1993/119/oj.

European Commission (EC), 2005. Council Regulation (EC) No 1/2005 of 22 December 2004 on the protection of animals during transport and related operations and amending Directives 64/432/EEC and 93/119/EC and Regulation (EC) No 1255/97. Official Journal of the European Union L3: 1-44. Available at: http://data.europa.eu/eli/reg/2005/1/oj.

European Commission (EC), 2019. Pilot project on best practices for animal transport – final report. Available at: https://ec.europa.eu/food/animals/welfare/practice/transport_en.

European Food Safety Authority (EFSA), 2004. Opinion of the Scientific Panel on Animal Health and Welfare (AHAW) on a request from the Commission related to welfare aspects of the main systems of stunning and killing the main commercial species of animals. EFSA Journal 2: 45. https://doi.org/10.2903/j.efsa.2004.45

European Food Safety Authority (EFSA), 2013. Scientific opinion on monitoring procedures at slaughterhouses for pigs. EFSA Journal 11: 3523. https://doi.org/10.2903/j.efsa.2013.3523

European Food Safety Authority (EFSA), 2020. Scientific opinion on the welfare of pigs at slaughter. EFSA Journal 18: 6148. https://doi.org/10.2903/j.efsa.2020.6148

Fàbrega, E., Manteca, X., Font, J., Gispert, M., Carrión, D., Velarde, A., Ruiz-de-la-Torre, J.R. and Diestre, A., 2002. Effects of halothane gene and pre-slaughter treatment on meat quality and welfare from two pig crosses. Meat Science 62: 463-472.

Fàbrega, E., Velarde, A., Cros, J., Gispert, M., Suarez, P., Tibau, J. and Soler, J., 2010. Effect of vaccination against gonodotropingreleasing hormone, using Improvac, on growth performance, body composition, behaviour and acute phase proteins. Livestock Science 132: 53-59.

Faucitano, L and Velarde, A., 2021. Optimising pig welfare during transport, lairage and slaughter. In: Edwards, S. (ed.) Understanding the behaviour and improving the welfare of pigs. Burleigh Dodds Science Publishing, Cambridge, UK, pp. 247-278.

Faucitano, L. and Geverink, N.A., 2008. Effects of preslaughter handling on stress response and meat quality in pigs. In: Faucitano, L. and Schaefer, A.L. (eds.) The welfare of pigs – from birth to slaughter. Wageningen Academic Publishers, Wageningen, the Netherlands, pp. 197-224.

Faucitano, L. and Goumon, S., 2018. Transport of pigs to slaughter and associated handling. In: Špinka, M. (ed.) Advances in pig welfare. Woodhead Publishing, London, UK, pp. 261-293.

Faucitano, L. and Lambooij, E., 2019. Transport of pigs. In: Grandin, T.A. (ed.) Livestock transport and handling. CABI Publishing, Wallingford, UK, pp. 307-327.

Faucitano, L. and Pedernera, C., 2017. Reception and unloading. In: Raj, M. and Velarde, A. (eds.) Animal welfare at slaughter: a practitioner guide. 5M Publishing, Sheffield, UK, pp. 33-50.

Faucitano, L., 2001. Causes of skin damage to pig carcasses. Canadian Journal of Animal Science 81: 39-45.

Faucitano, L., 2010. Effects of lairage and slaughter conditions on animal welfare and pork quality. Canadian Journal of Animal Science 90: 461-469.

Faucitano, L., 2018. Preslaughter handling practices and their effects on animal welfare and pork quality. Journal of Animal Science 96: 728-738.

Faucitano, L., Chevillon, P. and Ellis, M., 2010. Effects of feed withdrawal prior to slaughter and nutrition on stomach weight, and carcass and meat quality in pigs. Livestock Science 127: 110 114.

Faucitano, L., Conte, S., Pomar, C., Paiano, D., Duan, Y., Zhang, P., Drouin, G., Rina, S., Guay, F. and Devillers, N., 2020. Application of extended feed withdrawal time preslaughter and its effects on animal welfare and carcass and meat quality of enriched-housed pigs. Meat Science 167: 108163. https://doi.org/10.1016/j.meatsci.2020.108163

Faucitano, L., Saucier, L., Correa, J.A., Méthot, S., Giguère, A., Foury, A., Mormède, P. and Bergeron, R., 2006. Effects of feed texture, meal frequency and pre-slaughter fasting on carcass and meat quality, and urinary cortisol in pigs. Meat Science 74: 697-703.

Fitzgerald, R.F., Stalder, K.J., Matthews, J.O., Schultz-Kaster, C.M. and Johnson, A.K., 2009. Factors associated with fatigued, injured, and dead pig frequency during transport and lairage at a commercial abattoir. Journal of Animal Science 87: 1156-1166.

Food Safety and Inspection Service (FSIS), 2011. Market swine condemned ante-mortem for deads at USDA inspected packing plants for the calendar year 2010. Freedom of Information Act Case # 10-148. FSIS, Washington, DC, USA.

Food Safety and Inspection Service (FSIS), 2013. Market swine condemned ante-mortem for deads at USDA inspected packing plants for the calendar year 2012. Freedom of Information Act Case #2013-100. FSIS, Washington, DC, USA.

Food Safety and Inspection Service (FSIS), 2014. Market swine condemned ante-mortem for deads at USDA inspected packing plants for the calendar year 2013. Freedom of Information Act Case #2014-200. FSISD, Washington, DC, USA.

Food Safety and Inspection Service (FSIS), 2015. Market swine condemned ante-mortem for deads at USDA inspected packing plants for the calendar year 2014. Freedom of Information Act Case #2015-188. FSIS, Washington, DC, USA.

Food Safety and Inspection Service (FSIS), 2016. Market swine condemned ante-mortem for deads at USDA inspected packing plants for the calendar year 2015. Freedom of Information Act Case #2016-113. FSIS, Washington, DC, USA.

Fox, J., Widowski, T., Torrey, S., Nannoni, E., Bergeron, R., Gonyou, H.W., Brown, J.A., Crowe, T., Mainau, E. and Faucitano, L., 2014. Water sprinkling market pigs in a stationary trailer. 1. Effects on pig behaviour, gastrointestinal tract temperature and trailer micro-climate. Livestock Science 160: 113-123.

Franck, M., Svensson, M., Von Seth, G., Josell, A., Figwer, Ph., Poirel, M.T. and Monin, G., 2003. Effect of stunning conditions on occurence of PSE defects in hams of rn+/RN$^-$ pigs. Meat Science 64: 351-355.

Fraqueza, M.J., Roseiro, L.C., Almeida, J., Matias, E., Santos, C. and Randall, C., 1998, Effects of lairage temperature and holding time on pig behaviour and on carcass and meat quality. Applied Animal Behaviour Science 60: 317-330.

Garcia, A. and McGlone, J.J., 2015. Loading and unloading finishing pigs: effects of bedding types, ramp angle, and bedding moisture. Animals 5: 13-26.

Geverink, N.A., Buhnemann, A., Van de Burgwal, J.A., Lambooij, E., Blokhuis, H.J. and Wiegant V.M., 1998. Responses of slaughter pigs to transport and lairage sounds. Physiology & Behaviour 63: 667-673.

Geverink, N.A., Engel, B., Lambooij, E. and Wiegant, V.M., 1996. Observations on behavior and skin damage of slaughter pigs and treatment during lairage. Applied Animal Behaviour Science 50: 1-13.

Giersing, M., Lundström, K. and Andersson, A., 2000. Social effects and boar taint: significance for production of slaughter boars (*Sus scrofa*). Journal of Animal Science 78: 296-305.

Gispert, M., Faucitano, L., Oliver, M.A., Guàrdia, M.D., Coll, C., Siggins, K., Harvey, K. and Diestre, A., 2000. A survey of pre-slaughter conditions, halothane gene frequency, and carcass and meat quality in five Spanish pig commercial abattoirs. Meat Science 55: 97-106.

Gispert, M., Font i Furnols, M., Gil, M., Velarde, A., Diestre, A., Carrión, D., Sosnicki, A. and Plastow, G.S., 2007. Relationships between carcass quality parameters and genetic types. Meat Science 77: 397-404.

Gispert, M., Oliver, M.A., Velarde, A., Suárez, P., Pérez, J. and Font i Furnols, M., 2010. Carcass and meat quality characteristics of immunocastrated male, surgically castrated male, entire male and female pigs. Meat Science 85: 664-670.

Gonyou, H.W. and Brown, J., 2012. Reducing stress and improving recovery from handling during loading and transport of market pigs. Final report submitted to Alberta Livestock and Meat Agency, 40 pp.

Goumon, S. and Faucitano, L., 2017. Influence of loading handling and facilities on subsequent response to pre-slaughter stress in pigs. Livestock Science 200: 6-13.

Goumon, S., Brown, J.A., Faucitano, L., Bergeron, R., Crowe, T., Connor, M.L. and Gonyou, H.W., 2013b. Effect of rest duration on recovery from repeated exercise in near market-weight pigs. Journal of Animal Science 91: 5859-5867.

Goumon, S., Brown, J.A., Faucitano, L., Bergeron, R., Widowski, T.M., Crowe, T., Connor, M.L. and Gonyou, H.W., 2013c. Effects of transport duration on maintenance behavior, heart rate and gastrointestinal tract temperature of market-weight pigs in 2 seasons. Journal of Animal Science 91: 4925-4935.

Goumon, S., Faucitano, L., Bergeron, R., Crowe, T., Connor, M.L. and Gonyou, H.W., 2013a. Effect of ramp configuration on easiness of handling, heart rate and behavior of near-market pigs at unloading. Journal of Animal Science 91: 3889-3898.

Government of Canada, 2020. Health of animals regulations. Regulations amending the health of animals regulations. Available at: https://tinyurl.com/m3hxv9yv

Grandin, T.A. and Vogel, K.D., 2011. Restaurant animal welfare and humane slaughter audits in U.S. federally inspected beef and pork slaughter plants. Available at: http://www.grandin.com/survey/2011.restaurant.audits.html

Grandin, T.A., 2002. The welfare of pigs during transport and slaughter. Available at: https://tinyurl.com/xvr3tdyy

Grandin, T.A., 2010. The importance of measurement to improve the welfare of livestock, poultry and fish. In: Grandin, T.A. (ed.) Improving animal welfare – a practical approach. CABI Publishing, Wallingford, UK, pp. 1-20.

Grandin, T.A., 2016. Transport fitness of cull sows and boars: a comparison of different guidelines of fitness for transport. Animals 6: 77. https://doi.org/10.3390/ani6120077

Grandin, T.A., 2017. Recommended animal handling guidelines and audit guide: a systematic approach to animal welfare. North American Meat Institute, Washington, DC, USA, 131 pp. Available at: http://animalhandling.org/producers/guidelines_audits

Gregory, N.G., 1998. Physiology of stress, distress, stunning and slaughter. In: Gregory, N.G. (ed.) Animal welfare and meat science. CABI Publishing, Wallingford, UK, pp. 64-92.

Gregory, N.G., Mohan Raj, A.B.M., Audsley, A.R.S. and Daly, C.C., 1990. Effect of carbon dioxide on man. Fleischwirtschaft 70: 1173-1174.

Guàrdia, M.D., Estany, J., Balash, S., Oliver, M.A., Gispert, M. and Diestre, A., 2004. Risk assessment of PSE condition due to pre-slaughter conditions and RYR1 gene in pigs. Meat Science 67: 471-478.

Guàrdia, M.D., Estany, J., Balash, S., Oliver, M.A., Gispert, M. and Diestre, A., 2005. Risk assessment of DFD meat due to pre-slaughter conditions in pigs. Meat Science 70: 709-716.

Guàrdia, M.D., Estany, J., Balash, S., Oliver, M.A., Gispert, M. and Diestre, A., 2009. Risk assessment of skin damage due to pre-slaughter conditions and RYR1 gene in pigs. Meat Science 81: 745-751.

Guàrdia, M.D., Gispert, M. and Diestre, A., 1996. Mortality rates during transport and lairage in pigs for slaughter. Meat Focus International 10: 362-366.

Guay, K., Salgado, G., Thompson, G., Backus, B., Sapkota, A., Chaya, W. and McGlone, J.J., 2013. Behavior and handling of physically and immunologically castrated market pigs on farm and going to market. Journal of Animal Science 91: 5410-5417.

Haley, C., Dewey, C.E., Widowski, T. and Friendship, R., 2008a. Association between in-transit losses, internal trailer temperature, and distance travelled by Ontario market hogs. Canadian Journal of Veterinary Research 72: 385-389.

Haley, C., Dewey, C.E., Widowski, T. and Friendship, R., 2010. Relationship between estimated finishing-pig space allowance and in transit loss in a retrospective survey for 3 packing plant in Ontario in 2003. Canadian Journal of Veterinary Research 74: 178-184.

Haley, C., Dewey, C.E., Widowski, T., Poljak, Z. and Friendship, R., 2008b. Factors associated with in-transit losses of market hogs in Ontario in 2001. Canadian Journal of Veterinary Research 72: 377-384.

Hambrecht, E., Eissen, J.J., Newman, D.J., Smits, C.H.M., Den Hartog, L.A. and Verstegen, M.W.A., 2005. Negative effects of stress immediately before slaughter on pork quality are aggravated by suboptimal transport and lairage conditions. Journal of Animal Science 83: 440-448.

Hamilton, D.N., Ellis, M., Bertol, T.M. and Miller, K.D., 2004. Effects of handling intensity and live weight on blood acid-base status in finishing pigs. Journal of Animal Science 82: 2405-2409.

Herskin, M.S., Aaslyng, M.D., Anneberg, I., Thomsen, P.T., Gould, L.M. and Thodberg, K., 2020a. Significant variation in the management of cull sows before transport for slaughter: results from a survey of Danish pig farmers. Veterinary Record 186: 1-8.

Herskin, M.S., Christensen, S.W. and Rousing, T., 2020b. Handling and moving cull sows upon arrival at the slaughterhouse – effects of small versus larger groups of sows. Applied Animal Behaviour Science 232: 105-113.

Hoenderken, R., 1978. Electrical stunning of slaughter pigs. PhD-thesis, University of Utrecht, Utrecht, the Netherlands.

Hoenderken, R., Van Logtestjin, J.G., Sybesma, W. and Spanjaard, W.J.M., 1979. Kohlendioxid-betabung von schlachtschweinen. Fleischwirtschaft 59: 1572-1578.

Hoffman, L.C. and Fisher, P., 2010. Comparison of the effects of different transport conditions and lairage times in a Mediterranean climate in South Africa on the meat quality of commercially crossbred Large white x Landrace pigs. Journal of the African Veterinary Association 81: 225-227.

Holmes, R., Gerritzen, M.A., Herskin, M.S., Schwarzlose, I. and Ruis, M.A.W., 2020. Review on arrival and lairage management at pig slaughterhouses. European Reference Centre for Animal Welfare (EURCAW) – pigs. Available at: https://www.eurcaw.eu/en/eurcaw-pigs/output/scientific-output.htm

Honkavaara, M., 1989. Influence of lairage on blood composition of pig and on the development of PSE pork. Journal of Agriculture Science in Finland 61: 433-440.

Humane Farm Animal Care, 2018. Humane farm animal care standards. Available at: www.certifiedhumane.org

Huynh, T.T.T., Aarnink, A.J.A., Truong, C.T., Kemp, B. and Verstegen, M.W.A., 2006. Effects of tropical climate and water cooling on growing pigs' responses. Livestock Science 104: 278-291.

Iulietto, M.F., Sechi, P., Mansi Gaudenzi, C., Grispoldi Ceccarelli, M., Barbera, S. and Cenci-Goga, B.T., 2018. Noise assessment in slaughterhouses by means of a smartphone app. Italian Journal of Food Safety 7: 7053. https://doi.org/10.4081/ijfs.2018.7053

Kavanagh, L., 2010. The interaction of group size and alley width on the movement of near-market pigs. MSc-thesis, University of Guelph, Guelph, Canada.

Kephart, K.B., Harper, M.T. and Raines, C.R., 2010. Observations of market pigs following transport to a packing plant. Journal of Animal Science 88: 2199-2203.

Knecht, D. and Duzinski, K., 2016. The effect of sex, carcass mass, back fat thickness and lean meat content on pork ham and loin characteristics. Archives of Animal Breeding 59: 51-57.

Knowles, T.G., Brown, S.N., Edwards, J.E. and Warriss, P.D., 1998. Ambient temperature below which pigs should not be continuously showered in lairage. Veterinary Record 143: 575-578.

Lambooij, E. and Engel, B., 1991. Transport of slaughter pigs by road over a long distance: some aspects of loading density and ventilation. Livestock Production Science 28: 163-174.

Lambooij, E., 1988. Road transport of pigs over a long distance: some aspects of behaviour, temperature and humidity during transport and some effects of the last two factors. Animal Production 46: 257-263.

Lambooij, E., 2014. Transport of pigs. In: Grandin, T.A. (ed.) Livestock handling and transport. CABI Publishing, Wallingford, UK, pp. 280-297.

Lambooij, E., Garssen, G.J., Walstra, P., Mateman, F. and Merkus, G.S.M., 1985. Transport of pigs by car for two days: some aspects of watering and loading density. Livestock Production Science 13: 289-299.

Lammens, V., Peeters, E., De Maere, H., De Mey, E., Paelinck, H., Leyten, J. and Geers, R., 2007. A survey of pork quality in relation to pre-slaughter conditions, slaughterhouse facilities, and quality assurance. Meat Science 74: 381-387.

Lewis, C.R.G. and McGlone, J.J., 2007. Moving finishing pigs in different group sizes: cardiovascular responses, time and ease of handling. Livestock Science 107: 86-90.

Lindahl, C., Sindhøj, E., Hellgren, R.B., Berg, C. and Wallenbeck, A., 2020. Responses of pigs to stunning with nitrogen filled high-expansion foam. Animals 10: 2210-2236.

Llonch, P., Dalmau, A., Rodríguez, P., Manteca, X. and Velarde, A., 2012a. Aversion to nitrogen and carbon dioxide mixtures for stunning pigs. Animal Welfare 21: 33-39.

Llonch, P., Rodríguez, P., Gispert, M., Dalmau, A., Manteca, X. and Velarde, A., 2012b. Stunning pigs with nitrogen and carbon dioxide mixtures: effects on animal welfare and meat quality. Animal 6: 668-675.

Llonch, P., Rodríguez, P., Jospin, M., Dalmau, A., Manteca, X. and Velarde, A., 2013. Assessment of unconsciousness in pigs during exposure to nitrogen and carbon dioxide mixtures. Animal 7: 492-498.

Long, V.P. and Tarrant, P.V., 1990. The effect of pre-slaughter showering and post-slaughter rapid chilling on meat quality in intact pork sides. Meat Science 27: 181-195.

McGlone, J.J. and Curtis, S.E., 1985. Behavior and performance of weanling pigs in pens equipped with hide areas. Journal of Animal Science 60: 20-24.

McGlone, J.J., Johnson, A.K., Sapkota, A. and Kephart, R.K., 2014a. Establishing trailer ventilation (boarding) requirements for finishing pigs during transport. Animals 4: 515-523.

McGlone, J.J., Johnson, A.K., Sapkota, A. and Kephart, R.K., 2014b. Transport of market pigs: improvements in welfare and economics. In: Grandin, T.A. (ed.) Livestock handling and transport. CABI Publishing, Wallingford, UK, pp. 298-314.

McGlone, J.J., McPherson, R.L. and Anderson, D.L., 2004. Case study: moving devices for finishing pigs: efficacy of electric prod, board, paddle, or flag. The Professional Animal Scientist 20: 518-523.

McKeegan, D.E.F., Reimert, H.G.M., Hindle, V.A., Boulcott, P., Sparrey, J.M., Wathes, C.M., Demmers, T.G.M. and Gerritzen, M.A., 2013. Physiological and behavioral responses of poultry exposed to gas-filled high expansion foam. Poultry Science 92: 1145-1154.

Ministério da Agricultura, Pecuária e Abastecimento (MAPA), 2020. Decreto No. 9.013, de 29 de março de 2017, alterado pelo Decreto No. 10.468, de 18 de agosto de 2020. Regulamento da Inspeção Industrial e Sanitária de Produtos de Origem Animal. Diário Oficial da União: seção 1, Brasília, n. 159, pp. 5.

Miranda-de La Lama, G.C., Bermejo-Poza, R., Formoso-Rafferty, N., Mitchell, M., Barreiro, P. and Villaroel, M., 2021. Long-distance transport of finisher pigs in the Iberian peninsula: effects of season on thermal and enthalpy conditions, welfare indicators and meat pH. Animals 11: 2410. https://doi.org/10.3390/ani11082410

Moak, K., 2021. Impact of two novel trailer designs on trailer microclimate, animal welfare, and meat quality during short distance transportation of pigs to slaughter under Canadian summer and winter conditions. MSc-thesis, University of Guelph, Guelph, Canada.

Morris, B.K., Davis, R.B., Brokesh, E., Flippo, D.K., Houser, T.A., Najar-Villarreal, F., Turner, K.K., Williams, J.G., Stelzleni, A.M. and Gonzalez, J.M., 2021. Measurement of the three-axis vibration, temperature, and relative humidity profiles of commercial transport trailers for pigs. Journal of Animal Science 99: skab027. https://doi.org/10.1093/jas/skab027

Moss, B.W., 1978, Some observations on the activity and aggressive behavior of pigs when penned prior to slaughter, Applied Animal Ethology 4: 323-339.

Mota-Rojas, D., Becerril, M., Lemus, C., Sanchez, P., Gonzalez, M., Olmos, S.A., Ramirez, R. and Alonso-Spilsbury, M., 2006. Effects of mid-summer transport duration on pre- and post-slaughter performance and pork quality in Mexico. Meat Science 73: 404-412.

Mota-Rojas, D., Bolas-Lopez, D., Mendez, M.C., Ramirez-Tellez, J., Roldan-Santiago, P., Flores-Peinado, S. and Mora-Pedina, P., 2012. Stunning swine with CO2 gas: controversies related to animal welfare. International Journal of Pharmacology 8: 141-151.

Nanni Costa, L., Lo Fiego, D.P., Dall'Olio, S., Davoli, R. and Russo, V., 1999. Influence of loading method and stocking density during transportation on meat and dry-cured ham quality in pigs with different halothane genotypes. Meat Science 51: 391-399.

Nanni Costa, L., Lo Fiego, D.P., Dall'Olio, S., Davoli, R. and Russo, V., 2002. Combined effects of pre-slaughter treatments and lairage time on carcass and meat quality in pigs of different halothane genotype. Meat Science 61: 41-47.

Nannoni, E., Liuzzo, G., Serraino, A., Giacometti, F., Martelli, G., Sardi, L., Vitali, M., Romagnoli, L., Moscardini, E. and Ostanello, F., 2016. Evaluation of pre-slaughter losses of Italian heavy pigs. Animal Production Science 57: 2072-2081.

Nannoni, E., Widowski, T., Torrey, S., Fox, J., Rocha, L.M., Gonyou, H.W., Weschenfelder, A.V., Crowe, T. and Faucitano, L., 2014. Water sprinkling market pigs in a stationary trailer. 2. Effects on selected exsanguination blood parameters and carcass and meat quality variation. Livestock Science 160: 124-131.

National Agricultural Statistics Service (NASS), 2014. Agricultural statistics. NASS, USDA, Washington, DC, USA.

National Farm Animal Care Council (NFACC), 2014. Code of practice for the care and handling of pigs. NFACC, Ottawa, Canada. Available at: https://www.nfacc.ca/pdfs/codes/pig_code_of_practice.pdf

National Pork Board, 2016. Transport quality assurance (TQA) handbook, version 6. Available at: http://www.pork.org/tqa-certification/tqa-program-materials/

Newman, D., Young, J., Carr, C., Ryan, M. and Berg, E., 2014. Effect of season, transport length, deck location, and lairage length on pork quality and blood cortisol concentrations of market hogs. Animals 4: 627-642.

Nielsen, N.J., 1982. Recent results from investigations of transportation of pigs for slaughter. In: Moss, R. (ed.). Transport of animals intended for breeding, production and slaughter. Martinus Njhoff Publishing, The Hague, the Netherlands, pp. 115-124.

Niewold, T.A., Van Essen, G.J., Nabuurs, M.J., Stockhofe-Zurwieden, N. and Van der Meulen, J., 2000. A review of porcine pathophysiology: a different approach to disease. Veterinary Quarterly 22: 209-212.

Nodari, S., Polloni, A., Giacomelli, S., Vezzoli, F. and Galletti, G., 2014. Assessing pig welfare at stunning in Northern Italy commercial abattoirs using electrical method. Large Animal Review 20: 87-91.

Nonneman, D.J., Brown-Brandl, T., Jones, S.A., Wiedmann, R.T. and Rohrer, G.A., 2012. A defect in dystrophin causes a novel porcine stress syndrome. BMC Genomics 13: 233. https://doi.org/10.1186/1471-2164-13-233

Nowak, B., Mueffling, T.V. and Hartung, J., 2007. Effect of different carbon dioxide concentrations and exposure times in stunning of slaughter pigs: impact on animal welfare and meat quality. Meat Science 75: 290-298.

Olesen, L.S., Nygaard, C.M., Friend, T.H., Bushong, D., Knabe, D.A., Vestergaard, K.S. and Vaughan, R.K., 1996. Effect of partitioning pens on aggressive behavior of pigs regrouped at weaning. Applied Animal Behaviour Science 46: 167-174.

Passafaro, T.L., Van de Stroet, D., Bello, N.M., Williams, N.H. and Rosa, G.J.M., 2019. Generalized additive mixed model on the analysis of total transport losses of market-weight pigs. Journal of Animal Science 97: 2025-2034.

Pedersen-MacNab, M.K., Tokareva, M. and Seddon, Y.M., 2019. Progress made towards the animal welfare research priorities for pigs: a review of research 2012-2019 for the Pig Code Technical Committee. In: National Farm Animal Care Council (ed.) Five-year review summary report code of practice for the care and handling of pigs. Available at: https://tinyurl.com/5ynu99ps

Peeters, E., Deprez, K., Bethers, F., De Baerdemaeker, J., Auberts, A.E. and Geers, R., 2008. Effect of a driver and driving style on the stress response of pigs during a short journey by trailer. Animal Welfare 17: 189-196.

Pereira, T., Titto, E.A., Conte, S., Devillers, N., Sommavilla, R., Diesel, T., Dalla Costa, F., Guay, F., Friendship, R., Crowe, T. and Faucitano, L., 2018. Use of fan-misters bank for cooling pigs kept in a stationary trailer before unloading: effects on trailer microclimate, and pig behavior and physiological response. Livestock Science 216: 67-74.

Peterson, E., Remmenga, M., Hagerman, A.D. and Akkina, J.E., 2017. Use of temperature, humidity, and slaughter condemnation data to predict increases in transport losses in three classes of swine and resulting foregone revenue. Frontiers in Veterinary Science 4: 67. https://doi.org/10.3389/fvets.2017.00067

Pilcher, C.M., Ellis, M., Rojo-Gómez, A., Curtis, S.E., Wolter, B.F., Peterson, C., Ritter, M.J. and Brinkmann, J., 2011. Effects of floor space during transport and journey time on indicators of stress and transport losses of market-weight pigs. Journal of Animal Science 89: 3809-3818.

Rabaste, C., Faucitano, L., Saucier, L., Foury, D., Mormède, P., Correa, J.A., Giguère, A. and Bergeron, R., 2007. The effects of handling and group size on welfare of pigs in lairage and its influence on stomach weight, carcass microbial contamination and meat quality variation. Canadian Journal of Animal Science 87: 3-12.

Raj, A.B.M. and Gregory, N.G., 1995. Welfare implications of the gas stunning of pigs 1. Determination of aversion to the initial inhalation of carbon dioxide or argon. Animal Welfare 4: 273-280.

Raj, A.B.M. and Gregory, N.G., 1996. Welfare implications of gas stunning pigs. 2. Stress of induction of anaesthesia. Animal Welfare 5: 71-78.

Raj, A.B.M., 1999. Behaviour of pigs exposed to mixtures of gases and the time required to stun and kill them: welfare implications. Veterinary Record 144: 165-168.

Raj, A.B.M., 2008. Welfare of pigs during stunning and slaughter. In: Faucitano, L. and Schaefer, A.L. (eds.) Welfare of pigs – from birth to slaughter. Wageningen Academic Publishers, Wageningen, the Netherlands, pp. 225-242.

Raj, A.B.M., Johnson, S.P., Wotton, S.B. and McKinstry, J.L.M., 1997. Welfare implications of gas stunning pigs: 3. the time to loss of somatosensory evoked potentials and spontaneous electrocorticogram of pigs during exposure to gases. The Veterinary Journal 153: 329-339.

Raj, A.B.M., Smith, C. and Hickman, G., 2008. Novel method for killing poultry in houses with dry foam created using nitrogen. Veterinary Record 162: 722-723.

Randall, J.M., 1993. Environmental parameters necessary to define comfort for pigs, cattle and sheep in livestock transporters. Animal Science 57: 299-307.

Randall, J.M., Duggan, J.A. and Alami, M.A., 1995. Influence of motion and vibration of animals. Fleischwirtshaft 75: 201-204.

Renaudeau, D., Gourdine, J. and St-Pierre, N., 2011. A meta-analysis of the effects of high ambient temperature on growth performance of growing-finishing pigs. Journal of Animal Science 89: 2220-2230.

Rioja-Lang, F.C., Brown, J.A., Brockhoff, E.J. and Faucitano, L., 2019. A review of swine transportation research on priority welfare issues: a Canadian perspective. Frontiers in Veterinary Science 6: 36. https://doi.org/10.3389/fvets.2019.00036

Ritter, M.J., 2007. Effects of animal handling and transportation factors on the welfare, stress responses, and incidences of transport losses in market weight pigs at the packing plant. PhD-thesis, University of Illinois, Urbana-Champaign, IL, USA.

Ritter, M.J., Ellis, M., Benjamin, M., Berg, E., DuBois, P., Marchant-Forde, J., Green, A., Matzat, P., Mormede, P., Moyer, T., Pfalzgraf, K., Siemens, M., Sterle, J., Whiting, T. and Wolter, B., 2005. The fatigued pig syndrome. Journal of Animal Science 83, Suppl. 1: 258.

Ritter, M.J., Ellis, M., Berry, N.L., Curtis, S.E., Anil, L., Berg, E., Benjamin, M., Butler, D., Dewey, C., Driessen, B., DuBois, P., Hill, J.D., Marchant-Forde, J.N., Matzat, P., McGlone, J.J., Mormede, P., Moyer, T., Pfalzgraf, K., Salak-Johnson, J., Siemens, M., Sterle, J., Stull, C., Whiting, T., Wolter, B., Niekamp, S.R. and Johnson, A.K., 2009. Transport losses in market weight pigs: I. A review of definitions, incidence, and economic impact. The Professional Animal Scientist 25: 404-414.

Ritter, M.J., Ellis, M., Bertelsen, C.R., Bowman, R., Brinkmann, J., DeDecker, J.M., Keffaber, K.K., Murphy, C.M., Peterson, B.A., Schlipf, J.M. and Wolter, B.F., 2007. Effects of distance moved during loading and floor space on the trailer during transport on losses of market weight pigs on arrival at the packing plant. Journal of Animal Science 85: 3454-3461.

Ritter, M.J., Ellis, M., Bowman, R., Brinkmann, J., Curtis, S.E., DeDecker, J.M., Mendoza, O., Murphy, C.M., Orellana, D.G., Peterson, B.A., Rojo, A., Schlipf, J.M. and Wolter, B.F., 2008a. Effects of season and distance moved during loading on transport losses of market-weight pigs in two commercially available types of trailer. Journal of Animal Science 86: 3137-3145.

Ritter, M.J., Ellis, M., Brinkmann, J., DeDecker, J.M., Keffaber, K.K., Kocher, M.E., Peterson, B.A., Schlipf, J.M. and Wolter, B.F., 2006. Effect of floor space during transport of market-weight pigs on the incidence of transport losses at the packing plant and the relationships between transport conditions and losses. Journal of Animal Science 84: 2856-2864.

Ritter, M.J., Ellis, M., Hollis, G.R., McKeith, F.K., Orellana, D.G., Van Genugten, P., Curtis, S.E. and Schlipf, J.M., 2008b. Frequency of the HAL-1843 mutation of the ryanodine receptor gene in dead and nonambulatory-noninjured pigs on arrival at the packing plant. Journal of Animal Science 86: 511-514.

Ritter, M.J., Ellis, M., Murphy, C.M., Peterson, B.A. and Rojo, A., 2008c. Effects of handling intensity, distance moved, and transport floor space on the stress responses of market weight pigs. Journal of Animal Science 86, e-Suppl. 3: 43.

Ritter, M.J., Yoder, C.L., Jones, C.L., Carr, S.N. and Calvo-Lorenzo, M.S., 2020. Transport losses in market weight pigs: II. U.S. incidence and economic impact. Translational Animal Science 4: 1103-1112.

Rocha, L.M., Bridi, A.M., Mormède, P., Weschenfelder, A.V., Devillers, N., Bertoloni, W. and Faucitano, L., 2013. Effects of ractopamine administration and castration method on the response to pre-slaughter stre ss and carcass and meat quality in pigs of two Piétrain genotypes. Journal of Animal Science 91: 3965-3977.

Rocha, L.M., Devillers, N., Maldague, X., Kabemba, F.Z., Fleuret, J., Guay, F. and Faucitano, L., 2019. Validation of anatomical sites for the measurement of infrared body surface temperature variation in response to handling and transport. Animals 9: 425. https://doi.org/10.3390/ani9070425

Rocha, L.M., Dionne, A., Saucier, L., Nannoni, E. and Faucitano, L., 2015. Hand-held lactate analyzer as a tool for the real-time measurement of physical fatigue before slaughter and pork quality prediction. Animal 9: 707-714.

Rocha, L.M., Velarde, A., Dalmau, A., Saucier, L. and Faucitano, L., 2016. Can the monitoring of animal welfare parameters predict pork meat quality variation through the supply chain (from farm to slaughter)? Journal of Animal Science 94: 359-376.

Rodríguez, P., Dalmau, A., Ruiz-De-La-Torre, J.L., Manteca, X., Jensen, E., Rodríguez, B., Litvan, H. and Velarde, A. 2008. Assessment of unconsciousness during carbon dioxide stunning in pigs. Animal Welfare 17: 341-349.

Rydhmer, L. and Canario, L., 2014. Behavioral genetics in pigs and relations to welfare. In: Grandin, T.A. and Deesing, M. (eds.) Genetics and the behavior of domestic animals. Academic Press, London, UK, pp. 397-434.

Rydhmer, L. and Lundheim, N., 2008. Breeding pigs for improved welfare. In: Faucitano, L. and Schaefer, A.L. (eds.) The welfare of pigs – from birth to slaughter. Wageningen Academic Publishers, Wageningen, the Netherlands, pp. 243-270.

Rydhmer, L., Hansson, M., Lundström, K., Brunius, C. and Andersson, K., 2013. Welfare of entire male pigs is improved by socialising piglets and keeping intact groups until slaughter. Animal 7: 1532-1541.

Rydhmer, L., Zamaratskaia, G., Andersson, H.K., Algers, B., Guillemet, R. and Lundström, K., 2006. Aggressive and sexual behavior of growing and finishing pigs reared in groups, without castration. Acta Agricultura Scandinavica, Section A Animal Science 56: 109-119.

Santos, C., Almeida, J.M., Matias, E.C., Fraqueza, M.J., Roseiro, C. and Sardinha, L., 1997. Influence of lairage environmental conditions and resting time on meat quality in pigs. Meat Science 45: 253-262.

Sardi, L., Gastaldo, A., Borciani, M., Bertolini, A., Musi, V., Martelli, G., Cavallini, D. and Nannoni, E., 2020. Pre-slaughter sources of fresh meat quality variation: the case of heavy pigs intended for protected designation of origin products. Animals 10: 2386. https://doi.org/10.3390/ani10122386

Saucier, L., Bernier, D., Bergeron, R., Méthot, S., Giguère, A. and Faucitano, L., 2007. Effect of feed texture, meal frequency and pre-slaughter fasting on behaviour, stomach weight and microbial carcass contamination in pigs. Canadian Journal of Animal Science 87: 479-487.

Scheeren, M.B., Gonyou, H.W., Brown, J., Weschenfelder, A.V. and Faucitano, L., 2014. Effects of transport time and location within truck on skin bruises and meat quality of market weight pigs in two seasons. Canadian Journal of Animal Science 94: 71-78.

Schwartzkopf Genswein, K., Faucitano, L., Dadgar, S., Shand, P., Gonzàlez, L.A. and Crowe, T., 2012. Road transport of cattle, swine and poultry in North America and its impact on animal welfare, carcass and meat quality: a review. Meat Science 92: 227-243.

Scientific Committee on Animal Health and Animal Welfare (SCAHAW), 2002. The welfare of animals during transport (details for horses, pigs, sheep and cattle). Report of the Scientific Committee on Animal Health and Animal Welfare, European Commission, Brussels, Belgium, 130 pp.

Sindhøj, E., Lindahl, C. and Bark, L., 2021. Potential alternatives to high-concentration carbon dioxide stunning of pigs at slaughter, Animal 15: 100164. https://doi.org/10.1016/j.animal.2020.100164

Sommavilla, R., Faucitano, L., Gonyou, H., Seddon, Y., Bergeron, R., Widowski, T., Crowe, T., Connor, L., Schereen, M.B., Goumon, S. and Brown, J., 2017. Season, transport duration and trailer compartment effects on blood stress indicators in pigs: relationship to environmental, behavioural and other physiological factors, and pork quality traits. Animals 7: 8. https://doi.org/10.3390/ani7020008

Spencer, T. and Veary, C.M., 2010. A study of preslaughter pig handling and stunning in selected South African Highveld Region abattoirs. Journal of South African Veterinary Association 81: 102-109.

Squires, E.J., Bone, C. and Cameron, J., 2020. Pork production with entire males: directions for control of boar taint. Animals 10: 1665. https://doi.org/10.3390/ani10091665

Sunstrum, J., Dewey, C. and Haley, C., 2006. Clinical signs of stress of finisher pigs transported to market in the summer In: Proceedings of the 2006 American Association of Swine Veterinarians. Orlando, FL, USA, pp. 45.

Sutherland, M.A., McDonald, A. and McGlone, J.J., 2009. Effects of variations in the environment, length of journey and type of trailer on the mortality and morbidity of pigs being transported to slaughter. Veterinary Record 165: 13-18.

Sutherland, M.A., Niekamp, S.R., Rodriguez-Zas, S.L. and Salak-Johnson, J.L., 2006. Impacts of chronic stress and social status on various physiological and performance measures in pigs of different breeds. Journal of Animal Science 84: 588-596.

Teixeira, D.L. and Boyle, L.A., 2014. A comparison of the impact of behaviours performed by entire male and female pigs prior to slaughter on skin lesion scores of the carcass. Livestock Science 170: 142-149.

Teixeira, D.L., Lykke, L. and Boyle, L.A., 2020. The effect of a novel transport system on the welfare and meat quality of slaughter pigs. Journal of Applied Animal Welfare Science 24(3): 260-271. https://doi.org/10.1080/10888705.2020.1790370

Terlouw, E.M., Deiss, V. and Astruc, T., 2021a. Comparing gas and electrical stunning: effects on meat quality of pigs when pre-stunning physical activity is minimal. Foods 10: 319. https://doi.org/10.3390/foods10020319

Terlouw, E.M., Deiss, V. and Astruc, T., 2021b. Stunning of pigs with different gas mixtures: behavioural and physiological reactions. Meat Science 175: 108452. https://doi.org/10.1016/j.meatsci.2021.108452

Thodberg, K., Fogsgaard, K.K. and Herskin, M.S., 2019. Transportation of cull sows – deterioration of clinical condition from departure and until arrival at the slaughter plant. Frontiers in Veterinary Science 6: 28. https://doi.org/10.3389/fvets.2019.00028

Torrey, S., Bergeron, R., Faucitano, L., Widowski, T., Lewis, N., Crowe, T., Correa, J.A., Brown, J. and Gonyou, H.W., 2013b. Transportation of market-weight pigs 2. Effect of season and animal location in the truck on behavior with a 8 hour transport. Journal of Animal Science 91: 2872-2878.

Torrey, S., Bergeron, R., Gonyou, H.W., Widowski, T., Lewis, N., Crowe, T., Correa, J.A., Brown, J. and Faucitano, L., 2013a. Transportation of market-weight pigs 1. Effect of season and truck type on behavior with a 2 hour transport. Journal of Animal Science 91: 2863-2871.

Troeger, K. and Woltersdorf, W., 1991. Gas anaesthesia of slaughter pigs. 1. Stunning experiments under laboratory conditions with fat pigs of known halothane reaction type: meat quality, animal protection. Fleischwirtschaft 71: 1063-1068.

Urrea, V.M., Bridi, A.M., Ceballos, M.C., Paranhos da Costa, M.J.R. and Faucitano, L., 2021. Behavior, blood stress indicators, skin lesions and meat quality in pigs transported to slaughter at different loading densities. Journal of Animal Science 99(6): skab119. https://doi.org/10.1093/jas/skab119

Van de Perre, V., Permentier, L., De Bie, S., Verbeke, G. and Geers, R., 2010. Effect of unloading, lairage, pig handling, stunning and season on pH of pork. Meat Science 86: 931-937.

Van Staaveren, N., Teixeira, D.L., Hanlon, A. and Boyle, L.A., 2015. The effect of mixing entire male pigs prior to transport to slaughter on behaviour, welfare and carcass lesions. PLoS ONE 10: e0122841. https://doi.org/10.1371/journal.pone.0122841

Velarde, A. and Dalmau, A., 2012. Animal welfare assessment at slaughter in Europe: moving from inputs to outputs. Meat Science. 92: 244-251.

Velarde, A., Cruz, J., Gispert, M., Carrión, D., De la Torre Ruiz, J.L., Diestre, A. and Manteca X., 2007. Aversion to carbon dioxide stunning in pigs: effect of carbon dioxide concentration and halothane genotype. Animal Welfare 16: 513-522.

Velarde, A., Gispert, M., Faucitano, L., Alonso, P., Manteca, X. and Diestre, A., 2001. Effects of the stunning procedure and the halothane genotype on meat quality and incidence of haemorrhages in pigs. Meat Science 58: 313-319.

Velarde, A., Gispert, M., Faucitano, L., Manteca, X. and Diestre, A., 2000a. Survey of the effectiveness of stunning procedures used in Spanish pig abattoirs. Veterinary Record 146: 65-68.

Velarde, A., Gispert, M., Faucitano, L., Manteca, X. and Diestre, A., 2000b. The effect of stunning method on the incidence of PSE meat and haemorrhages in pork carcasses. Meat Science 55: 309-314.

Vermeulen, L., Van de Perre, V., Permentier, L., De Bie, S., Verbeke, G. and Geers, R., 2015. Sound levels above 85 dB pre-slaughter influence pork quality. Meat Science 100: 269-274.

Vitali, A., Lana, E., Amadori, M., Bernabucci, Nardone, A. and Lacetera, N., 2014. Analysis of factors associated with mortality of heavy slaughter pigs during transport and lairage. Journal of Animal Science 92: 5134-5141.

Vogel, K.D., Badtram, G., Claus, J.R., Grandin, T., Turpin, S., Weyker, R.E. and Voogd, E., 2011. Head-only followed by cardiac arrest electrical stunning is an effective alternative to head-only electrical stunning in pigs. Journal of Animal Science 89: 1412-1418.

Von Borell, E., Baumgartner, J., Giersing, M., Jäggin, N., Prunier, A., Tuyttens, F.A.M. and Edwards, S.A., 2009. Animal welfare implications of surgical castration and its alternatives in pigs. Animal 3: 1488-1496.

Warriss, P.D., 1982. Loss of carcass weight, liver weight and liver glycogen, and the effects on muscle glycogen and ultimate pH in pigs fasted preslaughter. Journal of Science of Food and Agriculture 33: 840-846.

Warriss, P.D., 1996. Guidelines for the handling of pig ante-mortem: interim conclusion from ECAIR-Project CT920262. In: Proceedings of the EU-Seminar: New information on welfare and meat quality of pigs as related to handling, transport and lairage conditions, Landbauforschung Völkenrode, pp. 217-224.

Warriss, P.D., 1998. The welfare of slaughter pigs during transport. Animal Welfare 7: 365-381.

Warriss, P.D., 2003. Optimal lairage times and conditions for slaughter pigs: a review. Veterinary Record 153: 170-176.

Warriss, P.D. and Brown, S.N., 1985. The physiological responses of fighting between in pigs and the consequences for meat quality. Journal of Science of Food and Agriculture 36: 87-92.

Warriss, P.D. and Brown, S.N., 1994. A survey of mortality in slaughter pigs during transport and lairage. Veterinary Record 134: 513-515.

Warriss, P.D., Brown, S.N., Adams, S.J.M. and Corlett, I.K., 1994. Relationships between subjective and objective assessments of stress at slaughter and meat quality in pigs. Meat Science 38: 329-340.

Warriss, P.D., Brown, S.N., Bevis, E.A. and Kestin, S.C., 1990. The influence of pre-slaughter transport and lairage on meat quality in pigs of two genotypes. Animal Production 50: 165-172.

Warriss, P.D., Brown, S.N., Knowles, T.G., Wilkins, L.J., Pope, S.J., Chadd, S.A., Kettlewell, P.J. and Green, N.R., 2006. Comparison of the effects of fan-assisted and natural ventilation of vehicles on the welfare of pigs being transported to slaughter. Veterinary Record 158: 585-588.

Webster, A.J.F., 1994. Animal welfare: a cool eye towards Eden. Wiley-Blackwell Science Ltd., Oxford, UK.

Weeks, C.A., 2008. A review of welfare in cattle, sheep, and pig lairages, with emphasis on stocking rates, ventilation and noise. Animal Welfare 17: 275-284.

Weeks, C.A., Brown, S.N., Lane, S., Heasman, L., Benson, T. and Warriss, P.D., 2009. Noise levels in lairages for cattle, sheep, and pigs in abattoirs in England and Wales. Veterinary Record 165: 308-314.

Weschenfelder, A.V., Maldague, X., Rocha, L.M., Schaefer, A.L., Saucier, L. and Faucitano, L., 2013b. The use of orbital infra-red thermography for the measurement of the physiological condition of pigs just prior to slaughter and pork quality prediction. Meat Science 95: 616-620.

Weschenfelder, A.V., Torrey, S., Devillers, N., Crowe, T., Bassols, A., Saco, Y., Piñeiro, M., Saucier, L. and Faucitano, L., 2012. Effects of trailer design on animal welfare parameters and carcass and meat quality of three Pietrain crosses being transported over a long distance. Journal of Animal Science 90: 3220-3231.

Weschenfelder, A.V., Torrey, S., Devillers, N., Crowe, T., Bassols, A., Saco, Y., Piñeiro, M., Saucier, L. and Faucitano, L., 2013a. Effects of trailer design on animal welfare parameters and carcass and meat quality of three Pietrain crosses being transported over a short distance. Livestock Science 157: 234-244.

Wesoly, R., Jungbluth, I., Stefanski, V. and Weiler, U., 2015. Pre-slaughter conditions influence skatole and androstenone in adipose tissue of boars. Meat Science 99: 60-67.

Wotton, S.B. and Gregory, N.G., 1986. Pig slaughtering procedures: time to loss of brain responsiveness after exsanguination or cardiac arrest. Research in Veterinary Science 40: 148-151.

Wotton, S.B. and O'Callaghan, M., 2002. Electrical stunning of pigs: the effect of applied voltage on impedance to current flow and the operation of a fail-safe device. Meat Science 60: 203-208.

Wu, F. Vierck, K.R., DeRouchey, J.M., O'Quinn, T.G., Tokach, M.D., Goodband, R.D., Dritz, S.S. and Woodworth, J.C., 2017. A review of heavy weight market pigs: status of knowledge and future needs assessment. Translational Animal Science 1: 1-15.

Xiong, Y, Green, A. and Gates, R.S., 2015. Characteristics of trailer thermal environment during commercial swine transport managed under U.S. industry guidelines. Animals 5: 226-244.

Zou, B., He, G., Zhao, D., Yan, J., Zhang, Z., Xu, X., Zhou, G. and Lo, C., 2020. Proteomic analysis of the potential mechanism for improving pork quality by preslaughter mild driving. Food Science 41: 29-37.

Zurbrigg, K., Van Dreumel, T., Rothschild, M.F., Alves, D., Friendship, R. and O'Sullivan, T., 2017a. Pig-level risk factors for in-transit losses in swine: a review. Canadian Journal of Animal Science 97: 339-346.

Zurbrigg, K., Van Dreumel, T., Rothschild, M.F., Alves, D., Friendship, R. and O'Sullivan, T., 2017b. Postmortem lesions and heart weights of in-transit-loss market pigs in Ontario. Journal of Animal Science 95: 5532-5536.

Zurbrigg, K., Van Dreumel, T., Rothschild, M.F., Alves, D., Friendship, R. and O'Sullivan, T., 2018. Cardiac weights and weight ratios as indicators of cardiac lesions in pigs: a study of pig hearts from an Ontario abattoir. Canadian Journal of Veterinary Research 82: 198-202.

Photo: Leslie Moffat, Eyes on Animals, The Netherlands

Chapter 6
Poultry

M. Gerritzen[1], T. Shynkaruk[2], K. Buchynski[2], K. Schwean-Lardner[2] and T.G. Crowe[2]*

[1]Wageningen University & Research, P.O. Box 338, 6700 AH Wageningen, the Netherlands; [2]University of Saskatchewan, Saskatoon, SK, S7N 5A2, Canada; marien.gerritzen@wur.nl

Abstract

Before being slaughtered birds are subjected to a number events that are stressful and can lead to physical injuries, thermal stress, weight loss and in extremes to death. Catching and crating of birds is the major factor for introducing injuries like leg and wing bruising, broken or dislocated wings. Catching method, catching crew and the type and condition of the crates will all contribute to the risk on injuries. Feed withdrawal prior to transport is applied with the objectives of reducing wetness of birds during transport and allow the gastro-intestinal tract to empty prior to processing. However, the process can create a negative energy balance within the birds during the transport phase, leading to prolonged hunger and reduced ability to thermoregulate. Transport conditions, such as loading density, ventilation, vehicle design and protection from external weather conditions contribute to heat or cold stress. Thermal stress, both hot and cold, has been identified as a risk factor for increased mortality levels.

After arrival, poultry in the containers are placed in a lairage area. In lairage birds should be protected from adverse climatic conditions. Before stunning, birds are taken out of the containers either by hand or automatically depending on the type of container and stunning system. Electrical stunning involves shackling of birds in inverted position before stunning. Factors, such as electrical frequency and current, influence the success of the stun and carcass quality. High frequencies with low currents lead to ineffective stunning whereas low frequencies with higher currents lead to haemorrhages.

Gas stunning methods are an alternative to electrical stunning. Different type of birds, like chickens, turkeys, ducks, geese, are effectively stunned by exposure to different gas mixtures. However, induction of unconsciousness is not immediate and will induce respiratory distress before loss of consciousness.

Keywords: catching and crating, poultry, slaughter, stunning, transport

6.1 Introduction

Worldwide over 66 billion chickens are reared for meat production resulting in 109 million tonnes of poultry meat (https://www.ciwf.org.uk/media/5235303/Statistics-Broiler-chickens.pdf). Before being slaughtered birds are subjected to a number of subsequent events as feed withdrawal, catching and crating, transport, unloading and handling at the slaughter plant and stunning and slaughter. All these events in a relatively short period time includes many stressors and risk factors for animal welfare. Human proximity and handling, physical injuries, heat or cold stress, prolonged hunger, exhaustion can lead to pain, fear, distress, weight loss and in extremes to death. In this chapter, common practice and the major risk factors of the pre-transport phase, transport and handling and stunning at the slaughter plant are presented.

6.2 Pre-pairing and handling for transport

The process of transporting birds to the processing plant can be a stressful event for the birds. It is also a time period where injury can occur. Jacobs *et al.* (2017a,b) found that approximately 1.4% of birds had leg bruising, and 3.7% wing/breast bruising due to the catching and loading process. These authors also reported on the importance of training for catching crews, which can have a direct influence on injury level in the flock.

Bird welfare during transport can also be impacted by a number of factors birds are exposed to prior to the actual catching, loading and transportation process. For example, how healthy the flock is can affect the number of dead-on-arrival (DOA) upon reaching the processing plant. Wet feathers and poor feather cover have a significant effect on thermoregulation ability, which is particularly important during cold weather. Feed withdrawal, a process which improves food safety, can result in weight loss, and reduced ability to thermoregulate during the transport process.

6.2.1 Flock health

Health of a commercial poultry flock obviously has a significant impact on bird welfare during the production phase. However, it also can play a significant role on the number of DOA during transport to the processing plant. Research has indicated that higher DOA numbers during transport can be associated to poor chick quality upon arrival (Jacobs *et al.*, 2017a), or during the grow-out and pre-slaughter period (Chauvin *et al.*, 2011; Jacobs *et al.*, 2017a; Kittelsen *et al.*, 2017). Jacobs *et al.* (2017a) did show that appropriate culling on farm by transporting healthy, fit-for-transport animals only can reduce this number. For end-of-lay hens, poor feathering, low body weights, and higher flock mortality during the production phase have been associated with increased DOA levels during transport (Weeks *et al.*, 2012).

6.2.2. Feather quality and quantity

Feathers play a significant role in both skin protection and thermoregulation in poultry. When feathers become wet, which can happen when ventilation in the barn is reduced during cold periods, leading to higher humidity in the barn, the physiological ability for thermoregulation during cold weather is significantly reduced, and core body temperature may not be maintained (Hunter *et al.*, 1999). In addition, feather cover itself is important. End-of-lay hens with poor feather cover, for example, have a significantly more difficult time with thermoregulation during cold weather than do those with a good feather cover base (Beaulac *et al.*, 2020; Frerichs, 2021), resulting in a higher percentage of body weight lost during the transport process (Frerichs *et al.*, 2021).

6.2.3 Feed withdrawal

Feed withdrawal prior to transport is a common practice, conducted with the objectives of reducing wetness of birds during the transport, and allowing the gastro-intestinal tract to empty prior to processing. The latter reduces potential contamination during slaughter, leading to improved food safety. The length of time prior to slaughter that feed is removed is important in terms of health and potentially affective states. However, the process can create a negative energy balance within the birds during the transport phase, as energy is still required for basic maintenance and thermoregulation. Since bird types differ significantly with respect to muscle mass, body size and metabolic heat production, it is important to consider these separately.

6.2.3.1 Broilers

The effect of feed withdrawal on broilers has been studied over various parameters. With respect to time to empty gastrointestinal tracts, Warriss *et al.* (2004) found that 12 h of withdrawal resulted in the maximum effect on the emptying of the gastrointestinal tract, which is ideal in a food safety perspective. As a result of this emptying, there is an impact on the energy balance of birds during transport (Nijdam *et al.*, 2005). Delezie *et al.* (2007) removed feed for a 3 or a 10 h period from 42- or 43-day old broilers, respectively. A group of birds were left uncrated (not physically transported) and compared to broilers that were crated, exposed to feed withdrawal for 10 h and transported for 3 h. The authors noted reductions in a number of parameters, including a decrease in triglycerides, uric acid and triiodothyronine concentrations, which supports the negative energy balance that exists with feed withdrawal. They also noted that the effects of feed withdrawal and transport were additive, with larger declines in the parameters measured when transport was added to the feed withdrawal phase. Warriss *et al.* (1993) found reductions in blood parameters, including glucose, when broilers were exposed to 6 h of feed withdrawal, in addition to reductions in liver weight and glycogen content. Rodrigues *et al.* (2017) tested fasted (8 h) and transported (for 2 h) birds held in lairage for differing durations of time (0, 2, 4 or 6 h) prior to slaughter. These authors found that longer lairage, which results in longer feed withdrawal periods, produced a larger reduction in plasma glucose, meaning less energy available for homeostasis during the lairage period. Their recommendation was that 18 h should be a maximum feed withdrawal period before slaughter.

The effect of feed withdrawal on bird affect is not as well understood. Warriss *et al.* (1988) found reductions in liver glycogen levels with increased lengths of feed withdrawal (up to 36 h), and hypothesised that as that occurs, levels of hunger increase. Knowles *et al.* (1995) tested 24 h feed withdrawal on Ross broilers (7 week) and found a rise in blood corticosteroid levels, suggesting the 24 h period of withdrawal was stressful for the birds. Rodrigues *et al.* (2017) reported increased heterophil/lymphocyte (H/L) ratios in fasted birds, with the highest levels being found with 18 h feed withdrawal as compared to 12, 14 or 16 h. Frustration in broilers may also occur, particularly if the feed is visible (Khalil *et al.*, 2010; Zulkifli *et al.*, 2006). Interestingly, fear levels, as measured by tonic immobility, actually reduce in fasted birds (Sherwin *et al.*, 1993). Typically, tonic immobility is thought to be suggestive of fear, but in this case could be related to hunger itself.

6.2.3.2 End-of-lay hens

Very limited research has been conducted with end-of-lay hens overall, and to the authors' knowledge, the effect of length of feed withdrawal has not been studied. However, it may be possible to draw inferences from research conducted on molting procedures in hens.

In a study designed to examine stress in laying hens, Beuving and Vonder (1978) withdrew feed and water for 7 h from 34 week old hens, or deprived 40 week old chickens from food for 5 days and water for 2.5 days. The authors noted that the blood level of corticosteroid increased with feed withdrawal, suggesting a stress response. However, blood corticosteroid levels decreased when water was supplied to the birds. Interestingly, these authors also found that handling and crating the birds resulted in higher blood corticosteroid production than did feed and water withdrawal, suggesting that the birds find handling even more stressful than feed withdrawal.

Other measures of stress, including the H/L ratio, a measure of chronic stress in birds, have also been utilised in moulting feed withdrawal studies. Work by Gross and Siegel (1983) and Zulkifli *et al.* (2006) found this ratio to increase, indicative of stress response, when feed was withdrawn over a range of time lengths (24-48 h).

6.3 Catching and loading

Catching birds involves moving the birds from inside the poultry barn to the transport module. The methodologies of catching and loading vary by bird type, geographical location, etc. In many areas of the world, broilers and end-of-lay hens are caught by hand, often by specialised catching crews, and manually moved to crates and modules. Training of these crews is an important factor in determining the impact that catching and loading will have on bird welfare when manual loading is utilised (Kettlewell and Mitchell, 1994). In other cases, mechanical harvesters are utilised to remove the impact of human contact for the birds. The module systems may be lifted off of the truck and brought into the bird house to reduce the distance for carrying. Because of their size, commercial turkeys are generally moved in a different manner. They are often gently herded to a moving belt, then ride the belt up to the truck crates or modules.

Regardless of methodology, catching and loading is thought to be one of the most stressful times of a commercial bird's life (Kettlewell and Mitchell, 1994). Marchewka *et al.* (2013), for example, reported that the stress turkeys endure in the pre-transport phase significantly reduces their well-being, with effects ranging from stress to death. In addition, injury can also occur during this time. Griffiths (1985) reported that approximately 40% of bruising can occur during the catching process. There is also a link between injury, such as wing fractures (Jacobs *et al.*, 2017a,b), occurring during the catching and loading phase, and DOA on transport (Chauvin *et al.*, 2011; Kittelsen *et al.*, 2018). The risk of bone fractures is greater during the catching and loading of end-of-lay hens since they typically have reduced bone quality (Gregory and Wilkins, 1989). However, injury can be at least partially mitigated by careful handling, and perhaps by loading type.

6.3.1. Manual catching

Manual catching and carrying of birds to the transport module or crate can be conducted in a number of different formats. This can include carrying by one leg, by two legs, or in an upright position. Catchers can move more birds per trip with a one-leg catch, and the least number of birds per trip with an upright movement.

The data showing the effects of catching on the welfare of birds are limited, and there appears to be no comparison with one-hand catching compared to upright or data focused on time to load (direct impact on feed withdrawal, time birds remain on a stationary vehicle with limited air flow) and worker fatigue to catch and load large-scale barns with a one-hand carry versus upright carrying. Kittelsen *et al.* (2018) compared hand catching by two legs (2,010 birds) to birds moved in an upright catching technique (two birds caught by hand under the abdomen two birds at a time; 996 birds). In this study, the abdominal carry method was faster (231 vs 252 s to fill the module drawer). It should be noted though that the number of birds tested was small compared to the number that would be required on a commercial farm, and catcher fatigue may have a negative effect on bird welfare during the catching phase (Langkabel *et al.*, 2015; De Lima *et al.*, 2019).

Work by De Lima *et al.* (2019) tested different conditions during the manual catching and loading phase. This included studying lighting conditions, methods of catching (one or two upright at a time) and loading onto the module, and position of the module or crate on the truck. Their results suggested that bird welfare was best when birds were carried one bird at a time, carefully placed into the crate, with low lighting intensity (curtains closed), and with the crate set 21 cm above the ground. The applicability of these procedures in a large commercial barn was not tested.

6.3.2. Mechanical catching

Mechanical catching involves using machinery to physically catch birds and place them into the crates or modules. The efficiency of a number of 'mechanical harvester' types has been assessed, including pneumatic or vacuum-type systems, scoops or mats, or large combine-type machinery (Lacy and Czarick, 1998). The rationale behind this equipment is that bird stress and injury may be reduced with elimination of human handling and that cost would be lower and working conditions improved for those people working in the barn.

The effects on bird welfare are still unclear. With respect to injury levels, some research has indicated fewer injuries, body damage or bruises occurring when using mechanical harvesting equipment as compared to manual (Delezie *et al.*, 2005; Prescott *et al.*, 2000). Mönch *et al.* (2020) found no effect on the incidence of severe wing injuries, but reported more DOAs in mechanically caught birds. However, this difference may be explained by manual catchers identifying and not crating dead birds prior to placing in crates.

Mechanical harvesting may reduce stress levels in birds. Although no difference in heart rates between manually and mechanically caught broilers was found (Duncan *et al.*, 1986; Marchewka *et al.*, 2013), heart rate of the mechanically caught birds returned to baseline quicker, indicating lower stress levels (Duncan *et al.*, 1986). Furthermore, tonic immobility, a measure of fear in birds, was higher in manually caught birds (Duncan *et al.*, 1986). Delezie *et al.* (2005) found no differences in stress levels between manually and mechanically caught broilers, but the incidence of leg injuries was lower. The difference in the model of mechanical equipment that was used may be partially responsible for the conflicting results between studies.

Delezie *et al.* (2006) conducted a survey of consumer preferences for mechanical versus manual catching and loading of poultry. The authors found that if consumers were aware about the advantages and disadvantages of the equipment, they tended to be in favour of mechanically harvesting of birds. However, if they were not, they preferred manual catching. This consumer preference, along with cost effectiveness and functionality of this equipment, may be important in determining future directions with respect to catching.

6.4 Transport

6.4.1 Transport containers (crates and modules)

Birds are secured on transport vehicles either in crates or modules (Figure 6.1). These containers are designed and constructed to safely hold birds, without sharp protrusions that could injure birds or openings that could ensnare birds' body parts. The containers are often made from perforated plastic or wire mesh. These containers are designed with a large amount of perforations, typically on all six sides, to support air movement and allow inspection of birds after they have been loaded (Kettlewell and Mitchell, 1994).

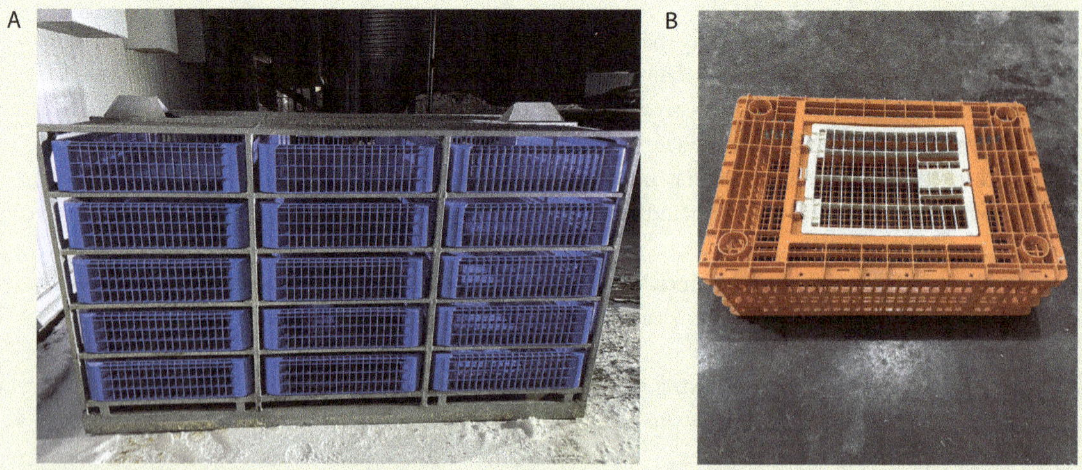

Figure 6.1. Broiler transport containers, (A) transport module produced by Anglia Autoflow and (B) example of a poultry transport crate (T. Crowe, University of Saskatchewan, Canada).

Both the module and crate are designed to be a discrete unit to contain multiple birds while in transit. Crates are typically manipulated by hand, and they can be moved from the truck into barns, sometimes using conveyors. If crates are moved into the barn to be filled with birds, the filled crates must then be reloaded onto the truck. Filled crates can become heavy, and extra care must be taken to ensure that the crates remain horizontal after birds are inserted. Alternatively, birds are caught and carried to the truck where they are inserted into crates, while the crates remain on the truck. This process, as compared to the process with modules, often requires that birds are carried further distances and may be subject to hand-offs from one person to another.

Modules are designed to contain multiple compartments (drawers or tiers) of birds and require a forklift to remove the modules from the truck, place them in or near the rearing site, then reload the module on the truck. Modules allow a greater degree of mechanisation and have the potential to reduce the amount and time of human handling of birds. By reducing time of human handling, worker fatigue is reduced, which can thereby reduce the potential for injury (Langkabel *et al.*, 2015; De Lima *et al.*, 2019).

The dimensions of crate/module interior coupled with the number and size of birds loaded into the container will dictate the amount of space that each bird is allowed. The stocking density of birds (bird mass per floor area) along with the preferred amount of headspace within the containers are not well understood (Section 6.4.3). Codes of practice (National Farm Animal Care Council, 2016) and standard operating practices within industry are largely guided and informed by experience and historical practice.

Wichman *et al.* (2010) reported that turkey behaviour, while birds are in transport containers, is affected by the amount of head space allowed, with taller crates allowing turkeys more freedom to move. With taller crates, birds tended to stand, which increased the potential for scratching and bruising other birds. Conversely, broilers tend to sit when the transport vehicle is moving (Vinco *et al.*, 2016). This study also showed that when greater headspace was available, broilers were observed to stand, flap their wings and tended to trample over other birds, which likely led to the higher incidence of injuries. Stress levels were also elevated as demonstrated by greater H/L ratios in the blood.

6.4.2 Trailer design and the influence on transport conditions

Conditions experienced by birds during transit can affect a number of outcomes, including bird behaviour, welfare and meat quality. Exposure conditions vary greatly, often within the same trailer, and are influenced by a number of factors. In some cases, especially for small chicks and poults, transport trailers incorporate active ventilation and environment-control equipment. However, most poultry transported to processing plants in North America are moved in passively ventilated, unheated trailers.

Passive ventilation takes advantage of the pressure envelope created around the periphery of a moving trailer to move the air through the load. The pressure envelope around a moving trailer is a function of the travel velocity combined with the direction and velocity of prevailing winds.

Generally, higher travel velocities and greater air velocities over the surface of the trailer will cause more distinct separation zones and greater pressure differentials. Air will move within the trailer from regions of relatively higher pressure to regions of relatively lower pressure.

The transfer of heat and moisture from within the trailer to the exterior can be achieved through ventilation, where the warm moist air within the load is exchanged for cooler and dryer ambient air. Ventilation is important to help maintain proper air quality by removing excess heat, water vapor and gases that can otherwise accumulate within the load.

Temperatures experienced by birds have been reported as a major factor in determining bird welfare (Mitchell and Kettlewell, 1998). The acceptable temperature ranges depend on the poultry species and the age of the animals. Air temperature that birds experience, as a single metric, is important, but it is important to recognise that the effects of temperature extremes can be compounded by a combination with humidity, stocking density and vibration and accelerations.

The objectives of incorporating ventilation within a trailer are not unlike the objectives of ventilating a barn. However, there are greater challenges to provide ventilation in a trailer, because the load is very dense with containers and birds. Any air movement within the load must move through perforations in the sides of containers, between containers and between containers and the trailer envelope. It is difficult for air to penetrate or to move through the load. The limited air exchange inside a passively ventilated trailer causes the internal conditions to vary significantly (Kettlewell and Mitchell, 1994). Knezacek *et al.* (2010) reported a difference of more than 56 °C between the ambient temperature and the warmest recorded temperature inside a truck during a cold Canadian transport event. Conversely, during colder ambient conditions, when tarps are installed along the sides of the load, there are also cold conditions, only slightly warmer than ambient, near the outside of the load and near inlets. Birds within a single transport vehicle during cold ambient conditions may experience heat stress or cold stress, depending on their location within the load. Figure 6.2 illustrates these points with a warm core near the midline at the front of the front trailer and colder regions located along the outside and at the rear of the front trailer.

Actively ventilated trailers have been described in published literature (Dadgar *et al.*, 2010; Kettlewell and Mitchell, 2001; Spurio *et al.*, 2016). Active ventilation requires fans to move air through the load. Further, some systems have the ability to condition the air by either heating or cooling it before it is introduced to the birds within the containers. The requirement for an electrical power source in addition to ventilation equipment and controllers makes actively ventilated trailers more complicated than passively ventilated trailers. Tarps must always be installed along the sides of actively ventilated trailers for the ventilation system to work. Kettlewell *et al.* (2001) reported that a well-managed actively ventilated trailer has the potential to improve the welfare of birds while in transport.

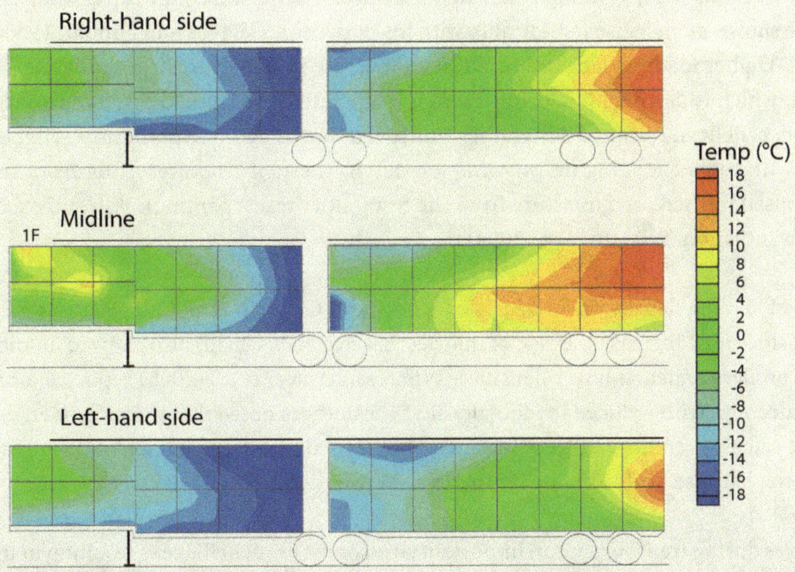

Figure 6.2. Temperature distribution within a two-trailer road train (Burlinguette *et al.*, 2012).

6.4.3 Loading density

The scientific literature regarding loading density for poultry transportation cannot provide an explicit ideal loading density. Rather, several factors need to be considered when determining loading density, including environmental conditions, i.e. ambient temperature, relative humidity, and weather conditions) and bird factors, i.e. sex, age, and live weight (Bayliss and Hinton, 1990). Ambient conditions have to be considered when determining loading densities for each transport load in order to minimise bird exposure to thermal stress. High loading densities limit bird movement and negatively impact the crate microclimate (Bedanova *et al.*, 2006). Loading densities need to be reduced in hot and/or humid weather to allow birds to perform thermoregulatory behaviours, allow better ventilation to dissipate heat, and reduce metabolic heat and humidity. In cold weather, birds need enough space to move around and access warmer areas of the crate and maintain body heat (Strawford *et al.*, 2011), while densities that are too low may reduce the available heat within the crate (Weeks *et al.*, 1997).

In Canada, the recommend maximum loading densities for broiler chickens and end-of-lay hens are 63 kg/m^2 at moderate temperatures and 54 kg/m^2 at hot temperatures (Poultry Industry Council, 2010). For turkeys, the recommended densities are 98 kg/m^2 at moderate temperatures and 83 kg/m^2 during hot conditions. These values do not take into account wind chill or humidity.

6.4.3.1 Broilers

Numerous studies have evaluated the impact of crating density on broiler chickens. Higher crating densities have been associated with higher numbers of DOAs (Chauvin *et al.*, 2011; Hussnain *et al.*,

2020a; Nijdam *et al.*, 2004; Whiting *et al.*, 2007). Conversely, during cold weather transportation, studies have shown an increase in mortality with lower densities (Baylis and Hinton, 1990; Caffrey *et al.*, 2017). Higher loading densities result in an increase in metabolic heat production (Caffrey *et al.*, 2017), which is detrimental during warmer temperatures as it could result in hyperthermia, however it is beneficial during cold weather transportation to avoid hypothermia. Higher DOAs associated with higher loading density may be due to thermal stress resulting from increased humidity caused by increased moisture from the respiratory tract (panting), skin and excreta, and inadequate ventilation (Nijdam *et al.*, 2004).

The impact of loading density on bird injuries is not clear. Chauvin *et al.* (2011) suggested that low loading densities may increase the risk of injuries. Saraiva *et al.* (2020) also showed an increase in bruising in broilers crated at lower densities. Whereas Petracci *et al.* (2005) reported a reduction in downgraded legs with reduced loading density. The authors noted that bruising often occurs on farm during catching, crating, and loading and hypothesised that the reduction in downgraded legs may have been because birds were easier to crate at lower densities.

High densities during transport are an important stressor for broiler chickens resulting in increased H/L ratios (Bedanova *et al.*, 2006), heat shock protein 70 and corticosterone concentrations, and rectal temperature (Delezie *et al.*, 2007).

Hussnain *et al.* (2020a,b) assessed the impacts of crating density on broilers transported during warm conditions and cold conditions. During warm transport conditions (27.3 to 33.6 °C) higher densities (0.033 m^2/bird) resulted in higher body weight loss, higher DOAs, and increases in serum catalase concentration which is an indicator of oxidative stress. Whereas, during cold transport conditions (3.6 to 9.5 °C), lower loading density (0.050 m^2/bird) increased body weight losses. Petracci *et al.* (2005) also reported higher live weight losses in broilers transported at low loading density (<55 kg/m^2) between September and May in Italy. However, transporting broilers at high stocking density during cold conditions may result in the bird's inability to use behavioural adaptions to cope with their environment (Strawford *et al.*, 2011; Watts *et al.*, 2011). Whereas lower loading densities would allow birds to adjust their location within the crate depending on the location of cold spots and huddle and/or burrow to maintain body heat.

6.4.3.2 Pullets and end-of-lay hens

There is a substantial lack of scientific literature related to loading density for pullets and end-of-lay hens. Weeks *et al.* (1997) reported the use of lower densities in end-of-lay hens compared to pullets as hens are more fragile and therefore may be more likely to sustain injury. However, these birds were then exposed to colder temperatures as the low stocking density reduced the amount of heat available within the load.

6.4.3.3 Turkeys

Relatively few studies have examined the impacts of loading density on turkeys. Mallia *et al.* (2000) found that turkeys transported at higher loading densities (3 to 20 birds per crate) presented a lower incidence of cyanosis. It was suggested that during cold transport, higher densities may cause birds to experience lower thermal loss and therefore reduced utilisation of their energy reserves for maintenance of body heat. Di Martino *et al.* (2017) reported that at temperatures ≤2 °C high loading density near to or just above the maximum density in EC Reg. 1/2005 (EC, 2005), the DOA risk was greater than for loads with densities of 10 kg/m^2 less than the EC maximum. These results may be explained by the fact that lower densities allow birds to move away from cold spots and to huddle in warmer locations within the crate (Strawford *et al.*, 2011). No effect of loading density was found for carcass downgrading, including trimming, bruises, scratches, and leg oedema (McEwen and Barbut, 1992).

6.4.4 Thermal stress

Thermal stress is one of the primary stressors in commercial transport. Ambient temperature can play a large role in the health and welfare of poultry during transport. However, there are multiple other factors that may contribute to the trailer thermal microclimate conditions to which an individual bird is exposed, such as trailer design, crating density, relative humidity, weather, distance and speed of travel. In addition, both time of year/season and geography can greatly impact whether birds are more likely to be exposed to cold vs heat stress.

6.4.4.1 Broilers

Thermal stress, both hot and cold, has been identified as a risk factor for increased DOA levels for broiler chickens (Caffrey *et al.*, 2017; Nijdam *et al.*, 2004). Vecerek *et al.* (2016) found that DOAs were higher in the winter months compared with summer months, specifically when temperatures were between -6 and -3.1 °C, with levels of DOAs decreasing at temperatures above -3.0 °C. Whereas, a study conducted in the UK reported higher levels of DOAs in the summer months, particularly when temperatures exceeded 17 °C (Warriss *et al.*, 2005).

When specifically evaluating the stress response, transport during ambient temperatures between -5 and 5 °C have also been associated with increased blood corticosterone concentration (Vosmerova *et al.*, 2010). However, increased H/L ratios have also been recorded in broilers that were crated and exposed to 34 °C compared with broilers exposed to 22 and 15 °C (Akşit *et al.*, 2006).

Carcass and meat quality can be impacted by both hot and cold exposure. Dadgar *et al.* (2011) noted that broilers exposed to cold conditions (0 to -17 °C) had higher live shrink levels than broilers exposed to neutral conditions (20 °C). Cold exposure also typically results in higher muscle pH, water-holding capacity/cooking loss, and dark colour (L* values below 46), all traits indicative of dark, firm, and dry (DFD) meat, resulting from the depletion of muscle glycogen stores (Barbut *et al.*, 2005; Dadgar *et al.*, 2011, 2012). On the other hand, an increased incidence of pale, soft, exudative (PSE) meat, as shown by lower muscle pH, water-holding capacity, and light colour (L* values above 57.7), has been reported in broilers exposed to hot conditions (Barbut *et al.*, 2005; Dadgar *et al.*, 2010; Sandercock *et al.*, 2001).

6.4.4.2 End-of-lay hens

End-of-lay hens may experience greater difficulty coping with thermal stress as they are metabolically stressed due to the demands of egg production and are typically poorly feathered which affects their ability to thermoregulate (Leeson and Morrison, 1978; Richards *et al.*, 2012). There is considerably less research on the transportation of end-of-lay hens as they are not transported as frequently as broilers. However, the majority of the published research focuses on DOAs. Despite some evidence of increased DOAs in summer (Petracci *et al.*, 2006), the number of DOAs more frequently increases during the winter months (Vecerkova *et al.*, 2019; Weeks *et al.*, 2012). Voslářová *et al.* (2007) found the highest levels of DOAs when the ambient temperature was below 10 °C. This was also seen in a large study conducted in Turkey, where DOAs were highest in the winter and spring months when temperatures ranged from 3.5 to 17.5 °C (Cavusoglu and Petek, 2021). A large study in the UK assessed risk factors associated with mortality during transport for end-of-lay hens and found that cooler temperatures resulted in higher mortality levels, with poor-feathered hens being less likely to survive compared to their well-feathered counterparts (Weeks *et al.*, 2012). Frerichs (2021) also observed higher mortality in poor-feathered white-strain end-of-lay hens compared to well-feathered hens when exposed to cold (-10 °C). Interestingly, the results of a simulated transport study of brown end-of-lay hens found that hens that were exposed to cold temperatures (-10 °C) for periods of 8 or 12 h experienced higher mortality levels than those exposed to more thermoneutral or warm conditions (21 or 30 °C), regardless of feather cover (Beaulac *et al.*, 2020). While it is unknown why brown hens were unable to cope regardless of feather cover, there is potential that it may be related to strain differences. Cavusoglu and Petek (2021) reported overall lower DOA rates for white-feathered hens (1.37%) compared with brown-feathered hens (1.64%). An increased incidence of PSE meat has also been found in end-of-lay hens transported under cold conditions (-10 °C) (Beaulac *et al.*, 2020).

6.4.4.3 Turkeys

Turkeys respond to thermal stress slightly differently when compared with chickens as turkeys tend to tolerate cold exposure better than heat exposure (-18 vs 20 °C; Henrikson *et al.*, 2018). Temperatures greater than 12 °C have been associated with increased DOAs, with temperatures above 26 °C further increasing DOAs (Di Martino *et al.*, 2017). A study conducted in Italy found that DOA levels in turkeys were highest in the summer months compared with spring, autumn, and winter with temperature ranging from 16.3 to 27.7 °C (Petracci *et al.*, 2006). Vermette *et al.* (2017) evaluated the effects of heat stress during simulated transport, where turkeys were crated and exposed to 35 °C for an 8 h period and found that turkeys exposed to those conditions presented increased core body temperature and H/L ratio, both indicators of heat stress and more elevated breast and thigh muscle pH values.

6.4.5 Transport duration or distance

Duration or distance of transport can have a large impact on bird welfare, with long durations decreasing bird welfare to a larger extent than short durations. Caffrey *et al.* (2017) highlighted that increasing transport duration increases bird risk of mortality due to: (1) inability to cope with transport as a result of previous health status; (2) injury during catching or loading; and (3) inability to cope with thermal stressors potentially worsened by lack of feed and water. As most results related to the effects of transport duration originate from retrospective studies, the effects of transport distance have been studied more than duration. While this information is useful, it is important to consider that duration of transport over the same distance can be extremely variable.

6.4.5.1 Broilers

Transport duration for broiler chickens is typically shorter than that of turkeys or end-of-lay hens, as there are a larger number of slaughter plants that process broilers which results in closer proximity of the farm to the slaughter facility. However, similar to other bird types, DOA levels of broilers increase as transport duration or distance increases.

A study conducted in the Czech Republic found that longer transport distance resulted in increased DOA numbers starting from 0.15% for broilers transported under 50 km up to 0.68% for broilers transported over 300 km (Voslářová *et al.*, 2007). Similarly, another study found that DOAs increased in a stepwise manner from approximately 0.45 to 0.75% with increasing transport distance from 101 km to more than 300 km (Vecerek *et al.*, 2016).

Bianchi *et al.* (2005) found that mortality was lower for broilers transported less than 3.5 h (0.24%) compared with those transported for 3.5-5 (0.41%) or >5 h (0.45%). Kittelsen *et al.* (2017) took a different approach and grouped broilers into high DOA (over 0.30%) and normal DOA (up to 0.30%) groups and found that the group with high DOA rates was transported on average 1.5 h longer than those in the normal group.

While it is well known that transportation is a stressful procedure, increased duration may exacerbate the stress response. Yalçin and Güler (2012) reported higher H/L ratios in broilers transported over longer distances (115-165 km) compared with broilers transported over shorter distances (65 km). Corticosterone levels have not typically been shown to increase in relation to transport duration. The effects of crating alone have been evaluated, with corticosterone concentrations being elevated in birds that had been crated compared with the control (uncrated birds). However, prolonged duration of crating had no further effect on stress levels (Bedanova *et al.*, 2014; Chloupek *et al.*, 2008). Vosmerova *et al.* (2010) found that corticosterone was actually highest in broilers prior to transport and attributed this to the catching and loading process.

Transport duration may also impact carcass and meat quality. Increased transport duration has been shown to increase live shrink, with broilers losing 1.27, 1.87, and 2.09% body weight for durations of ≤3.5, 3.5-5.0, or >5 h, respectively (Bianchi *et al.*, 2005). Yalçin and Güler (2012) found that, regardless of ambient temperature or season, long distance transport (165 km) resulted in PSE meat, but only in heavier broilers (>2.4 kg).

6.4.5.2 End-of-lay hens

End-of-lay hens are typically hauled for longer durations and distances as demand is low and there are few slaughter plants equipped to process them (Newberry *et al.*, 1999; Weeks *et al.*, 2012). In addition, geography may play a large role in the effects of transport durations on the welfare of end-of-lay hens. A paper discussing the management of spent hens stated that the average time hens spend on the trailer ranged from 16-26 h, depending on the origins of the birds (Newberry *et al.*, 1999). The authors also noted that as time spent on the trailer increased from less than 12 h to more than 24 h, DOAs and bird/carcass condemnation rates also increased (0.7 to 2.3% and 6.5 to 9.9%, respectively; Newberry *et al.*, 1999). The transport duration applied in the latter Canadian study was substantially longer than that reported in an Italian study on 3,241 loads of end-of-lay hens, where hens were transported for 1-8 h (Di Martino *et al.*, 2017). In this latter study, the DOA rate increased during hauls longer than 2 h, reaching a peak in DOA rate at 8 h of transport (Di Martino *et al.*, 2017). Simulated transport studies also reported greater dehydration and live shrink values with increasing transport time, i.e. 12 vs 4 h (Beaulac *et al.*, 2020; Frerichs *et al.*, 2021).

Similar results have been reported in studies assessing the effects of transport distance. Vecerkova *et al.* (2019) reported a lower mortality rate (0.34%) within loads of hens transported for shorter distances (≤50 km) compared with those transported for longer distances (0.80%; 201-300 km). A large retrospective study analysed 3,256 loads of end-of-lay hens in Turkey and found that DOA rates were lowest in loads transported for distances under 400 km (1.03-1.13%), while the highest rate was recorded at travel distances greater than 800 km (2.96%; Cavusoglu and Petek, 2021). In addition, body weight loss was again lowest for hens transported under 400 km (3.04-3.05%) and highest for those transported greater than 800 km (6.84%). These authors also found a significant and positive correlation between transport distance and body weight loss (r=0.40) as well as DOAs (r=0.29).

Ambient conditions can exacerbate the effects of transport duration on DOAs. A study looking at simulated transport of end-of-lay hens found that mortality of hens was 10 times higher when they were exposed to cold conditions (-10 °C) for 12 h (33.3%) vs 4 h (3.3%) (Beaulac *et al.*, 2020).

6.4.5.3 Turkeys

Transport durations for turkeys also vary by geographical location. Furthermore, similar to broilers and end-of-lay hens, exposing turkeys to longer transport increases DOA rates and carcass shrink, and reduces meat quality. An Italian study registered the highest DOA rate in loads transported for more than 60 min (Di Martino *et al.*, 2017). In a Canadian survey of 1,005 truck loads, McEwen and Barbut (1992) found that the average time turkeys spent on trucks was 13.4±3.8 h for toms and 14.9±3.3 h for hens and reported an association between increased travel duration and carcass downgrading due to bruised drum and half wing trims. Boukhris *et al.* (2017) found that turkeys that were transported longer (2-2.5 h) compared with 15-30 min exhibited detrimental effects on meat quality as exhibited by PSE meat characteristics and tougher meat. Two studies carried out in the Czech Republic have observed the lowest mortality in turkeys with transport distances under 50 km, while the highest mortality rates were found at distances greater than 200 km (Machovcova *et al.*, 2016; Voslářová *et al.*, 2007).

6.5 Lairage and unloading

6.5.1 Lairage time

After arrival at the slaughterhouse, poultry in the containers are placed in a lairage area. The most commonly used procedure is to unload the containers from the truck and placing them in the lairage area where they stay until the time of slaughter. In some cases containers stay on the trucks and the trucks are placed in the lairage area. The duration in lairage varies largely between slaughter plants.

In a survey in Italy, Grilli *et al.* (2015) estimated the lairage time for 233 different batches of broilers as ranging between 0.2 and 9.4 h with a mean of 4 h. A Belgian study undertaken at six slaughter plants, found lairage durations varying from 15 min to 9.3 h, with a mean at 4h35 (Jacobs *et al.*, 2016). In Brazil, Rodrigues *et al.* (2017) investigated the effects of different lairage times (0, 2, 4 and 6 h) and the position of chicken crates during transport to the slaughterhouse on the biochemical and haematological profile. They found an increase in the blood H/L ratio in broilers that were subjected to 6 h of lairage at the abattoir. This indicates that increasing the lairage time, including the additive effect of fasting, generates stress-related metabolic disorders in broiler chickens. In another study in Brazil, Vieira *et al.* (2010) evaluated 215 poultry transport trucks from a commercial poultry slaughterhouse from 2006 to 2007. Data were obtained from the holding area acclimatised by fans and water misting in the mornings, afternoons and at night. Lairage time was considered in the analysis besides the thermal variables (i.e. temperature and relative humidity), transport distance, and density of birds per cage). The results indicated that the effects of lairage time varying from 1 to 5 h, along with those of travel distance, were important in the variation of rectal temperature and on the number of dead birds per truck. Lairage time in a controlled environment had no effect in winter, whereas in summer increased lairage time decreased the mortality rate.

From these different published observations it can be concluded that lairage conditions and durations can have an important impact on a bird welfare. Considering the climatic conditions and the waiting time which the animals have been submitted to during transport and lairage itself, the hazards identified during this process are: 'too high or too low effective temperature', 'too long food and water deprivation', 'unexpected loud noise' and 'insufficient space allowance'. These hazards can cause heat or cold stress, prolonged hunger, prolonged thirst, fear and restriction of movements, as welfare consequences to the birds (EFSA, 2019). This implies that duration in lairage should take into account the total period of fasting and transport and should be kept within limits to prevent animals from being without feed and water too long.

Furthermore, birds in lairage should be protected from adverse climatic conditions and preventive measure should be taken to avoid thermal stress. If effective temperature are still above thermoneutral zone (TNZ = 23-29 °C; Meltzer, 1983), ventilation should be provided and in certain cases misting and nebulisation can be applied to bird in order to cool them down (Jiang *et al.*, 2015). However, the TNZ in poultry is not the same in all circumstances, it depends on genetics and physiological status, body weight, the degree of feather cover, acclimatisation, feed and dehydration status (Dawson and Whittow, 2000) as well as other environmental factors (Lara and Rostagno, 2013). If chickens are kept in the truck before unloading at ambient temperatures above the critical upper temperature

(29 °C; Meltzer, 1983), death may occur because of hyperthermia (Webster *et al.*, 1993). It is thus recommended not to leave a truck stationary for more than 45 min when the temperature of the load exceeds 30 °C, unless the truck is kept in shade and air circulation is provided (CARC, 2003). Alternatively, to cool-off birds during the stationary situation fan banks with or without water sprinklers can be also used before unloading. The combination of water sprinkling or misting and ventilation prevents for humidity rise in the environment and results in increased evaporative cooling allowing birds to lose body heat. Research showed that water sprinkling upon arrival followed by ventilation during 1 h lairage in the truck reduced the mortality of chickens and the production of PSE breasts (Guarnieri *et al.*, 2004; Vieira *et al.*, 2010).

6.5.2 Unloading

The stunning process of poultry often requires for the birds to be taken out of the containers to be further handled to be restrained on a shackle line (electrical stunning) or placed on a conveyor belt (gas stunning). Containers are moved from lairage area to the stunning area and emptied by hand or automatically depending on the type of container and stunning system. Most common types of containers are: (1) plastic crates that can be opened at the top and or side to take out birds manually containing up to 12 broilers; and (2) containers with 4 or 5 layers containing up to approximately 400 broiler chickens.

The type and design of crates and containers has an impact on the ease of removing birds from the containers. Small openings and sharp edges will increase the risk of animals being injured during handling or emptying. Similarly, when birds are difficult to reach, the catching of birds located away from the openings can be complicated, and the operator may tend to catch the animal roughly by the head, neck or wings, increasing the risk of injury (EFSA, 2019). Rough handling can cause pain and fear due to injuries. These would appear as main carcass defects and can be seen as post-mortem lesions, such as skin lesions, wing fractures, and bruising on wings and breasts (Jacobs *et al.*, 2017b).

Containers with different layers or drawers can be emptied manually by removing the drawers from the rack manually or automated. The drawers are placed on a conveyor belt and birds can be either shackled for electrical stunning or the drawers can go with birds into the controlled atmosphere stunner (CAS). When containers are emptied automatically, a hydraulic platform operated by an employee tilts the entire container to dump the live birds onto a conveyor that runs into the shackle room (Tinker *et al.*, 2005). The manually and mechanical unloading of birds from the crates or containers are inducing risk factors for animal welfare, including injuries, such as bruises and wing damage, that will affect carcass quality. One problem that can occur with dump modules is unloading the birds too fast. This results in birds falling on top of each other on the conveyor belts and pileups. When heavier birds are unloaded from the dump modules, the risk of broken wings is higher compared to lighter birds (AVMA, 2016). Negative aspects of dumping birds on a conveyor belt can be ameliorated by avoiding dumping birds before the dumping belt is empty by slowing down the operation.

Preventing unloading birds form the containers or crates by placing the containers or drawers of the containers into the stunner will reduce these risks. For this reason, more and more gas stunning systems are designed to stun birds in the containers or drawers.

6.6 Stunning methods

Stunning of animals prior to slaughter is mandatory in the EU and many countries around the world have set standards for welfare practices with regard to humane slaughter, and the OIE also includes standards for the humane conduct of slaughter in Chapter 7 of its Terrestrial Animal Health Code (OIE, 2019). These standards and regulations are based on the fact that animals, including poultry, are sentient beings that would experience pain, fear and distress when slaughtered while conscious.

Stunning is performed to render animals unconscious for a period that is long enough to guarantee that they do not recover during the bleeding phase. The overall goal should be to minimise or eliminate anxiety, pain, and distress prior to loss of consciousness. Therefore, both the induction of unconsciousness and handling prior to slaughter must be considered (AVMA, 2016).

Electrical stunning is the most commonly used method for stunning poultry worldwide, however, CAS is more and more applied for broiler chickens and turkeys. Besides the traditional electrical water bath and two-phase CAS stunning new and improved methods are being developed or under research.

6.6.1 Shackling

An unavoidable aspect of electrical water bath stunning is that the birds are inverted onto moving shackles prior to stunning. Shackling is probably the most important aspect affecting animal welfare during the process of electrical stunning. Different studies clearly indicates that inversion and more specific hanging upside down is leading to increasing plasma corticosterone levels (Debut *et al.*, 2005; Kannan *et al.*, 1997), vigorous wing flapping and escape attempts (Sparrey and Kettlewell, 1994). The reaction to shackling increase by rough handling, unsuitable shackles, bright lights and noise (Gregory and Bell, 1987). Shackling is proven to be painful, especially when hanging the birds on to the shackles is performed by workers, at high speed, often resulting in a considerable number of bruised legs due to compression of the legs or broken bones (Gentle and Tilston, 2000; Sparrey and Kettlewell, 1994; Veerkamp and De Vries, 1983). The pain due to shackling is likely to be worse in birds with leg abnormalities (e.g. lame broilers; Butterworth, 1999; Danbury *et al.*, 2000). Furthermore, the stress effect is depending on the duration of being shackled (Bedanova *et al.*, 2007); therefore, a maximum shackle duration of 1 min is recommended or even required in EU regulation (EC, 2009). The negative effects of shackling cannot be prevented as handling and inversion is a prerequisite of the stunning method, but they can be mitigated by reducing the duration of being inverted to a minimum, by handling animals with care and by using the right shackles. Furthermore, calming the animals and reduce the level of wing flapping by applying breast support will help to mitigate pain and fear during the inversion period. Furthermore, the size of the shackles should be appropriate to the size of the birds and birds' legs. The moving shackle line may include sharp curves, inclinations, drops and bunchy transitions, causing irregular movements that can increase the force the shackles exert on the legs of the animals. The irregular movements and increased force on the legs can lead to painful compression and a fear response (EFSA, 2019).

The only way to prevent shackling prior to stunning at high throughput rates at the moment is by replacing electrical water bath stunning by a CAS method.

6.6.2 Electrical stunning

The process of electrical stunning involves unloading the birds from the containers or crates, shackling them in inverted position and leading them trough an electrified water bath. The heads of the birds are immersed in the electrified water bath whereby a current then flows through the body of the bird. Depending on the dimensions of the water bath up to 30 birds are submerged simultaneously in a constant flow in the water. A metal strip in the base of the water bath forms the positive electrode and the shackles are earthed and form the negative electrode, so that the electric current passes through the bird from head to legs. The water bath is electrically live so that each bird is stunned from the moment it makes contact with water (Bilgili, 1999; Fernandez, 2004). This current causes generalised epileptiform activity in the brain of the birds, rendering them unconscious (Berg and Raj, 2015). Depending on the electrical frequency (Hz), the current may also affect the heart of the bird, causing permanent cardiac arrest, in which case the stunning is irreversible. It should be mentioned that there is a wide range of current levels, waveforms and frequencies used commercially, and that there is a complex relationship between these input variables (Berg and Raj, 2015).

The birds in the stunning system form a parallel pathway of resistance and together with the shackles and framework this will result in the overall current that is generated in the system. The resistance of every bird varies due to differences in body conditions and shackle conditions (Woolley *et al.*, 1986), i.e. degree of fouling and contact area with bird. These variations in resistance, which can range from 650 to 2,630 Ω (Hindle *et al.*, 2010: Sparrey *et al.*, 1993) influence the efficacy of the stun, so that some birds receive insufficient current to facilitate unconsciousness and insensibility. This can lead to problems with either bird welfare, in terms of failure to lose consciousness or rapid recovery, or product quality, i.e. blood splashing and bone fractures (Kranen *et al.*, 1999).

The combination of factors, such as frequency, voltage, current, wave form, resistance of the apparatus, resistance of the birds, and dimensions of the water bath all influence the success of the stun (Hindle *et al.*, 2010; Kranen *et al.*, 1999). It has been claimed that in practice only about one-third of birds are effectively stunned, one-third are inadequately stunned, and the remainder undergo cardiac arrest (Woolley *et al.*, 1986). The large between bird variation in effective stunning is confirmed by Hindle *et al.* (2010). Research expressed that effective stunning depends strongly on the received current in close relation with the applied frequency. Increasing frequencies demand higher currents to induce unconsciousness (Hindle *et al.*, 2010). In Figure 6.3 the probability of effective stunning is presented at three different frequency settings. For a probability on effective stunning, at a 50 Hz frequency a current of on average 150 mA is required, while at 400 Hz the required current is 250 mA. At a frequency of 1000 Hz a current of an average 450 mA will not

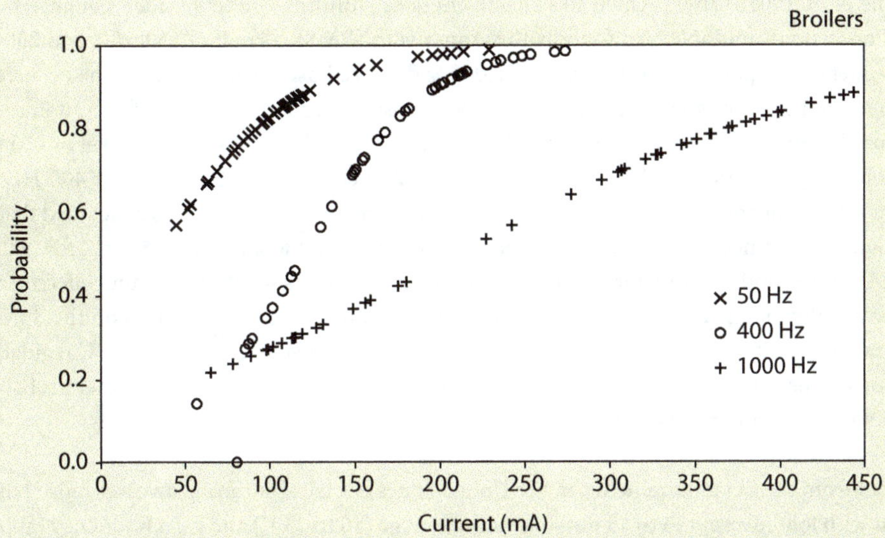

igure 6.3. Probability of an effective stun in relation to current (mA) based on observations from individual stuns with broilers t frequency levels of 50 Hz (×), 400 Hz (O), and 1000 Hz (+) (Hindle *et al.*, 2010).

even result in a probability of more than 0.8 to result in effective stunning. Based on these results maximum electrical frequency should not exceed 450 Hz.

Looking at the results from the study of Hindle *et al.* (2010) and at the minimum currents from the EU regulation (EC, 2009; Table 6.1), it appears that the required mA in the EU regulation are at a low level at the different frequencies and especially at the higher frequencies.

However, Raj *et al.* (2006) investigated the effectiveness of electrical water bath stunning of individual broilers for 1 s with a constant root mean square (RMS) current of 100, 150 or 200 mA, delivered using a variable voltage/constant current stunner with 200, 400, 600, 800, 1000, 1,200 or 1,400 Hz sine wave alternating current (AC). Based on the results of this study, it was recommended that effective water bath stunning of broilers with a minimum constant current of 100, 150 and 200 mA could be achieved with electrical frequencies of up to 200, 600 and 800 Hz, respectively. In addition, it was suggested that electrical frequencies of above 800 Hz would require a minimum current of greater than 200 mA to induce epileptiform activity in the EEGs of broilers.

able 6.1. Minimum average currents for water bath stunning of poultry (EC, 2009).

Frequency (Hz)	Chickens	Turkeys	Ducks and geese	Quails
<200	100 mA	250 mA	130 mA	45 mA
200-400	150 mA	400 mA	not permitted	not permitted
400-1,500	200 mA	400 mA	not permitted	not permitted

Girasole *et al.* (2016) investigated the effectiveness of stunning broilers under slaughterhouse conditions using a multiple bird water bath stunner with (RMS) current of 150, 200, and 250 mA and frequencies of 200, 400, 600, 800, and 1,200 Hz delivered using sinewave AC. The occurrence of corneal reflex, spontaneous eye blinking, and a positive response to a painful stimulus were monitored as indicators of consciousness. The results indicated that at a current level of 150 mA, the probability of a successful stun was over 90% at 200 Hz, approximately 40% at 400 Hz, and below 5% for frequencies greater than 600 Hz. Based on the results, the authors concluded that the minimum current necessary to achieve effective stunning in 90% of birds is 150 mA for 200 Hz, 200 mA for 400 and 250 mA for 600 Hz. The necessary currents for 90% stunning effectiveness, i.e. a probability of 0.9, is in line with the results from Hindle *et al.* (2010). Based on the different research projects, it can be concluded that the minimum required currents from the EU regulations will probably be sufficient to effectively stun 90% of the broilers, but to achieve a higher probability considerable higher currents will be necessary.

Different from the EU requirements, in the United States electrical stunning involves pulsed direct current with low current (25 to 45 mA/bird), low voltage (10 to 25 V) and high frequency (approx. 500 Hz) (McNeal *et al.*, 2003). A study involving two-phase electrical water bath stunning, with a first phase consisting of low-voltage (12 and 15 V), high-frequency (550 Hz) pulsed direct current for 10 s and a second phase consisting of sinusoidal wave AC (50 Hz at 40, 50, and 60 V for 5 s) showed that 22% of the birds had positive corneal reflexes, 18% showed spontaneous blinking, and <10% wing flapping (Prinz *et al.*, 2012). These behaviours are considered as signs of consciousness and eventually pain and fear. However, other results of studies of birds stunned with the US low-voltage model indicate that the birds are unresponsive to stimuli. However, the physiologic data are contradictory and it is unclear whether birds truly reach a state of unconsciousness (AVMA, 2016).

One of the most common problems in water bath stunning is that birds are receiving not enough current due too to low voltage settings or too high frequency for the applied voltage to be effective. This will lead to birds not rendered unconscious or a rapid return to consciousness after stunning. Another problem is birds missing the stunner water bath because they are extremely small compared to the average size of the flock. These birds are too short to have direct contact with the water bath. Also birds that pull up their heads or body will make no or intermediate contact with the water leading to pre-stun shocks or mis stuns. Pre-stun shocks will also occur frequently as birds are entering the stunner and the wing tips reach the water bath before the bird's head is in direct contact with the water. To prevent birds receiving pre-stun shocks or missing contact with the water completely the height of the water-bath stunner must be adjusted so that the birds cannot pull themselves up and avoid the stunner. Also a well-designed entrance ramp into the water bath can prevent pre-stun shocks.

After stunning, birds should be checked for unconsciousness by assessing indicators of consciousness/ unconsciousness at key stages (EFSA, 2013). Effective electrical stunning is characterised by the appearance of tonic/clonic seizures (showing as fast, uncontrolled muscle contractions), absence of breathing, absence of eye reflexes and spontaneous blinking (EFSA, 2013). However, due to the high line speeds, up to 15,000 birds per hour in the EU, it is not possible to check all animals on the line. Randomised checking of birds by taking them off the line will give a clear indication of the stun quality and should therefore be carried out systematically.

6.6.3 Head-only electrical stunning

Alternative electrical stunning method to the multi-bird water bath stunning can be head-only electrical stunning. Principle of head-only electric stunning methods is that electrodes are placed at both sides of the head, such that they span the brain. The birds can be restrained either manually, in case of slow speed slaughter, or in a cone (Lambooij *et al.*, 2014) or by being shackled prior to stunning. However, the latter are still not commercially applied yet.

As the birds are individually exposed, a constant current could be delivered to each bird (Berg and Raj, 2015). Ideally, head-only electrical stunning should be performed using a constant current source and a minimum current of 240 mA should be applied for broilers to ensure unconsciousness that will last long enough to prevent animals to recovery before or during the bleeding process. However, most of the head-only electrical stunning equipment used around the world is supplied with a constant voltage, and in some countries, this is limited to 110 V for operators' safety reasons. Such a low voltage may not be adequate to deliver the minimum currents necessary to induce immediate loss of consciousness. The electrical requirements for head-only electrical stunning equipment are, expressed as minimum currents, 240 mA for hens and broiler chickens, and 400 mA for turkeys (EC, 2009).

To control effective stunning the same indicators as being applied when looking for signs of consciousness in birds stunned using electrical water bath stunning can also be applied to head-only stunning. Such indicators are tonic seizures, absence of breathing, and spontaneous blinking, vocalisations and eye reflexes (EFSA, 2013; Grilli *et al.*, 2015).

6.6.4 Controlled atmosphere stunning

Gas stunning methods are developed as an alternative to electrical stunning since the 1990's. Different type of birds, like chickens, turkeys, ducks, geese, have been reported to be effectively stunned under commercial slaughter conditions by exposure to different gas mixtures (Gerritzen *et al.*, 2007, 2013; McKeegan *et al.*, 2007). Different developments and still on-going improvements have taken place using different gasses or gas mixtures, like anoxic methods using argon (Ar) or nitrogen (N_2; Raj *et al.*, 1991, 1998), carbon dioxide (CO_2; Gerritzen *et al.*, 2000, 2004, 2013; McKeegan *et al.*, 2006, 2007; Prinz *et al.*, 2012) or a mixture containing up to a maximum of 40% CO_2 mixed with inert gases (Ar and N_2) leading to anoxia (Raj *et al.*, 1992a,b; Raj and Gregory, 1994).

Gas stunning methods are often used to improve meat quality aspects (Section 6.6.6), but also to avoid the pain associated with shackling conscious birds in the case of electrical stunning. Since CAS stunning does not induce immediate unconsciousness, the use of CAS is limited to birds in containers or on conveyors. Several commercially available CAS systems differ in the way birds are placed in the stunning system. Crates, containers or de-stacked drawers can be conveyed through a

CAS system. In these processes, no handling of conscious birds is required. Shackling and handling are performed after the birds are stunned with a gas mixture (AVMA, 2016).

According to EU regulation 1099/2009 (EC, 2009), it is not allowed to expose birds to a high concentration (>40%) of CO_2, while still conscious. This is because exposure of conscious birds to more than 40% CO_2 will cause painful stimulation of the nasal mucosa and aversive reactions (McKeegan *et al.*, 2007). Therefore, gas stunning systems based on CO_2 are working with a minimum of two phases: a first phase with a gas concentration below 40% CO_2 to render the birds unconscious and a second phase above 40% CO_2 to secure deep levels of unconsciousness. The most common CAS stunning method is a two phase-controlled atmosphere stunner or a multi-step CO_2 stunning system (MCAS), but systems containing 30% CO_2 with 70% N_2 are also used in practice. Exposure to high CO_2 concentrations or to very low O_2 concentrations as with N_2 or Ar stunning systems induce serious uncontrolled muscle contractions (convulsions). These convulsions can lead to wing damage, like bruises and broken or dislocated wings, leading to economic losses. Convulsive movements can be prevented by a gradual exposure to increasing CO_2 concentrations. In studies containing multi-stepwise increase of CO_2 the appearance of convulsions was not observed or only at a very moderate level (Gerritzen *et al.*, 2013). During exposure to the gas mixture, birds gradually lose consciousness. The speed of induction, depth and duration of unconsciousness induced with gas mixtures depend on both exposure time and gas concentration. Higher concentrations of CO_2 require shorter exposure times to induce a sufficient level of unconsciousness than lower CO_2 concentrations. Exposure times and gas concentrations are therefore two crucial parameters to control during gas stunning. In practice in a two-phase CAS stunning system birds are generally exposed for 1 min to approximately 30% CO_2 with 25% O_2 in air followed by a second phase of 2 min in >60% CO_2 in air. Since CAS stunning does not induce immediate unconsciousness, animal welfare aspects during the induction phase (phase 1) should be assessed. The exposure to gas mixtures will induce a series of behaviours that are typical and will have more or less impact on animal welfare. However, the interpretation of the effect is not always easy, but monitoring the birds during this phase is important to avoid unnecessary suffering and to enact appropriate interventions (e.g. adjusting gas concentrations or mixtures), if required. During the induction phase, gasping (deep breathing with a stretched or arched neck) and head shaking will occur (Abeyesinghe *et al.*, 2007; Gerritzen *et al.*, 2007; McKeegan *et al.*, 2006). Gasping is inherent to CO_2 stunning and is indicative of breathlessness. However, agents identified as being less aversive (e.g. Ar or N_2 gas mixtures) can still produce overt signs of behavioural distress (e.g. open-mouth breathing) for extended periods of time prior to loss of consciousness under certain conditions of administration (e.g. gradual displacement; Webster and Collett, 2012). Head shaking can be a reaction to different situations and is often perceived as an alarming or alerting response, but it is also associated with a reaction to unpleasant, irritating or painful inhalation of gas. After some time, birds will lose balance, frequently being corrected by wing flapping or jumping up. However, this should not be interpreted as escape attempts, but rather as corrective measures from the birds to keep their balance or keep their posture. Loss of posture and thus being unable to maintain body position indicates the moment of loss of consciousness (Benson *et al.*, 2012; Gerritzen *et al.*, 2004; McKeegan *et al.*, 2011). After loss of posture, when animals are unconscious, gasping will continue for some period of time and convulsions (uncontrolled muscular movements) can occur. The interpretation of visual animal-based measures (ABMs) during the exposure to gas mixtures depends strongly on the moment that behavioural expressions start. For example, continuous wing flapping that starts directly after exposure can indicate escape attempts and is seen as an aversive reaction to high CO_2, whereas short wing flapping to regain or maintain posture does not necessarily indicate an aversive response to high CO_2. In addition, for head shaking, the interpretation depends on the intensity and level of head shaking. Severe or vigorous head shaking can be interpreted as a sign of painful or irritating inhalation of gas mixtures.

At the end of controlled atmosphere stunning, birds should be checked at key stages for deep unconsciousness using appropriate indicators as described by EFSA (2013) to ensure that animals will not recover before or during bleeding. In most cases, birds will be dead when exiting the CAS system. In any case, it is important to ensure that all birds are dead before processing them. Death can be confirmed from permanent absence of breathing, absence of corneal or palpebral reflex, dilated pupils and relaxed carcass (EFSA, 2013).

6.6.5 LAPS

A new development in CAS stunning is stunning poultry by exposure to a Low Atmospheric Pressure Stunning method (LAPS). In this method, broilers are placed while in the transport containers into the decompression chamber and exposed to gradual decompression with a reduction of available oxygen to less than 5% (Holloway and Pritchard, 2017; Mackie and McKeegan, 2016; Martin *et al.*, 2016a,b,c; Vizzier-Thaxton *et al.*, 2010). This method was found to be 'acceptable' based on the published scientific evidence for broiler chickens only up to 4 kg live weight (EFSA, 2017). However, EFSA recently identified several hazards during LAPS stunning, namely 'too fast decompression', 'expansion of gases in the body cavity', and 'too short exposure time'. These hazards can cause persistence of consciousness, pain and respiratory distress (EFSA, 2019).

6.6.6 Product quality

Gas mixtures and electrical stunning methods showed different effects on meat quality (Uijttenboogaart, 1997) and carcass quality (Kranen *et al*, 2000; Raj *et al.*, 1992a). Blood clotting incidence was lower in CO_2 gas stunning systems compared to electrical stunning. Haemorrhaging, if present in the valuable breast meat in particular, results in a decrease in quantity (due to trimming) and quality of poultry products, and hence causes economical losses to the poultry industry. However, haemorrhages is not only induced by stunning, but its cause can be multifactorial, like catching and crating, handling and shackling (Kranen *et al.*, 1996). Raj *et al.* (1992a) reported a higher incidence of wing damage after gas stunning compared to electrical stunning. Wing damage in that study was due to the severe convulsions caused by gas stunning. Gas killing of broilers would substantially reduce the incidence of broken bones and haemorrhaging in breast muscles in association with or without the broken bones (Uijttenboogaart, 1997).

Haem content in the liver and breast meat is also affected by the way of stunning. The haem levels in the liver tend to be lower in the CO_2 stunned birds, resulting in darker, more red and less yellow liver compared to electrical stunned birds (Uijttenboogaart, 1999). Whereas, breast meat of gas stunned birds have significant higher levels compared to electrical stunned birds, resulting in lighter breast meat than from electrical. Colour of breast meat is also paler in CO_2 gas stunned birds compared to Ar gas stunned birds.

When compared to the two the electrical and CO_2 gas stunning systems, the pH of breast meat presented a sharper decrease in Ar gas stunned turkeys (Raj, 1994). The faster pH fall was probably due to the convulsions that occur during the Ar gas stun/kill operation compared to no or less convulsions induced by the other two stunning systems. However, the difference in pH value between stunning systems disappeared after 24 h of chilling. Differences in the rate of post-mortem glycolysis, induced by other stunning procedures, like gaseous stunning, have been reported to result in differences in meat quality traits, like more tender meat (Raj, 1994).

6.7 Bleeding

Bleeding following stunning is performed to prevent animals from regaining consciousness before death occurs (EC, 2009). Since the main stunning methods are reversible accurate neck cutting, i.e. severing both carotid arteries supplying oxygenated blood to the brain, should be performed as soon as possible. Neck cutting can be performed automatically by a single or twin-bladed neck-cutting machine or manually with a sharp knife.

It is expected that unconsciousness induced by stunning should last longer than the time interval between the end of stunning/neck cutting and the induction of death due to blood loss following neck cutting (Figure 6.4). In broilers, the time to onset of an isoelectric EEG indicative of brain death following effective waterbath electrical stunning and severance of two carotid arteries and two jugular veins has been reported to be about 22 s (Raj *et al.*, 2006). This implies that unconsciousness induced by the stunning method should last at least 22 s. Therefore, neck cut needs to be applied direct after stunning. The duration of unconsciousness after gas stunning depends strongly on the gas(mixture), gas concentration and the exposure time. For the most common applied gas stunning method, e.g. two-phase CO_2, the stunning method is in many cases irreversible and, if not, the period of unconsciousness exceeds 2 min. This means that the stun to stick interval with gas stunning can be longer than for electrical stunning. However, due to the variation between animals and application of the methods close surveillance of unconsciousness is required.

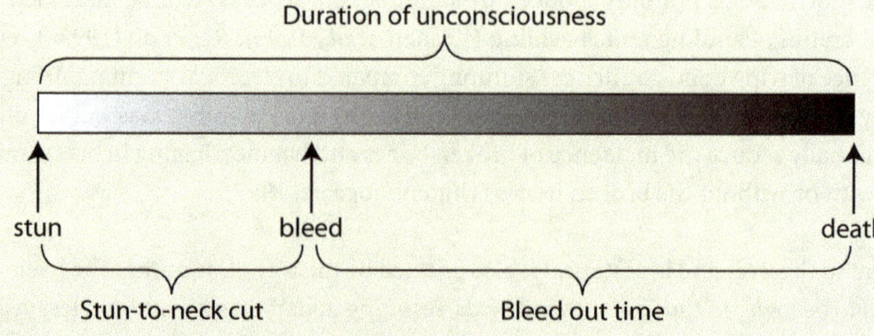

Figure 6.4. Duration of stun-to-neck-cutting and bleeding intervals (EFSA, 2004).

6.8 Conclusions

Before being slaughtered birds are subjected to a number of subsequent events, such as feed withdrawal, catching and crating, transport, unloading and handling at the slaughter plant and stunning. All these events in a relatively short period time contain many stressors and risk factors that can lead to physical injuries, thermal stress, weight loss and in extremes to death.

Feed withdrawal prior to transport can create a negative energy balance within the birds during the transport phase, leading to prolonged hunger and reduced ability to thermoregulate. Thermal stress, both hot and cold, has been identified as a risk factor for increased DOA levels.

Transport conditions, such as loading density, ventilation, vehicle design and protection from external weather conditions contribute to heat or cold stress.

The stunning process of poultry often requires for the birds to be taken out of the containers to be further handled to be restrained on a shackle line (electrical stunning) or placed on a conveyor belt (gas stunning). The manually and mechanical unloading of birds from the crates or containers are inducing risk factors for animal welfare including injuries, such as bruises and wing damage, that will affect carcass quality.

An unavoidable aspect of electrical water bath stunning is that the birds are inverted onto moving shackles prior to stunning. Shackling is probably the most important aspect affecting animal welfare during the process of electrical stunning.

During exposure to the gas mixture, birds gradually lose consciousness. The speed of induction, depth and duration of unconsciousness induced with gas mixtures depend on both exposure time and gas concentration. Higher concentrations of CO_2 require shorter exposure times to induce a sufficient level of unconsciousness than lower CO_2 concentrations. Exposure times and gas concentrations are therefore two crucial parameters to control during gas stunning.

Carcass and meat quality are affected by the pre- and slaughter process as well as by the stunning and bleeding method.

References

Abeyesinghe, S.M., McKeegan, D.E., McLeman, M.A., Lowe, J.C., Demmers, T.G.M., White, R.P., Kranen, R.W., Van Bemmel, H., Lankhaar, J.A.C. and Wathes, C.M., 2007. Controlled atmosphere stunning of broiler chickens. I. Effects on behaviour, physiology and meat quality in a pilot scale system at a processing plant. British Poultry Science 4: 406-423.

Akşit, M., Yalçin, S., Özkan, S., Metin, K. and Özdemir, D., 2006. Effects of temperature during rearing and crating on stress parameters and meat quality of broilers. Poultry Science 85: 1867-1874.

American Veterinary Medical Association (AVMA), 2016. AVMA guidelines for the humane slaughter of animals: 2016 edition. American Veterinary Medical Association, Schaumburg, IL, USA. Available at: https://tinyurl.com/8fets2vs.

Barbut, S., Zhang, L. and Marcone, M., 2005. Effects of pale, normal, and dark chicken breast meat on microstructure, extractable proteins, and cooking of marinated fillets. Poultry Science 84: 797-802.

Bayliss, P.A. and Hinton M.H., 1990. Transportation of broilers with special reference to mortality rates. Applied Animal Behaviour Science 28: 93-118.

Beaulac, K., Crowe, T.G. and Schwean-Lardner, K., 2020. Simulated transport of well- and poor-feathered brown-strain end-of-cycle hens and the impact on stress physiology, behaviour, and meat quality. Poultry Science 99: 6753-6763.

Bedáňová, I., Voslářová E., Večerek V., Pištěková V. and Chloupek, P., 2006. Effects of reduction in floor space during crating on haematological indices in broilers. Berkiner und Münchener tierärztliche Wochenschrift 119: 17-21.

Bedanova, I., Voslarova, E., Chloupek, P., Pistekova, V., Suchy, P., Blahova, J., Dobsikova, R. and Vecerek, V., 2007. Stress in broilers resulting from shackling. Poultry Science 86: 1065-1069.

Bedanova, I., Voslarova, E., Zelinska, G., Blahova, J., Marsalek, P. and Chloupek, J., 2014. Neopterin and biopterin as biomarkers of immune system activity associated with crating in broiler chickens. Poultry Science 93: 2432-2438.

Benson, E.R., Alphin, R.L., Rankin, M.K., Caputo, M.P., Kinney, C.A. and Johnson, A.L., 2012. Evaluation of EEG based determination of unconsciousness vs. loss of posture in broilers. Research in Veterinary Science 93: 960-964.

Berg, C. and Raj, M., 2015. A review of different stunning methods for poultry – animal welfare aspects (stunning methods for poultry). Animals 5: 1207-1219.

Beuving, G. and Vonder, G.M.A., 1978. Effect of stressing factors on corticosterone levels in the plasma of laying hens. General and Comparative Endocrinology 35: 153-159.

Bianchi, M., Petracci, M. and Cavani, C., 2005. Effects of transport and lairage on mortality, liveweight loss, and carcass quality in broiler chickens. Italian Journal of Animal Science 4: 516-518.

Bilgili, S.F., 1999. Recent advances in electrical stunning. Poultry Science 78: 282-286.

Boukhris, H., Damergi, C., Najar, T. and Samet, A., 2017. Transport stress impact on postmortem metabolisms of turkey meat quality. Journal of New Sciences 35: 2049-2054.

Butterworth, A., 1999. Infectious components of broiler lameness: a review. World's Poultry Science Journal 55: 327-352.

Burlinguette, N.A., Strawford, M.L., Watts, J.M., Classen, H.L., Shand, P.J. and Crowe, T.G., 2012. Broiler trailer thermal conditions during cold climate transport. Canadian Journal of Animal Science 92: 109-122.

Caffrey, N.P., Dohoo, I.R. and Cockram, M.S., 2017. Factors affecting mortality risk during transportation of broiler chickens for slaughter in Atlantic Canada. Preventative Veterinary Medicine 147: 199-208.

Canadian Agri-Food Research Council (CARC), 2003. Recommended code of practice for the care and handling of farm animals – chickens, turkeys and breeders from hatchery to processing plant. Canadian Agri-Food Research Council, Ottawa, Canada. Available at: https://tinyurl.com/3djrz3yy.

Cavusoglu, E. and Petek, M., 2021. Effects of season, plumage colour, and transport distance on body weight loss, dead-on-arrival, and reject rate in commercial end-of-lay hens. Animals 11: 1827. https://doi.org/10.3390/ani11061827.

Chauvin, C., Hillion, S., Balaine, L., Michel, V., Peraste, J., Petetin, I., Lupo, C. and Le Bouquin, S., 2011. Factors associated with mortality of broilers during transport. Animal 5: 287-293.

Chloupek, P., Vecerek, V., Voslarova, E., Bedanova, I., Suchy, P., Pistekova, V. and Kozak, A., 2008. Effects of differ crating periods on selected biochemical indices in broiler chickens. Berliner und Münchener Tierärztliche Wochenschrift 121: 132-136.

Dadgar, S., Lee, E.S., Leer, T.L.V., Burlinguette, N., Classen, H.L., Crowe, T.G. and Shand, P.J., 2010. Effect of microclimate temperature during transportation of broiler chickens on quality of the pectoralis major muscle. Poultry Science 89: 1033-1041.

Dadgar, S., Lee, E.S., Leer, T.L.V., Crowe, T.G., Classen, H.L. and Shand, P.J., 2011. Effect of acute cold exposure, age, sex, and lairage on broiler breast meat quality. Poultry Science 90: 444-457.

Dadgar, S., Lee, E.S., Crowe, T.G., Classen, H.L. and Shand, P.J., 2012. Characteristics of cold-induced dark, firm, dry broiler chicken breast meat. British Poultry Science 53: 351-359.

Danbury, T.C., Weeks, C.A., Chambers, J.P., Waterman-Pearson, A.E., Kestin, S.C. and Chambers, J.P., 2000. Self-selection of the analgesic drug carprofen by lame broiler chickens. Veterinary Record 146: 307-311.

Dawson, W.R. and Whittow, G.C., 2000. Regulation of body temperature. In: Sturkie, P.D. (ed.) Avian physiology. Academic Press, New York, NY, USA, pp. 343-390.

Debut, M., Berri, C., Arnould, C., Guemene, D., Sante-Lhoutellier, V., Sellier, N., Baéza, E., Jehl, N., Jégo, Y., Beaumont, C. and Le Bihan-Duval., E., 2005. Behavioural and physiological responses of three chicken breeds to pre-slaughter shackling and acute heat stress. British Poultry Science 46: 527-535.

Delezie, E., Lips, D., Lips, R. and Decuypere, E., 2005. Mechanical catching of broiler chickens is a viable alternative for manual catching from an animal welfare point of view. Animal Science Papers and Reports 23 (Suppl. 1): 257-264.

Delezie, E., Verbeke, W., De Tavernier, K. and Decuypere, E., 2006. Consumers' preferences toward techniques for improving manual catching of poultry. Poultry Science 85: 2019-2027.

Delezie, E., Swennen, Q., Buyse, J. and Decuypere, E., 2007. The effect of feed withdrawal and crating density in transit on metabolism and meat quality of broilers at slaughter weight. Poultry Science 86: 1414-1423.

De Lima, V.A., Ceballos, M.C., Gregory, N.G. and Paranhos da Costa, M.J.R., 2019. Effect of different catching practices during manual upright handling on broiler welfare and behavior. Poultry Science 98: 4282-4289.

Di Martino, G., Capello, K., Russo, E., Mazzucato, M., Mulatti, P., Ferrè, N., Garbo, A., Brichese, M., Marangon, S. and Bonfanti, L., 2017. Factors associated with pre-slaughter mortality in turkeys and end of lay hens. Animal 11: 2295-2300.

Duncan, I.H.J., Slee, G.S., Kettlewell, P., Berry, P. and Carlisle, A.J., 1986. Comparison of the stressfulness of harvesting broiler chickens by machine and by hand. British Poultry Science 27: 109-114.

European Commission (EC), 2005. Council Regulation (EC) No 1/2005 of 22 December 2004 on the protection of animals during transport and related operations and amending Directives 64/432/EEC and 93/119/EC and Regulation (EC) No 1255/97. Official Journal of the European Union L3: 1-44. Available at: http://data.europa.eu/eli/reg/2005/1/oj.

European Commission (EC), 2009. Council Regulation (EC) No 1099/2009 of 24 September 2009 on the protection of animals at the time of killing. Official Journal of the European Union L303: 1-30. Available at: http://data.europa.eu/eli/reg/2009/1099/oj.

European Food Safety Authority (EFSA), 2004. Opinion of the Scientific Panel on Animal Health and Welfare (AHAW) on a request from the Commission related to welfare aspects of the main systems of stunning and killing the main commercial species of animals. EFSA Journal 2: 45. https://doi.org/10.2903/j.efsa.2004.45.

European Food Safety Authority (EFSA), 2013. Scientific Opinion on monitoring procedures at slaughterhouses for poultry. EFSA Journal 11: 3521. https://doi.org/10.2903/j.efsa.2013.3521.

European Food Safety Authority (EFSA), 2017. Low atmospheric pressure system for stunning broiler chickens. EFSA Journal 15: 5056. https://doi.org/10.2903/j.efsa.2017.5056.

European Food Safety Authority (EFSA), 2019. Scientific opinion on slaughter of animals: poultry. EFSA Journal 17: 5849. https://doi.org/10.2903/j.efsa.2019.5849.

Fernandez, X., 2004. A short overview of the welfare implications of pre-slaughter stunning in poultry. Proceedings of the International Society for Animal Hygiene. Saint-Malo, France, pp. 41-42.

Frerichs, C., 2021. The effects of simulated transportation on white-feathered laying hens. M.Sc. thesis, University of Saskatchewan, Saskatoon, Canada.

Frerichs, C., Beaulac, K., Crowe, T.G. and Schwean-Lardner, K., 2021. The effects of simulated transport on the muscle characteristics of white-feathered end-of-cycle hens. Poultry Science 100: 101280, https://doi.org/10.1016/j.psj.2021.101280.

Gentle, M.J. and Tilston, V.L., 2000. Nociceptors in the legs of poultry: implications for potential pain in pre-slaughter shackling. Animal Welfare 9: 227-236.

Gerritzen, M.A, Lambooij E., Hillebrand S.J.W., J.A.C. Lankhaar J.A.C. and C. Pieterse, C., 2000. Behavioral responses of broilers to different gaseous atmospheres. Poultry Science 79: 928-933.

Gerritzen, M., Lambooij, B., Reimert, H., Stegeman, A. and Spruijt, B., 2004. On-farm euthanasia of broiler chickens: effects of different gas mixtures on behavior and brain activity. Poultry science 83: 1294-1301.

Gerritzen, M.A., Lambooij, B., Reimert, H., Stegeman, A. and Spruijt, B., 2007. A note on behaviour of poultry exposed to increasing carbon dioxide concentrations. Applied Animal Behaviour Science 108: 179-185.

Gerritzen, M.A., Reimert, H.G.M., Hindle, V.A., Verhoven, M.T.W. and Veerkamp, W.B., 2013. Multistage carbon dioxide gas stunning of broilers. Poultry Science 92: 41-50.

Girasole, M., Marrone, R., Anastasio, A., Chianese, A., Mercogliano, R. and Cortesi, M.L., 2016. Effect of electrical water bath stunning on physical reflexes of broilers: evaluation of stunning efficacy under field conditions. Poultry Science 95: 1205-1210.

Gregory N.G. and Bell J.C., 1987. Duration of wing flapping in chickens shackled before slaughter. Veterinary Record 121: 567-569.

Gregory, N.G. and Wilkins, L.J., 1989. Broken bones in domestic fowl: handling and processing damage in end-of-lay battery hens. British Poultry Science 3: 555-562.

Griffiths, G.L., 1985. Ageing bruises on chicken legs. Proceedings of the 6th Australasian Poultry and Stockfeed Convention, Melbourne, Australia, pp. 269-299.

Gross, W.B. and Siegel, H.S., 1983. Evaluation of the heterophil/lymphocyte ratio as a measure of stress in chickens. Avian Diseases 27: 972-979.

Grilli, C., Loschi, A.R., Rea, S., Stocchi, R., Leoni, L. and Conti, F., 2015. Welfare indicators during broiler slaughtering. British Poultry Science 56: 1-5.

Guarnieri, P.B., Soares, A.L, Olivo, R., Schneider, J.P., Macedo, R.M. and Ida, E.I., 2004. Preslaughter handling with water shower spray inhibits PSE (pale, soft, exudative) broiler breast meat in a commercial plant. Journal of Food Biochemistry 28: 269-277.

Henrikson, Z.A., Vermette, C.J., Schwean-Lardner, K. and Crowe, T.G., 2018. Effects of cold exposure on physiology, meat quality, and behavior of turkey hens and toms crated at transport density. Poultry Science 97: 347-357.

Hindle, V.A., Lambooij, E., Reimert, H.G., Workel, L.D. and Gerritzen, M.A., 2010. Animal welfare concerns during the use of the waterbath for stunning broilers, hens, and ducks. Poultry Science 89: 401-412.

Holloway P.H. and Pritchard D.G., 2017. Effects of ambient temperature and water vapor on chamber pressure and oxygen level during low atmospheric pressure stunning of poultry. Poultry Science 96: 2528-2539.

Hunter, R.R., Mitchell, M.A. and Carlisle, A.J., 1999. Wetting of broilers during cold weather transport: a major source of physiological stress? British Poultry Science 40: S48-S49.

Hussnain, F., Mahmud, A., Mehmood, S. and Jaspal, M.H., 2020a. Effect of transportation distance and crating density on preslaughter losses and blood biochemical profile in broilers during hot and humid weather. Turkish Journal of Veterinary and Animal Sciences 44: 418-426.

Hussnain, F., Mahmud, A., Mehmood, S. and Jaspal, M.H., 2020b. Effect of broiler crating density and transportation distance on preslaughter losses and physiological response during the winter season in Punjab, Pakistan. Brazilian Journal of Poultry Science 22: 1-10.

Jacobs, L., Delezie, E., Duchateau, L., Goethals, K. and Tuyttens, F.A.M., 2016. Impact of the separate pre-slaughter stages on broiler chicken welfare. Poultry Science 96: 266-273.

Jacobs, L., Delezie, E., Duchateau, L., Goethals, K. and Tuyttens, F.A.M., 2017a. Broiler chickens dead on arrival: associated risk factors and welfare indicators. Poultry Science 96: 259-265.

Jacobs, L., Delezie, E., Duchateau, L., Goethals, K. and Tuyttens, F.A., 2017b. Impact of the separate pre-slaughter stages on broiler chicken welfare. Poultry Science 96: 266-273.

Jiang, N.N., Xing, T., Wang, P., Xie, C. and Xu, X.L., 2015. Effects of water-misting Sprays with forced ventilation after transport during summer on meat quality, stress parameters, glycolytic potential and microstructures of muscle in broilers. Asian-Australasian Journal of Animal Sciences 28: 1767-1773.

Kannan, G., Heath, J.L., Wabeck, C.J., Souza, M.C., Howe, J.C. and Mench J.A., 1997. Effects of crating and transport on stress and meat quality characteristics in broilers. Poultry Science 76: 523-529.

Kettlewell, P.J. and Mitchell, M.A., 1994. Catching, handling and loading of poultry for road transportation. World's Poultry Science Journal 50: 54-56.

Kettlewell, P.J. and Mitchell, M.A., 2001. Comfortable ride: concept 2000 provides climate control during poultry transport. Resource magazine, Engineering and Technology for a Sustainable World 8: 13-14.

Kettlewell, P.J., Hoxey, R.P., Hampson, C.J., Green, N.R., Veale, B.M. and Mitchell, M.A., 2001. AP – Animal production technology: design and operation of a prototype mechanical ventilation system for livestock transport vehicles. Journal of Agricultural Engineering Research 79: 429-439.

Khalil, A.M., Matsui, K. and Takeda, K., 2010. Responses to abrupt changes in feeding and illumination in laying hens. Turkish Journal of Veterinary and Animal Sciences 34: 433-439.

Kittelsen, K.E., Moe, R.O., Hoel, K., Kolbjørnsen, Ø., Nafstad, O. and Granquist, E.G. 2017. Comparison of flock characteristics, journey duration and pathology between flocks with a normal and a high percentage of broilers 'dead-on-arrival' at abattoirs. Animal 11: 2301-2308.

Kittelsen, K.E., Granquist, E.G. Aunsmo, A.L., Moe, R.O. and Tolo, E., 2018. An evaluation of two different broiler catching methods. Animals 8: 141. https://doi.org/10.3390/ani8080141

Knezacek, T.D., Olkowski, A.A., Kettlewell, P.J., Mitchell, M.A. and Classen, H.L., 2010. Temperature gradients in trailers and changes in broiler rectal and core body temperature during winter transportation in Saskatchewan. Canadian Journal of Animal Science 90: 321-330.

Knowles, T.G., Warriss, P.D., Brown, S.N., Edwards, J.E. and Mitchell, M.A., 1995. Response of broilers to deprivation of food and water for 24 hours. British Veterinary Journal 151: 197-202.

Kranen, R.W., C.H. Veerkamp, C.H., Lambooy, E, Van Kuppevelt, T.H. and Veerkamp, J.H., 1996. Hemorrhages in muscles of broiler chickens: the relationships among blood variables at various rearing temperature regimens. Poultry Science 75: 570-576.

Kranen, R, Van Kuppevelt, T.H., Goedhart, H.A., Veerkamp, C.H., Lambooy, E. and Veerkamp, J.H., 1999. Hemoglobin and myoglobin content in muscles of broiler chickens. Poultry Science 78: 467-476.

Kranen, R.W., Lambooy, E., Veerkamp, C.H., Van Kuppevelt, T.H. and Veerkamp, J.H., 2000. Histological characterization of hemorrhages in muscles of broiler chickens. Poultry Science 79: 110-116.

Lacy, M.P. and Czarick, M., 1998. Mechanical harvesting of broilers. Poultry Science 77: 1794-1797.

Lambooij, E., Reimert, H.G., Verhoeven, M.T. and Hindle, V.A., 2014. Cone restraining and head-only electrical stunning in broilers: effects on physiological responses and meat quality. Poultry Science 93: 512-518.

Langkabel, N., Baumann, M.P.O., Fieler, A., Sanguankiat, A. and Fries, R., 2015. Influence of two catching methods on the occurrence of lesions in broilers. Poultry Science 94: 1735-1741.

Lara, L.J. and Rostagno, M.H., 2013. Impact of heat stress on poultry production. Animals 3: 356-369.

Leeson, S. and Morrison, W.D., 1978. Effect of feather cover on feed efficiency of laying birds. Poultry Science 57: 1094-1096.

Machovcova, Z., Vecerek, V., Voslarova, E., Malena, M., Conte, F., Bedanova, I. and Vecerkova, L., 2016. Pre-slaughter mortality among turkeys related to their transport. Animal Science Journal 88: 705-711.

Mackie, N. and McKeegan, E.F., 2016 Behavioural responses of broiler chickens during low atmospheric pressure stunning. Applied Animal Behaviour Science 74: 90-98.

Mallia, J.G., Vaillancourt, J.P., Martin, S.W. and McEwen, S.A., 2000. Risk factors for abattoir condemnation of turkey carcasses due to cyanosis in Southern Ontario. Poultry Science 79: 831-837.

Marchewka, J., Watanabe, T.T.N., Ferrante, V. and Estevez, I., 2013. Review of the social and environmental factors affecting the behavior and welfare of turkeys (*Meleagris gallopavo*). Poultry Science 92: 1467-1473.

Martin, J.E., Christensen, K., Vizzier-Thaxton, Y., Mitchell, M.A. and McKeegan, D.E.F., 2016a. Behavioural, brain and cardiac responses to hypobaric hypoxia in broiler chickens. Physiology & Behavior 163: 25-36.

Martin, J.E., Christensen, K., Vizzier-Thaxton, Y. and McKeegan, D.E.F., 2016b. Effects of analgesic intervention on behavioural responses to low atmospheric pressure stunning. Animal Behaviour Science 180: 157-165.

Martin, J.E., Christensen, K., Vizzier-Thaxton, Y. and McKeegan, D.E.F., 2016c. Effects of light on responses to low atmospheric pressure stunning in broilers. British Poultry Science 57: 585-600.

McEwen, S.A. and Barbut, S., 1992. Survey of turkey downgrading at slaughter: carcass defects and associations with transport, toenail trimming, and type of bird. Poultry Science 71: 1107-1115

McKeegan, D.E.F., McIntyre, J., Demmers, T.G.M., Wathes, C.M. and Jones, R.B., 2006. Behavioural responses of broiler chickens during acute exposure to gaseous stimulation. Applied Animal Behaviour Science 99: 271-286.

McKeegan, D.E., Abeyesinghe, S., McLeman, M., Lowe,J.C., Demmers, T., White,R, Kranen, R.W., Van Bemmel, H., Lankhaar, J.A.C. and Wathes, C., 2007. Controlled atmosphere stunning of broiler chickens: II. Effects on behaviour, physiology and meat quality in a commercial processing plant. British Poultry Science 48: 430-442.

McKeegan, D.E., Sparks, N.H., Sandilands, V., Demmers, T.G.M., Boulcott, P. and Wathes, C.M., 2011. Physiological responses of laying hens during whole-house killing with carbon dioxide. British Poultry Science 52: 645-657.

McNeal, W.D., Fletcher, D.L. and Buhr, R.J., 2003. Effects of stunning and decapitation on broiler activity during bleeding, blood loss, carcass, and breast meat quality. Poultry Science 82: 163-168.

Meltzer, A., 1983 Thermoneutral zone and resting metabolic rate of broilers. British Poulty Science 24: 471-476.

Mitchell, M.A. and Kettlewell, P.J., 1998. Physiological stress and welfare of broiler chickens in transit: solutions not problems! Poultry Science 77: 1803-1814.

Mönch, J., Rauch, E., Hartmannsgruber, S., Erhard, M., Wolff, I., Schmidt, P., Schug, A.R. and Louton, H., 2020. The welfare impacts of mechanical and manual broiler catching and of circumstances at loading under field conditions. Poultry Science 99: 5233-5252.

National Farm Animal Care Council, 2016. Code of practice for the care and handling of hatching eggs, breeders, chickens and turkeys. Available at: https://www.nfacc.ca/codes-of-practice/chickens-turkeys-and-breeders.

Newberry, R.C., Webster, A.B., Lewis, N.J. and Van Arnam, C., 1999. Management of spent hens. Journal of Applied Animal Welfare Science 2: 12-29.

Nijdam, E., Arens, P., Lambooij, E., Decuypere, E. and Stegeman, J.A., 2004. Factors influencing bruises and mortality of broilers during catching, transport, and lairage. Poultry Science 83: 1610-1615.

Nijdam, E., Delezie, E., Lambooij, E., Nabuurs, M.J.A., Decuypere, E. and Stegeman, J.A., 2005. Feed withdrawal of broilers before transport changes plasma hormone and metabolite concentrations. Poultry Science 84: 1146-1152.

Office International des Epizooties (OIE), 2019. Terrestrial animal health code. 28th Edition, World Organisation for Animal Health, Paris, France. Available at: https://tinyurl.com/35ufd3ka.

Petracci, M., Bianchi, M. and Cavani, C., 2005. Pre-slaughter factors affecting mortality, live weight loss, and carcass quality in broiler chickens. Proceedings of the XVII European Symposium on the Quality of Poultry Meat, Doorwerth, the Netherlands, pp. 23-26.

Petracci, M., Bianchi, M., Cavani, C., Gaspari, P. and Lavazza, A., 2006. Preslaughter mortality in broiler chickens, turkeys, and spent hens under commercial slaughtering. Poultry Science 85: 1660-1664.

Poultry Industry Council, 2010. Should this bird be loaded? A guide for preparing, loading, and transporting poultry. Available at: https://www.poultryindustrycouncil.ca/downloads/handbook.pdf.

Prescott, N.B., Berty, P.S., Haslam, S. and Tinker, D.B., 2000. Catching and crating turkeys: effects on carcass damage, heart rate, and other welfare parameters. Journal of Applied Poultry Research 9: 424-432.

Prinz, S., Coenen, A., Ehinger, F., Van Oijen, G. and Bessei, W., 2012. Stunning effectiveness of broiler chickens using a two-phase stunner: pulsed direct current followed by sine wave alternating current. Archiv für Geflügelkunde 76: 63-71.

Raj, A.B.M., 1994. Effect of stunning method, carcase chilling temperature and filleting time on the texture of turkey breast meat. British Poultry Science 35: 77-89.

Raj, A.B.M. and Gregory, N.G., 1994. An evaluation of humane gas stunning methods for turkeys. Veterinary Record 135: 222-223.

Raj, A.B.M., Gregory, N.G. and Wotton, S.B., 1991. Changes in the somatosensory evoked potentials and spontaneous electroencephalogram of hens during stunning in argon-induced hypoxia. British Veterinary Journal 147: 322-330.

Raj, A.B.M., Gregory, N.G. and Wilkins, L.J., 1992a. Survival rate and carcass downgrading after the stunning with carbon dioxide-argon mixtures. Veterinary Record 130: 325-328.

Raj, A.B.M. Wotton, S.B. and Whittington, P.E., 1992b. Changes in the spontaneous and evoked electrical activity in the brain of hens during stunning with 30 per cent carbon dioxide in argon with 5 per cent residual oxygen. Research in Veterinary Science 53: 126-129.

Raj, A.B.M., Wotton, S.B. McKinstry, J.L., Hillebrand, S.J.W. and Pieterse, C., 1998. Changes in the somatosensory evoked potentials and spontaneous electroencephalogram of broiler chickens during exposure to gas mixtures. British Poultry Science 39: 686-695.

Raj, A.B.M., O'Callaghan, M. and Knowles, T.G., 2006. The effects of amount and frequency of alternating current used in water bath stunning and of slaughter methods on electroencephalograms in broilers. Animal Welfare 15: 7-18.

Richards, G.J., Wilkins, L.J., Weeks, C.A., Knowles, T.G. and Brown, S.N., 2012. Evaluation of the microclimate in poultry transport module drawers during the marketing process of end-of-lay hens from farm to slaughter. Veterinary Record 19: 474-481.

Rodrigues, D.R., Café, M.B., Jardim Filho, R.M., Oliveira, E., Trentin, T.C., Martins, D.B. and Minafra, C.S., 2017. Metabolism of broilers subjected to different lairage times at the abattoir and its relationship with broiler meat quality. Arquivo Brasileiro de Medicina Veterinária e Zootecnia 69: 733-741.

Sandercock, D.A., Hunter, R.R., Nute, G.R., Mitchell, M.A. and Hocking, P.M., 2001. Acute heat stress-induced alterations in blood acid-base status and skeletal muscle membrane integrity in broiler chickens at two ages: implications for meat quality. Poultry Science 80: 418-425.

Saraiva, S., Esteves, A., Oliveira, I., Mitchell, M. and Stilwell, G., 2020. Impact of pre-slaughter factors on welfare of broilers. Veterinary and Animal Science 10: 100146.

Sherwin, C.M., Kestin, S.C., Nicol, C.J., Knowles, T.G., Brown, S.N., Reed, H.J. and Warriss, P.D., 1993. Variation in behavioural indices of fearfulness and fatigue in transported broilers. British Veterinary Journal 149: 571-578.

Sparrey, J.M. and Kettlewell, P.J., 1994. Shackling of poultry: is it a welfare problem? World's Poultry Science Journal 50: 167-176.

Sparrey, J.M., Kettlewell, P.J., Paice, M.E.R. and Whetlor, W.C., 1993. Development of a constant current water bath stunner for poultry processing. Journal of Agricultural Engineering Research 56: 267-274.

Spurio, R.S., Soares, A.L., Carvalho, R.H., Silveira Junior, V., Grespan, M., Oba, A. and Shimokomaki, M., 2016. Improving transport container design to reduce broiler chicken PSE (pale, soft, exudative) meat in Brazil. Animal Science Journal 87: 277-283.

Strawford, M.L., Watts, J.M., Crowe, T.G., Classen, H.L. and Shand, P.J., 2011. The effect of simulated cold weather transport on core body temperature and behavior of broilers. Poultry Science 90: 2415-2424.

Tinker, D., Berry, P., White, R., Prescott, N., Welch, S. and Lankhaar, J., 2005. Improvement in the welfare of broilers by changes to a mechanical unloading system. Journal of Applied Poultry Research 14: 330-337.

Uijttenboogaart, T.G., 1997. Effects of gas and electrical stunning methods on meat quality. Proceedings of the Satellite Symposium 'Developments of new humane stunning and related processing methods for poultry to improve product quality and consumer acceptability'. European poultry meat and egg quality symposia '97, WPSA, Poznan, Poland.

Uijttenboogaart, T.G., 1999. Effects of gas and electrical stunning methods on meat quality. Proceedings of the Simpósio Internacional sobre Tecnologia de Processamento e Qualidade, Concórdia, Brazil.

Vecerek, V., Voslarova, E., Conte, F., Vecerkova, L., and Bedanova, I., 2016. Negative trends in transport-related mortality rates in broiler chickens. Asian Australasian Journal of Animal Science 29: 1796-1804.

Vecerkova, L., Vecerek, V. and Voslarova, E., 2019. Welfare of end-of-lay hens transported for slaughter: effects of amibient temperature, season, and transport distance on transport-related mortality. Poultry Science 98: 6217-6224.

Veerkamp, C.H. and De Vries, A.W., 1983. Influence of electrical stunning on quality of broilers. In: Eikelenboom, G. (ed.) Stunning animals for slaughter. Martinus Nijhoff, The Hague, the Netherlands, pp. 187-199.

Vermette, C.J., Henrikson, Z.A., Schwean-Lardner, K.V. and Crowe, T.G., 2017. Influence of hot exposure on 12-week-old turkey hen physiology, welfare, and meat quality and 16-week-old turkey tom core body temperature when crated at transport density. Poultry Science 96: 3836-3843.

Vieira, F.M.C., Silva, I.J.O., Barbosa Filho, J.A.D. and Vieira, A.M.C., 2010. Productive losses on broiler preslaughter operations: effects of the distance from farms to abattoirs and of lairage time in a climatized holding area. Revista Brasileira de Zootecnia 39, https://doi.org/10.1590/S1516-35982010001100021

Vinco, L.J., Archetti, I.L., Giacomelli, S. and Lombardi, G., 2016. Influence of crate height on the welfare of broilers during transport. Journal of Veterinary Behavior 14: 28-33.

Vizzier-Thaxton, Y., Christensen, K.D., Schilling, M.W., Buhr, R.J. and Thaxton, J.P., 2010. A new humane method of stunning broilers using low atmospheric pressure. Journal of Applied Poultry Research 19: 341-348.

Voslářová, E., Janáčková, B., Rubešová, L., Kozák, A., Bedáňová, I., Steinhauser, L. and Večerek, V., 2007. Mortality rates in poultry species and categories during transport for slaughter. Acta Veterinaria Brno 76: S101-S108.

Vosmerova, P., Chloupek, J., Bedanova, I., Chloupek, P., Kruzikova, Blahova, J. and Vecerek, V., 2010. Changes in selected biochemical indices related to transport of broilers to slaughterhouse under different ambient conditions. Poultry Science 89: 2719-2725.

Warriss, P.D., Kestin, S.C., Brown, S.N. and Bevis, E.A., 1988. Depletion of glycogen reserves in fasting broiler chickens. British Poultry Science 29: 149-154.

Warriss, P.D., Kestin, S.C., Brown, S.N., Knowles, T.G., Wilkins, L.J., Edwards, J.E., Austin, S.D. and Nicol, C.J., 1993. The depletion of glycogen stores and indices of dehydration in transported broilers. British Veterinary Journal 149: 391-398.

Warriss, P.D., Wilkins, L.J., Brown, S.N., Phillips, A.J. and Allen, V., 2004. Defecation and weight of the gastrointestinal tract contents after feed and water withdrawal in broilers. British Poultry Science 45: 61-66.

Warriss, P.D., Pagazaurtundua, A. and Brown, S.N., 2005. Relationship between maximum daily temperature and mortality of broiler chickens during transport and lairage. British Poultry Science 46: 647-651.

Watts, J.M., Graff, L.J., Strawford, M.L., Crowe, T.G., Burlinguette, N.A., Classen, H.L. and Shand, P.J., 2011. Heat and moisture production by broiler during simulated cold weather transport. Poultry Science 90: 1890-1899.

Webster, A.J.F., Tuddenham, A., Saville, C.A. and Scott, G.B., 1993 Thermal stress on chickens in transit. British Poultry Science 34: 267-277.

Webster, A.B. and Collett, S.R., 2012. A mobile modified-atmosphere killing system for small-flock depopulation. Journal of Applied Poultry Research 21: 131-144.

Weeks, C.A., Webster, A.J.F. and Wyld, H.M., 1997. Vehicle design and thermal comfort of poultry in transit. British Poultry Science 38: 464-474.

Weeks, C. A., Brown, S.N., Richards, G.J., Wilkins, L.J. and Knowles, T.G., 2012. Levels of mortality in hens by end of lay on farm and in transit to slaughter in Great Britain. Veterinary Record 25: 647-650.

Whiting, T.L., Drain, M.E. and Rasali, D.P., 2007. Warm weather transport of broiler chickens in Manitoba. II. Truck management factors associated with death loss in transit to slaughter. Canadian Veterinary Journal 48: 148-154.

Wichman, A., Norring, M., Pastell, M., Algers, B., Pösö, R., Valros, A., Saloniemi, H. and Hänninen, L., 2010. Effect of crate height during short-term confinement on the welfare and behaviour of turkeys. Applied Animal Behaviour Science 126: 134-139.

Woolley, S.C., Borthwick, F.J.W. and Gentle, M.J., 1986. Tissue resistivities and current pathways and their importance in preslaughter stunning of chickens. British Poultry Science 27: 301-306.

Yalçin, S. and Güler, H.C., 2012. Interaction of transport distance and body weight on preslaughter stress and breast meat quality of broilers. British Poultry Science 53: 175-182.

Zulkifli, I., Norazlina, I., Htin, N.N. and Juriah, K., 2006. Physiological and behavioural responses of laying hens to repeated feed deprivation. Archiv Fur Geflugelkunde 70: 22-27.

Photo: Michael Cockram, University of Prince Edward Island, Charlottetown, Canada

Chapter 7
Sheep

M.S. Cockram[1] and A. Velarde[2]*

[1]Sir James Dunn Animal Welfare Centre, Department of Health Management, Atlantic Veterinary College, University of Prince Edward Island, 550 University Avenue, Charlottetown, PEI, C1A 4P3, Canada; [2]Animal Welfare Program, IRTA, Veïnat de Síes, 17121 Girona, Spain; mcockram@upei.ca

Abstract

Preslaughter handling, transport and slaughter can potentially affect animal welfare and the quality of sheep meat (lamb from young sheep and mutton from older sheep). A reduction in the muscle glycogen concentration at slaughter can reduce post-mortem muscle acidification. Bruising can occur following trauma and inappropriate handling. However, the effects of each preslaughter stage on animal welfare and meat quality are variable. Road transportation is normally stressful, but the other risks to welfare can be affected by the manner of loading and the quality of the journey, e.g. stocking density and ventilation. The rumen contents can provide energy and water during long periods without access to feed and water, but sheep can show signs of hunger after long journeys. A slaughter plant lairage provides an area where sheep can rest, drink, be inspected and assembled before slaughter. After lairage, the sheep are moved to the point of slaughter through a combination of passageways, races and forcing pens to a moving restrainer or circular pen. The main method used for stunning is head-only electrical stunning. Tongs are placed on the head to apply an electrical current to the brain that causes temporary unconsciousness. After stunning, the sheep must be stuck and bled without delay before it recovers consciousness. The major blood vessels in the neck are cut to cause rapid blood loss and death. To avoid the risk of recovery, electrical stunning can be carried out by passing a current simultaneously through the brain and the heart (head-to-body electrical stunning) to induce cardiac arrest and death. Captive bolt stunning can induce immediate loss of consciousness from brain concussion caused by the impact of the bolt on the skull. Some religious slaughter is conducted without prior stunning. If sheep are conscious during sticking and bleeding, they can experience stress, pain, and fear.

Keywords: handling, lamb, transport, slaughter, welfare

7.1 Introduction to pre-slaughter handling and transport

Preslaughter handling and transport have the potential to affect animal welfare by causing fear and distress, injury and pain, and hunger and discomfort. They can also influence the quality of sheep meat by affecting its tenderness, juiciness, flavour and colour (Ferguson *et al.*, 2014). The severity of any animal welfare and meat quality issues as a result of preslaughter handling depend on the way in which the sheep are managed. There are fewer meat quality issues associated with the handling and transport of sheep, and they also cope better with transportation than many other types of livestock, e.g. pigs (Knowles, 1998). A reduction in the muscle glycogen concentration of sheep at the time of slaughter can impair post-mortem muscle acidification (Pethick *et al.*, 1995) and increase the risk of dark cutting or dark, firm, dry meat (Ponnampalam *et al.*, 2017). Muscle glycogen concentration can be reduced by a combination of factors during the preslaughter period

(Gardner *et al.*, 2014). In some circumstances, the preslaughter treatment of the sheep is stressful, and this can affect meat quality. For example, one study showed that wethers slaughtered on-farm using captive bolt stunning had a lower muscle pHu value than those transported for 2 h, lairaged for 24 h and then electrically stunned (Gardner *et al.*, 1999). However, the effects of each stage of preslaughter management on meat quality are variable, and inconsistent effects on meat quality have been reported. Other than general guidance to reduce preslaughter stress, it is difficult to prescribe specific key preslaughter management procedures that will provide consistent meat quality (Young *et al.*, 2005). Jacob *et al.* (2005a) reported a small effect of journey duration (0.5 to 9 h) in increasing pHu in *Semimembranosus*, but not in *Semitendinosus* muscle; on-farm fasting without water (4 to 24 h) increased muscle pHu values, but there was no effect of lairage duration (0, 1 or 2 days). There were no significant differences between the muscle glycogen concentrations measured on-farm and those measured at the time of slaughter. The administration of adrenaline or cortisol before slaughter can increase the plasma concentration of these hormones at slaughter, but they do not necessarily have a significant effect on muscle pHu values (Pighin *et al.*, 2013). Similarly, Stewart *et al.* (2018) did not find a significant correlation between plasma indicators of stress (e.g. cortisol) at the time of slaughter and lamb loin tenderness.

Other potential factors that can affect meat quality are bruising from trauma and inappropriate handling, feed and water restriction, and body temperature (Ferguson *et al.*, 2014). The body temperature of sheep can increase after handling events during each stage of preslaughter management, e.g. during assembly before transport, penning, weighing, loading, transport, unloading, lairage, and assembly before slaughter. A significant negative correlation has been reported between elevations in body temperature after exercise and muscle glycogen concentrations at slaughter (Pighin *et al.*, 2014).

There are several guides to good practices for the handling and transport of sheep, e.g. the sheep handling guide from New Zealand (https://www.worksafe.govt.nz/topic-and-industry/agriculture/working-with-animals/working-with-sheep/safe-sheep-handling-gpg/) and the EU sheep transport guide (http://www.animaltransportguides.eu).

7.1.1 Types of sheep transported to slaughter

Lamb is produced from young sheep, typically less than a year old, and mutton is produced from older sheep, usually about 3 years of age. There is variation between countries in sheep production systems and in whether the sheep are kept primarily for meat, wool or milk production. The type of system affects when the sheep are sent for slaughter. Weaned lambs from meat breeds can be slaughtered at 4 to 12 months of age when they have reached the desired weight and condition. In some systems, lighter lambs are produced for slaughter that are weaned at about 1 month of age and slaughtered at about 70 days of age. Other lambs are weaned just before their transport to slaughter. These are either lambs that have been suckling their mother at pasture until they reach 3-5 months of age or lambs from milk breeds that have been reared on milk replacer until they are 30 to 45 days of age. Mutton is mainly the meat from cull ewes that are no longer required for breeding. However, it can also come from castrated males (wethers) (about 1 year of age) or entire rams.

7.2 Pre-transport handling

This stage can involve collection and assembling, handling, sorting and preparation of the sheep, including selection of fitness for transport, feed and water provision or a holding period without feed.

7.2.1 Fitness

Before loading, sheep must be assessed for their fitness for the intended journey. Sheep can die or become unfit during journeys if they have health issues before loading (Knowles *et al.*, 1994) or the conditions during the journey are extremely poor. In one survey, 0.008% of lambs transported to slaughter in the UK were dead-on-arrival (Knowles *et al.*, 1994); at inspections at a control post in Italy, 0.084% were dead-on-arrival, and 0.019% were unfit to continue their journey (Padalino *et al.*, 2018), and on arrival at their destination in the EU after a long journey (>8 h), 0.2% of the sheep were dead-on-arrival or non-ambulatory, and 0.09% were severely lame (Messori *et al.*, 2015b). Some non-ambulatory sheep show signs of hypocalcaemia and hypoglycaemia (Shorthose and Shaw, 1977). Sheep should arrive at their destination in a similar condition to that assessed before loading with minimal deterioration during their journey. Sheep with conditions that affect their physical fitness and could result in exhaustion, debility, or fatigue should not be transported (Cockram, 2019). Sheep that are not fit for transport should be treated or euthanised on the farm. Lameness is common in sheep (Kaler and Green, 2009; Winter, 2008) and can be detected using locomotion scoring (Kaler *et al.*, 2009). To be fit for transport, a sheep must be able to stand, bear weight on all of its legs and be able to adjust its footing to maintain its balance during the journey. It must also be able to walk up and down ramps at the start and the end of the journey. However, sheep with a disease known as footrot are more sensitive to the mechanical stimulation of their feet than non-lame sheep (Ley *et al.*, 1989), and some in this condition are transported to slaughter (König *et al.*, 2011). As they have to make frequent foot changes to maintain stability during transport, they likely experience pain each time that they move their feet.

To assess fitness for transport, sheep should be observed to evaluate signs of illness, such as behaviour (e.g. differences from other sheep, greater or reduced activity and reduced responses to external stimuli), posture, gait, respiration and discharges, and, if necessary, they should be caught and examined in detail (Lovatt, 2010). Signs of pain associated with lesions, such as footrot and mastitis, can be observed using changes in facial expression, such as orbital and masseter tightening, abnormal ear position, and changes in the shape of the lips, jaw, nostril and philtrum (McLennan *et al.*, 2016). Some sheep that are assessed as fit for transport and transported to slaughter will nevertheless have pathological conditions that may or may not have been clinically apparent before loading. Condemnation statistics for carcasses assessed as not fit for human consumption indicate that some lambs sent to slaughter have pathology, e.g. pneumonia, nephritis/nephrosis and arthritis (Edwards *et al.*, 1999; Green *et al.*, 1995), that might or might not have been apparent on the farm (Green *et al.*, 1997) and might or might not have affected their ability to respond to the challenges of transportation. However, in many cases, the presence of pathology in these sheep will have been associated with systemic disease and inflammatory processes (Kostro *et al.*, 2009) that likely affected their physiological fitness for transport. For example, they may have had reduced respiratory function, reduced exercise tolerance (Scott and Gessert, 1998), impaired ability to walk and maintain posture, and reduced capacity to thermoregulate and respond to periods of feed and water restriction (Cockram, 2019). Common health issues in cull ewes that could affect their fitness for transport, especially if they experience prolonged marketing at auction markets, are emaciation, lameness, mastitis and 'broken mouth'/dental issues (Herrtage *et al.*, 1974; McLaren *et al.*, 2020; Ridler and West, 2010). Unfortunately, sheep with lameness, respiratory disease and in poor body condition can be observed at auction markets (Van Metre *et al.*, 2009). Ewes in poor body condition are likely to be less able to cope with prolonged periods of feed restriction and have greater plasma free fatty acids (FFA) concentrations and a lower plasma glucose concentration than those in better body condition (Caldeira *et al.*, 2007). Recently shorn sheep are susceptible to wind-chill (Alexander, 1974) during transport and are therefore not fit to be transported in cold conditions. There are several guidance documents to assist in the assessment of the fitness of sheep and to determine whether sheep are compromised but could potentially be transported on short journeys with additional provisions (e.g. Alberta Farm Animal Care Association and Alberta Lamb Producers, 2019).

7.2.2 Feed and water withdrawal

Some sheep have their feed removed for a period before transport to empty their gastrointestinal tract. In New Zealand, this period of pre-transport fasting can range from 3 to 12 h (Fisher *et al.*, 2012). However, the total duration between the start of on-farm fasting and the time of slaughter can be two or more days. The reasons for this practice include a reduction in defecation during transport that can prevent wool staining, reduce faecal and bacterial contamination of the fleece and skin before slaughter, and reduce the risk of carcass contamination from gastrointestinal tract rupture during evisceration after slaughter (Byrne *et al.*, 2007; Fisher *et al.*, 2012). However, prolonged periods of fasting can increase the proliferation of bacteria in the gastrointestinal tract that are pathogenic to humans, thereby increasing the risk of food safety issues (Hadley *et al.*, 1997; Pointon *et al.*, 2012). Sheep should have access to drinking water facilities before transport (http://www.animaltransportguides.eu). However, in some circumstances, they might not have access to water for several hours before loading (Pearce *et al.*, 2008).

Fasting before transport may not always produce the desired reduction in food safety risks and is likely to cause negative effects on live weight, meat quality and animal welfare. In ewes removed from pasture and then fasted for 0, 9, 18 or 30 h before a journey lasting 1 h, washed on arrival at the slaughter plant and then lairaged for about 3 h before slaughter, no effect of duration of fasting on carcass weight or *Longissimus* muscle pHu value was identified (Fisher *et al.*, 2015). However, bacterial contamination on the carcasses of sheep was greater after 9 h fasting before transport, and serum concentrations of FFA, β-hydroxybutyrate, triglycerides and urea increased with fasting duration (Fisher *et al.*, 2015), indicating that the sheep had to mobilise their body energy reserves. Lambs (27 kg) that were kept on pasture and then fasted for 0, 24 or 48 h (with and without access to water) before transport to slaughter, lost live weight during the first 24 h of fasting (10% loss) and this loss increased over the next 24 h (11% loss over 48 h) with no effect of access to water on weight loss. Significant carcass weight loss as a result of water and tissue loss occurred after 24 h regardless of whether or not the sheep had access to water. The weight of the contents of the gastrointestinal tract decreased after 24 h of fasting, but the dry matter concentration of the contents also decreased. Therefore, although the gastrointestinal tract was not as full, any puncture of the tract during evisceration could have led to increased spread of contents onto the carcass (Kirton *et al.*,1968). Fasting sheep for 3 days with occasional access to water compared with 14 h of fasting increased losses in live and carcass weights, but there was no greater loss in the weight of rumen contents (Kirton *et al.*, 1967).

7.2.3 Weaning

Lambs that are weaned just before transport are more likely to experience stress, have difficulty in obtaining drinking water, and be more susceptible to fasting because of their reduced rumen contents, compared with weaned lambs. In lambs, slaughtered at 4 months of age, those weaned just before 0.75 h transport and 12 h lairage had greater plasma cortisol concentration at slaughter than those that had been weaned at 1.5 or 2.5 months of age, but there was no effect of weaning on plasma total protein concentration and meat quality (Ekiz *et al.* 2012b).

7.2.4 On-farm handling

It is preferable to avoid excessive handling and exercise before loading. If preslaughter handling (yarding or segregating from the rest of the flock, weighing and crutching) is undertaken within 3 days of slaughter, the *Longissimus* muscle pHu value can be greater than if the handling procedures are undertaken 8-10 days before slaughter (Devine *et al.*, 2006). Sheep can be stressed by movement from their normal pen and mixing before or at the time of loading (Miranda-de la Lama *et al.*, 2010b). The movement of sheep down a race by a handler using shouts and arm movements and separation of the sheep into groups using a swing gate (drafting) can increase the plasma cortisol concentration (Hargreaves and Hutson, 1990). Separating sheep according to size, condition, and social grouping, and avoiding any mixing of breeding ewes and rams will facilitate the loading of the sheep into separate compartments on the vehicle. Although social mixing of lambs may not necessarily increase the frequency of aggressive interactions between the sheep (Ruiz-de-la-Torre and Manteca, 1999), mixing some males can increase the frequency of aggressive interactions and cause raised plasma cortisol concentration (Miranda-de la Lama *et al.*, 2012a).

In Australia, transporters reported that sheep that had been handled and transported previously were easier to handle than those that had never been handled and that the breed of sheep affected the ease of handling (Burnard *et al.*, 2015). Sheep move more readily along familiar handling routes than along those that they have not used previously (Hutson, 1980a). However, the temperament of weaned lambs as assessed on-farm by the flight speed test in a weighing crate was not found to be related to subsequent meat quality (Dodd *et al.*, 2014).

During on-farm handling, sheep can be moved using trained dogs, plastic bags, sticks, bottles with stones inside, and other rattling instruments to produce noise (Gallo *et al.*, 2018). In Australia, transporters reported that electric prods and sheep rattles were used, but sticks and polyethene pipes were not used during on-farm handling (Burnard *et al.*, 2015). Sheep driving aids may sometimes not be used appropriately, e.g. some sheep can be hit with driving aids, and some may be pulled by the wool, ears and legs (Gallo *et al.*, 2018). If sheep are moved too quickly without sufficient space to move, they are at risk of bruising from sheep riding behaviour (Figure 7.1), where one sheep places its forelegs on the back of the sheep in front (Jarvis and Cockram, 1994).

7.2.5 Loading

In Australia, transporters reported that yards and ramps on farms were often in poor repair or poorly designed, and this had a major effect on ease of handling (Burnard *et al.*, 2015). In some countries, lambs are loaded manually. One person catches the lamb and carries it to the vehicle by supporting the lamb with two hands under the abdomen, and they might then hand the lamb to a second person on the vehicle (Ekiz *et al.*, 2012a; Gallo *et al.*, 2018). Loading sheep via a ramp is less stressful, as shown by the lower plasma cortisol concentration, than loading by shouting, pushing and lifting the sheep manually onto the vehicle (Yardimci *et al.*, 2013). Loading of sheep onto a single deck vehicle does not always cause a rise in the plasma cortisol concentration (Cockram *et al.*, 1996). However, loading of sheep onto multi-deck vehicles can involve loading via an external ramp to reach the middle deck, an internal ramp (22° slope) to reach the upper deck, or an internal ramp (41° slope) to reach the belly compartment (Miranda-de la Lama *et al.*, 2018). Loading sheep via an 18° angle ramp with horizontal cleats compared with using a hydraulic lift can increase plasma noradrenaline concentration, and both methods can increase heart rate, but increases in the plasma concentrations of cortisol and adrenaline concentration after loading may not occur (Parrott *et al.*, 1998). The time taken by a group of sheep to walk up a ramp increases with the angle of the ramp. For narrow ramps (0.5 m wide), sheep move slower up a ramp angle of 10° than on a level surface

Figure 7.1. Sheep riding behaviour (M. Cockram, University of Prince Edward Island, Canada).

and slow down considerably at a ramp angle of 45° compared with one at 30°. For wider ramps (1.5 m wide) the speed of movement up a ramp does not slow down until the angle of the ramp is 45°. The floor surface of the ramp can affect the speed of movement up a wide, steep ramp with the sheep moving faster on surfaces with cross battens at 30 cm intervals than on one composed of steps (22 cm high and 22 cm deep; Hitchcock and Hutson, 1979b).

7.3 Transport

7.3.1 Destinations

Although some sheep are transported to slaughter between countries by sea in specialised vessels, only the transport of sheep in vehicles by road and ferry will be considered here. Journeys from the farm to slaughter can be single direct journeys, or they can consist of multiple journeys, where the sheep remain on the vehicle, such as journeys involving multiple collections from several farms or periods on-board a ferry, and those involving unloading at various types of staging points or for immediate transfer from one vehicle to another (Murray *et al.*, 2000). Long journeys may be broken into stages where sheep are unloaded at staging points to provide periods of rest, feed and water. For example, in the EU, for journeys longer than 8 h, sheep must, after 14 h of travel, be given a rest period of at least 1 h sufficient for them in particular, to be given liquid and, if necessary, fed. After this rest period, they may be transported for a further 14 h (EC, 2005). Many sheep are sold at auction markets (Figure 7.2) before they are finally transported to slaughter. Auction markets can increase the risk of some animal welfare issues. The additional handling that sheep receive at markets (Jarvis and Cockram, 1995a) can result in more bruising than that seen in sheep transported directly to slaughter from farms (Jarvis *et al.*, 1996). At a market, the sheep can be kept in pens that do not provide sufficient space to lie down, and they can easily be disturbed by surrounding animal and human activity. During the day, the sheep may not have ready access to drinking water, feed and bedding. Even if they are kept overnight at a market, there can be a reluctance to provide feed for sheep destined for slaughter (Lambton *et al.*, 2017).

Sheep might also be transported from the farm to holding areas where they are assembled into larger loads or sorted and where necessary kept until they reach the desired weight or condition for slaughter. In Spain, the duration that weaned rams (100 days of age) were kept at a preslaughter classification centre for weighing and sorting into groups until they reached the desired live weight of 25 kg affected meat quality (Miranda-de la Lama *et al.*, 2009). Those kept for 28 days because their live weight was only 16 kg on arrival produced tougher meat than those that reached 25 kg on arrival or after 7 days. The authors suggested that social mixing at the classification centre had an important influence. However, there was no effect on *Longissimus* muscle pHu or bruising. In a second study, Miranda-de la Lama *et al.* (2010a) reported increased plasma cortisol and lactate concentrations at slaughter in lambs that had stayed at the classification centre for 7 or 28 days, but no effect on *Longissimus* muscle pHu or bruising. In a third study, Miranda-de la Lama *et al.* (2012b) found no effect of the classification centre on *Longissimus* muscle pHu or other meat quality traits. Miranda-de la Lama *et al.* (2012c) found no effect on the plasma cortisol concentration, creatine kinase (CK) activity, bruising or *Longissimus* muscle pHu between transporting young male lambs

Figure 7.2. Sheep handling at auction market showing flocking behaviour (M. Cockram, University of Prince Edward Island Canada).

for 3 h direct to slaughter compared with sending the lambs to a classification centre where they were handled for 1 h.

7.3.2 Vehicle design and maintenance

In some countries, the design of livestock vehicles is regulated to provide appropriate conditions for the welfare of the sheep (e.g. EC, 2005). Vehicle designs vary from small trailers to multi-deck vehicles with mechanical ventilation, temperature monitoring and navigation systems, and facilities to offer feed and water. Some vehicles have a roof, whereas others are open. Some designs incorporate a ramp or hydraulic lift for loading and unloading the sheep. Adjustable partitions are used to provide compartments to segregate groups of animals and to provide some lateral protection (Lapworth, 2008).

7.3.3 Stress, motion and road surface

Road transport is normally stressful for sheep. The plasma cortisol concentration peaks after about 3 h of transport and can then decline to pre-loading values by the end of a 12 h journey (Cockram *et al.*, 1996). Both lambs that have been weaned and consuming dry feed and those that had been suckling until the start of transportation can show raised plasma concentrations of cortisol (Ergül Ekiz *et al.*, 2019) and adrenaline (Bórnez *et al.*, 2009) in response to transportation. Sheep can also show an increased body temperature during transport (Parrott *et al.*, 1999) and increased acute phase proteins (i.e. serum amyloid A and haptoglobin) 48 h post-transport (Fazio *et al.*, 2015; Piccione *et al.*, 2012). However, in one study, wethers that were transported for 1.5 h did not show increases in the plasma concentrations of ACTH and cortisol, but heart rate and body temperature increased. The sheep were assessed by observers as 'alert', 'anxious', and 'aware', but these behavioural and physiological effects were reduced after repeated journeys (Wickham *et al.*, 2012). It is likely that sheep find movement stressful as sheep placed in an oscillating pen to simulate transport showed a plasma cortisol and adrenaline response (Parrott *et al.*, 1994). The plasma cortisol response of sheep to transportation can be reduced by the administration of a sedative (xylazine; Ali *et al.*, 2006) and by the restriction of visual stimuli of the moving environment (da Cunha Leme *et al.*, 2012). When a livestock vehicle is in motion, sheep are exposed to noise, vibration and irregular accelerations in vertical, horizontal and lateral directions. The characteristics of these stimuli can depend on the vehicle, e.g. the type of suspension, the road surface and the driving style (Gebresenbet *et al.*, 2011). When sheep stand during a journey, they attempt to maintain balance in response to changes in motion and experience a greater frequency of potentially traumatic events, mainly losses of balance and slips during transport than during stationary confinement (Cockram *et al.*, 1996). Although losses of balance are common, falls during a journey do not often occur (Cockram *et al.*, 2004).

Differences in driving style can affect the frequency of losses of balance by sheep and the degree of disturbance to the sheep when they are attempting to rest during a journey. In one study (Cockram *et al.*, 2004), about 80% of the losses of balance were considered to have been caused by driving events, such as acceleration, braking, stopping, cornering, gear changes and uneven road surfaces that preceded the event. However, only about 22% of driving events were followed by a loss of balance. It was likely that driving events were also responsible for many interruptions to both lying behaviour and rumination. Clear benefits of motorway (highway) driving compared with single carriageway driving were fewer losses of balance, more lying down, more rumination and fewer disturbances amongst the sheep. Transporting sheep on rough compared with smooth journeys is more stressful (Bradshaw *et al.*, 1996). Ruiz-de-la-Torre *et al.* (2001) transported young lambs for 15 h on either smooth highways or on rougher secondary roads. More sheep remained standing on the rough roads, plasma cortisol concentration and *Longissimus* muscle pHu values were also greater than on smooth roads, but there was no effect on plasma CK activity. Miranda-de la Lama *et al.* (2011) transported young lambs for 3 h on paved, two-lane roads that provided smooth driving conditions and on unpaved rural secondary roads that provided rougher conditions. After a 0.5 h lairage, the lambs transported in rough conditions had greater values for plasma concentrations of cortisol and lactate, plasma CK activity, bruising, and *Longissimus* muscle pHu and toughness than those that had been transported in smooth conditions.

7.3.4 Space allowance

Floor space allowance (m^2/sheep) or its inverse loading density (kg/m^2) are important factors affecting how sheep respond to transport. For this reason, guidelines on appropriate ranges for different sizes and types of sheep have been produced. For example, the European Regulation (EC, 2005) specifies a space allowance for shorn sheep and lambs that are between 26 and 55 kg of 0.2 to 0.3 m^2/sheep and >0.3 m^2/sheep for those >55 kg. For unshorn sheep, a space allowance of 0.3

to 0.4 m^2/sheep for those <55 kg and >0.4 m^2/sheep for those >55 kg is recommended. Allometric equations to estimate space allowances have been provided in the form of space allowance = k liveweight$^{2/3}$, where k is a constant of 0.020 for short journeys and 0.027 for longer journeys, where all sheep require room to lie down (Petherick and Phillips, 2009).

Transporting sheep at too high a stocking density can prevent them from lying down, achieving adequate thermoregulation and making positional adjustments to maintain their balance when the vehicle is moving (Warriss *et al.*, 2003). Increased stocking density can also increase *Longissimus* muscle pHu (Chulayo and Muchenje, 2013). In another study, increased *Longissimus* muscle pHu values and meat toughness have been reported in lambs transported for 3 h at 0.2 and 0.3 m^2/sheep compared with those transported at 0.4 m^2/sheep (Akin *et al.*, 2018).

After 3 h of a 12 h journey, lambs (30-40 kg) transported at 0.22 m^2/sheep laid down less than those at space allowances of 0.27, 0.31 and 0.41 m^2/sheep. During the first 12 h after transport, the sheep that had been transported at the 0.22 m^2/sheep space allowance laid down more than those that had been transported at the 0.41 m^2 per sheep space allowance. During transport, the frequency of losses of balance and the frequency of slips were lower at the 0.22 m^2/sheep space allowance than at the 0.27 and 0.41 m^2 per sheep space allowances (Cockram *et al.*, 1996). Jones *et al.* (2010) studied the effects of space allowance during a 6 h journey on behaviour in shorn and fleeced lambs and ewes. In shorn lambs (33 kg), the frequencies of losses of balance and of trampling were lower at 1.04 m^2/sheep than at 0.23 m^2/sheep, and all sheep were able to lie down at 1.04 m^2/sheep, but only a maximum of 46% laid down at 0.23 m^2/sheep. At higher space allowances, sheep were able to immediately regain a standing posture if they went down, but at 0.23 m^2/sheep, it took on average 435 s to recover their standing posture. In fleeced lambs (40 kg), the frequency of losses of balance was lower at 1.6 m^2/sheep than at 0.40 m^2/sheep. At higher space allowances sheep, were able to stand straight away if they went down, but at 0.42 m^2/sheep it took on average 172 s. In shorn ewes (68 kg), the frequencies of losses of balance and of trampling were lower at 1.56 m^2/sheep than at 0.30 m^2/sheep. The frequency of falls was lower at 1.56 and 0.69 m^2/sheep than at 0.35 and 0.30 m^2/sheep. 31% of the sheep were able to lie down at 1.56 m^2/sheep, but only 5% at 0.35 m^2/sheep. At higher space allowances sheep were able to stand straight away if they went down, but at 0.30 m^2/sheep it took on average 174 s. In fleeced ewes (65 kg), the frequency of losses of balance was lower at 1.56 m^2/sheep than at 0.42 m^2/sheep. The frequency of trampling was lower at 0.78 m^2/sheep than at 0.40 m^2/sheep. At higher space allowances sheep were able to stand straight away if they went down, but at 0.40 m^2/sheep it took on average 367 s. On the basis of these results, Jones *et al.* (2010) proposed that the constant (k) for use in calculations of space allowance for shorn ewes should be k=0.026 (i.e. 0.44 m^2 for 67 kg), for fleeced ewes and lambs, k=0.033 (i.e. 0.56 m^2 for 65 kg and 0.4m^2 for 40.5 kg), and for shorn lambs, k=0.029 (i.e. 0.3 m^2 for 32.5 kg).

In suckling lambs transported for 0.5 h, those provided with a space allowance of 0.125 m^2/lamb stood more than those at 0.25 m^2/lamb (Ibáñez *et al.*, 2002). During transport on highways, but not on rough roads, more lambs laid down at 0.25 than at 0.12 m^2/lamb (De la Fuente *et al.*, 2012). Another study in suckling lambs (13 kg) found no effect of space allowance (0.12, 0.20 and 0.25 m^2/lamb) on plasma cortisol concentration, plasma CK and lactate dehydrogenase (LDH) activities, carcass quality and meat quality. Muscle pHu as assessed in the *Psoas Major* was lower at 0.12 than at 0.20 or 0.25 m^2/lamb, although all values were within the normal range (De la Fuente *et al.*, 2010). In suckling lambs (13 kg) transported for 3 h at either 0.08, 0.12 or 0.20 m^2/lamb, there was no effect of space allowance on live weight, carcass weight, muscle pHu or other meat quality traits (Sanchez-Sanchez *et al.*, 2013). In young lambs (29 kg) transported for 2.25 h at 0.20 m^2/lamb and then lairaged for 1 h, the plasma cortisol concentration and CK activity were greater than those that had been transported at 0.27 m^2/lamb. However, there was no effect of space allowance on muscle pHu or toughness (Teke *et al.*, 2014). In young lambs (28 kg), transported for 5.5 h and then lairaged for 18 h with and without feed, there was no effect of space allowance during transport (0.16, 0.20 and 0.30 m^2/lamb) on muscle pH_u or toughness (Vergara *et al.*, 2017).

Another important factor that determines an appropriate stocking density is whether adequate ventilation can be provided for the number of sheep carried in the vehicle (Randall and Patel, 1994). Reducing stocking density can reduce the rises in internal temperature and the temperature-humidity index that occur during stationary periods (Fisher *et al.*, 2002).

7.3.5 Heat stress

The thermoneutral zone of sheep depends on many factors, including the fleece depth, body size, air movement, time since the last feed, floor type and the proximity of other animals (Randall, 1993). The main factors that affect the risk of heat stress within a vehicle are the external conditions, the type of ventilation, internal airflow patterns and the total heat and moisture production from the sheep (Norton *et al.*, 2013). During transport, the sheep may be standing or lying on a metal floor or a well-bedded floor, and this can affect heat exchange between the sheep and the vehicle. When stationary, parking the vehicle in the shade to avoid solar radiation will also reduce the risk of heat stress (http://www.animaltransportguides.eu.). European Regulation (EC, 2005) requires ventilation systems to be designed, constructed and maintained in such a way that, at any time during the journey, when the vehicle is stationary or moving, it is capable of maintaining a range of temperatures from 5 to 30 °C within the vehicle for all animals, with a ±5 °C tolerance, depending on the outside temperature. Ideally, sheep should not be exposed to temperatures higher than about 25 °C (Silanikove, 2000). During heat stress, the sheep have an increased respiration rate that is followed by open-mouthed breathing (panting) (Hales and Brown, 1974). Transport of rams for 3 h in an open vehicle in very hot conditions (42 °C) can increase plasma cortisol and catecholamine concentrations, *Longissimus* muscle pHu and toughness (Kadim *et al.*, 2009). If the ambient temperature is near to body temperature, sheep have to rely on evaporative cooling. As humidity increases, the effectiveness of evaporative cooling decreases, and therefore there is an increased risk of heat distress at high temperature and high humidity (Randall, 1993). A monitoring device for temperature and humidity inside the compartment where sheep are transported can be used to alert the driver and reduce the risk of heat stress (European Commission, 2000).

7.3.6 Cold stress

Shorn sheep with a short fleece or sheep with very wet fleece, and young lambs are susceptible to cold temperatures. If cold stress is a risk, extra bedding should be provided, and if the air temperature is ≤6 °C (Bennett, 1972), it will be necessary to protect the sheep from wind chill by reducing airflow directly onto the sheep from ventilation openings. The provision of feed before the start of the journey and at intervals during the journey can help sheep to maintain their energy metabolism (Alexander, 1974).

7.3.7 Ventilation

Efficient ventilation of the vehicle is required to remove the heat, moisture, and carbon dioxide produced by the sheep (Randall, 1993). If, during a long journey, ventilation is poor, ammonia from the urine and faeces produced by the sheep can accumulate (Fisher *et al.*, 2002) and, when its concentration exceeds 45 ppm, it can cause mucosal irritation (Phillips *et al.*, 2012). Most vehicles used to transport sheep rely on passive ventilation. Adjustable openings on the sides of each deck of the vehicle and sometimes on the front and rear provide air exchange between the external and internal environment (Figure 7.3). When the vehicle is moving, external pressure changes cause air to enter the vehicle through rear openings, move forward within the vehicle over the animals and leave through the openings at the front. If there is air movement over the sheep, it can also cause convective cooling. However, when the vehicle is stationary, natural ventilation has to depend on wind and, to a smaller extent, on temperature differences between the inside and outside of the vehicle (Kettlewell *et al.*, 2001). The control of natural ventilation is achieved by the driver manually opening and closing the ventilation openings at the start of the journey and sometimes when the vehicle is stopped during the journey. The efficiency of the ventilation is affected by the percentage of

Figure 7.3. Multi-deck sheep vehicle with adjustable ventilation openings (M. Cockram, University of Prince Edward Island Canada).

the floor surface area represented by air openings, the deck height and the loading density (Randall and Patel, 1994). For small enclosed trailers with a rear tailboard vent, airflow can be better on the top than on the bottom deck (Gilkeson *et al.*, 2016). If the floor space allowance and the deck height are too low, the airflow through ventilation openings along the sides of a vehicle can be blocked by the sheep (Randall, 1993). For 65 kg ewes, a deck height of at least 0.96 m is recommended to allow the sheep to stand in a normal position during transport (Jones *et al.*, 2002) and for air to flow above the sheep. On a multi-deck pot-belly trailer, lambs (47 kg) that were placed in either the bottom deck or the front compartment of the middle deck where the ventilation would be expected to be worse than in other areas (Mitchell and Kettlewell, 2008) had increased body temperature, plasma cortisol concentration and CK activity compared with those in the central areas of the middle and top decks and those in the rear of the vehicle (Miranda-de la Lama *et al.*, 2018).

When vehicles loaded with sheep are stationary, such as when on a ferry, the natural ventilation is less efficient, resulting in increased temperature and humidity within the vehicle (Fisher *et al.*, 2002, 2005; Knowles *et al.*, 1996). This is particularly the case in hot environments within the vehicle, where sheep have increased water loss, primarily from the respiratory system (Mitchell and Kettlewell, 2008). In this situation, larger vents and fewer animals are beneficial (Randall, 1993). Mechanical ventilation systems on vehicles can be beneficial in hot environments, and situations where the vehicle is stationary for long periods, such as traffic or border controls and driver breaks. Forced or mechanically ventilated systems are used in some sheep vehicles that have been adapted for long-distance transportation. Extraction fans are located at regions of low pressure, and air inlets and outlets are located to allow the air to pass over all of the animals within the livestock compartment (Kettlewell *et al.*, 2001).

7.3.8 Bedding

The floors of livestock vehicles are normally made of metal with ridges or grids to provide traction against slipping. Bedding, such as straw and sawdust, is sometimes used (Messori *et al.*, 2015b; Miranda-de la Lama *et al.*, 2010b) to provide thermal insulation and a comfortable surface for lying behaviour. It can also absorb urine and slurry and improve stability. However, Rodríguez *et al.* (2017) found no significant effects of type of bedding (rice husk and two depths of sawdust) on lambs transported for 5.5 h on the serum cortisol concentration, serum CK activity and plasma concentrations of adrenaline and noradrenaline.

7.3.9 Journey duration

Some journeys to slaughter can be long, e.g. over 3 days (Gallo *et al.*, 2018) consisting of one or more road journeys, sometimes with ferry transport. The necessity for long distance transportation of live sheep to slaughter has been questioned. It could be considered more appropriate to slaughter animals close to their source of production and transport the carcass (such as the use of mobile slaughter; Hultgren, 2022 – Chapter 10 of this book) rather than put the welfare of live animals at risk. Many countries have legal regulations (e.g. EC, 2005) that restrict journey durations often by specifying the intervals at which the sheep must be offered feed, water and rest. There is a strong argument that too much emphasis has been placed on journey durations and that greater focus should be placed on the quality of the journey. However, if there is non-compliance with regulations or industry standards, the sheep are placed at risk during long journeys (Cockram, 2007). If transport conditions are good and the sheep are fit, they can be transported on long journeys without showing obvious signs of fatigue, dehydration and exhaustion of body energy reserves. In ewes transported for 30 or 48 h, Fisher *et al.* (2010) reported that body weight decreased and total plasma protein concentration tended to increase compared with those transported for 12 h. However, there were no significant effects of journey duration on dehydration (serum osmolality), plasma β-hydroxybutyrate concentration and lying behaviour during the first 18 h post-transport. In another study (Li *et al.*, 2011), ewes transported for 48 h showed metabolic effects associated with fasting and muscular exertion that were greater than those seen after a 12 h journey. The sheep transported for 48 h mobilised and utilised their adipose reserves, and became ketogenic. However, the ewes recovered during a post-transport rest period, when they had access to feed and water.

No consistent effects of journey length on meat quality have been reported. Short journeys of up to 6 h, followed by 1 h lairage with a total fasting period of 21 h did not significantly affect *Semitendinosus* muscle pHu values of lambs compared with no transport before slaughter (Warriss *et al.*, 1990). In 13 kg dairy breed weaned lambs, a journey of 24 h increased faecal cortisol metabolites, the redness and the toughness of the *Longissimus* muscle compared with a 1 h journey, but there were no effects on *Longissimus* muscle pHu or other measures of meat quality. In 17 kg lambs from a meat breed, there were no effects of journey duration on faecal cortisol metabolites or meat quality traits (Dalmau *et al.*, 2014). However, in wethers, a significant positive relationship between journey distance and *Longissimus* muscle pHu has been reported (Chulayo and Muchenje, 2013).

7.3.10 Intervals for feed, water and rest

During the preslaughter period, sheep can potentially be exposed to long periods without feed and water. The feed and water in the rumen of sheep can provide them with a source of energy and water for several days. However, if feed and/or water is restricted or the sheep do not eat and/or drink enough before transport, during a journey, while at an intermediate stop such as an auction market, and during lairage before slaughter, the cumulative effect is that by the time of slaughter some sheep will be dehydrated and some will show signs of significant mobilisation of body energy reserves. Dehydration of lambs before slaughter can cause the meat to appear darker, and it reduces meat yield (Ferguson *et al.*, 2014). In one study, the accumulative effects of time without ready access to drinking water during long journeys and while at auction markets followed by a period of overnight lairage caused raised plasma osmolality at the time of slaughter compared with sheep transported directly from farms (Jarvis *et al.*, 1996). After 24 h of transport, plasma total protein concentration and osmolality can increase, but they recover rapidly during lairage when water was available (Knowles *et al.*, 1996). If the environmental conditions are not extreme, sheep do not show clear physiological signs of dehydration during a 12 h journey, but post-transport, their water intake and plasma osmolality can be greater than non-transported sheep with constant access to food and water (Cockram *et al.*, 1996). After a long journey, e.g. of 15 h, the immediate priority for sheep is normally to eat. The consumption of dry feed such as hay increases their water intake, and they spend less time than normal lying down (Cockram *et al.*, 1999). For long journeys, drinking water can be provided from nipple drinkers and/or water bowls connected to a storage tank on the vehicle (Messori *et al.*, 2015b). Messori *et al.* (2017) transported ewes (42 kg) at 0.27 m^2/sheep for 29 h in a vehicle that provided water via nipple drinkers and did not find a significant change in the serum total protein concentration.

After transport for 12 h without feed, the plasma concentration of free fatty acids of lambs can be greater, and after 24 h without feed, the plasma concentration of β-hydroxybutyrate can be greater than that in controls with access to feed and water (Cockram *et al.*, 1997; Knowles *et al.*, 1996). Fasting for 24 h can decrease carcass weight (Thompson *et al.*, 1987) and reduce liver glycogen concentration to minimal values, but it does not have a major and consistent effect on muscle glycogen concentration and meat quality (Ferguson *et al.*, 2014; Warriss *et al.*, 1989). The greatest live weight losses occur within the first 24 h of fasting, but these losses can continue for several days (Thompson *et al.*, 1987). In wethers (46 kg), fasting for 2 or 4 days reduced live weight, but no effects on carcass weight, *Longissimus, Semimembranosus* and *Semitendinosus* muscle glycogen concentration and pHu value were found (Daly *et al.*, 2006).

The length of journey before it becomes necessary to provide sheep with an opportunity to rest in response to signs of fatigue and to drink to prevent signs of dehydration is in many circumstances longer than many advocates concerned about animal welfare have suggested. However, sheep show signs of hunger after they have been transported on long journeys without access to feed, but it can be problematic to attempt to offer them feed during the journey. If they consume large quantities of dry feed without an opportunity to drink, they can become dehydrated by a transfer of water from the circulation into the rumen in response to an increased post-prandial rumen osmolality (Ternouth, 1967). When sheep are offered hay during and after a 24 h journey, eating is a priority (Cockram *et al.*, 1997). Post-transport, the increase in the time spent eating compared with that before transport can last for 5 h in sheep that have been transported for 24 h without a rest stop, but it is shorter in those offered feed during the journey (Cockram *et al.*, 1997). If sheep are provided with a rest period or either 3 or 12 h during a 24 h journey, the initial increase in the plasma concentrations of β-hydroxybutyrate and FFA that occurs during the first 12 h of the journey before the rest period decreases during the mid-journey rest period when hay is offered, and then the concentrations increase again during the second 12 h of the journey when the feed is no longer available. The mean time taken by sheep to take their first drink during a mid-journey rest period after 12 h of transport was about 1 h. However, some sheep were not observed to drink during a 3 h lairage period or a 3 h rest period on the vehicle (Cockram *et al.*, 1997). After sheep had consumed hay during a 3 h rest period, their plasma osmolality was greater than in those that had been transported for 24 h without access to food and water (Cockram *et al.*, 1997). Krawczel *et al.* (2007) compared the effects of a continuous 22 h journey of weaned lambs with a journey consisting of an 8 h period of transport followed by unloading and access to concentrates and water for 6 h, a second 8 h journey followed by unloading and access to concentrates and water for 24 h and then a 6 h journey. In the lambs transported for 22 h without feed, the plasma glucose concentration was lower between 10 and 22 h of transport than in those provided with concentrations after each 8 h journey segment. The lambs provided with a 6 h rest period after their first 8 h journey had a lower plasma cortisol concentration than those that were transported continuously. There was no benefit of providing the two rest periods on the plasma total protein concentration and no detrimental effect on the plasma

CK activity from the unloading and reloading of the lambs. Most of the lambs laid down during the journeys, and after unloading, they ate the concentrates rather than drinking or lying down.

In ewes transported for 29 h with access to water via nipple drinkers, the plasma total protein concentration increased and then decreased after an 8 h rest period either on or off of the vehicle when the sheep had access to hay, concentrates and water. Ewes that remained on the vehicle during the 8 h rest period spent less time eating and lying than those unloaded into a pen. However, no difference in plasma CK activity was found between the group unloaded for the mid-journey rest period compared with those that remained on the vehicle (Messori *et al.*, 2015a). In another study, Messori *et al.* (2017) transported ewes for 29 h with access to water from nipple drinkers on the vehicle and a 1 h rest stop. The sheep were then unloaded and provided with feed and water for 8, 16 or 24 h before reloading and a subsequent 6 h journey. There were no significant effects of the duration of the feed and water provision following the 29 h journey on the serum total protein concentration, serum CK activity and salivary cortisol concentration.

Sheep that stand for most of a long journey (especially if additional muscular activity is required to deal with frequent postural instability) are at risk of developing fatigue. However, observational studies of lying behaviour have shown that sheep do not show signs of fatigue after long journeys of up to 24 h (Cockram *et al.*, 1997). Studies that have used gentle walking exercise for several hours on a treadmill to attempt to induce behavioural and physiological signs of fatigue in sheep have shown that sheep are capable of prolonged exercise before they show any signs of fatigue (Cockram *et al.*, 2012).

7.4 Handling at the slaughter plant

7.4.1 Unloading

There can sometimes be a delay between the arrival of a vehicle at a slaughter plant and unloading (Miranda-de la Lama *et al.*, 2010b). To avoid heat stress, it is preferable to minimise this delay, e.g. by improving the scheduling of arrival of the vehicles. A slaughter plant should have an unloading ramp or a bridge/bay that allows the sheep to walk off the vehicle on a flat surface. If no ramp or other unloading facility is present that allows the sheep to walk off of the vehicle, the sheep have to be manually unloaded (Figure 7.4B), or in some cases, they have to jump off the vehicle at considerable risk of injury (Gallo *et al.*, 2018). If supervision is inadequate, handling practices can be inappropriate with some sheep made to jump from the vehicle, sometimes from upper decks, and some might be thrown from an upper or lower deck (Rickenbacker, 1962). Unloading is assisted by the use of a ramp (Figure 7.4A), but the time taken by a group of sheep to move down a ramp increases with the angle of the ramp. For narrow ramps (0.5 m wide), sheep slow down considerably at a ramp angle of 30° compared with one at 20°. For wider ramps (1.5 m wide), the speed of movement down a ramp slows down when the angle of the ramp is 30-45°. The floor surface of the ramp can affect the speed of movement down a wide, steep ramp with the sheep moving faster on surfaces with cross battens than on one composed of steps (Hitchcock and Hutson, 1979b). When Messori *et al.* (2015b) observed sheep that had been transported on long journeys to destinations in the EU, they observed that between 0.3 and 62% of the sheep fell during unloading. A significant

Figure 7.4. Sheep unloading (A) down vehicle ramps or (B) by manual handling (photo (A) M. Cockram, University of Prince Edward Island, Canada; photo (B) H. Li, Inner Mongolia Academy of Agricultural & Husbandry Sciences, China).

positive correlation was found between the ramp angle and the percentage of falling events per vehicle. Slipping and reluctance to move were observed in most consignments. When the ramp was totally covered with straw, there was less falling and slipping.

7.4.2 Lairage

The lairage at a slaughter plant provides a holding area where ante-mortem inspection of the sheep for food safety and animal welfare issues can be conducted. It also provides an area where sheep can be held to enable the slaughter line to run at a constant and efficient rate. If sheep arrive wet, lairage also provides an opportunity for the fleece to dry before the sheep are slaughtered (Weeks, 2008). In many countries, sheep are slaughtered on the day of arrival, but in others, it is common practice to keep them overnight in the lairage (with water and sometimes with access to feed). Whether it is beneficial for animal welfare and for meat quality for sheep to be lairaged at a slaughter plant rather than slaughtered as soon as possible is controversial. A period of lairage can allow a raised plasma cortisol concentration following transport and handling to decrease. For example, suckling lambs transported for 0.5 h and then lairaged for 3 h without feed and water had a lower plasma cortisol concentration than those lairaged for 1.5 h (Ibáñez *et al.*, 2002). However, lairage does not necessarily improve meat quality, and there can be deterioration in some aspects of carcass and meat quality with prolonged lairage. Prolonged lairage can also increase the risk of fleece contamination and the spread of pathogens between the sheep (Grau and Smith, 1974). From an animal welfare perspective, immediate slaughter is preferable to being kept in a lairage, especially in an environment that does not enable the sheep to rest and drink.

On arrival in a lairage, some sheep will start to lie down and rest. They show increased lying behaviour with duration in a lairage, and after overnight lairage, most sheep are lying down (Kim *et al.*, 1994). Lying behaviour is facilitated by the provision of adequate floor space and by keeping the sheep in the same group in which they were transported. However, resting is disturbed when sheep are moved to a different pen, by a person entering the pen to provide feed, by the addition of sheep to the pen and by general activity in the lairage, such as the movement of animals and especially people in adjacent passageways (Jarvis and Cockram 1995b; Kim *et al.*, 1994). Sheep in lairage appear to be more responsive to human vocalisation and to mechanical noise such as metal banging and hosing than to noises of animal origin (Kim *et al.*, 1994). A space allowance of at least 1 m^2/sheep can provide an opportunity for the sheep to lie down. At lower space allowances, sheep were observed to walk over other sheep, and movement can be restricted by other sheep lying down (Kim *et al.*, 1994). For sheep of 35-45 kg, a space allowance of 0.3 m^2/sheep was observed to restrict lying and drinking behaviour during the first 8 h in a lairage. Whereas a space allowance of 0.6 or 1 m^2/sheep provided sufficient space for more than 90% of the sheep in a pen to lie down (Jongman *et al.*, 2008). Entire rams can show butting behaviour during lairage after transport (Ekiz *et al.*, 2012c). Groups of entire rams, ewes and a mixed-sex group all showed a cortisol response to transport that decreased during 2 h lairage. There was more butting in the ram group than in the ewe group, and it was intermediate in the mixed-sex group (Ekiz *et al.*, 2013).

Several studies have shown that even if plasma cortisol concentration reduces during lairage, this was not accompanied by improved meat quality (Jacob *et al.*, 2005b; Liste *et al.*, 2011). Other studies have shown improved meat quality after lairage (Ekiz *et al.*, 2012a; Li *et al.*, 2018; Najafi *et al.*, 2020). However, prolonged fasting during lairage can produce negative effects, including increased live weight loss and reduced liver glycogen concentration (Díaz *et al.*, 2014). Although sheep may not normally be offered feed in a lairage, if it is offered after a long journey, they are likely to eat for several hours (Cockram *et al.*, 1996). After a 5.5 h journey, provision of concentrates and water during 18 h lairage resulted in greater plasma glucose concentrations than in those without access to feed (Cozar *et al.*, 2016). In ewes transported for 4.5 h, the plasma cortisol and FFA concentrations were increased. Post-transport, they showed increased eating and drinking, but spent less time lying down. The mean latencies to eat, drink and lie down were 0.1, 0.9 and 2 h, respectively (Pascual-Alonso *et al.*, 2017). In wether lambs fasted for 48 h with or without access to water, serum β-hydroxybutyrate concentration and serum total protein concentration increased. Water deprivation increased the calculated serum osmolality and serum urea concentration. It also caused water to move out of the muscle. This dehydration caused the meat to be darker, the drip and cooking losses were reduced, and the *Longissimus* muscle pHu was lowered (Jacob *et al.*, 2006a). Lairage can provide an opportunity for sheep to drink, and this can avoid prolonged thirst and dehydration. However, not all sheep drink during lairage. Some sheep can become dehydrated during their journey to slaughter and are still dehydrated at the time of slaughter (Ferguson *et al.*, 2014). This is a particular problem for recently weaned lambs that may not be familiar with drinking water from troughs and other drinkers (Jacob *et al.*, 2006b). In lambs transported for 1.25 h, a lairage for 18 h resulted in greater plasma total protein concentration than after a shorter lairage of 0.5 h. There was no noticeable consumption of drinking water from a trough (Ekiz *et al.*, 2012a).

7.4.3 Handling

After unloading, the sheep are moved along passageways to lairage pens and after a period of lairage to the point of slaughter through a combination of passageways, races and forcing pens to a restrainer where the sheep are slaughtered. A common handling system consists of a series of pens that funnel the sheep into an ascending single file race that leads to a V-shaped conveyor restrainer that moves groups of sheep in a continuous flow to the point of slaughter. Circular rotating crowd pens can also be used to reduce handling and to encourage voluntary loading into the restrainer (Bates *et al.*, 2014). Grandin and Deesing (2008) provide details of facility designs suitable for slaughter plants. Factors in facility design that are considered to reduce the risk of injury and handling difficulties are a well-maintained and drained, non-slip floor, absence of projections and impact points, no right-angled bends or dead ends, no shadows or contrasts and lateral protection from jumping (Gregory *et al.*, 2009). Sheep run faster through races 1.5 m wide (three sheep wide) than narrower races 0.5 m wide (one sheep wide) because they are better able to move as a group. In both narrow and wide races, sheep run faster when the sides of a race are covered than when they are open. A corner of 180° within a race can, in some circumstances, impede movement. In wide races, the time taken to run through a race increases as the angle of a corner increases from 0 to 135°. However, in narrow races, a corner can increase the speed of movement (Hutson and Hitchcock, 1978). In wide races, the time taken by sheep to move through a race is faster when forced by a handler (Hutson, 1980b). In some countries, e.g. New Zealand, sheep are washed by moving them through a swim bath or an area where they are exposed to a shower of water spray before slaughter. Although washing can remove visible contamination, it can increase the bacterial load on the carcass (Biss and Hathaway, 1996). Repeated washing of sheep can increase muscle pHu values (Bray *et al.*, 1989; Geesink *et al.*, 2011; Petersen, 1983) and the risk of bruising (Petersen, 1978).

The handling facilities available, how efficiently handling is undertaken, and how handlers interact with the sheep are variable, and this can influence the risk of stress and injury. Sheep are at risk of bruising from wool-pulls, falling, hitting structures and from other sheep placing their forelegs on their back (Cockram and Lee, 1991; Jarvis and Cockram, 1994). Bruising as a result of injury can require excision from the carcass and loss of value (Rickenbacker, 1962; Wroth and Suiter, 1978). The types of handling issues that can occur during preslaughter handling are refusal to move forward, baulking (arrested movement due to distraction), slips, falls, impacts, and undesired movement of the group (Gregory *et al.*, 2009). In Australia, handlers have been observed to interact with sheep physically by touching and pushing, slapping and hitting and by using an electric goad/prod, auditorily by talking, whistling, shouting and use of artificial noises, such as shaking metallic rattles and banging on pen fittings, and visually by waving (Coleman *et al.*, 2012). If there is a lack of skilled handlers, poor design of facilities and inadequate supervision, examples of poor handling technique can be observed. Types of poor handling that have been observed include beating the body and sometimes the head with an object, such as a stick, pushing with the use of hands, pulling with a rope, sometimes pulling the head/horns, or tail, slapping with the hands or kicking (Bekele *et al.*, 2020), and wool pulling (Cockram and Lee, 1991). Wool pulling has been observed during unloading when the first sheep will not readily run off the vehicle (Gregory *et al.*, 2009). Handlers who feel under time pressure to move sheep quickly are more likely to physically touch and push the sheep, whereas those who feel under control are less likely to use an electric goad (Coleman *et al.*, 2012). An electric goad/prod is not a recommended handling tool for sheep as it does not consistently improve the flow of sheep in a race (McCutchan *et al.*, 1992).

Key aspects of sheep behaviour used to facilitate handling are a major aversion to social isolation, flocking in response to fear, and behavioural synchrony that facilitates following behaviour. Following behaviour is used when sheep are driven as a group and moved through passageways and races. Sheep movement along a race is facilitated by the sight of a sheep at the end of the race (Franklin and

Hutson 1982c). Sheep (Judas sheep) can be trained to initiate following behaviour in other sheep, and this can be used to assist unloading and movement (Grandin, 1980; Hinch, 2017). Any handling that involves separating or disrupting groups is contrary to their natural behaviour (Hargreaves and Hutson, 1997). Sheep become distressed when visually isolated from other sheep, their heart rate increases, they vocalise and become difficult to handle, making frequent escape attempts to rejoin other sheep (Baldock and Sibly, 1990; Syme and Elphick, 1982). Sheep must therefore be handled in groups. Sheep move away from humans or dogs when their flight zone is penetrated (Hargreaves and Hutson, 1997). The flight distance of sheep is reduced by prior regular human contact and handling (gentling) (Hargreaves and Hutson, 1990). Distractions, such as shadows and noises, can influence the ease of sheep handling, but visual and olfactory cues from previous sheep movement through a race were not found to affect sheep movement compared with a clean race (Franklin and Hutson 1982a). Sheep movement along a race was also not affected by sheep vocalisations (Franklin and Hutson 1982b). The first sheep in a group is likely to move more readily into a well-lit area than a dark area (Hitchcock and Hutson, 1979a). Sheep have a total visual field ranging from 190 to 306° with a binocular field ranging from 4 to 77°. Conformation of the head, presence or absence of horns, position of the ears and length of wool on the face affect the field of view (Hutson, 1980b). They can distinguish between colours, contrasting shades and patterns (Sugnaseelan *et al.*, 2013; Tanaka *et al.*, 1989). Sheep have good depth perception (Walk and Gibson, 1961), but this can cause sheep to baulk if there is a change in the floor surface (Hargreaves and Hutson, 1997).

Sheep can be stressed by handling procedures and show increased heart rate in response to approach by a human and a human with a dog and following movement by a man and a dog (Baldock and Sibly, 1990). Sheep show more signs of fear when a dog is present than when a human is present (Beausoleil *et al.*, 2005) and have increased plasma cortisol concentration when gathered by a human and even greater concentration when gathered by a dog (Thurley and McNatty, 1973). Increased use of dogs to move sheep in a slaughter plant was found to be related to the plasma cortisol concentration of the sheep at slaughter (Hemsworth *et al.*, 2011). Hemsworth *et al.* (2019) found a significant correlation between aggressive dog behaviour when lambs were in a slaughter plant handling system and the plasma cortisol concentration at exsanguination. Exposure to barking dogs for 5 min before slaughter increased plasma cortisol concentration at the time of slaughter, but there was no effect on *Longissimus* muscle pHu or toughness (Zimerman *et al.*, 2013).

Handling systems before slaughter should ideally minimise stress, excitement and exercise. Hemsworth *et al.* (2019) found significant correlations between the jumping and escape behaviour of lambs in a slaughter plant handling system and the plasma cortisol concentration at slaughter, and between the time that lambs remained in the handling system and the plasma cortisol concentration at slaughter. Prolonged restraint and isolation can cause raised plasma concentrations of cortisol and adrenaline, reduce the muscle glycogen concentration and increase *Longissimus* muscle pHu (Apple *et al.*, 1995). Adrenaline release associated with exercise can also cause the utilisation of muscle glycogen (Pethick *et al.*, 1991). Intensive exercise and movement by a dog during the preslaughter period can increase muscle pHu compared with those not handled intensively. However, plasma cortisol was not affected by the intensity of the handling treatment. Drip loss and cooking loss was reduced, and the meat (as assessed in the *Longissimus* muscle) was more tender in sheep that had been handled intensively (Sutherland *et al.*, 2016). Exercise immediately before slaughter can increase *Semimembranosus* muscle pHu and reduce toughness (Chrystall *et al.*, 1982; Warner *et al.*, 2005). However, exercise does not have a consistent effect on muscle glycogen concentration, muscle pHu (Apple *et al.*, 1994) and muscle toughness (Bond *et al.*, 2004). The magnitude of the reduction in muscle glycogen concentration in response to exercise varies between different muscles. The recovery of muscle glycogen concentration after exercise depends on the availability of feed. In those fasted after exercise, liver glycogen concentration fell to minimal concentrations after 48 h of fasting, but in those offered feed it increased. After exercise, the muscle glycogen in fasted sheep did not change, but in those offered feed, the glycogen concentration in some muscles increased after 24-48 h (Jacob *et al.*, 2009).

7.5 Stunning and slaughter

7.5.1 Stunning

To avoid pain, distress and suffering during the slaughtering process, sheep are stunned before neck cutting. The European Council Regulation No. 1099/2009 defines stunning as any intentionally induced process that causes loss of consciousness and sensibility without pain, including any process resulting in instantaneous death (EC, 2009). During loss of consciousness, the animal is unable to perceive external stimuli and control its voluntary mobility and, therefore, does not respond to normal stimuli, including pain (EFSA, 2004). The most widely used methods for stunning sheep in commercial slaughterhouses are electrical and, to a small extent, mechanical stunning.

7.5.1.1 Electrical stunning

Electrical stunning is the most widely used method in sheep. It involves the application of an electric current of sufficient magnitude to the brain such that a generalised epileptiform activity is induced similar to that recorded in humans during grand mal epileptic seizures (Croft, 1952; Hoenderken, 1978). This seizure-like state, induced by an increase in the release of several excitatory neurotransmitters (e.g. glutamate and aspartate) into the extracellular space (Cook *et al.*, 1992), is suggestive of an immediate loss of consciousness and appears to be associated with a lack of sensory awareness (Anil, 1991). The loss of consciousness after electrical stunning lasts about 30 s (Velarde *et al.*, 2002).

Effective stunning induces tonic and clonic (kicking) seizures, which are the outward signs of grand mal epilepsy. Following the stun, the hind legs are flexed under the abdomen and the forelegs fully extended. The body is tense and tonic (rigid), breathing is absent, and the eyeballs may be rotated to such a great extent that the pupils may not be visible. The clonic phase, which can be either a galloping, cantering or erratic kicking action, usually follows immediately after the tonic phase (Anil, 1991; Gregory, 1998). To avoid recovery, neck sticking should be performed while the animal is in the tonic phase.

The main determinant of the epileptiform activity and the subsequent loss of consciousness is the amount of current flowing through the cerebral cortex (Hoenderken, 1978). Several studies have concluded that effective electrical stunning in sheep should be induced by using minimum currents of 1.0 A (Cook *et al.*, 1995; Gregory and Wotton, 1984). Based on these results, the European Council Regulation (1099/2009 CE) on animal welfare at slaughter sets out the minimum current intensity of 1.0 A for electrical stunning of small ruminants (EC, 2009). The administered current is a function of the amount of voltage and inversely proportional to the electrical resistance of the stunning tongs and tissues (i.e. presence of wool, dryness of skin, thickness of subcutaneous fat, and skull bone density and thickness) in the current pathway between both stunning electrodes (Velarde *et al.*, 2000). The most common electrical stunning methods use a 50 Hz sine wave, alternating current (AC) frequency. Higher frequencies require higher voltages to induce epilepsy, and the duration of epilepsy is also shorter (EFSA, 2004).

Head-only electrical stunning is normally applied by placing the tongs between the eyes and the base of the ears (Figure 7.5) on both sides of the head (Velarde *et al.*, 2000). If the electrodes are positioned on the neck too far away from the head (i.e. in caudal direction), the electrodes do not span the brain to induce immediate unconsciousness in spite of good electrical contact. Factors leading to wrong placement of the electrodes are variation in the size of animal and presence and direction of horns in some breeds hindering the accurate placement of electrodes. Ineffective stunning might also occur if the pressure of the stunning electrodes on the head is not sufficient to facilitate the current flow or the exposure to the electrical current is too short to result in epileptiform activity in the brain.

To avoid recovery, the electrical stunning can be carried out by passing a current simultaneously through the brain and the heart (head-to-body electrical stunning) to induce cardiac ventricular fibrillation that severely impairs cardiac output and normal blood circulation (Figure 7.6). Consequently, it induces hypoxia in the brain and myocardium, which prolongs the period of unconsciousness induced by the head-only electrical stun and leads to death (Von Mickwitz *et al.*, 1989), even if the sheep is not bled out. As head-to-body electrical stunning provokes death, stun-

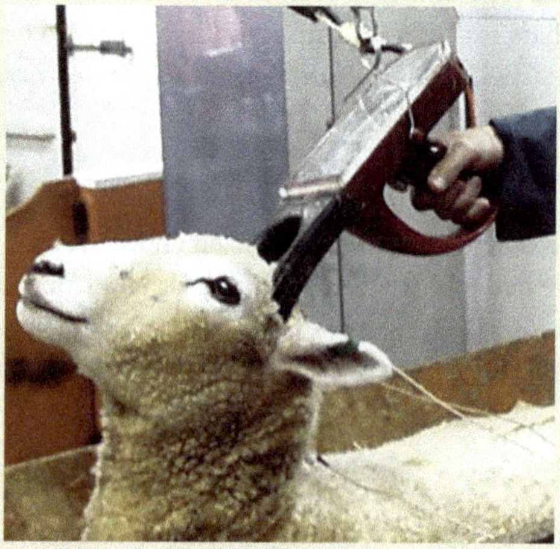

Figure 7.5. Head-only electrical stunning (A. Velarde, IRTA, Spain).

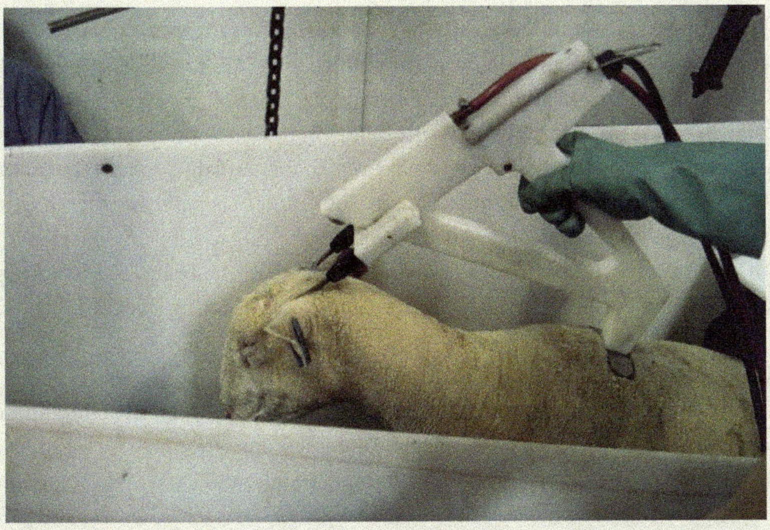

Figure 7.6. Head-to-body electrical stunning (photo courtesy of T. Grandin, Colorado State University, CO, USA).

to-stick interval is not critical in this case. To induce cardiac ventricular fibrillation, a 50 Hz sine wave alternating current should be applied.

Head-only electrical stunning induces an increase in blood pressure (Kirton *et al.*, 1978), which, when accompanied by strong skeletal muscle activity during the tonic and clonic phases, can lead to petechial haemorrhages and ecchymoses (often referred to as 'blood splash') in the carcass (Gregory, 2007; Vergara *et al.*, 2005) and reduced carcass quality (Velarde *et al.*, 2003). In head-to-body electrical stunning, the current flows from the electrodes located on the head to a third one placed on the spinal cord, caudal to the position of the heart, thereby causing the current to pass through the heart and the spinal nerves. The inhibition of the spinal nerve function produces a reduction in clonic convulsions (Gilbert *et al.*, 1984). This reduces the intensity of muscle contractions and limits the increase in blood pressure, reducing the incidence and severity of haemorrhages (Petersen *et al.*, 1986).

For the correct application of the electrodes and the uninterrupted flow of current, sheep should be restrained. Restraining facilities may vary depending on the volume of throughput. Small abattoirs may use group stunning pens where several sheep are held in a pen, and the slaughterman individually restrains each animal for stunning. The size of the pen varies, but should not permit too many animals to be held in it and will often include a crowding gate which allows the size of the stun-pen to be reduced such that, as the number of animals declines, there is less room for the remainder to move away from the operator. Larger slaughterhouses may have automatic V-type restrainers or conveyor belt system to move the sheep up to the point of stunning, in which the animals are placed with their legs straddling, not touching the ground, with their body weight supported under the brisket and belly.

7.5.1.2 Mechanical stunning

Mechanical stunning induces immediate loss of consciousness as the impact of the bolt on the skull results in brain concussion and disruption of normal brain function (EFSA, 2004). The most used method to stun sheep mechanically is the penetrative captive bolt powered by a cartridge. These guns are designed to fire a retractable steel bolt that penetrates the cranium and enters the brain. While the substantial damage sustained by vital centres in the caudal regions of the brain, including the brain stem rapidly renders the animal unconscious, it is the major haemorrhages caused by rupture of the arteries supplying the brain that ensures a long-lasting unconsciousness (EFSA, 2004). Death may occur depending on the degree of injury to the brain, but is not a guaranteed outcome (Lambooij and Algers, 2016). Therefore, captive bolt stunning shall be followed as quickly as possible by bleeding.

Successful induction of brain concussion manifests as an immediate collapse of the animal and onset of apnoea (absence of breathing) followed by the onset of a tonic seizure, which can be recognised from its head extended, hind legs rigidly flexed under the body, and fixed eyes. The forelegs may be flexed initially and then gradually straighten out. This period lasts for 10 to 20 s and is followed by a period of kicking movement. Ineffective or unsuccessful captive bolt stunning can be recognised from the failure to collapse, the presence of breathing (including laboured breathing) and the absence of tonic seizure.

The bolt diameter, velocity and penetration depth are important parameters to ensure the efficacy of the stun. The gun powder content (strength) of the cartridge should be selected according to the manufacturers' instructions to suit the animal type. Sheep have extremely variable skull morphology, particularly with regard to the presence, size and internal complexity of the frontal sinuses, that occupies the entire frontal bone (Farke, 2010). These are air-filled paranasal spaces, located within the expanded frontal sinuses which occasionally extends up into the horncores (Farke, 2010). It has been suggested that the enlarged frontal sinuses of horned sheep may be an adaptation for head-to-head combat and that these structures may have a shock-absorbing function – protecting the brain from impacts to the horns (Farke, 2008).

The shooting position depends on whether the animal is polled (hornless) or horned. For polled sheep, the muzzle of the stunner should be placed on the highest point of the head, and on the mid-line, aiming straight down (Figure 7.7A; HSA, 2016). Whereas, for horned sheep, the muzzle of the stunner should be placed on the mid-line, behind the ridge between the horns, and aimed towards the base of the tongue (Figure 7.7B; HSA, 2016).

7.5.2 Slaughter

Slaughter is the process of bleeding to induce death as a result of the permanent absence of nutrients and oxygen in the brain supplied by the common carotid arteries. Bleeding of sheep is carried out by neck cutting close to the head using a blade at least 120 mm long to sever both carotid arteries and both jugular veins, i.e. a cut across the throat (Figure 7.8; cut position 2). Although not very common in sheep, chest sticking with incision of the brachiocephalic trunk and the major blood vessels, which arise from the heart, can also be used (Figure 7.8; cut position 1).

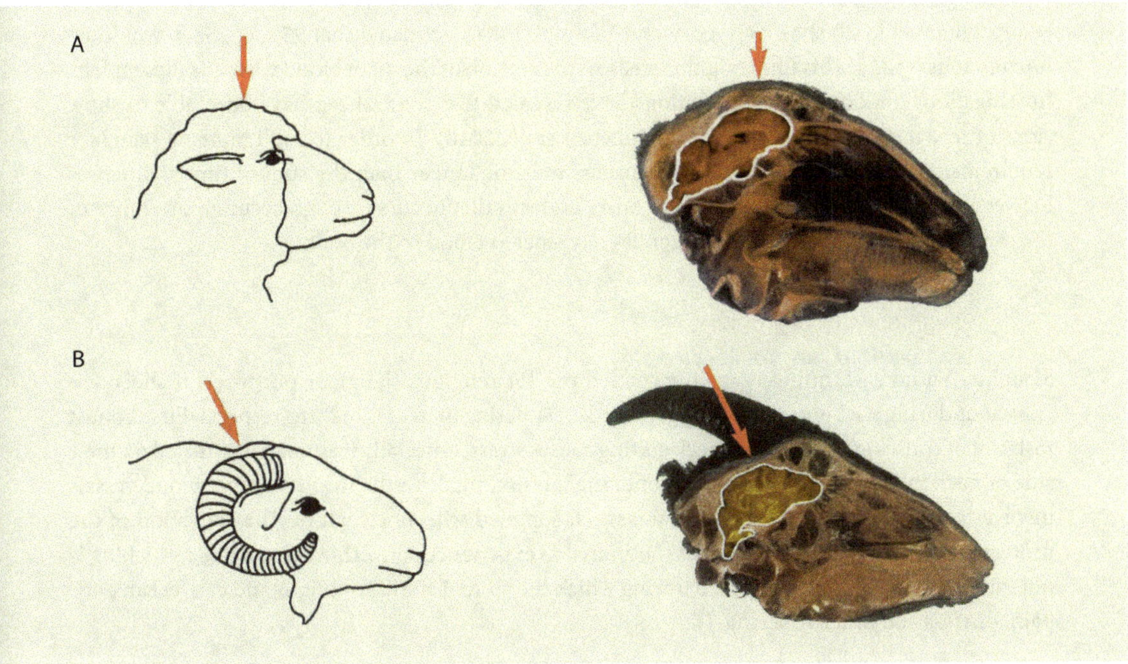

Figure 7.7. Captive bolt shooting position for (A) polled and (B) horned sheep (HSA, 2016).

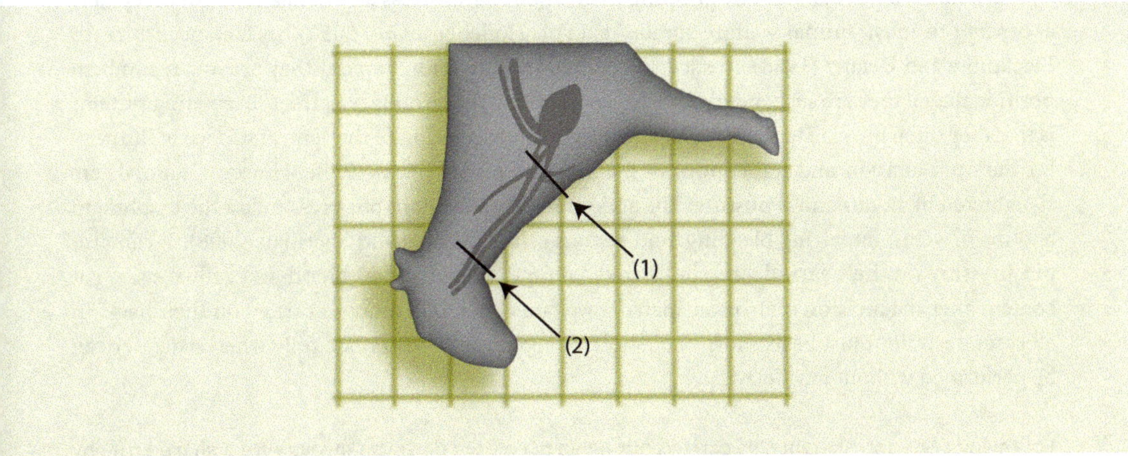

Figure 7.8. Cutting position for sheep (HSA, 2016). (1) chest sticking; (2) cut across the throat.

The weight of blood in a sheep is about 8% of the liveweight of the animal, and about 18% of total cardiac output flows through the brain at any one time (EFSA, 2004). About 50% of the total blood is removed by exsanguination. If not previously stunned, sheep are gradually rendered unconscious as brain perfusion becomes insufficient to sustain normal function. Consciousness is lost when 30-40% of the total blood volume is lost or if blood pressure drops to below 35 to 50 mmHg (Gregory, 2005). The mean time to loss of consciousness is generally quoted as being 14±1 s. Based on the results obtained in 20 sheep, Gregory and Wotton (1984) estimated that 95% of sheep would be unconscious by 22 s. This time might increase if the section of the main blood vessels is incomplete. Investigations under practical conditions have revealed that some sheep have been able to show signs of recovery for up to 2 min (von Holleben *et al.*, 2010). In order to fulfil humane slaughter requirements, the duration of unconsciousness must be longer than the sum of time that lapses between the end of stun and the time to onset of death that occurs through accurate severance of the carotid arteries in the neck that supplies oxygenated blood to the brain.

7.5.2.1 Slaughter without stunning

Slaughter without stunning is mainly carried out for religious slaughter purposes. If sheep are conscious during sticking and bleeding, the risk of suffering increases in three respects. First, robust restraint of conscious animals for neck cutting causes stress, especially if animals are turned on their side or back in rotary casting pens. Second, the incision made in the throat to sever blood vessels involves substantial tissue damage in areas well supplied with nociceptors. The activation of the protective nociceptive system induces the animal to experience pain. Third, death due to sticking is not immediate, and there is a period during which the animal is still conscious and can feel anxiety, pain, distress and other suffering.

For religious slaughter, sheep are restrained either in an upright position, lying on their side or rotated by 180° on their back (Levinger, 1995). Restraining conscious sheep by suspending their hind legs or holding or lifting them by grasping their wool, causes stress and pain and is not acceptable according to most animal welfare standards (Von Holleben *et al.*, 2010). In fact, according to Blackmore and Delany (1988), in sheep bleeding is slightly more rapid if they are in a recumbent position than if they are suspended in a vertical position. The advantage of the 180° rotating systems is that they often provide to the slaughterman a good presentation of the ventral surface of the neck for the neck incision and full control of the pressure placed on the knife. However, animals are slaughtered in an unnatural posture. The abdominal organs might put pressure on the diaphragm and major veins, impairing bleeding, and venous reflux (Adams and Sheridan, 2008). It can also put pressure on the heart, decreasing stroke volume. Aspiration of blood and refluxing of gut content after the incision is also considered a welfare concern in animals turned on their back. In all cases, restraint must be as short as possible. Sheep must be restrained only when slaughter can be performed without any delay.

Following restraint, slaughter is carried out by an accurate cut to the throat with a sharp knife by means of an incision below the angle of the jaw that involves severing major blood vessels in the ventral neck region (skin and vessels cut simultaneously). The two carotid arteries and jugular veins are severed at the same time as the oesophagus, trachea and vagus nerves. Although this practice (i.e. cutting of the oesophagus and trachea) has been suggested as not being optimal for hygiene reasons in the EU (Regulation No 853/2004), it is allowed in ritual slaughter (EC, 2004).

The neck has a high density of nociceptors (Lambooij and Kijlstra, 2008). As loss of unconsciousness is not immediate after the cut, it must be assumed that the cut and consequences of the cut are painful, as nociceptors in the area of the incision are stimulated (Kavaliers, 1989). During spot visits (Velarde *et al.*, 2014), it has been observed that sheep struggle and sometimes shiver after the cut.

Several factors can increase the perception of pain, such as performing multiple cuts (more nociceptors are affected), changes of direction of the cut, blunt knife (extension of the wound), insufficient length of the knife, wound manipulation, presentation of the neck in a position disturbing a good cut or flow of the blood (i.e. increased flexibility of the skin due to insufficient tension) and any other circumstance that delays the loss of consciousness in the animal and increases the inflammatory pain (Brooks and Tracey, 2005; Woolf, 2004). The number of cuts performed affects the magnitude of pain perception, as each time the knife touches the surface of the wound, or more tissue is damaged, more nociceptors are activated. The number of cuts can range from 1 to 6 sweeps of the knife (Velarde *et al.*, 2014).

To reduce pain perception, animals should be neck cut by a single swift or continuous back and forward movement of the knife without interruption. The knife used must be sufficiently long for each type of animal to minimise the need for multiple cuts. The knife must be sharp for each animal. The thickness of the wool might also affect the quality of the cut, and it is recommended that the wool be parted before the cut. Otherwise, this would constrain the blade during cutting and could cause blunting of the knife.

During neck cutting, the head of the sheep may be stretched manually in addition to the mechanical restraining of the body, and it is recommended that the head continues to be supported during the early stages of bleeding. The head restraint must be such that it provides good access to the neck for effective neck cutting and bleeding out. It must also be set with sufficient neck tension to make sure that the neck can be stretched such that an optimum cut is possible, and the wound site remains open to enable profuse bleeding, while preventing further stimulation of the wound. The head restraint must be designed to avoid mechanical stimuli (such as physical contact or scraping) and chemical stimuli (such as contamination with stomach contents) on the surface of the wound during the conscious period. When inspecting the wound, unnecessary contact with the severed edge of the skin must be avoided. Thus, visual inspection is preferable.

7.5.2.2 Monitoring of unconsciousness state

According to the EU Regulation (EC) No 1099/2009, ruminants shall be released from the restraint only when unconsciousness has been ensured, and carcass dressing shall begin only after the onset of death (EC, 2009). Therefore, it is important to carry out systematic checks to ensure that the animals do not present any signs of consciousness or sensibility before being released from restraint and do not present any sign of life before undergoing dressing.

Under laboratory conditions, the induction and maintenance of unconsciousness following stunning can be ascertained by recording the brain activity using electroencephalography (EEG) or electrocorticography (ECoG). Induction of unconsciousness is normally recognised from the predominance of slow-wave (low frequency) and high-amplitude EEG activity followed by a quiescent EEG (EFSA, 2013a). When stunning-induced EEG or ECoG changes are ambiguous, abolition of auditory, somatosensory or visual evoked potentials in the brain has been used to ascertain the brain responsiveness to these external stimuli (EFSA, 2013a). Under field conditions, the effectiveness of stunning can be recognised from the characteristic changes in the behaviour of animals (e.g. loss of posture), physical signs (e.g. onset of seizures, cessation of breathing, fixed eye) and from the presence or absence of responses to physiological reflexes (e.g. response to an external stimulus, such as a blinking response to touching of the cornea (corneal reflex), response to pain stimuli, such as a nose prick or toe pinching. In the scientific literature, these physical signs and reflexes have been referred to as indicators of unconsciousness or consciousness and used to monitor the welfare at slaughter of animals (EFSA, 2004). The EFSA AHAW panel developed toolboxes of welfare indicators for developing monitoring procedures at slaughterhouses for sheep stunned with head-only electrical method (EFSA, 2013b). In particular, it proposed welfare indicators together with their corresponding outcomes of consciousness, unconsciousness or death. The toolbox of indicators is proposed to be used to assess consciousness at three key stages of monitoring: (1) after stunning and during shackling and hoisting; (2) during sticking; and (3) during bleeding (Figure 7.9).

The recommended indicators for monitoring the state of consciousness after head-only electrical stunning application (in Toolbox 1) are tonic/clonic seizures, breathing and the corneal or palpebral reflex. Additionally, spontaneous blinking, posture and vocalisations may be used. At sticking (Toolbox 2), the recommended indicators to be used are breathing, tonic/clonic seizures and muscle tone. Additionally, the corneal or palpebral reflex, spontaneous blinking and vocalisations may be used. During bleeding (Toolbox 3), the recommended indicators are breathing and muscle tone. Additionally, vocalisations, the corneal or palpebral reflex and spontaneous blinking may be used. The most common intervention to apply if there is a risk of consciousness is first to re-stun and second to evaluate the risk factors for ineffective stunning (EFSA, 2013b; Figure 7.9).

The EFSA AHAW panel also developed toolboxes of welfare indicators (Figure 7.10) for developing monitoring procedures at slaughterhouses for small ruminants slaughtered without stunning (EFSA, 2013b).

The signs used to assess the onset of unconsciousness in sheep are:

- Cessation of rhythmic breathing. Rhythmic breathing can be recognised from regular flank movements. Therefore, sustained absence of rhythmic breathing (absence of a respiratory cycle – inspiratory and expiratory movements) can be used as an outcome of unconsciousness.
- Cessation of breathing can also be used as an outcome of unconsciousness in animals. However, since the trachea is also severed at the time of neck cutting at slaughter without stunning, absence of breathing cannot be assessed from the air movement at the external nostrils.
- Loss of muscle tone, resulting in a relaxed body. This may be used to recognise the onset of unconsciousness. However, involuntary muscle jerks may occur occasionally. Depending upon the type and severity of restraint loss of tone in neck and leg muscles could be used as an alternative when all of the body cannot be assessed.
- Sustained loss of posture (collapse or loss of balance). In conscious animals, the cerebral cortex integrates posture and movement. After sticking, collapse, which manifests when a free-standing animal falls to the ground, is the earliest indication of approaching unconsciousness after the

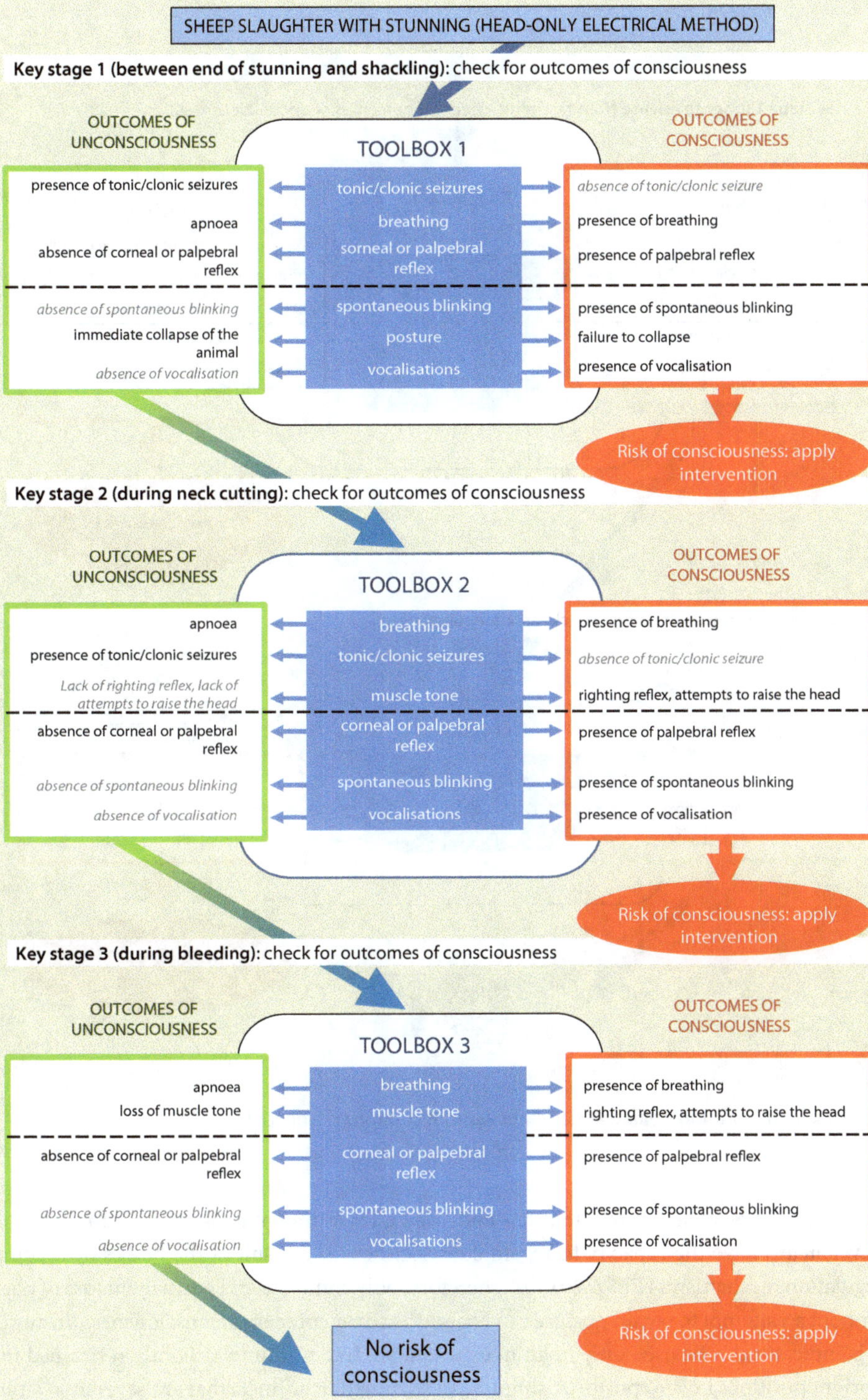

Figure 7.9. Toolbox for monitoring welfare at slaughter of sheep following head-only electrical stunning (EFSA, 2013b).

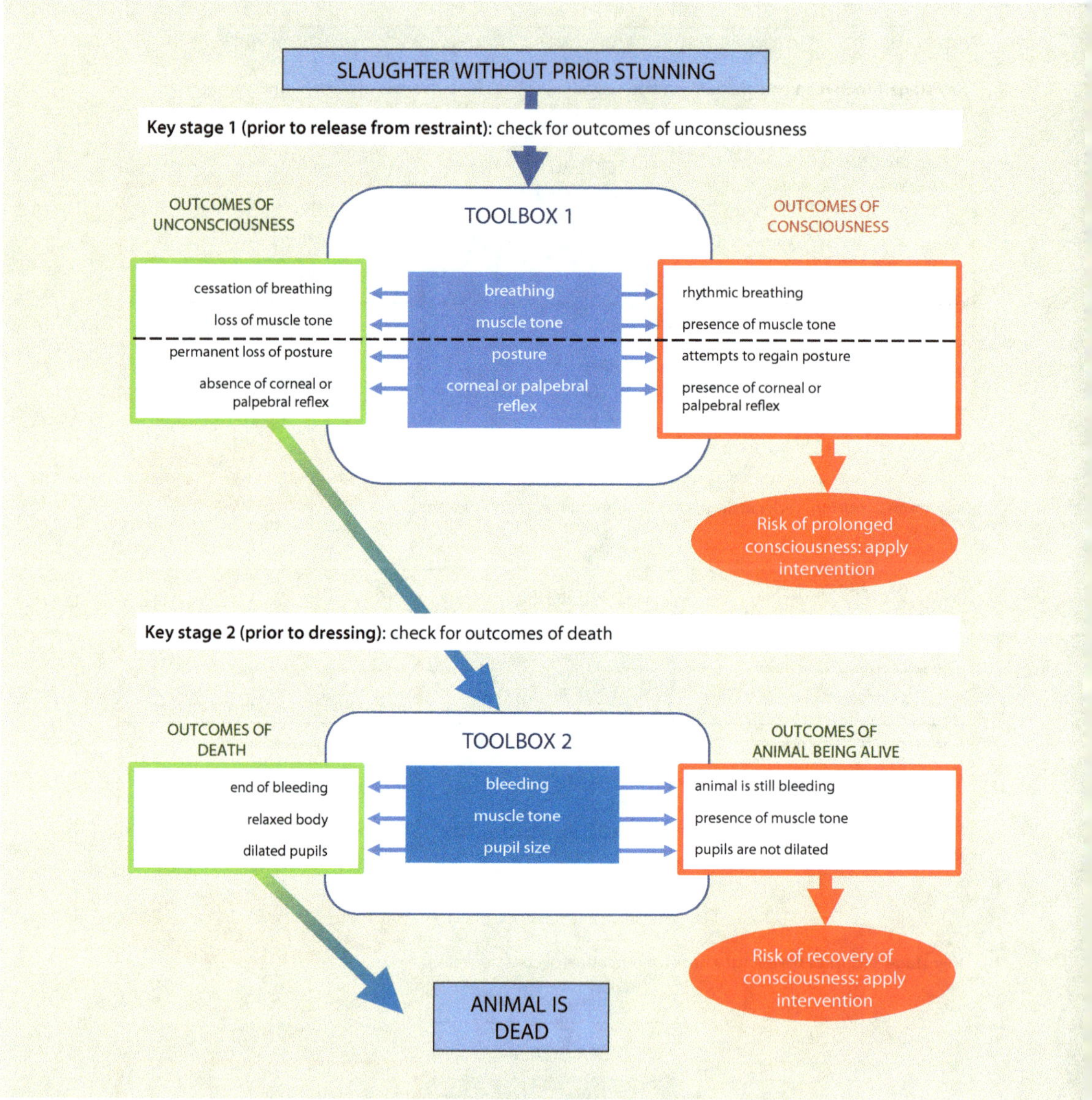

Figure 7.10. Toolbox for monitoring welfare at slaughter without stunning of sheep (EFSA, 2013b).

neck cut. However, an animal that has collapsed after a dramatic loss of blood pressure may nevertheless have the capacity to regain consciousness as a result of the body's own counter-regulation mechanisms (EFSA 2013b). Therefore, only complete and permanent loss of posture without an attempt to regain posture can be used as an outcome of unconsciousness. In any case, this indicator can be used only in animals which are free-standing or lightly restrained in the upright position. Loss of posture cannot be determined in animals that are severely restrained and/or rotated.

- Absence of corneal or palpebral reflex. The blinking response elicited by touching or tapping the cornea, the inner/outer eye canthus or eyelashes, will inevitably be absent in unconscious sheep.

According to EFSA, breathing and muscle tone are the preferred indicators, while loss of posture, corneal and palpebral reflexes can be additional indicators that should not be relied upon alone in ruminants (Figure 7.10).

In the event of inefficient bleeding or prolonged consciousness being exhibited during repeated checks after neck cutting, animals should be stunned with a suitable method as soon as possible, even if this requires the religious authorities to declare the animal as non-kosher or haram (unlawful according to the Islamic rules). Optimally, this should be done within 30 s (Velarde *et al.*, 2010).

When the function of the brain stem is sufficiently impaired as a result of the blood loss, respiration, and cardiac activity will cease over time (EFSA 2013b). The main clinical signs of death are (Figure 7.10):

- End of bleeding. Slaughter leads eventually to cessation of bleeding, with only minor dripping, from the neck cut wound.
- Relaxed body. Complete and irreversible loss of muscle tone leads to this state, which can be recognised from the limp carcass.
- Dilated pupils (mydriasis), which requires close examination of the eyes.

7.6 Conclusions

There are multiple factors that can affect the welfare of sheep during handling, transport and slaughter. An initial assessment is required of the fitness of the sheep for the intended journey together with consideration of any factors that will increase the vulnerability of particular animals to handling and transport. A risk assessment of the potential welfare issues that might arise during the journey together with a contingency plan to deal with these issues if they occur represents best practice. Attention to the manner in which sheep are handled and handling facilities that are designed to utilise the behaviour of sheep will minimise injury and stress and facilitate optimal movement of the sheep. Provision of an unloading bridge or ramps that are not steep or slippery will expedite loading and unloading. Avoiding overcrowding and providing sufficient ventilation to prevent heat stress during transport are important aspects of transport. Driving style, road conditions, vehicle design and stocking density can affect the stability of the sheep and their opportunities for rest during a journey. The effects of journey duration on sheep welfare are affected by the quality of the journey and the duration without access to feed and water. The rumen provides a store of water and digesta, and if sheep are provided with feed, water and rest, they can recover rapidly after a long journey. However, sheep can show signs of hunger, and they mobilise body energy reserves during a long journey. These effects are even more apparent if the sheep are fasted before and after transport. Provision of drinking water before, at intervals during, and after transport is important, especially in hot conditions to avoid excessive thirst. Increased sophistication in vehicle design and environmental monitoring together with journey planning to provide opportunities for rest, feed and water is required for long journeys in some countries. The influence of handling, transport, lairage and slaughter on meat quality in sheep is not as apparent as in some other non-ruminant food animals. However, bruising and reduced muscle glycogen at the time of slaughter can affect meat quality. Electrical stunning that passes sufficient current through the brain (head-only stunning) is an effective means of inducing immediate loss of consciousness in sheep. However, to avoid recovery, death must then be caused by cardiac arrest produced by passing an electrical current across the heart (head-to-body electrical stunning) or by prompt and accurate exsanguination following severance of the major blood vessels in the neck or chest. Mechanical stunning using a penetrative captive bolt is another option to effectively stun sheep. Slaughter without stunning leads to severe pain, fear and distress before losing consciousness. The stunning equipment used must be maintained and tested at regular intervals. The effectiveness of stunning and exsanguination must be assessed by monitoring the sheep for signs of consciousness. If any signs of consciousness are observed, the sheep must be re-stunned immediately, and corrective action taken to avoid ineffective stunning.

References

Adams, D.B. and Sheridan, A.D., 2008. Specifying the risk to animal welfare associated with livestock slaughter without induced insensibility. Available at: https://tinyurl.com/5y463znz.

Akin, P.D., Yilmaz, A. and Ekiz, B., 2018. Effects of stocking density on stress responses and meat quality characteristics of lambs transported for 45 minutes or 3 hours. Small Ruminant Research 169: 134-139.

Alberta Farm Animal Care Association and Alberta Lamb Producers, 2019. Humane handling guidelines for sheep. Standards for the care of compromised and unfit animals. Available at: https://tinyurl.com/533bs5tb.

Alexander, G., 1974. Heat loss from sheep. In: Monteith, J.L. and Mount, L.E. (eds.) Heat loss from animals and man. Butterworth-Heinemann, London, UK, pp. 173-203.

Ali, B.H., Al-Qarawi, A. and Mousa, H.M., 2006. Stress associated with road transportation in desert sheep and goats, and the effect of pretreatment with xylazine or sodium betaine. Research in Veterinary Science 80: 343-348.

Anil, M.H., 1991. Studies on the return of physical reflexes in pigs following electrical stunning. Meat Science 30: 13-21.

Apple, J.K., Dikeman, M.E., Minton, J.E., McMurphy, R.M., Fedde, M.R., Leith, D.E. and Unruh, J.A., 1995. Effects of restraint and isolation stress and epidural blockade on endocrine and blood metabolite status, muscle glycogen metabolism, and incidence of dark-cutting longissimus muscle of sheep. Journal of Animal Science 73: 2295-2307.

Apple, J.K., Minton, J.E., Parsons, K.M., Dikeman, M.E. and Leith, D.E., 1994. Influence of treadmill exercise on pituitary-adrenal secretions, other blood constituents, and meat quality of sheep. Journal of Animal Science 72: 1306-1314.

Baldock, N.M. and Sibly, R.M., 1990. Effects of handling and transportation on the heart rate and behaviour of sheep. Applied Animal Behaviour Science 28: 15-39.

Bates, L.S.W., Ford, E.A., Brown, S.N., Richards, G.J., Hadley, P.J., Wotton, S.B. and Knowles, T.G., 2014. A comparison of handling methods relevant to the religious slaughter of sheep. Animal Welfare 23: 251-258.

Beausoleil, N.J., Stafford, K.J. and Mellor, D.J., 2005. Sheep show more aversion to a dog than to a human in an arena test. Applied Animal Behaviour Science 91: 219-232.

Bekele, T., Szonyi, B., Feleke, A. and Grace, D., 2020. Assessment of small ruminant welfare in Ethiopia – an abattoir-based study. Journal of Applied Animal Welfare Science 23: 356-365.

Bennett, J.W., 1972. The maximum metabolic response of sheep to cold: effects of rectal temperature, shearing, feed consumption, body posture, and body weight. Australian Journal of Agricultural Research 23: 1045-1058.

Biss, M.E. and Hathaway, S.C., 1996. Effect of pre-slaughter washing of lambs on the microbiological and visible contamination of the carcases. Veterinary Record 138: 82-86.

Blackmore, D.K. and Delany, M.W., 1988. Slaughter of stock – a practical review and guide. Veterinary Continuing Education, Massey University, Palmerston North, New Zealand, 137 pp.

Bond, J.J., Can, L.A. and Warner, R.D., 2004. The effect of exercise stress, adrenaline injection and electrical stimulation on changes in quality attributes and proteins in Semimembranosus muscle of lamb. Meat Science 68: 469-477.

Bórnez, R., Linares, M.B. and Vergara, H., 2009. Haematological, hormonal and biochemical blood parameters in lamb: effect of age and blood sampling time. Livestock Science 121: 200-206.

Bradshaw, R.H., Hall, S.J.G. and Broom, D.M., 1996. Behavioural and cortisol response of pigs and sheep during transport. Veterinary Record 138: 233-234.

Bray, A.R., Graafhuis, A.E. and Chrystall, B.B., 1989. The cumulative effect of nutritional, shearing and preslaughter washing stresses on the quality of lamb meat. Meat Science 25: 59-67.

Brooks, J. and Tracey, I., 2005. From nociception to pain percepton: imaging the spinal and supraspinal pathways. Journal of Anatomy 207: 19-33.

Burnard, C.L., Pitchford, W.S., Hocking Edwards, J.E. and Hazel, S.J., 2015. Facilities, breed and experience affect ease of sheep handling: the livestock transporter's perspective. Animal 9: 1379-1385.

Byrne, B., Dunne, G., Lyng, J. and Bolton, D.J., 2007. The development of a 'clean sheep policy' in compliance with the new Hygiene Regulation (EC) 853/2004 (Hygiene 2). Food Microbiology 24: 301-304.

Caldeira, R.M., Belo, A.T., Santos, C.C., Vazques, M.I. and Portugal, A.V., 2007. The effect of body condition score on blood metabolites and hormonal profiles in ewes. Small Ruminant Research 68: 233-241.

Chrystall, B.B., Devine, C.E., Snodgrass, M. and Ellery, S., 1982. Tenderness of exercise-stressed lambs. New Zealand Journal of Agricultural Research 25: 331-336.

Chulayo, A.Y. and Muchenje, V., 2013. The effects of pre-slaughter stress and season on the activity of plasma creatine kinase and mutton quality from different sheep breeds slaughtered at a smallholder abattoir. Asian-Australasian Journal of Animal Sciences 26: 1762-1772.

Cockram, M.S. and Lee, R.A., 1991. Some preslaughter factors affecting the occurrence of bruising in sheep. British Veterinary Journal 147: 120-125.

Cockram, M.S., 2007. Criteria and potential reasons for maximum journey times for farm animals destined for slaughter. Applied Animal Behaviour Science 106: 234-243.

Cockram, M.S., 2019. Fitness of animals for transport to slaughter. Canadian Veterinary Journal 60: 423-429.

Cockram, M.S., Baxter, E.M., Smith, L.A., Bell, S., Howard, C.M., Prescott, R.J. and Mitchell, M.A., 2004. Effect of driver behaviour, driving events and road type on the stability and resting behaviour of sheep in transit. Animal Science 79: 165-176.

Cockram, M.S., Kent, J.E., Goddard, P.J., Waran, N.K., McGilp, I.M., Jackson, R.E., Muwanga, G.M. and Prytherch, S., 1996. Effect of space allowance during transport on the behavioural and physiological responses of lambs during and after transport. Animal Science 62: 461-477.

Cockram, M.S., Kent, J.E., Jackson, R.E., Goddard, P.J., Doherty, O.M., McGilp, I.M., Fox, A., Studdert-Kennedy, T., McConnell, T.I. and O'Riordan, T., 1997. Effect of lairage during 24 h of transport on the behavioural and physiological responses of sheep. Animal Science 65: 391-402.

Cockram, M.S., Kent, J.E., Waran, N.K., McGilp, I.M., Jackson, R.E., Amory, J.R., Southall, E.L., O'Riordan, T., McConnell, T.I. and Wilkins, B.S., 1999. Effects of a 15h journey followed by either 12h starvation or *ad libitum* hay on the behaviour and blood chemistry of sheep. Animal Welfare 8: 135-148.

Cockram, M.S., Murphy, E., Ringrose, S., Wemelsfelder, F., Miedema, H.M. and Sandercock, D.A., 2012. Behavioural and physiological measures following treadmill exercise as potential indicators to evaluate fatigue in sheep. Animal 6: 1491-1502.

Coleman, G.J., Rice, M. and Hemsworth, P.H., 2012. Human-animal relationships at sheep and cattle abattoirs. Animal Welfare 21: 15-21.

Cook, C.J., Devine, C.E., Gilbert, K.V., Smith, D.D. and Maasland, S.A., 1995. The effect of electrical head-only stun duration on electroencephalographic-measured seizure and brain amino acid neurotransmitter release. Meat Science 40: 137-147.

Cook, C.J., Devine, C.E., Tavener, A. and Gilbert, K.V., 1992. Contribution of amino acid transmitters to epileptiform activity and reflex suppression in electrically head stunned sheep. Research in Veterinary Science 52: 48-56.

Cozar, A., Rodriguez, A.I., Garijo, P., Calvo, L. and Vergara, H., 2016. Effect of space allowance during transport and fasting or nonfasting during lairage on welfare indicators in Merino lambs. Spanish Journal of Agricultural Research 14: e0501. https://doi.org/10.5424/sjar/2016141-8313

Croft, P.G., 1952. Problem of electrical stunning. Veterinary Record 64: 255-258.

da Cunha Leme, T. M., Titto, E.A.L., Titto, C.G., Amadeu, C.C.B., Fantinato Neto, P., Vilela, R.A. and Pereira, A.M.F., 2012. Influence of transportation methods and pre-slaughter rest periods on cortisol level in lambs. Small Ruminant Research 107: 8-11.

Dalmau, A., Di Nardo, A., Realini, C.E., Rodríguez, P., Llonch, P., Temple, D., Velarde, A., Giansante, D., Messori, S. and Villa, P.D., 2014. Effect of the duration of road transport on the physiology and meat quality of lambs. Animal Production Science 54: 179-186.

Daly, B.L., Gardner, G.E., Ferguson, D.M. and Thompson, J.M., 2006. The effect of time off feed prior to slaughter on muscle glycogen metabolism and rate of pH decline in three different muscles of stimulated and non-stimulated sheep carcasses. Australian Journal of Agricultural Research 57: 1229-1235.

De la Fuente, J., De Chávarri, E.G., Sánchez, M., Vieira, C., Lauzurica, S., Díaz, M.T. and Pérez, C., 2012. The effects of journey duration and space allowance on the behavioural and biochemical measurements of stress responses in suckling lambs during transport to an abattoir. Applied Animal Behaviour Science 142: 30-41.

De la Fuente, J., Sanchez, M., Perez, C., Lauzurica, S., Vieira, C., Gonzalez De Chavarri, E. and Diaz, M.T., 2010. Physiological response and carcass and meat quality of suckling lambs in relation to transport time and stocking density during transport by road. Animal 4: 250-258.

Devine, C.E., Lowe, T.E., Wells, R.W., Edwards, N.J., Edwards, J.E.H., Starbuck, T.J. and Speck, P.A., 2006. Pre-slaughter stress arising from on-farm handling and its interactions with electrical stimulation on tenderness of lambs. Meat Science 73: 304-312.

Díaz, M.T., Vieira, C., Pérez, C., Lauzurica, S., De Chávarri, E.G. Sánchez, M. and De la Fuente, J., 2014. Effect of lairage time (0h, 3h, 6h or 12h) on glycogen content and meat quality parameters in suckling lambs. Meat Science 96: 653-660.

Dodd, C.L., Hocking Edwards, J.E., Hazel, S.J. and Pitchford, W.S., 2014. Flight speed and agitation in weaned lambs: genetic and non-genetic effects and relationships with carcass quality. Livestock Science 160: 12-20.

Edwards, D.S., Christiansen, K.H., Johnston, A.M. and Mead, G.C., 1999. Determination of farm-level risk factors for abnormalities observed during post-mortem meat inspection of lambs: a feasibility study. Epidemiology and infection 123: 109-119.

Ekiz, B., Ekiz, E.E., Yalcintan, H., Yilmaz, A., Kocak, O. and Gunes, H., 2013. Effect of ram-ewe mixed transportation on certain welfare parameters in red Karaman and Imroz sheep. Istanbul Universitesi Veteriner Fakultesi Dergisi 39: 155-167.

Ekiz, B., Ergul Ekiz, E., Kocak, O., Yalcintan, H. and Yilmaz, A., 2012a. Effect of pre-slaughter management regarding transportation and time in lairage on certain stress parameters, carcass and meat quality characteristics in Kivircik lambs. Meat Science 90: 967-976.

Ekiz, B., Ergul Ekiz, E., Yalcintan, H., Kocak, O. and Yilmaz, A., 2012b. Effects of suckling length (45, 75 and 120 d) and rearing type on cortisol level, carcass and meat quality characteristics in Kivircik lambs. Meat Science 92: 53-61.

Ekiz, B., Ergül Ekiz, E., Yalçintan, H., Koçak, O., Yilmaz, A. and Güneş, H., 2012c. The effects of transport stress on certain welfare parameters and behaviours in Red Karaman, Imroz, Sakiz and Karakul Rams. Istanbul Universitesi Veteriner Fakultesi Dergisi 38: 15-28.

Ergül Ekiz, E.E., Yalcintan, H., Kocak, O. and Ekiz, B., 2019. The effects of weaning status and transport duration on some physiological and behavioural responses to transportation in Kivircik lambs. Small Ruminant Research 181: 51-56.

European Commission (EC), 2000. Report to the Council and the European Parliament on the experience acquired by Member States since the implementation of Council Directive 95/29/EC amending Directive 91/628/EEC concerning the protection of animals during transport COM(2000) 809 final. Available at: https://www.europarl.europa.eu/doceo/document/A-5-2001-0347_EN.html.

European Commission (EC), 2004. Regulation (EC) No 853/2004 of the European Parliament and the Council, 2004. Laying down specific hygiene rules for the hygiene of foodstuffs. Official Journal of the European Union L139: 55-205. Available at: http://data.europa.eu/eli/reg/2004/853/oj.

European Commission (EC), 2005. Council Regulation (EC) No 1/2005 of 22 December 2004 on the protection of animals during transport and related operations and amending Directives 64/432/EEC and 93/119/EC and Regulation (EC) No 1255/9. Official Journal of the European Union L3: 1-44. Available at: http://data.europa.eu/eli/reg/2005/1/oj.

European Commission (EC), 2009. EU Council Regulation (EC) No. 1099/2009 EC of 24 September 2009 on the protection of animals at the time of killing. Official Journal of the European Union L303: 1-30. Available at: http://data.europa.eu/eli/reg/2009/1099/oj.

European Food Safety Authority (EFSA), 2004. The welfare aspects of the main systems of stunning and killing the main commercial species of animals. EFSA Journal 45: 1-29.

European Food Safety Authority (EFSA), 2013a. EFSA panel on animal health and welfare (AHAW): guidance on the assessment criteria for studies evaluating the effectiveness of stunning interventions regarding animal protection at the time of killing. EFSA Journal 11: 3486. https://doi.org/10.2903/j.efsa.2013.3486

European Food Safety Authority (EFSA), 2013b. EFSA panel on animal health and welfare (AHAW): scientific opinion on monitoring procedures at slaughterhouses for sheep and goats. EFSA Journal 11: 3522. https://doi.org/10.2903/j.efsa.2013.3522

Farke, A., 2008. Frontal sinuses and head-butting in goats: a finite element analysis. Journal of Experimental Biology 211: 3085-3094.

Farke, A., 2010. Evolution and functional morphology of the frontal sinuses in *Bovidae* (Mammalia: Artiodactyla), and implications for the evolution of cranial pneumaticity. Zoological Journal of the Linnean Society 159: 988-1014.

Fazio, F., Ferrantelli, V., Cicero, A., Casella, S. and Piccione, G., 2015. Utility of acute phase proteins as biomarkers of transport stress in ewes and beef cattle. Italian Journal of Food Safety 3: 4210. https://doi.org/10.4081/ijfs.2015.4210

Ferguson, D.M., Schreurs, N.M., Kenyon, P.R. and Jacob, R.H., 2014. Balancing consumer and societal requirements for sheep meat production: An Australasian perspective. Meat Science 98: 477-483.

Fisher, A.D., Niemeyer, D.O., Lea, J.M., Lee, C., Paull, D.R., Reed, M.T. and Ferguson, D.M., 2010. The effects of 12, 30, or 48 hours of road transport on the physiological and behavioral responses of sheep. Journal of Animal Science 88: 2144-2152.

Fisher, A.D., Stewart, M., Duganzich, D.M., Tacon, J. and Matthews, L.R., 2005. The effects of stationary periods and external temperature and humidity on thermal stress conditions within sheep transport vehicles. New Zealand Veterinary Journal 53: 6-9.

Fisher, A.D., Stewart, M., Tacon, J. and Matthews, L.R., 2002. The effects of stock crate design and stocking density on environmental conditions for lambs on road transport vehicles. New Zealand Veterinary Journal 50: 148-153.

Fisher, M.W., Gregory, N.G. and Muir, P.D., 2012. Current practices on sheep and beef farms in New Zealand for depriving sheep of feed prior to transport for slaughter. New Zealand Veterinary Journal 60: 171-175.

Fisher, M.W., Muir, P.D., Gregory, N.G., Thomson, B.C., Smith, N.B., Johnstone, P.D. and Bicknell, N., 2015. The effects of depriving feed to facilitate transport and slaughter in sheep – a case study of cull ewes held off pasture for different periods. New Zealand Veterinary Journal 63: 260-264.

Franklin, J.R. and Hutson, G.D., 1982a. Experiments on attracting sheep to move along a laneway: I. Olfactory stimuli. Applied Animal Ethology 8: 439-446.

Franklin, J.R. and Hutson, G.D., 1982b. Experiments on attracting sheep to move along a laneway. II. Auditory stimuli. Applied Animal Ethology 8: 447-456.

Franklin, J.R. and Hutson, G.D., 1982c. Experiments on attracting sheep to move along a laneway. III. Visual stimuli. Applied Animal Ethology 8: 457-478.

Gallo, C., Tarumán, J. and Larrondo, C., 2018. Main factors affecting animal welfare and meat quality in lambs for slaughter in Chile. Animals 8: 165. https://doi.org/10.3390/ani8100165

Gardner, G.E., Kennedy, L., Milton, J.T.B. and Pethick, D.W., 1999. Glycogen metabolism and ultimate pH of muscle in Merino, first-cross, and second-cross wether lambs as affected by stress before slaughter. Australian Journal of Agricultural Research 50: 175-181.

Gardner, G.E., McGilchrist, P. and Pethick, D.W., 2014. Ruminant glycogen metabolism. Animal Production Science 54: 1575-1583.

Gebresenbet, G., Aradom, S., Bulitta, F.S. and Hjerpe, E., 2011. Vibration levels and frequencies on vehicle and animals during transport. Biosystems Engineering 110: 10-19.

Geesink, G., Sujang, S. and Koohmaraie, M., 2011. Tenderness of pre- and post rigor lamb longissimus muscle. Meat Science 88: 723-726.

Gilbert, K.Y., Devine, C.E., Hand, R., Ellery, S., 1984. Electrical stunning and stillness of lambs. Meat Science 11: 45-58.

Gilkeson, C.A., Thompson, H.M., Wilson, M.C.T. and Gaskell, P.H., 2016. Quantifying passive ventilation within small livestock trailers using computational fluid dynamics. Computers and Electronics in Agriculture 124: 84-99.

Grandin, T. and Deesing, M., 2008. Humane livestock handling. Storey Publishers, North Adams, MA, USA, 227 pp.

Grandin, T., 1980. Livestock behavior as related to handling facilities design. International Journal for the Study of Animal Problems 1: 33-52.

Grau, F.H. and Smith, M.G., 1974. *Salmonella* contamination of sheep and mutton carcasses related to pre-slaughter holding conditions. Journal of Applied Microbiology 37: 111-116.

Green, L.E., Berriatua, E. and Morgan, K.L., 1997. The relationship between abnormalities detected in live lambs on farms and those detected at post mortem meat inspection. Epidemiology and Infection 118: 267-273.

Green, L.E., Berriatua, E., Cripps, P.J. and Morgan, K.L., 1995. Lesions in finished early born lambs in southwest England and their relationship with age at slaughter. Preventive Veterinary Medicine 22: 115-126.

Gregory, N. G., and Wotton, S. B., 1984. Sheep slaughtering procedures. 2. Time to loss of brain responsiveness after exsanguination or cardiac-arrest. British Veterinary Journal 140: 354-360.

Gregory, N.G., 1998. Stunning and slaughter. In: Gregory N.G. (ed.) Animal welfare and meat science. CAB International, Wallingford UK, pp. 223-240.

Gregory, N.G., 2005. Recent concerns about stunning and slaughter. Meat Science 70: 481-491.

Gregory, N.G., 2007. Animal welfare and meat production. CABI Publishing, Wallingford, UK, 400 pp.

Gregory, N.G., Benson, T., Smith, N. and Mason, C.W., 2009. Sheep handling and welfare standards in livestock markets in the UK. Journal of Agricultural Science 147: 333-344.

Hadley, P.J., Holder, J.S. and Hinton, M.H., 1997. Effects of fleece soiling and skinning method on the microbiology of sheep carcases. Veterinary Record 140: 570-574.

Hales, J.R.S. and Brown, G.D., 1974. Net energetic and thermoregulatory efficiency during panting in the sheep. Comparative Biochemistry and Physiology Part A: Physiology 49: 413-422.

Hargreaves, A.L. and Hutson, G.D., 1990. The stress response in sheep during routine handling procedures. Applied Animal Behaviour Science 26: 83-90.

Hargreaves, A.L. and Hutson, G.D., 1997. Handling systems for sheep. Livestock Production Science 49: 121-138.

Hemsworth, P.H., Rice, M., Borg, S., Edwards, L.E., Ponnampalam, E.N. and Coleman, G.J., 2019. Relationships between handling, behaviour and stress in lambs at abattoirs. Animal 13: 1287-1296.

Hemsworth, P.H., Rice, M., Karlen, M.G., Calleja, L., Barnett, J.L., Nash, J. and Coleman, G.J., 2011. Human-animal interactions at abattoirs: relationships between handling and animal stress in sheep and cattle. Applied Animal Behaviour Science 135: 24-33.

Herrtage, M.E., Saunders, R.W. and Terlecki, S., 1974. Physical examination of cull ewes at point of slaughter. Veterinary Record 95: 257-260.

Hinch, G.N., 2017. Understanding the natural behaviour of sheep. In: Ferguson, D.M., Lee, C. and Fisher, A. (eds.) Advances in sheep welfare. Elsevier, Duxford, UK, pp. 1-15.

Hitchcock, D.K. and Hutson, G.D., 1979a. Effect of variation in light intensity on sheep movement through narrow and wide races. Australian Journal of Experimental Agriculture and Animal Husbandry 19: 170-175.

Hitchcock, D.K. and Hutson, G.D., 1979b. The movement of sheep on inclines. Australian Journal of Experimental Agriculture and Animal Husbandry 19: 176-182.

Hoenderken, R., 1978. Elektrische bedwelming van slachtvarkens. PhD. Dissertation, University of Utrecht, Utrecht, the Netherlands.

HSA (Humane Slaughter Association), 2016. Captive-bolt stunning of livestock. Available at: https://tinyurl.com/4fp89r8p.

Hultgren, J., 2022. Avoiding live-animal transport to slaughter: mobile abattoirs. In: Faucitano, L. (ed.) Preslaughter handling and slaughter of meat animals. Wageningen Academic Publishers, Wageningen, the Netherlands, pp. 390-434.

Hutson, G.D. and Hitchcock, D.K., 1978. The movement of sheep around corners. Applied Animal Ethology 4: 349-355.

Hutson, G.D., 1980a. The effect of previous experience on sheep movement through yards. Applied Animal Ethology 6: 233-240.

Hutson, G.D., 1980b. Visual field, restricted vision and sheep movement in laneways. Applied Animal Ethology 6: 175-187.

Ibáñez, M., De la Fuente, J., Thos, J. and González de Chavarri, E., 2002. Behavioural and physiological responses of suckling lambs to transport and lairage. Animal Welfare 11: 223-230.

Jacob, R.H., Gardner, G.E. and Pethick, D.W., 2009. Repletion of glycogen in muscle is preceded by repletion of glycogen in the liver of Merino hoggets. Animal Production Science 49: 131-138.

Jacob, R.H., Pethick, D.W. and Chapman, H.M., 2005a. Muscle glycogen concentrations in commercial consignments of Australian lamb measured on farm and post-slaughter after three different lairage periods. Australian Journal of Experimental Agriculture 45: 543-552.

Jacob, R.H., Pethick, D.W., Clark, P., D'Souza, D.N., Hopkins, D.L. and White, J., 2006a. Quantifying the hydration status of lambs in relation to carcass characteristics. Australian Journal of Experimental Agriculture 46: 429-437.

Jacob, R.H., Pethick, D.W., Ponnampalam, E., Speijers, J. and Hopkins, D.L., 2006b. The hydration status of lambs after lairage at two Australian abattoirs. Australian Journal of Experimental Agriculture 46: 909-912.

Jacob, R.H., Walker, P.J., Skerritt, J.W., Davidson, R.H., Hopkins, D.L., Thompson, J.M. and Pethick, D.W., 2005b. The effect of lairage time on consumer sensory scores of the M. longissimus thoracis et lumborum from lambs and lactating ewes. Australian Journal of Experimental Agriculture 45: 535-542.

Jarvis, A.M. and Cockram, M.S., 1994. Effects of handling and transport on bruising of sheep sent directly from farms to slaughter. Veterinary Record 135: 523-527.

Jarvis, A.M. and Cockram, M.S., 1995a. Handling of sheep at markets and the incidence of bruising. Veterinary record 136: 582-585.

Jarvis, A.M. and Cockram, M.S., 1995b. Some factors affecting resting behaviour of sheep in slaughterhouse lairages after transport from farms. Animal Welfare 4: 53-60.

Jarvis, A.M., Cockram, M.S. and McGilp, I.M., 1996. Bruising and biochemical measures of stress, dehydration and injury determined at slaughter in sheep transported from farms or markets. British Veterinary Journal 152: 719-722.

Jones, T.A., Look, A., Guise, H.J. and Lomas, M.J., 2002. Head height requirements, and assessing stocking density, for sheep in transit. Veterinary Record 150: 49-50.

Jones, T.A., Waitt, C. and Dawkins, M.S., 2010. Sheep lose balance, slip and fall less when loosely packed in transit where they stand close to but not touching their neighbours. Applied Animal Behaviour Science 123: 16-23.

Jongman, E.C., Edge, M.K., Butler, K.L. and Cronin, G.M., 2008. Reduced space allowance for adult sheep in lairage for 24 hours limits lying behaviour but not drinking behaviour. Australian Journal of Experimental Agriculture 48: 1048-1051.

Kadim, I.T., Mahgoub, O., Al-Marzooqi, W., Khalaf, S., Al-Sinawi, S. and Al-Amri, I., 2009. Effects of transportation during the hot season and low voltage electrical stimulation on histochemical and meat quality characteristics of sheep longissimus muscle. Livestock Science 126: 154-161.

Kaler, J. and Green, L.E., 2009. Farmers' practices and factors associated with the prevalence of all lameness and lameness attributed to interdigital dermatitis and footrot in sheep flocks in England in 2004. Preventive Veterinary Medicine 92: 52-59.

Kaler, J., Wassink, G.J. and Green, L.E., 2009. The inter- and intra-observer reliability of a locomotion scoring scale for sheep. Veterinary Journal 180: 189-194.

Kavaliers, M., 1989. Evolutionary aspects of the neuro-modulation of nociceptive behaviors. American Zoology 29: 1345-1353.

Kettlewell, P.J., Hoxey, R.P., Hampson, C.J., Green, N.R., Veale, B.M. and Mitchell, M.A., 2001. Design and operation of a prototype mechanical ventilation system for livestock transport vehicles. Journal of Agricultural and Engineering Research 79: 429-439.

Kim, F.B., Jackson, R.E., Gordon, G.D.H. and Cockram, M.S., 1994. Resting behaviour of sheep in a slaughterhouse lairage. Applied Animal Behaviour Science 40: 45-54.

Kirton, A.H., Bishop, W.H. and Mullord, M.M., 1978. Relationship between time of stunning and time of throat cutting and their effect on blood pressure and blood splash in lambs. Meat Science 2: 199-206.

Kirton, A.H., Clarke, J.N. and Carter, A.H., 1967. Effect of pre-slaughter fasting on liveweight, carcass weight, and carcass composition of Southdown ram lambs. New Zealand Journal of Agricultural Research 10: 43-55.

Kirton, A.H., Quartermain, A.R., Uljee, A.E., Carter, W.A. and Pickering, F.S., 1968. Effect of 1 and 2 days' ante-mortem fasting on live weight and carcass losses in lambs. New Zealand Journal of Agricultural Research 11: 891-902.

Knowles, T.G., 1998. A review of the road transport of slaughter sheep. Veterinary Record 143: 212-219.

Knowles, T.G., Maunder, D.H., Warriss, P.D. and Jones, T.W., 1994. Factors affecting the mortality of lambs in transit to or in lairage at a slaughterhouse, and reasons for carcase condemnations. Veterinary Record 135: 109-111.

Knowles, T.G., Warriss, P.D., Brown, S.N., Kestin, S.C., Edwards, J.E., Perry, A.M., Watkins, P.E. and Phillips, A.J., 1996. Effects of feeding, watering and resting intervals on lambs transported by road and ferry to France. Veterinary Record 139: 335-339.

König, U., Nyman, A.J. and De Verdier, K., 2011. Prevalence of footrot in Swedish slaughter lambs. Acta Veterinaria Scandinavica 53: 27-31.

Kostro, K., Jarosz, L., Gruszecki, T., Junkuszew, A. and Lipecka, C., 2009. Utility of haptoglobin assay for sheep welfare and health status evaluation in pre- and postslaughter period. Bulletin of the Veterinary Institute in Pulawy 53: 111-116.

Krawczel, P.D., Friend, T.H., Caldwell, D.J., Archer, G. and Ameiss, K., 2007. Effects of continuous versus intermittent transport on plasma constituents and antibody response of lambs. Journal of Animal Science 85: 468-476.

Lambooij, B. and Algers, B., 2016. Mechanical stunning and killing methods. In: Velarde, A. and Raj, M. (eds.) Animal welfare at slaughter. 5M Publishing Ltd., Sheffield, UK, pp. 91-110.

Lambooij, B. and Kijlstra, A. 2008. Ritueel slachten en het welzijn van herkauwers en pluimvee. Rapport 161, Animal Sciences Group van Wageningen UR, Lelystad, the Netherlands.

Lambton, S.L., Brouwer, A., Knowles, T.G., Richards, G., Statham, P.T.E., Walton, J. and Weeks, C.A., 2017. Factors affecting the ability of sheep to rest during time in markets in Great Britain. Animal Welfare 26: 83-93.

Lapworth, J.W., 2008. Engineering and design of vehicles for long distance road transport of livestock: the example of cattle transport in northern Australia. Veterinaria Italiana 44: 215-222.

Levinger, I.M., 1995. Shechita in the light of the year 2000. Critical view of the scientific aspects of methods of slaughter and shechita. Maskil L'David, Jerusalem, Israel, 204 pp.

Ley, S.J., Livingston, A. and Waterman, A.E., 1989. The effect of chronic clinical pain on thermal and mechanical thresholds in sheep. Pain 39: 353-357.

Li, J., Wijffels, G., Yu, Y., Nielsen, L.K., Niemeyer, D.O., Fisher, A.D., Ferguson, D.M. and Schirra, H.J., 2011. Altered fatty acid metabolism in long duration road transport: an NMR-based metabonomics study in sheep. Journal of Proteome Research 10: 1073-1087.

Li, X., Xia, A., Chen, L., Du, M., Chen, L., Kang, N. and Zhang, D., 2018. Effects of lairage after transport on post mortem muscle glycolysis, protein phosphorylation and lamb meat quality. Journal of Integrative Agriculture 17: 2336-2344.

Liste, G., Miranda-de la Lama, G.C., Campo, M.M., Villarroel, M., Muela, E. and Maria, G.A., 2011. Effect of lairage on lamb welfare and meat quality. Animal Production Science 51: 952-958.

Lovatt, F.M., 2010. Clinical examination of sheep. Small Ruminant Research 92: 72-77.

McCutchan, J.C. and Freeman, R.B., 1992. Failure of electrical prompting to improve sheep movement in single file races. Proceedings of the Australian Society of Animal Production 19: 455.

McLaren, A., McHugh, N., Lambe, N.R., Pabiou, T., Wall, E. and Boman, I.A., 2020. Factors affecting ewe longevity on sheep farms in three European countries. Small Ruminant Research 189: 106145. https://doi.org/10.1016/j.smallrumres.2020.106145

McLennan, K.M., Rebelo, C.J.B., Corke, M.J., Holmes, M.A., Leach, M.C. and Constantino-Casas, F., 2016. Development of a facial expression scale using footrot and mastitis as models of pain in sheep. Applied Animal Behaviour Science 176: 19-26.

Messori, S., Pedernera-Romano, C., Magnani, D., Rodriguez, P., Barnard, S., Dalmau, A., Velarde, A. and Dalla Villa, P., 2015a. Unloading or not unloading? Sheep welfare implication of rest stop at control post after a 29h transport. Small Ruminant Research 130: 221-228.

Messori, S., Pedernera-Romano, C., Rodriguez, P., Barnard, S., Giansante, D., Magnani, D., Dalmau, A., Velarde, A. and Dalla Villa, P., 2017. Effects of different rest-stop durations at control posts during a long journey on the welfare of sheep. Veterinaria Italiana 53: 121-129.

Messori, S., Sossidou, E., Buonanno, M., Mounaix, B., Barnard, S., Vousdouka, V., Dalla Villa, P., De Roest, K. and Spoolder, H., 2015b. A pilot study to develop an assessment tool for sheep welfare after long journey transport. Animal Welfare 24: 407-416.

Miranda-de la Lama, G. C., Rodríguez-Palomares, M., Cruz-Monterrosa, R., Rayas-Amor, A., Pinheiro, R.S.B., Galindo, F.M. and Villarroel, M., 2018. Long-distance transport of hair lambs: effect of location in pot-belly trailers on thermo-physiology, welfare and meat quality. Tropical Animal Health and Production 50: 327-336.

Miranda-de la Lama, G.C., Monge, P., Villarroel, M., Olleta, J.L., García-Belenguer, S. and María, G.A., 2011. Effects of road type during transport on lamb welfare and meat quality in dry hot climates. Tropical Animal Health and Production 43: 915-922.

Miranda-de la Lama, G.C., Rivero, L., Chacón, G., Garcia-Belenguer, S., Villarroel, M. and Maria, G.A., 2010a. Effect of the pre-slaughter logistic chain on some indicators of welfare in lambs. Livestock Science 128: 52-59.

Miranda-de la Lama, G.C., Villarroel, M. and María, G.A., 2012a. Behavioural and physiological profiles following exposure to novel environment and social mixing in lambs. Small Ruminant Research 103: 158-163.

Miranda-de la Lama, G.C., Villarroel, M., Campo, M.d.M., Olleta, J.L., Sañudo, C. and María, G. A., 2012b. Effects of double transport and season on sensorial aspects of lamb's meat quality in dry climates. Tropical Animal Health and Production 44: 21-27.

Miranda-de la Lama, G.C., Villarroel, M., Liste, G., Escós, J. and María, G.A., 2010b. Critical points in the pre-slaughter logistic chain of lambs in Spain that may compromise the animal's welfare. Small Ruminant Research 90: 174-178.

Miranda-de la Lama, G.C., Villarroel, M., Olleta, J.L., Alierta, S., Sañudo, C. and Maria, G.A., 2009. Effect of the pre-slaughter logistic chain on meat quality of lambs. Meat Science 83: 604-609.

Miranda-de la Lama, G.C., Salazar-Sotelo, M., Pérez-Linares, C., Figueroa-Saavedra, F., Villarroel, M., Sañudo, C. and Maria, G.A., 2012c. Effects of two transport systems on lamb welfare and meat quality. Meat Science 92: 554-561.

Mitchell, M.A. and Kettlewell, P.J., 2008. Engineering and design of vehicles for long distance road transport of livestock (ruminants, pigs and poultry). Veterinaria Italiana 44: 201-213.

Murray, K.C., Davies, D.H., Cullinane, S.L., Eddison, J.C. and Kirk, J.A., 2000. Taking lambs to the slaughter: marketing channels, journey structures and possible consequences for welfare. Animal Welfare 9: 111-122.

Najafi, M.H., Mohammadi, Y., Najafi, A., Shamsolahi, M. and Mohammadi, H., 2020. Lairage time effect on carcass traits, meat quality parameters and sensory properties of Mehraban fat-tailed lambs subjected to short distance transportation. Small Ruminant Research 188: 106122.

Norton, T., Kettlewell, P. and Mitchell, M., 2013. A computational analysis of a fully-stocked dual-mode ventilated livestock vehicle during ferry transportation. Computers and Electronics in Agriculture 93: 217-228.

Padalino, B., Tullio, D., Cannone, S. and Bozzo, G., 2018. Road transport of farm animals: Mortality, morbidity, species and country of origin at a southern Italian control post. Animals 8: 155. https://doi.org/10.3390/ani8090155

Parrott, R.F., Hall, S.J.G. and Lloyd, D.M., 1998. Heart rate and stress hormone responses of sheep to road transport following two different loading procedures. Animal Welfare 7: 257-267.

Parrott, R.F., Lloyd, D.M. and Brown, D., 1999. Transport stress and exercise hyperthermia recorded in sheep by radiotelemetry. Animal Welfare 8: 27-34.

Parrott, R.F., Misson, B.H. and De La Riva, C.F., 1994. Differential stressor effects on the concentrations of cortisol, prolactin and catecholamines in the blood of sheep. Research in Veterinary Science 56: 234-239.

Pascual-Alonso, M., Miranda-de la Lama, G.C., Aguayo-Ulloa, L., Villarroel, M., Mitchell, M. and María, G.A., 2017. Thermophysiological, haematological, biochemical and behavioural stress responses of sheep transported on road. Journal of Animal Physiology and Animal Nutrition 101: 541-551.

Pearce, K.L., Masters, D.G., Jacob, R.H., Hopkins, D.L. and Pethick, D.W., 2008. Effects of sodium chloride and betaine on hydration status of lambs at slaughter. Australian Journal of Experimental Agriculture 48: 1194-1200.

Petersen, G.V., 1978. Factors associated with wounds and bruises in lambs. New Zealand Veterinary Journal 26: 6-9.

Petersen, G.V., 1983. The effect of swimming lambs and subsequent resting periods on the ultimate pH of meat. Meat Science 9: 237-246.

Petersen, G.V., Carr, D.H., Davies, A.S. and Pickett, B.T., 1986. The effect of different methods of electrical stunning of lambs on blood pressure and muscular activity. Meat Science 16: 1-15.

Petherick, J.C. and Phillips, C.J.C., 2009. Space allowances for confined livestock and their determination from allometric principles. Applied Animal Behaviour Science 117: 1-12.

Pethick, D.W., Miller, C.B. and Harman, N.G., 1991. Exercise in Merino sheep – the relationships between work intensity, endurance, anaerobic threshold and glucose metabolism. Australian Journal of Agricultural Research 42: 599-620.

Pethick, D.W., Rowe, J.B. and Tudor, G., 1995. Glycogen metabolism and meat quality. Recent Advances in Animal Nutrition in Australia 7: 97-103.

Phillips, C.J.C., Pines, M.K., Latter, M., Muller, T., Petherick, J.C., Norman, S.T. and Gaughan, J.B., 2012. Physiological and behavioral responses of sheep to gaseous ammonia. Journal of Animal Science 90: 1562-1569.

Piccione, G., Casella, S., Giannetto, C., Giudice, E. and Fazio, F., 2012. Utility of acute phase proteins as biomarkers of transport stress in ewes. Small Ruminant Research 107: 167-171.

Pighin, D.G., Brown, W., Ferguson, D.M., Fisher, A.D. and Warner, R.D., 2014. Relationship between changes in core body temperature in lambs and post-slaughter muscle glycogen content and dark-cutting. Animal Production Science 54: 459-463.

Pighin, D.G., Cunzolo, S.A., Zimerman, M., Pazos, A.A., Domingo, E., Pordomingo, A.J. and Grigioni, G., 2013. Impact of adrenaline or cortisol injection on meat quality development of Merino hoggets. Journal of Integrative Agriculture 12: 1931-1936.

Pointon, A., Kiermeier, A. and Fegan, N., 2012. Review of the impact of pre-slaughter feed curfews of cattle, sheep and goats on food safety and carcase hygiene in Australia. Food Control 26: 313-321.

Ponnampalam, E.N., Hopkins, D.L., Bruce, H., Li, D., Baldi, G. and Bekhit, A.E., 2017. Causes and contributing factors to 'dark cutting' meat: current trends and future directions: a review. Comprehensive Reviews in Food Science and Food Safety 16: 400-430.

Randall, J.M. and Patel, R., 1994. Thermally induced ventilation of livestock transporters. Journal of Agricultural Engineering Research 57: 99-107.

Randall, J.M., 1993. Environmental parameters necessary to define comfort for pigs, cattle and sheep in livestock transporters. Animal Production 57: 299-307.

Rickenbacker, J.E., 1962. Losses from handling sheep and lambs. Farmer Cooperative Service, US Department of Agriculture, Washington, DC, USA.

Ridler, A.L. and West, D.M., 2010. Examination of teeth in sheep health management. Small Ruminant Research; Special Issue: Sheep Diagnostic Medicine 92: 92-95.

Rodríguez, A.I., Cózar, A., Calvo, L. and Vergara, H., 2017. Effect of bedding materials during transport on welfare indicators and microbiological quality in lambs. Animal Production Science 57: 1924-1930.

Ruiz-de-la-Torre, J.L. and Manteca, X., 1999. Behavioural effects of social mixing at different stocking densities in prepubertal lambs. Animal Welfare 8: 117-126.

Ruiz-de-la-Torre, J.L., Velarde, A., Diestre, A., Gispert, M., Hall, S.J.G., Broom, D.M. and Manteca, X., 2001. Effects of vehicle movements during transport on the stress responses and meat quality of sheep. Veterinary Record 148: 227-229.

Sanchez-Sanchez, M., Vieira-Aller, C., Fuente-Vazquez, J., Perez-Marcos, C., Lauzurica-Gomez, S., Gonzalez-de-Chavarri, E. and Diaz-Diaz-Chiron, M.T., 2013. Effect of season and stocking density during transport on carcass and meat quality of suckling lambs. Spanish Journal of Agricultural Research 11: 394-404.

Scott, P.R. and Gessert, M.E., 1998. Ultrasonographic examination of the ovine thorax. The Veterinary Journal 155: 305-310.

Shorthose, W.R. and Shaw, F.D., 1977. Plasma constituents of 'downer' sheep slaughtered at an abattoir. Australian Veterinary Journal 53: 330-333.

Silanikove, N., 2000. Effects of heat stress on the welfare of extensively managed domestic ruminants. Livestock Production Science 67: 1-18.

Stewart, S.M., McGilchrist, P., Gardner, G.E. and Pethick, D.W., 2018. Lamb loin tenderness is not associated with plasma indicators of pre-slaughter stress. Meat Science 137: 147-152.

Sugnaseelan, S., Prescott, N.B., Broom, D.M., Wathes, C.M. and Phillips, C.J.C., 2013. Visual discrimination learning and spatial acuity in sheep. Applied Animal Behaviour Science 147: 104-111.

Sutherland, M.A., Worth, G.M., Stuart, A.D., Dobbie, P.M. and Clerens, S., 2016. Effect of pre-slaughter handling, exercise and the presence of a dog on lamb welfare and meat quality. Animal 10: 1360-1367.

Syme, L.A. and Elphick, G.R., 1982. Heart-rate and the behaviour of sheep in yards. Applied Animal Ethology 9: 31-35.

Tanaka, T., Asakawa, K., Kawahara, Y., Tanida, H. and Yoshimoto, T., 1989. Color discrimination in sheep. Japanese Journal of Livestock Management 24: 89-94.

Teke, B., Ekiz, B., Akdag, F., Ugurlu, M., Ciftci, G. and Senturk, B., 2014. Effects of stocking density of lambs on biochemical stress parameters and meat quality related to commercial transportation. Annals of Animal Science 14: 611-621.

Ternouth, J.H., 1967. Post-prandial ionic and water exchange in the rumen. Research in Veterinary Science 8: 283-293.

Thompson, J.M., O'Halloran, W.J., McNeill, D., Jackson-Hope, N.J. and May, T.J., 1987. The effect of fasting on liveweight and carcass characteristics in lambs. Meat Science 20: 293-309.

Thurley, D.C. and McNatty, K.P., 1973. Factors affecting peripheral cortisol levels in unrestricted ewes. European Journal of Endocrinology 74: 331-337.

Van Metre, D.C., Barkey, D.Q., Salman, M.D. and Morley, P.S., 2009. Development of a syndromic surveillance system for detection of disease among livestock entering an auction market. Journal of the American Veterinary Medical Association 234: 658-664.

Velarde, A, Rodríguez, P., Dalmau, A., Fuentes, C., Llonch, P., von Holleben, K.V., Anil, M.H., Malbooij, J.B., Pleiter, H., Yesildere, T. and Cenci-Coga B.T., 2014. Religious slaughter: evaluation of current practices in selected countries. Meat Science 96: 278-287.

Velarde, A., Gispert, M., Diestre, A. and Manteca, X., 2003. Effect of electrical stunning on meat and carcass quality in lambs. Meat Science 63: 35-38.

Velarde, A., Rodriguez, P., Fuentes, C., Llonch, P., von Holleben, K., von Wenzlawowicz, M., Anil, H., Miele, M., Cenci Goga, B., Lambooij, B., Zivotofsky, A., Gregory, N., Bergeaud-Blackler, F. and Dalmau, A., 2010. Dialrel project – improving animal welfare during religious slaughter: recommendations for good parctice. Available at: https://tinyurl.com/5e3d5dxn.

Velarde, A., Ruiz-de-la-Torre, L., Stub, C., Diestre, A. and Manteca, X., 2000. Factors affecting the effectiveness of head-only electrical stunning in sheep. Veterinary Record 147: 40-43.

Velarde, A., Ruiz-De-La-Torre, J. L., Rosello, C. Fàbrega, E, Diestre A. and Manteca X., 2002. Assessment of unconsciousness after electrical stunning in sheep. Animal Welfare 11: 333-341.

Vergara, H., Cózar, A., Rodríguez, A.I. and Calvo, L., 2017. Effect of space allowance during transport and fasting or non-fasting during lairage on carcass contamination and meat traits in Merino lamb. Spanish Journal of Agricultural Research 15: e0503. https//doi.org/10.5424/sjar/2017152-10227

Vergara, H., Linares, M.B., Berruga, M.I. and Gallego, L., 2005. Meat quality in suckling lambs: effect of pre-slaughter handling. Meat Science 69: 473-478.

Von Holleben, K., von Wenzlawowicz, M., Gregory, N., Anil, H., Velarde, A., Rodríguez, P., Cenci-Goga, B.T., Catanese, B. and Lambooij, E., 2010. Dialrel project – Report on good and adverse practices – animal welfare concerns in relation to slaughter practices from the viewpoint of veterinary sciences. Available at: https://www.dialrel.net/dialrel/images/veterinary-concerns.pdf.

Von Mickwitz, G., Heer, A., Demmler, T., Rehder, H. and Seidler, M., 1989. Slaughter of cattle, swine and sheep according to the regulations on animal welfare and disease control using an electric stunning facility. Deutsche Tierärztliche Wochenschrift 96: 127-133.

Walk, R.D. and Gibson, E.J., 1961. A comparative and analytical study of visual depth perception. Psychological Monographs: General and Applied 75: 1-44.

Warner, R.D., Ferguson, D.M., McDonagh, M.B., Channon, H.A., Cottrell, J.J. and Dunshea, F.R., 2005. Acute exercise stress and electrical stimulation influence the consumer perception of sheep meat eating quality and objective quality traits. Australian Journal of Experimental Agriculture 45: 553-560.

Warriss, P.D., Bevis, E.A., Brown, S.N. and Ashby, J.G., 1989. An examination of potential indices of fasting time in commercially slaughtered sheep. British Veterinary Journal 145: 242-248.

Warriss, P.D., Brown, S.N. and Knowles, T.G., 2003. Assessment of possible methods for estimating the stocking density of sheep being carried on commercial vehicles. Veterinary Record 153: 315-319.

Warriss, P.D., Kestin, S.C., Young, C.S., Bevis, E.A. and Brown, S.N., 1990. Effect of preslaughter transport on carcass yield and indices of meat quality in sheep. Journal of the Science of Food and Agriculture 51: 517-523.

Weeks, C.A., 2008. A review of welfare in cattle, sheep and pig lairages, with emphasis on stocking rates, ventilation and noise. Animal Welfare 17: 275-284.

Wickham, S.L., Collins, T., Barnes, A.L., Miller, D.W., Beatty, D.T., Stockman, C., Blache, D., Wemelsfelder, F. and Fleming, P.A., 2012. Qualitative behavioral assessment of transport-naïve and transport-habituated sheep. Journal of Animal Science 90: 4523-4535.

Winter, A.C., 2008. Lameness in sheep. Small Ruminant Research 76: 149-153.

Woolf, C.J., 2004. Pain: moving from symptom control toward mechanism-specific pharmacologic management. Annals of Internal Medicine 140: 441-451.

Wroth, R.H. and Suiter, R.J., 1978. Rejections of sheep and lamb carcasses are costly. Journal of the Department of Agriculture 19: 10-15.

Yardimci, M., Sahin, E.H., Cetingul, I.S., Bayram, I., Aslan, R. and Sengor, E., 2013. Stress responses to comparative handling procedures in sheep. Animal 7: 143-150.

Young, O.A., Hopkins, D.L. and Pethick, D.W., 2005. Critical control points for meat quality in the Australian sheep meat supply chain. Australian Journal of Experimental Agriculture 45: 593-601.

Zimerman, M., Domingo, E., Grigioni, G., Taddeo, H. and Willems, P., 2013. The effect of pre-slaughter stressors on physiological indicators and meat quality traits on Merino lambs. Small Ruminant Research 111: 6-9.

Photos: Louw Hoffman, The University of Queensland, Australia

Chapter 8

Species destined for non-traditional meat production: 1. African game species, cervids, ostriches, crocodiles and kangaroos

T. Needham[1] and L.C. Hoffman[2]*

[1]Department of Animal Science and Food Processing, Faculty of Tropical AgriSciences, Czech University of Life Sciences Prague, Kamýcká 129, 16500 Prague-Suchdol, Czech Republic; [2]Center for Nutrition and Food Sciences, Queensland Alliance for Agriculture and Food Innovation (QAAFI), The University of Queensland, Health and Food Sciences Precinct, 39 Kessels Rd, Coopers Plains 4108, Australia; louwrens.hoffman@uq.edu.au

Abstract

An overview of the preslaughter handling of various animals that are not conventionally considered as traditional species for meat production is described. The practical handling of species discussed include those derived from both wild harvesting, using hunting and culling, or farming, with on-farm or off-farm slaughtering. Particular emphasis is placed on the correct selection, handling, and processing of these animals within their unique production circumstances, to ensure that high welfare standards, meat hygiene, and food safety requirements are maintained. More specifically, this chapter focuses on non-traditional species of which their meat enters the commercial market through official and legislated channels, including cervids, antelope, ostriches, kangaroos, and crocodiles. Of the species included, only the ostrich follows typical livestock preslaughter handling procedures (herding, loading, transportation, and abattoir lairage), due to the substantial development of the ostrich farming industry within South Africa, initially driven by feather production for the fashion industry, with meat as a by-product until recently. Similarly, crocodiles are harvested for skin production, with meat as a by-product; however, these operations are typically performed on-site to minimise handling. While deer farming has increased, most deer are still slaughtered on-site at the farm, and the carcasses are transported to further processing units. Typically, only deer that are habituated to handling are transported to an abattoir for slaughter, to minimise unnecessary stress and losses of unhabituated animals. On the other hand, African game species and Australian kangaroos are harvested within their extensive systems or natural environment, using free bullet shooting (day or night-culling) without preslaughter handling. Recently, with the intensification of game farming in southern Africa, capture methods may be used preslaughter to allow for more efficient slaughtering (using free bullet) within enclosures, as well as rapidly harvesting utilising a helicopter in difficult terrain, or if the game carcass price warrants this approach.

Keywords: culling, hunting, game ranching, venison farming, wildlife

8.1 Introduction

Non-traditional meat production broadly encompasses the meat which originates from animals which do not fall within the typical classification of beef, lamb, pork, or poultry. Non-traditional meat may originate from a range of species, such as antelope, cervids, reptiles, ratites, marsupials, insects, goats, camelids and rodents. Together with the topic of non-traditional species and their meat, some jargon needs to be discussed and clarified first. Firstly, the terms 'game species/meat' and 'venison species/meat' are often used interchangeably in literature, and even for the commercial marketing of non-traditional meat products in some countries. In some cases, game is often used to describe wildlife species which are wild and free-ranging, while venison describes farmed wildlife species (Kudrnáčová *et al.*, 2018). However, venison is more appropriately used to describe the meat from intensively farmed cervid species, specifically for meat production, and often involves domesticated animals (Hoffman and Wiklund, 2006). African wildlife species used for game meat production are still considered as being wild and free-ranging, due to the current nature of management, which is described as 'game ranching' in southern Africa. Often, the term 'bush meat' is also used when describing non-traditional meat production and defines any terrestrial wild animal that is killed for subsistence or commercial purposes (Milner-Gulland and Bennett, 2003), but the terminology is typically associated with the illegal hunting of these animals and is thus not appropriate for this chapter. Bush meat, and the general overview of the wide array of wildlife species consumed in Africa, is succinctly reviewed by Hoffman and Cawthorn (2012), together with their various meat nutritional values.

Game meat and venison is considered to be a healthy alternative to meat from traditional species, due to its low fat content (<3%) and favourable fatty acid profile, with the polyunsaturated fatty acids:saturated fatty acids ratio above the recommended value of 0.4, and low n6:n3 ratios (Hoffman and Wiklund, 2006). Furthermore, the production of species, such as antelope and deer, in extensive grazing systems without intensive human intervention through handling, feeding, pharmaceutical application enhances this public opinion (Kuba *et al.*, 2015). Thus, game and venison meat are often considered to be natural, organic, and safe, in comparison to intensively farmed traditional livestock. Currently, this is particularly important for consumers, especially those in developed countries where the carbon footprint, sustainability, animal welfare, and meat residue content are considered when deciding which food products to purchase (Van Schalkwyk and Hoffman, 2016). Therefore, it is important to consider these factors when meat from non-traditional species is to be harvested and integrated into the commercial market. For this reason, maintaining sustainable harvesting and selection of these species, upholding high standards of animal welfare during preslaughter processing, and ensuring a high level of hygiene throughout, are essential.

While non-traditional animal species are usually considered as being 'wild' or 'undomesticated', this is not necessarily true when the definition of 'domestication' is analysed. Domestication may be defined simply as 'the process of hereditary reorganisation of wild animals and plants into domestic and cultivated forms according to the interests of people' (Encyclopedia Britannica, 2020), and thus many species which may be considered as non-traditional species cannot be deemed as undomesticated. For instance, the management of many deer and antelope species involves their enclosure into fenced areas, the selection of breeding animals into breeding herds, and the selection of culled animals for consumption, as well as the provision of supplementary feeding in some instances. In fact, deer were domesticated in China centuries ago (Chakanya *et al.*, 2016). This process has also enabled more intensive handling of non-traditional meat species, together with the implementation of slaughtering protocols. Under extensive production systems utilised for non-traditional meat species (often termed 'free-ranging' for those still maintained within some boundary fencing and 'free-roaming' or 'wild' for those not maintained within some boundary), animals are not handled intensively, or at all, and thus they are usually slaughtered through hunting, harvesting, or cropping. Typically, hunting implies a 'fair chase' situation, where an individual pursues the animal on foot, and may be accompanied by a few/no other people and may utilise dogs, or observes the animal from hides. This hunting can be done using firearms or bow-hunting. Usually only a few animals are targeted, for trophy purposes or self-consumption. During trophy hunting, headshots are not used, and thus significant damage to the carcass, and thus meat, results from the body shot placements (Van Schalkwyk and Hoffman, 2016). Typically, the meat generated from trophy hunting does not enter the commercial system, but rather serves as a source of food for local people. However, harvesting, cropping, or culling involves killing a large number of animals for the purpose of commercial meat production. Harvesting generally implies no selectivity when killing the animals, whereas cropping involves a degree of selection of the animal prior to the animal being removed from the herd (Van Schalkwyk and Hoffman, 2016).

To maintain the sustainable culling of free-ranging populations of non-traditional meat species, a number of universal principles should be considered when selecting these animals for slaughter, regardless of the slaughtering protocol. Firstly, an ideal system should allow for population control, while maintaining population structure and not disrupting population growth (Van Schalkwyk and Hoffman, 2016). Typically, this involves a greater selection pressure on males than females, leaving pregnant, lactating, and future breeding females in the herd, as only females who are not reproducing or show poor breeding potential are removed. This selection pressure is further exacerbated when trophy hunting is incorporated into the economic activities of the ranching enterprise, as often seen in African game species. The sustainable harvesting of these animals depends on the season, to allow for greater nutritional availability for the breeding herd during seasons of poor nutritional quality, and not to interrupt mating and birthing seasons. Selection should consider the natural population and social structure of the species, to ensure optimal welfare conditions. Furthermore, intensive selection pressure of a particular species should be avoided, particularly in game ranching systems where multiple species are maintained together, to ensure the ecological balance is conserved.

While keeping the aforementioned in mind, harvesting methods should also uphold high ethical standards. Depending on the species and the current production systems used to manage these species, various slaughter and preslaughter handling practises are used. On large areas of land, animals may be shot from vehicles or helicopters, but they may also be driven into a smaller enclosed area for either free bullet killing, or slaughter by exsanguination after restraint and stunning. The current preslaughter handling methods for various non-traditional meat species which are incorporated into the commercial market will be discussed in detail within this chapter, focusing on popular free-ranging species, as well as more intensively farmed species.

8.2 Free-ranging African game species

Wildlife farming, or rather game ranching, for meat purposes in Africa is currently possible in countries which allow for ownership rights over wildlife, including South Africa, Namibia, and Zimbabwe. African game species are primarily maintained in extensive systems, typically on marginal lands, with minimal human intervention. However, recently in countries such as Namibia and South Africa, the private wildlife sector has grown rapidly. Within these private game ranches, diverse economic activities are practiced, including ecotourism, live sales, trophy hunting, meat hunting (referred to as 'biltong' hunting in southern Africa) and culling. In addition to these commercial activities, game ranches often provide meat for their employees and surrounding communities, aiding in local food security (Taylor *et al.*, 2015). They also employ up to twice the number of people per unit area than livestock farms, thereby providing employment opportunities for local communities as well (Taylor *et al.*, 2020).

While trophy hunting currently generates a substantial proportion of the income from these farms, culling of game species for commercial meat production is rather underutilised considering its potential. The reasons for this being the different management objectives of the game ranches (i.e. trophy hunting), a high cost of compliance to current meat hygiene legislations, and a higher effort to harvest wild species compared to livestock species (Hoffman *et al.*, 2004; Taylor *et al.*, 2020). Because these game species are maintained on large areas of land, with little to no human interaction, and are not habituated to any form of handling or manipulation, they are flighty and thus cannot be handled in the same fashion as traditional livestock species. Therefore, when harvesting or culling is performed for meat production purposes, the animals are typically slaughtered on-farm and using free bullet killing (or shooting), without stunning. There are multiple protocols that can be followed for the shooting of game species, from walk-and-stalk hunting, to in-field harvesting from vehicles or helicopters, as well as herding and shooting within 'bomas'. Bomas are small pen-like structures, built to contain wild animals while allowing minimal movement. The chosen method depends on the type of species being culled, the terrain present, and the abattoir facilities available in the vicinity (Hoffman and Laubscher, 2010).

African game species can be divided into three categories: large game (i.e. elephant, hippopotamus, giraffe, and buffalo), medium game (i.e. wildebeest, kudu, eland, gemsbok, hartebeest, and zebra), and small game (i.e. impala, springbok, blesbok, and duiker). Currently, the meat from these large game species does not enter the commercial meat industry, as these species are currently hunted mostly for trophy hunting purposes, or for population control in areas where there is no infrastructure (abattoirs) available for the legal processing of the carcasses.

Medium and small game species are utilised regularly for meat production and includes numerous antelope species. Popular species for game meat production (including export) includes common eland (*Taurotragus oryx*), gemsbok (*Oryx gazella*), kudu (*Tragelaphus strepsiceros*), Hartmann's zebra (*Equus hartmannae*), springbok (*Antidorcas marsupialis*), impala (*Aepyceros melampus*), and red hartebeest (*Alcelaphus buchelaphus*) and blesbok (*Damaliscus pygargus phillipsi*) (Hoffman and Wiklund, 2006; Van Schalkwyk and Hoffman, 2016). The reason for the popularity of these species being their population numbers, fecundity, accessibility of their production regions and their proximity to processing facilities (Van Schalkwyk and Hoffman, 2016). A range of preslaughter handling options are available, but they are chosen based on the terrain and options available to the game rancher. Whilst it is possible to capture and load antelope onto a truck, transportation is utilised only for translocation of animals and not currently for slaughtering purposes. Considering the extremely nervous and stressed behaviour of antelope species unhabituated to human contact, loading and transportation without tranquilisers is not recommended (Bothma, 1996).

However, common eland antelope can be handled and slaughtered in a similar fashion to deer, after habituation to the handling system. Wirtu *et al.* (2005) habituated female common eland within zoological gardens to restraint within a hydrolytic crush, using bananas as positive reinforcement, and Musa *et al.* (2021) subsequently described the use of facilities similar to that of deer, designed for handling common eland. The system is modified to accommodate for the heavier eland and allows for the separation of animals before entering the squeeze chute (Figure 8.1).

The common eland in this study were kept within a cattle feedlot barn and selected for temperament for ~10 years. Habituation to the system was performed by allowing the animals to freely exit and enter the barn into the paddocked area, followed by a habituation program that stepped-up handling activities, while monitoring temperament. The eland were lured from the barn using concentrated feed, directly into the raceway system, and offered carrots in the squeeze chute during training. Musa *et al.* (2021) showed that common eland temperament improved after repeated routine handling (fortnightly for four months). Following these techniques, Needham *et al.* (2020a) were able to slaughter 18 common eland within the squeeze restraint system, using a captive bolt immediately as the animal was secured. The restraint door was opened, and the animal was exsanguinated using a thoracic stick. The animal was suspended by its heel during the bleeding process, using a rope attached to a digger-loader. This system allowed for immediate death and effective bleeding. No carcass bruising or injuries were recorded, and no indication of stress based on the evaluation of meat quality traits as assessed in the *Longissimus* and *Biceps femoris* muscles (i.e. pH, colour, and tenderness) was found. It is advisable that a bucket be used to collect the blood of the animals during exsanguination in these on-farm systems to not deter the other animals from entering the system,

Figure 8.1. The squeeze chute system used by Musa *et al.* (unpublished data) and Needham *et al.* (2020a) for the handling and slaughtering of common eland antelope after habituation (photo courtesy of A. Musa, Czech University of Life Sciences, Czech Republic).

and to maintain hygiene (T. Needham, personal observation). However, due to the nature of the current production systems used for game species, slaughtering in the field (day or night culling), or within a boma after mass capture, are currently utilised within the game ranching industry for the mass culling of game for meat purposes.

8.2.1 Day and night culling

Similar to sport hunting, culling may only be performed by qualified and competent shooters/marksmen. Animals culled during the day are usually shot from vehicles, helicopters, or within bomas. Head or upper neck shot placements should be used, with the intention of instantaneous death. Care must be taken to ensure the shot is high enough on the neck, so that the animal is not just paralysed, but rendered unconscious. Also, it must be ensured that the jaw of the animal is not shot, as this neither kills them nor renders them unconscious, and thus the heavily wounded animal flees. A jaw shot animal will lead a slow agonising death, as it will be unable to drink water or graze/browse (L. Hoffman, personal observation). Lung, heart, or body shots result in significant carcass losses (Figure 8.2). For example, 3, 20 and 50% of the carcass may be lost due to neck, shoulder or hindquarter shot placements, respectively (Bothma, 1996). Thoracic shot placement also leads to excessive bleeding and thus carcass contamination, as well as the attraction of flies onto the large wound. Abdominal shot placement should be avoided, to prevent carcass contamination, and to

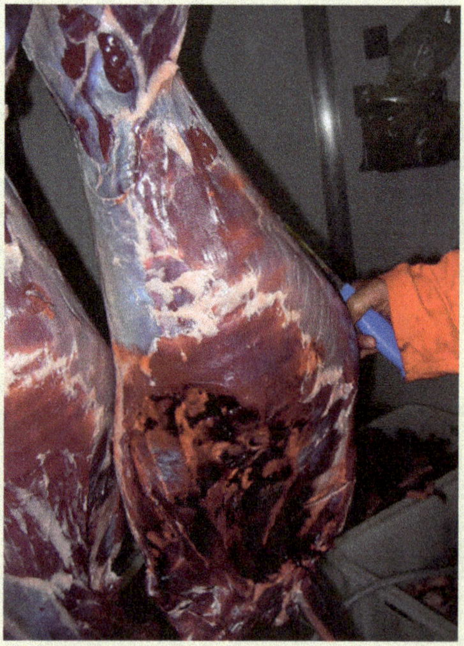

Figure 8.2. A carcass showing significant damage and loss as a result of thoracic shot placement (L. Hoffman, University of Queensland, Australia).

ensure the meat meets export requirements. Both these sources of carcass contamination and presence of flies under hot environmental conditions pose serious concerns for meat hygiene and safety.

Typically, for culling purposes, custom made vehicles are used. These are designed in a specialised manner for the driver, who is also the marksman, to have a rifle rest so as to ensure easy placement of the rifle for an accurate shot. The rifle will also have a sound suppressor as this not only deadens the sound (and thus not frighten and stress the herd), but also lessens the recoil onto the marksman's shoulder – it is not uncommon for a marksman to shoot >100 animals per day/night and excessive rifle recoil could result in the marksman flinching and missing or wounding the animals (L. Hoffman, personal observation). The culling vehicle should also have either hooks on the side for hanging the smaller carcasses or a sloped deck for larger carcasses with the necessary system to load the carcasses rapidly. There should also be standing room for at least two additional crew who would typically look for game, handle the spotlights (during night culling), exsanguinate and load the carcasses (L. Hoffman, personal observation).

While day culling allows for better visibility and shooting from further distances (up to 300 m), the herding or chasing of game during the day can result in stress and exhaustion, as animals are highly active and aware during daylight hours and environmental temperatures are generally higher, leading to various ethical and meat quality issues. Thus, night culling is preferred for the commercial harvest of many game species (Hoffman and Laubscher, 2009a, 2010, 2011). To ensure its success, night culling should be performed during moonless nights, from a vehicle. Spotlights are required (halogen, 1 million candle power), to blind the animals, and a telescopic sight should be used to identify age and sex (Conroy, 2005).

Repeated, and intensive shooting of individuals within the same herd results in the remaining animals becoming nervous, causing them to flee more readily when they see or hear vehicles approaching (Hoffman and Laubscher, 2009b). The use of two vehicles (during day-time culling), approaching at opposite ends of a herd can, however, aid in improving efficiency and decreasing harvesting stress. The use of helicopters also improves culling efficiency, and wounded animals are easily traced, but to make the operation financially feasible, a large number of game animals should be processed. Vehicles should be used to locate the shot animals as fast as possible, to exsanguinate them and take the carcass for processing as soon as possible (L. Hoffman, personal observation).

Carcasses should be bled within 10 min of being shot, but complete bleeding is seldom achieved, as the heart stops functioning after shooting – hanging the carcasses by their Achilles heel (smaller carcasses) or on a slope facilitates bleeding. It is important to ensure that knives are cleaned between animals, either with water (>82 °C) or with an approved chemical steriliser (e.g. 10 ppm free chlorine solution: Van Schalkwyk and Hoffman, 2016). Larger carcasses are eviscerated in the field with the brown offal being left in the field for scavengers to consume, whilst lighter carcasses are suspended with their viscera intact, or the latter can be removed if the field depot/abattoir is a long drive away from the killing site. The red offal/pluck is retained in the carcass for health inspection by a trained health officer or veterinarian (L. Hoffman, personal observation).

8.2.2 Mass game capture for boma culling

Various methods of game capture are available, from individual animal methods (hand capture, net guns, leg-hold traps, and cage traps) to mass game capture, and are classified as either physical or chemical capture. The chosen method depends on a number of factors, including species, tameness, condition, behaviour and herd structure, the safety and training of handlers, and the environment (O'Dell, 2016). With regards to animals intended for slaughter, chemical capture is not permitted, as chemical residues from the drugs used for these purposes remain within the meat. Thus, physical capture is preferred, primarily via mass capture in a boma, in which the animals are slaughtered. Habituated animals are easier to capture, as they can be approached closer than animals unhabituated to humans, but mature males remain dangerous in many large antelope species (and deer) and should be managed with caution (Ebedes *et al.*, 1996).

The plastic boma capture method (Oelofse method) is considered the best method for the mass capture of game species, with its primary advantage being its time-efficiency, but it lacks selectivity and can be expensive (Ebedes *et al.*, 1996). Thus, if a few individuals are to be selected, they are generally shot within the herd, using vehicle or helicopter culling. A variety of species can be successfully handled in this way, including antelope, giraffe, buffalo, ostriches, and warthogs (Ebedes *et al.*, 1996). The plastic boma consists of a large durable-plastic funnel, supported by cables and nets (Figure 8.3).

The boma sides should be secured to the ground for 'digging' species, and high enough for jumping species (impala, eland, kudu). As the animals are driven towards the boma, using vehicles or helicopters, a large funnel with plastic sides helps guide the animals into the boma, and a series of curtains are closed across the boma behind the herded animals, by people on foot, with the exception of dangerous species, where they are driven from pulling ropes outside of the boma (Ebedes *et al.*, 1996). Adult breeding males should be cut out of the herding drive, when possible, as they can injure other animals. It is not advised to make use of cattle prods to move the animals, as antelope tend to kick. When herding the animals towards the funnel, helicopters are generally the most successful, but

Figure 8.3. Entrance of a plastic boma game capture system, showing the funnel shape and curtains retracted, and ready to close behind the animals (L. Hoffman, University of Queensland, Australia).

alternatives, such as on horseback or motorbikes are possible as well (Ebedes *et al.*, 1996). The final section of the funnel is a solid wall crush, generally made from either metal, wood, or a conveyor belt. Here, animals can be loaded into a vehicle, or into smaller camps. It is essential that the first attempt at capturing the herd is successful, as recapture success is diminished.

Erecting the boma should take place either early morning on the day of capture, or late on the evening prior, as animals may become cautious and avoid it (Ebedes *et al.*, 1996). The site should be away from water sources and feeding areas, and should have a fair amount of vegetative cover, ideally close to a road if the animals are to be transported. The vegetation should, however, not be too dense (Ebedes *et al.*, 1996), such that that herding the animals become difficult, but it should also allow for a degree of camouflage. While large trees should be avoided at the entrance of the funnel, they are suitable for anchoring the cables supporting the walls. Rough terrain is generally not suitable for mass game capture (Ebedes *et al.*, 1996). The entrance of the funnel should be in the direction such that the wind is blowing into the funnel during capture, ideally, as this has a significant impact on its ultimate success. Should helicopters be used for herding, the chosen site should avoid power lines, telephone poles, perimeter fences and tall trees. Herding should be slow, to avoid scattering and exhaustion, e.g. 8 to 12 km per hour for common eland (du Toit, 2005), and should allow for rest and regrouping if necessary. Using this system of active capture, particularly with helicopters, is highly expensive and requires professional teams. However, it allows for shooting from the ground, in a small area, allowing for short harvest times and easy and quick accessibility to refrigeration units. Furthermore, wounded animals can be quickly attended to (L. Hoffman, personal communication).

It is also possible to capture large groups of game using passive capture, where temporary plastic bomas (for small to medium game) can be constructed close to a frequented area, and the animals are lured gradually into the boma with food. The entrances should be 8 to 12 m wide, and there should be at least two of them (Furstenburg, 2005). Once the selected animals have entered the system, pop-up, drop-down, or swing gates should be used to close them inside the boma. Generally, the use of a camera allows this action to be performed from a distance remotely, so that the entrance of the animals into the boma is not disturbed (Ebedes *et al.*, 1996). Such bomas can also be constructed in a way that they 'pop-up' or 'drop-down' around feeding or drinking sites. Pop-up bomas are constructed by rolling the plastic sheeting up (and submerging them into trenches, covered with some soil in some cases) and releasing them by pulling the sheeting up rapidly once the animals have entered the area (Ebedes *et al.*, 1996). The drop-down bomas function in the opposite way, where sheeting is rolled and suspended by poles and cables, and released rapidly downwards once the animals enter, whereafter it is secured to the ground. The preferred method is influenced by the target species, with pop-up bomas being more suitable for digging species (e.g. wildebeest), and drop-down for jumping species (e.g. kudu). These plastic structures should not remain erected for extended periods, as the components may weather, thereby increasing risk of failure in future captures.

More permanent structures for game capture should be constructed from wood, steel, and conveyor belt rubber, with typical dimensions of 3 m in height, and 30 m wide × 30 m long (Ebedes *et al.*, 1996). The pen should avoid sharp corners, so that submission animals do not get pinned by more dominant ones. Shelter can also provide areas for subordinate animals to escape, rest and recuperate. Providing a heaped structure, like branches, for more dominant males to redirect their attention towards can also prove to be beneficial in this regard (Ebedes *et al.*, 1996). Slaughter animals can be kept within these structures, to either rest and recover prior to slaughtering, or provide some supplementary feeding for 'finishing'. Feed and water should thus be provided, in the form of a total mixed ration for game species, together with the provision of a roughage source. Food should be provided at all feeding points simultaneously, to avoid excessive competition. While the game industry is intensifying in terms of its production system management, to include the provision of supplementary feeding and management in smaller camps, the effects of boma-finishing on the stress, yield and meat quality of game species remain poorly described. Initial investigations into an intensive boma-finishing of impala fed a total mixed ration for six months indicated that investigation into optimising the management of boma-finishing practises still needs to be performed (Needham *et al.*, 2020b). However, passive capture eliminates herding, and thus has the potential to produce meat of higher quality.

Once the game species have been captured in one of the boma systems, they should be allowed to rest prior to slaughter. This rest period is necessary for recovery purposes, restoring muscle glycogen, allowing for sufficient glycogen for post-mortem glycolysis and thus pH decline, thereby ensuring that meat quality is maintained. Typically, the game are rested for an hour or longer, with overnight resting occurring when the animals are captured in the late afternoon. However, no research to date has been conducted on the effect of rest period on stress levels and meat quality for African ungulates. If the animals are left overnight, depending on species (e.g. impala), males might have to be removed and separated to stop them killing each other (La Grange, 2006). After recovery, the game species may be accurately and carefully selected, to ensure sustainable culling, and moved into a small boma attached to the larger catching boma and shot from outside of the boma using a light calibre rifle with a sound suppressor, thereby minimising human interaction (L. Hoffman, personal observation; Figure 8.4). Shooting can thus be done within close proximity, minimising shot misplacement and thus wounding of animals. Furthermore, headshots can be more accurately

Figure 8.4. Blue wildebeest within a boma system, in the 'deaths pen', prior to shooting. Animals can easily be seen and shot from outside of the boma system (L. Hoffman, University of Queensland, Australia).

applied, and bleeding can be performed faster. The carcasses are normally removed from the small boma and transported a short distance away prior to exsanguination so as to minimise the smell of blood in the small killing boma.

8.2.3 Handling considerations for game capture and day-culling

The following basic guidelines should be considered when handling and harvesting game prior to slaughter (Ebedes *et al.*, 1996):

- Comprehensive planning of the entire operation, with the team of trained personnel involved, allocating clear duties to each person.
- The personnel involved in the operation should be highly trained and skilled, to minimise both injuries to animals and personnel.
- Schedule active captures during cooler periods (<25 °C) of the year (April-September in southern Africa), and day. Hot, humid, and rainy weather should be completely avoided, as animals slip on wet ground, increasing the risk of injuries.
- Avoid herding long distances and allow for rest and regrouping, if necessary.
- Avoid capturing or culling animals during critical periods of the reproductive cycle, and during periods of poor body condition. Harvesting animals during the breeding season can result in embryo resorption, while in the lambing/calving season, it can result in abortions, and separation of young from their mothers during the crucial and delicate phase of early lactation, resulting in losses of young animals.
- Family group structures should be respected and maintained as much as possible during capture, and aggressive animals should be separated, to avoid fighting and injury.
- Boma walls should be high enough to prevent jumping out (height depends on species, but typically 2-3 m), and secured on the ground to prevent digging under, which not only defeats the purpose of capture, but the escape process also predisposes the animals to injury that may result in mortality.
- Food and water should be placed into the bomas prior to the animals entering if the animals are to stay for longer (>1 day) periods before the culling starts, to prevent disturbance immediately after capture. For the animals fed, there is no feed withdrawal period as is normally found in domesticated animals transported to an abattoir (Vercauteren *et al.*, 2011).
- Adequate space should be provided, to allow space for resting, and to minimise competition for resources. No data exists for potential stocking densities; the space per animal will depend on the species and whether they are herd animals or not, also whether they are from the same herd/group. As discussed, when two or more groups are herded together in bomas, it is advisable to remove the males to minimise fighting of alpha males.
- The boma environment should be as quiet as possible, and visitors should be forbidden.
- Protection should be provided by shelters, draining should be adequate to ensure a dry environment, and the bomas should be cleaned, thus preventing respiratory diseases and footrot.

8.2.4 Effect of stress during herding and culling, and its consequences for game meat production

Considering the natural instinct of wild animals to flee until it has escaped danger or exhausted itself, a number of factors should be considered during game capture and culling. This stress is exacerbated by (Ebedes *et al.*, 1996):

- close proximity to humans, vehicles, and helicopters;
- sudden and excessive muscular activity;
- injuries, from the terrain, infrastructure, or other animals;
- new surroundings and unfamiliar food within the boma.

Thus, the effect of the above parameters should be minimised as much as possible and eliminated before the 'breaking point' of the animal is reached (exhaustion and/or death). The consequences of exceeding this breaking point are heat stress, capture myopathy, DFD (dark, firm, dry) meat and mortality (Ebedes *et al.*, 1996). Thus, signs of heat stress should always be monitored, including excessive sweating, panting, and lying down with the head tucked towards the back of the body, on the ground. Capture myopathy (or overstraining disease/white muscle disease) is a result of overexertion, irreparably damaging their muscles and internal organs, and is characterised by lameness, exhaustion, depression, and death (Hattingh *et al.*, 1984). Symptoms of capture myopathy are similar to heat stress, but also include shivering, coffee-coloured urine, and muscle paralysis (Hattingh *et al.*, 1984). The cause for capture myopathy is the extreme accumulation of lactic acid in the blood, as a result of excessive muscle glycogen breakdown, together with myoglobin release from the muscles, causing kidney damage and/or heart failure. Capture myopathy is not limited to game capture itself, but it also occurs when animals are chased excessively during helicopter and vehicle culling. Red hartebeest, springbok, female impala, kudu, and blesbok appear to be more susceptible to capture myopathy (Ebedes *et al.*, 1996). The consequence of capture myopathy on the carcass can be seen in the form of lesions, haemorrhaging, and bruising in the muscles, as well as the heart (Ebedes *et al.*, 1996). Muscles are pale, and in chronic cases, connective tissue forms at the damaged muscle sites. Similar to wild boars, in warthogs, PSE (pale, soft, exudative) meat is also possible when the pigs are exposed to stress (Hoffman, 2001a,b).

8.3 Farmed cervids

New Zealand pioneered modern deer farming in the early 1970's, as the first country to legalise the practice. Nowadays, New Zealand, Australia, China, and Canada utilise deer farming systems, with growing interest in Europe and the United States (Kudrnáčová *et al.*, 2018). Within these countries, the most popular farmed deer species are fallow deer (*Dama dama*) and red deer (*Cervus elaphus*), whereas reindeer (*Rangifer tarandus*) are popular in Scandinavian countries, while wapiti (*Cervus canadensis*) and chital deer (*Axis axis*) are commonly farmed in North America. While African game species are used for trophy hunting, skin and meat production, deer are also utilised for these aforementioned purposes, and their teeth, hair, and antlers (including velvet antlers in some countries) are also used. Deer farming systems range from large, enclosed areas similar to wild conditions and vegetation (especially for reindeer) to intensive grazing systems comparable to commercial livestock, where total mixed rations may be provided. Although they are raised under these 'wilder' conditions, deer are not free-roaming and are normally harvested using similar culling methods as the African game species (shooting in the field and transporting carcass to the abattoir for processing). There are, however, some systems where the deer are either loaded and transported to abattoirs (commonly found in New Zealand) or the deer are killed and slaughtered in specially designed abattoirs on-farm (e.g. as used in the UK; see Hultgren, 2022 – Chapter 10 of this book). In wild reindeer herds, capture is also possible by roping and shooting with a captive bolt (Wiklund *et al.*, 2001). The in-field slaughter of deer by free bullet shooting (Figure 8.5), without disturbing their natural behaviour, is considered to be the method which maintains the highest standard of animal welfare, provided it is performed by a competent marksman (preferably the deer handler). Providing supplementary feed can aid in delivering headshots to a number of animals in a short period of time (Figure 8.6). This method eliminates the stress of handling, transport or lairage, and prevents losses from bruising, injury, and mortalities (Goddard, 2014).

Wild deer can also be captured for slaughter, using a series of progressively smaller camps, following the topography of the terrain, and being hidden within the surroundings as best possible, into which the deer are herded. The deer can be similarly passively captured into this system, using food, and closed in (usually at night). Transportation of wild game is not encouraged without the use of tranquilisers, which is not permitted for slaughter animals. Similar to African game species, these animals are susceptible to capture myopathy. Thus, they should rather be slaughtered in-field, after a rest period. As the in-field slaughtering of deer follows that of African game species, this section will focus on (semi) intensively managed farmed deer, which can either be slaughtered on-farm using routine handling facilities or transported live to be slaughtered off-farm.

Figure 8.5. Shooting antelope in the field, without handling or transport intervention (L. Hoffman, University of Queensland, Australia).

Figure 8.6. Supplementary feeding of deer, with a hide for shooting (T. Needham, Czech University of Life Sciences, Czech Republic).

Similar to African game species, deer should not be culled during important phases of their reproductive cycle. Deer are seasonal breeders, and thus their reproductive behaviour determines their management. They have a prominent rutting period, where males grow their antlers and become more aggressive and lose body condition as they direct their attention towards the females. After parturition, the hind becomes protective of her young, until weaning. Thereafter, yearlings are generally harvested from the herd for meat production, depending on farm management and nutrition. The age at slaughter is particularly important for male slaughter animals, as the meat quality of mature males (especially around rut) has a distinct and objectionable odour and can be considered rather tough. Similar principles to game capturing of African species, e.g. seasonal effects, ambient temperature, appropriate equipment, trained and experienced staff, etc., also apply to deer species, with similar consequences on meat quality.

8.3.1 Herding deer for capture, and on-farm slaughter

The herding, driving through a raceway or transportation of deer is appropriate for animals which are somewhat habituated to human contact, otherwise animal welfare may be compromised, and significant losses may be caused by this process. Losses range from carcass bruising, wounds (which result in carcass and skin trimming), and fractures, to mortalities from fighting (especially when handling yearling stags, or so-called 'spikers'; Goddard, 2014).

Whilst considered flighty animals, deer can be trained and habituated to human contact, as well as to handling (Ceacero *et al.*, 2014; Goddard, 2014). Selection has also focused on temperament as an important parameter (Goddard, 2014), and thus, deer are no longer considered entirely undomesticated (also in light of their intensive farming). Examples of positive selection for temperament are red deer and fallow deer that have become easier to handle. However, fallow deer are naturally more nervous than red deer during handling (Goddard, 2014).

More suitable handling facilities have been developed, thereby further assisting the ease of handling deer. The design of these facilities and the routine handling of various deer species therein are well-described in literature and should encourage natural movement of the animals through the system (Goddard, 2014). However, despite these advances, some deer species (e.g. roe deer; Goddard,

2014) still remain unsuitable to intensive management systems. Similar to some African antelope, deer can jump up to 2.5 m, and wild animals have a long flight distance, and are highly reactive to disturbances (Goddard, 2014). Depending on the distance of the disturbance, they may either bunch, or scatter and then regroup later. Isolation of an individual during herding results in the animal becoming highly agitated, often leading to injury while adopting escape practices. During handling, when deer cannot adopt their natural flight distance, they tend to use different strategies to cope (Goddard, 2014). Typically, they may try to escape under the fence (Clift *et al.*, 1985), if not constructed correctly, or start exhibiting agonistic behaviour towards one another. Teeth-grinding and salivation are signs of such agitation (Bartos, 1985; Lincoln *et al.*, 1970), of which handlers should be vigilant in order to prevent inter-animal fighting or attacks directed towards the handler. It is thus important to keep these natural instincts in mind when handling deer, to ensure that the handling is performed safely and efficiently.

The flight distance response of the herd is largely determined by the lead animal, and thus handlers should focus on its behaviour to improve efficiency of handling. This flight response can be used to calmly herd the animals in the direction of the handling system, by applying the correct balance of pressure when entering their flight zone. Simply walking slowly behind a group of deer is often enough pressure for them to move (Goddard, 2014). Should motorbikes or horseback riding be used to move the herd, deer should be habituated to their presence beforehand. Surrounding fencing should be high enough (≥2 m) to prevent jumping and should not allow the head of the deer to push through (Goddard, 2014). Gates should be located in the corner of the camps, preferably with sturdier fences surrounded them, and they should have deer-proof latches, that are simple to operate. In the handling pen, a space allowance of approximately 0.5 m^2/100 kg is recommended to allow deer to gather before moving them into a raceway system (Goddard, 2014). Providing positive reinforcement by food during handling (often together with an auditory cue) will make animals more willing to enter the facilities (Goddard, 2014), thereby making handling at final slaughter easier. Deer should be handled by someone who they are familiar/ habituated to, and they should be handled in a calm and gentle manner. This is not only more effective, but a handler who is experienced with deer handling understands their behaviour and can recognise signs of agitation, aggression, and excessive stress (e.g. lying down). It is important for the handler to wear the appropriate protective clothing (i.e. thick, or padded, and loose material, and headgear) and protective boards while driving the animals through the system, to protect them from antlers and hooves, as deer tend to rear and box, as well as kick out, in confined spaces (Goddard, 2014). For handler safety, in the United Kingdom and New Zealand, stags who are to be slaughtered off-farm need to have their antlers removed prior to transport (MAFF, 1989), which would require them either being darted previously for removal, or handled within these facilities for removal, some time prior to slaughter. Raising boards, or hands, high above the head increases the apparent height of the handler and may be used to assert dominance against threatening behaviour. This display works less effectively in large deer, such as wapiti (Goddard, 2014).

After the short rest period, smaller groups of animals (approximately 6 per group) should be moved through the facilities at a time (Goddard, 2014), either to be loaded for transportation to an abattoir, or to be slaughtered on-site. The raceway should be constructed using sturdy materials, and provide little visual opportunities to deer for escape. Currently, these raceways are mostly made from wood, as metal fittings are noisy and may further agitate the animals. Deer should be moved through the alley (5-10 m wide) closely together (Goddard, 2014), and preferably through phases splitting the group on its way to slaughter. Such partitioning of animals not only eases the flow of the herd through the system preventing back-ups, but it also avoids overcrowding, and decreases the level of human contact (Goddard, 2014). These partitions need to be high enough that the animals will not attempt to jump out of the system. To reduce light conditions and reduce the inclination to jump, the system should be covered, as subdued light conditions has been shown to improve the handling of fallow and white-tailed deer (Kelly *et al.*, 1984; Langridge, 1992). The flooring should not be dusty to prevent animals from slipping and it should be made of sand or concrete to ease cleaning (Goddard, 2014). The raceway should be constructed in a U-shape or curved (Goble, 1991), giving the deer the impression that they are moving around a corner, and not into a dead-end. Furthermore, as deer prefer to move uphill, the system should be built on a slope. Movement through the system is also easier if the animals have been habituated to the raceway (Price *et al.*, 1993), by allowing free access from their usual camps, and/or regular handling. Luring with concentrated feed may also encourage movement through the system. The raceway should exit into a squeeze chute/crush (drop-floor, mechanical or pneumatic) or a loading bay for transportation (L. Hoffman, personal communication).

Deer can be slaughtered within the squeeze chute on-farm (Figure 8.7), using captive bolt stunning followed by exsanguination. Should drop-floor chutes be used, the sides should be padded to

Figure 8.7. Deer in a typical restraint box on farm, entering through a raceway system (T. Needham, Czech University of Life Sciences, Czech Republic).

support the animal's body weight, preventing panic in the deer. The squeeze chute may be placed in a brighter light area as deer tend to move towards light, encouraging them to enter the chute. Once in the chute, the captive bolt can be applied effectively. The chute should be opened swiftly, and the animal should be exsanguinated as quickly as possible. Easier bleeding is achieved by placing the deer on a table, slope, or suspending it by the hindleg.

8.3.2 Transportation, lairage, and off-farm slaughter

Loading and off-loading deer should be done using ramps that allow for them to move up in a group, using a slope lower than 20° (Goddard, 2014). Care should be taken that no gaps exist between the ramp and the vehicle deck to avoid injuries or attempts to escape. Allowing previous exposure to the loading ramp before transportation aids in future loading and off-loading. In many countries farming deer, regulations exist regarding their transportation, in terms of distance, duration, conditions, provision of food and water, resting, and emergency slaughter protocols (e.g. DEFRA, 1989 in the United Kingdom, and NAWAC, 2018 in New Zealand). Deer should be transported on livestock enclosed vehicles with sides high enough (in accordance with local regulations) to prevent them from jumping out and relatively small compartments to accommodate them in divided groups, with floor space allocation depending on the journey length. Space allowance over long journeys should allow for animals to lie down, and thus adjustable compartments are favourable so that the space allowance may be adjusted according to the journey length (Jago *et al.*, 1997). Furthermore, compartments height should be sufficient to allow comfortable standing, but not allow for rearing up on their hind legs. The following compartment heights are recommended (MAFF, 1989):

- 1.53 m for deer of 75-100 kg;
- 1.22 m for deer of 45-75 kg;
- 1.00 m for deer <45 kg.

Furthermore, compartments should not have any gaps through which the legs of deer may get stuck, resulting in injuries and bone fractures. However, they should be sufficiently ventilated to ensure that the animals do not overheat. Ventilation around the sides of the compartments may assist with air movement, and spraying with water may be used when deer are exhibiting signs of heat stress (Goddard, 2014).

The load should consist of animals of same sex and size, and familiar to each other. Stags with hardened antlers should be transported separately, while the transportation of deer with velvet antlers is typically prohibited. The following requirements need to be considered/adhered to when transporting deer (Goddard, 2014):

- A plan of the journey should be drafted, including how to proceed with delays or accidents.
- The personnel involved in the capture, loading and driving should be qualified and experienced with working (and driving) with deer. Responsibilities for each person must be assigned beforehand.
- Driving quality should be ensured, keeping in mind that animals may lose their footing early during the journey. Braking and steering on winding roads should be as gentle as possible.

- Vehicles suitable for deer should be used (with compartments), to decrease the risk of injury. Deer should be able to lie down and stand up without restriction (see MAFF, 1989 for space recommendations). Generally, six animals of 100 kg per compartment is acceptable.
- The animals should not be lame, in pain, or distress prior to transportation. Thus, inspection prior to loading is necessary.
- Feed and water should be available (see NAWAC, 2018 for examples on timing), and in-transit rest periods should be planned, depending on the condition of the animals, the duration, distance, and environmental conditions, considering the local legislation. For example, young animals should not travel for longer than 12 h, and 24 h for adult animals (NAWAC, 2018).
- The driver should be able to observe the animals throughout transport (by video camera), or at least stop and inspect them.

Some countries have codes of practice for the slaughter of deer species, which specify that stress should be minimised by using appropriate facilities and equipment, competent staff, minimum force, and efficient stunning and killing methods that ensure immediate unconsciousness and death (e.g. DEFRA, 1985). However, maintaining high standard of animal welfare during the necessary handling, transportation and lairage required for off-farm slaughter is challenging. These situations are novel and result in significant stress for deer species that generally manifest aggressive behaviour. For the slaughter of deer, the facilities (i.e. lairage, restraint and stunning box) should be designed with their natural behaviour in mind, and thus slaughtering deer in multi-species abattoirs is not always suitable.

After transportation, it is necessary for the animals to rest and recover in lairage, as with other species slaughtered off-site. Similar principles apply to other livestock, where mixing of unfamiliar groups and different species should be avoided. According to DEFRA (1985), deer should be allowed to settle and then slaughtered in the shortest time possible, preferably within three hours of off-loading. Overnight lairage is not advised, and results in increased carcass bruising due to fighting (Jago *et al.*, 1996). As for other livestock species, food and water should be provided during rest. However, deer tend to be nervous at the sight of novel objects, such as troughs. In hot environments, misters can be used to cool deer in lairage if signs of heat stress are evident (L. Hoffman, personal communication).

The raceway leading to the stunning area should follow the same principles previously described for the on-farm handling facilities, Movement of the animals through the raceway and into the stunning area should be as fast as possible, and captive-bolt stunning should be applied immediately. The ideal positioning for the captive bolt is in the middle of the forehead, at the crossing point of two imaginary lines drawn from the eyes to the top of the base of the opposite ears (HSA, 2016). Small deer may be stunned in a small pen, whereas larger deer should be restrained in a stunning box which typically has bars on the top to stop deer jumping out. Although electrical stunning of deer has been trialled, this procedure was not as successful as hoped, mainly due to the difficulty in positioning the electric tongs correctly on the head (Blackmore *et al.*,1993). After stunning, exsanguination should be done immediately.

8.4 Ostriches

Since ostrich farming started in South Africa in the 1880s, this species has increasingly gained prominence as a livestock animal, due to its dual potential to produce a skin/hide sought after by the fashion industry, as well as producing healthy red meat with a low-fat content (Hoffman and Fisher, 2001). Interestingly, feathers which were the original cause of the booming ostrich industry pre-WWI, is becoming more popular once again (Brand *et al.*, 2018). Ostrich products are popular for export from South Africa to the European Union. In fact, South Africa is considered the world leader in the global ostrich marketplace, with 90% of the produced commodities (meat, leather and feathers) exported. With this high amount of exports, South Africa comprises 75% of the global market share, with 45% skin, 45% meat and 10% feathers (Barends-Jones and Pienaar, 2020).

The ostrich production system can be classified as being horizontal because of sector autonomy, with each phase of the animal production chain being owned by a different entity/company, resulting in the frequent need to move ostriches from one production entity/system to another. This, in combination with the bird's physical nature, such as being bipedal, could lead to an increase in the incidence of bruising, injuries and mortalities during transport and lairage (Hoffman *et al.*, 2010). Ostriches are normally transported three times in their production cycle. The first transport is for the transfer of day-old chicks from the hatching facility to the juvenile rearing facility. The second transport is for the relocation of juveniles to a grow-out phase or pastoral system. Then, finally, the third transport is for shipping birds to slaughter at the abattoir. During each transport cycle, precautions should be taken to ensure the bird's safety and to limit exposure to any stressful stimuli (Reiner *et al.*, 1996).

8.4.1 Loading

All birds slaughtered in southern Africa require a two-week quarantine period, where they undergo tests to ensure that they are not infected by Avian influenza. The current legislature for the export of ostriches and related products (i.e. meat, leather, and feathers) from South Africa requires that feedlots, as well as quarantine facilities, must be surrounded by a 3 m vegetation-free zone on all sides. This will prevent or minimise the access of rodents and ticks that carry potential disease-causing bacteria and viruses, such as the Crimean Congo Haemorrhagic Fever virus. If any bird tests positive for Avian influenza or has any visible ticks, the whole consignment may not be transported for slaughter (Hoffman and Lambrechts, 2011). Quarantine paddocks should also preferably be cleaned on a monthly basis, i.e. remove faeces and any loose soil. Loading facilities should also be close to the quarantine camps to minimise the exposure of birds to potential vegetation or areas where ticks may occur.

Ostriches are farmed both intensively and extensively. The latter incorporates mustering and capturing of the birds for loading resulting in stress (Hoffman and Lambrechts, 2011). Ostriches should always be treated with the utmost care and respect, for they are quite unpredictable under certain circumstances, and can inflict serious damage; it should be remembered that unlike other animals, ostriches kick forwards and not backwards. For this reason, the loading of ostriches requires a correct design of the loading facilities, specific handling techniques and appropriate level of training of the stockmen handling the birds.

The loading facilities should be designed to allow the transport truck access to the area where the birds will exit the holding pens. Typically, the loading facility is situated in a corner of the pen, and consists of a crowd/forcing pen, a raceway and then the loading ramp (Figure 8.8). The crowd gate that closes behind the birds should be solid and should present no sharp corners or protrusions as these will cause injury and bruises (Hoffman and Lambrechts, 2011). The level of the raceway has to be at the same height as the floor of the truck to allow birds to walk into the truck compartments. All the sides should be solid from leg height upwards to ensure ventilation, as they keep the birds calmer and facilitate movement. The sides should be high enough (1.7-2.0 m) to prevent the birds from seeing distractions. A walkway on one or both sides will allow a stockman access to the birds (L. Hoffman, personal observation). The raceway is normally wide (3 m) as very few birds will allow themselves to be herded individually. The floor of the crowding pen and raceway should be coarse and slip-proof to minimise injuries should a bird fall, and is normally covered with sand, free of stones, etc. that may cause damage to the feet of the birds. Care should be taken, however, to ensure that the sand that is trodden out is replaced, as ostriches, especially those that are not used

Figure 8.8. A well-designed raceway and loading ramp for ostriches (L. Hoffman, University of Queensland, Australia).

to humans and being handled, have been noted to lie down so as to allow the crowding gate to pass over them (Hoffman and Lambrechts, 2011).

Loading facilities were designed in the past to incorporate a ramp, but this is not advisable for ostriches which do not like to walk uphill, especially when being hooded as they cannot see where they are going. When using a loading facility where a ramp forms part of the design, the ramp incline should not be too steep (<30° angle). When loading ramps are made out of metal, a covering, such as a thin layer of conveyor belting should be used to cover the surface of the ramp. This will absorb the noise that is created when a bird steps onto the ramp, which will help to minimise the noise stress that is experienced during handling. The conveyor belting will also prevent the metal from heating when loading activities are performed during a hot day (L. Hoffman, personal observation). The latter scenario is an exception to the rule, as normally loading should be limited to the cooler hours of the day to ensure the least amount of heat stress to ostriches (Hoffman and Lambrechts, 2011). Care should also be taken to ensure that the truck is parked correctly and that there is no space between the loading ramp and truck where a bird's feet can be caught.

At loading, ostriches are caught by means of the following procedures: one bird is caught and loaded at a time, and if two catching teams are used, this will ensure for a more rapid loading time. It is also advisable to place a hood over the bird's head to quiet it down as much as possible. The best procedure for the catching and handling of slaughter birds are as follows (Hoffman and Lambrechts, 2011):

- Single out a bird from the group; normally the catching team will be the only people in the pen. The bird is usually caught by means of a shepherd's hook, which has a long handle (about 2 m long) and a specially shaped hook that is designed to be wide enough to accommodate the width of the neck of the bird. Sometimes, if the handlers are experienced enough, they can jump up to catch the bird by the part of the neck just below the head.
- As soon as the bird is caught in the hook, another handler should take a position behind the tail end of the bird, to prevent the bird from running in a circle or reversing, which will make it difficult to remove the head from the hook.
- Immediately after the handler has taken position behind the bird, the head must be taken out of the hook, and a hood placed over the head to quieten the bird down. A hood is normally 15×30 cm in size and should have an opening large enough for the beak, which will also allow the bird to breath comfortably. The opening should, however, not be too big, otherwise the bird will still be able to observe its surroundings and start running when not being handled properly. Sometimes handlers use old socks as a hood, e.g. in the case of younger birds. This is inadvisable, though, as the elasticity of the sock material does not allow the birds to breathe comfortably.
- After the hood has been placed over the bird's head, two handlers position themselves behind the wings (i.e. on either side of the body) to help to steer the bird onto the truck. Ostriches do not like their neck being touched and doing so during a loading activity may result in a bird reversing off the ramp.

Ostriches are very sensitive to the ambient conditions in the handling/loading facility and will become difficult to handle if noise levels are excessively high. Handlers should always work quietly and calmly during the herding and loading of the birds.

The use of electric prodders or hitting the bird with a stick against the neck should not be allowed. Hitting against the neck can result in a bird losing its balance and falling over backwards, which can injure its neck to the extent that it has to be culled. The use of whips is also not advised, for they tend to get entangled in the bird's legs and cause them to fall, injuring themselves and possibly also the handlers (Hoffman and Lambrechts, 2011).

When loading slaughter birds, it is advisable to adhere to the compartment dimensions, and not to load more birds than the number allowed per compartment. Preferably load birds of similar size together in a compartment. Larger birds in a compartment may be more aggressive and are prone to kick or injure smaller birds in the compartment. The opposite is also applicable, i.e. do not load too few birds per compartment, for this may cause them to fall around during the journey and injure themselves. Falling of birds will cause bruising of the carcass and damage the skins, which will result in poor welfare as well as downgrading of both products, and a loss of income in the case of slaughter birds. Males and females should be transported separately. During hot and/or humid days it is advised to allocate fewer birds to a compartment to ensure that the birds will be able to dissipate body heat effectively. When birds are overweight, it is advisable to allocate fewer birds to a compartment to allow the birds to not overheat during the journey. Furthermore, it is not advisable to transport male birds in their breeding season as they are very aggressive and tend to injure themselves and other birds (Hoffman and Lambrechts, 2011).

When a handling facility without a ramp is used, i.e. where the loading surface is level with the floor of the truck compartment, care must be taken that the birds are not chased into the compartment. Birds should be allowed to walk into/out of the compartment at their own pace. Pressure can be applied at the back of a group to ensure that the speed at which compartments are filled/emptied is not too slow, as this may cause stress to the birds that were loaded first, especially on humid and/or hot days (Hoffman and Lambrechts, 2011).

8.4.2 Transportation

The design of trucks and trailers that are used to transport ostriches (Figure 8.9) must adhere to the following requirements to ensure the safe transport of ostriches (Hoffman and Lambrechts, 2011):

- The sides of the truck should be solid (i.e. with no gaps at the bottom) and there should be an opening between the roof and the side to ensure adequate ventilation. However, most trucks that are used in the South African industry to transport ostriches are open, so ventilation is generally not a problem.
- The floor of the vehicle must be solid, non-absorbent, and slip-free. Typically, metal gridding or rubber matting is used. This will prevent birds from slipping, and the latter will also provide protection against the cold metal surfaces, especially if long distances are going to be covered. Although rubber matting is less durable, observations and discussions with transporters indicate that this material is more suitable, as the metal mesh can hurt the toes of the birds.
- The corners of trucks need to be protected with a form of cushioning material to prevent the birds being injured by any sharp edges in the corners. The loading space may not have any other sharp angles, protrusions or holes that may injure the birds. No loose objects should be stored in the truck compartments.

Figure 8.9. A truck designed to transport ostriches. Note the handler in each compartment (L. Hoffman, University of Queensland, Australia)

- Partitions must be installed at every 3 m of loading space if the truck is longer than 4 m. The minimum required floor space per bird is 0.5 m^2.

The transport regulations pertaining to the transport of live ostriches are set in the World Ostrich Association Welfare Codes for Ostriches (http://www.ostrich.org.uk/industry/welfare.pdf). These regulations state that workers should be on the trucks with the ostriches. Their presence is a precautionary measure to ensure animal safety by decreasing the incidence of trampling and/or lying down, thereby decreasing bruising in the final transport phase. These consequences caused by transporting live animals could be seen as being unavoidable, as transport of animals between sectors is a necessity.

Transportation of ostriches could have an influence on meat quality and quantity, with the latter being affected by live weight losses and carcass losses due to the trimming of bruised tissues (Hoffman *et al.*, 2010). The longer the distance, the more weight is lost. Hoffman *et al.* (2012) showed that ostriches transported for 60 km lost 2.4% weight compared to birds that were transported for 600 km (8.1% weight loss). These losses can be explained by the exposure of ostriches to various stressors during transport, including nutrient deprivation, heat exposure, wind chill, social anxiety, equilibrium control and physical harm. The degree of impact of these stressors depends on the bird's predisposition to the specific stressor and the amount of time taken by the bird to return to physiological homeostasis after reacting to stress (Fasone and Priolo, 2005).

Nutrient deprivation, in terms of water and feed removal, during transport has various effects on the bird's physiological status, which takes time to correct. The limitation of energy caused by nutrient deprivation influences the para-sympathetic response as well as influencing the glucose levels in the tissue and organs of the animal. These ante-mortem changes have a direct influence on the quality and quantity of meat by influencing the glucose levels available in the muscle post-mortem. Nutritional supplementation pre-transport with electrolyte therapy, as a means of controlling/minimising this effect, is neither practical nor economical (Schaefer *et al.*, 1995).

Dehydration is a major concern in the transport of ostriches, especially during daytime heat. Extreme heat exposure can lead to an increase in dead-on-arrival (DOA) birds. The loss of water experienced by the animal during transport is caused by the bird's thermoregulation to counteract the increased core and skin temperatures and can lead to a significant decrease in body weight, but without a major change in muscle weight. This loss in weight in ostriches is speculated to be a result of water loss pre- and post-mortem (Hoffman *et al.*, 2012). In birds, temperature is regulated mainly through movement of air in the airways (panting), and excretion of metabolic waste products. Birds do not have sweat glands and thus cannot reduce body temperature by sweating through the skin. Panting and increased excretion cause loss of water which, in some circumstances, can cause dehydration. An increased breathing rate is induced when core temperatures increase, as the body attempts to release excess heat (Hoffman *et al.*, 2012). The evaporation of water in the airways is a key response to thermogenesis and the removal of the heat from the bird's core. This is an effective mechanism but causes increased moisture loss.

World Ostrich Association Welfare Codes for Ostrich (http://www.ostrich.org.uk/industry/welfare.pdf) state that when transporting ostriches for trips of approximately 12 h, feed should be removed 10 h before loading. This is to reduce the volume of wet droppings on the vehicle floor and thus decreases the instances of ostriches slipping and falling. No provision has been made in the regulations for availability of water or feed while transporting ostriches for periods longer than 12 h.

The most prevalent stressor affecting the physiological homeostasis of ostriches in sub-Saharan Africa is heat exposure. Excessive heat exposure (>40 °C) causes a stress response, dehydration, heat shock and feed refusal. These negatively affect the bird's health status and affect the quantity and quality of meat by lowering glycogen reserves in the muscle and inducing muscle protein denaturation (Hoffman and Lambrechts, 2011; Hoffman *et al.*, 2012).

Being a bipedal species, ostriches ability to keep its centre of gravity stable is challenged during transport resulting in difficult control of their balance when exposed to external horizontal and vertical forces. During transport these forces can be simultaneous or separate. The animal then exerts a force equal to or greater than the forces acting on it, to counteract the exerted forces and ensure that the centre of gravity is kept as stationary as possible. In-transit ostriches were also noted to align themselves either parallel or perpendicular to the direction of travel, most likely in the attempt to lower the amount of external force exerted on the body, or to counteract these forces more effectively (Hoffman and Lambrechts, 2011). These principles must be considered by the transporter for careful driving, i.e. during acceleration/braking and travelling at speed around corners (Hoffman *et al.*, 2012).

8.4.3 Lairage

The offloading of ostriches into lairage at the abattoir is very similar to the loading. The inclusion of lairage as a bridging phase between transport and slaughter has resulted in some division between farmers and abattoirs on the responsibility for weight or animal losses during this period. Lairage also helps ostriches to get familiar with their new surroundings and give them the time to return to a relaxed state before slaughter, thereby improving meat quality (Hoffman and Lambrechts, 2011).

In ostrich abattoirs, under normal conditions, time in lairage could range from 0 to 24 h. However, it can be extended up to 4 days due to stop of slaughter activities during holidays and weekends. In slaughter weight birds (approximately 93 kg live weight), 3.23 kg weight was lost in birds kept in lairage without food (water supplied *ad libitum*) for 56 h compared to a weight loss of 1.04 kg in birds held in lairage for 24 h. A large portion of this weight loss was attributed to emptying of the gut contents in the birds standing in the longer lairage (Van Schalkwyk *et al.*, 2005). However, in a later study it was shown that some of the weight lost during transport and lairage is also linked to weight loss in the muscles. Birds that were transported for 600 km lost 3% more carcass weight than birds transported for only 60 km; both groups of birds were maintained under the same lairage conditions (Hoffman *et al.*, 2012).

The main stress factors in lairage are social anxiety, lack of familiarisation with the new environment, nutrient deprivation and lairage conditions.

Social anxiety can be induced by the animals interacting with each other. To avoid this animal condition, ostriches from different backgrounds and production systems are normally kept separate from each other during lairage, although some interaction still occurs between adjacent pens (Hoffman and Lambrechts, 2011).

During lairage, animals are frequently moved between pens and then to the killing point. This movement of animals in the lairage area could stimulate the herding instinct of the animals causing handling and safety problems. The herding instinct induced by moving the animals increases the frequency of animals tripping, falling and being trampled on, leading to a loss in leather quality and carcass value due to bruising (Hofmann *et al.*, 2010). The exposure of unfamiliar noises and smells at different intensities during lairage also affects the animals. The noise generated by distressed animals, motorised mechanisms and humans could induce a fight or flight response from the animal (Hoffman and Lambrechts, 2011).

Hormones released by the animals when sexually active are known to induce physiological and behavioural changes in the opposite sex (Balthazart and Taziaux, 2009). Certain avian species are capable of aromatic communication and new consideration should be given to the role of odours in avian communication. This could be a problem when slaughtering breeding animals as the older birds might induce sexual behaviour in the younger birds or they themselves may be sexually stimulated. The breeding birds need not be in the same pen to stimulate this response. Ratite species, such as emu and ostriches, are dangerous and aggressive when sexually active. The sexual behaviour of male ostriches is marked by an increase in aggression towards intruders that enter their territory, and this increases the possibility of physical injuries to the animal and the handler (Hoffman and Lambrechts, 2011). In females, it can lead to egg production and sexual behaviour being induced. These behavioural changes, such as cuing and displays, further increase the sexual activity of the males in lairage (L. Hoffman, personal observation).

Ostrich behaviour has been described as being similar in complexity and variability as animals belonging to the most developed and complex social orders. This makes the identification of standard behaviour patterns and abnormal behaviour difficult to ascertain. Ostriches are driven by diurnal and annual patterns like most birds and are active during the daylight hours of the day, and sitting down and becoming inactive during the evening, unless disturbed. According to South African regulations, the space allowance in lairage per bird should be 1.2 m^2 (Department of Agriculture, 2000). The decrease of space sometimes associated with placing ostriches in lairage might incur abnormal behaviour and stress, due to movement restriction as a result of over population. However, the increased density encountered during lairage might also influence vigilance and lower stress as increased group numbers has been shown to lower the overall vigilance per bird allowing more time for feeding/resting (Bejaei and Cheng, 2014). Behaviours, such as panting, yawning, and stretching, commonly seen in ostriches during the lairage period help regulate the physiological balance (Hoffman and Lambrechts, 2011).

Animal behaviour during lairage is influenced by the animal's familiarity with humans and the handlers' level of experience and their understanding and application of animal welfare regulations (Bejaei and Cheng, 2014). The inappropriate and incorrect use of herding tools by handlers, such as prodders or whips, increases discomfort for the animals inducing a stress response, i.e. fleeing with the wings extended and running blindly into surrounding structures, and contributes to increased bruising ante-mortem, resulting in a decreased leather value due to haemorrhages under the skin and carcass yield losses due to trimming (Hofmann *et al.*, 2010). Improved training also increases handler and animal safety, as certain animal species can cause fatal injuries to themselves as well as the handlers, if not correctly handled (Hoffman and Lambrechts, 2011).

The design of the lairage facility is important in helping the handlers effectively move animals without extensive use of herding tools at abattoirs. As for other livestock species, the effective design of the lairage area would lead to improved management of ostriches during lairage, thereby increasing production output. Distances between offloading of animals to lairage pens and between these pens to slaughter should be as minimal as possible (Department of Agriculture, 2000), so as to decrease physical exertion which decreases the glycogen reserves prior to slaughter and affect meat quality. The lairage pens should also be sizeable as to accommodate a number of animals without making it unmanageable within regulation guidelines (Department of Agriculture, 2000). This eases lairage management as well as ensuring that the groups of animals transported together can be kept together, thereby lowering social anxiety. At ostrich slaughter plants, the typical lairage pen is of octagonal design, which prevents ostriches from flocking together in sharp corners limiting the risk of bruises and/or injuries. The lairage floors are either sand or cement with grooves and metal grids (Figure 8.10) to prevents slips, although no evidence of the effects of these different flooring types on the birds' behaviour and on skin and meat quality exists.

As adequate water should always be available to birds at all times (World Ostrich Association Welfare Code, 2007), drinking troughs are placed outside the pens at an acceptable height (at least 1000 mm from the ground) to allow the birds to drink comfortably while preventing injury (Department of Agriculture, 2000). These welfare requirements also include adequate protection against the weather elements, by providing a structure with a roof to protect the animals against the sun and rain, with open sides to allow air circulation (Hoffman and Lambrechts, 2011).

8.4.4 Stunning and slaughter

Restraint and stunning of ostriches require specialised facilities and procedures because of their long neck, head anatomy, and physiology.

Stunning is traditionally performed using hand-held tongs on the head of the restrained bird. An electrical current in excess of 400 mA at 50 Hz AC applied only to the head prevents recovery in more than 90% of the ostriches when bled within 60 s of stunning (Hoffman, 2012). The South African legislation requires a stunning current of 400-600 mA, 90-110V for a duration of 4-6 s (Hoffman, 2012).

The retraining area is often a V-shaped structure, high enough (1.2 m) that the stunning operator is not kicked. After (and sometimes during) stunning, the birds are rocked backwards, and a rubberised leg clamp placed over the legs at the tarso-metatarsal bone, which immobilises them, and allows the ring/chain shackling via the big toes. Birds are then hoisted onto a 3.4 m overhead rail and manually conveyed to another area for exsanguination. This conventional stunning procedure has been replaced in many abattoirs with a new restraining and stunning mechanism (stunning box) that completely encompasses the ostrich in a padded clamp holder and allows for the whole bird to be restrained without affecting the quality of the feathers and the skin.

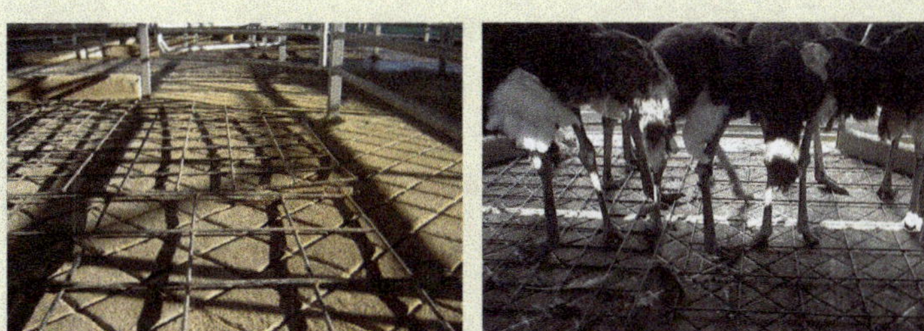

Figure 8.10. Typical lairage flooring used for ostriches (L. Hoffman, University of Queensland, Australia).

The new Divac Ostrich Stunning box (Divac, Knysna, Republic of South Africa) is built from a combination of galvanised mild steel and stainless steel, which encapsulates the ostrich in a padded clamp type holder, ensuring no physical damage occurs to the bird (Figure 8.11). The bird is gently pushed into the box which is then closed manually around the bird. The double-padded sides restrain the bird by holding the upper thighs. A rubberised pneumatic foot clamp restrains the bird's feet, whilst the bird's head is placed manually into the stunning clamp. As the bird is stunned, the whole stunning box rotates through 180°, thereby positioning the bird for the toe clamps respecting the safety measures for the operator. After the stunning is completed the box is opened and the bird hoisted for exsanguination. The time from stunning to exsanguination has also been reduced to lower than 20 s. Once released from the clamp, the unit rotates to its initial position, ready for the next ostrich.

The post-stunning first stages of recovery in the birds are accompanied by rhythmic breathing movements (Hoffman and Lambrechts, 2011), whose identification, however, may be difficult as they can be confused with rhythmic body movements in response to spinal reflexes and limb muscle contractions.

At bleeding, the ostrich head is normally held between two horizontal metal bars to minimise blood spillage on the feathers and skin. Slaughter occurs by bleeding through a complete ventral cut to the neck and/or by thoracic sticking. In a pilot investigation (Hoffman *et al.*, 2009), where the efficiency of the two bleeding methods was compared, a lower bleed-out percentage (defined a weight of blood expressed as percentage of body weight) was observed when the ventral throat cut alone was performed compared with the combined ventral cut and thoracic sticking (2.8±1.03%

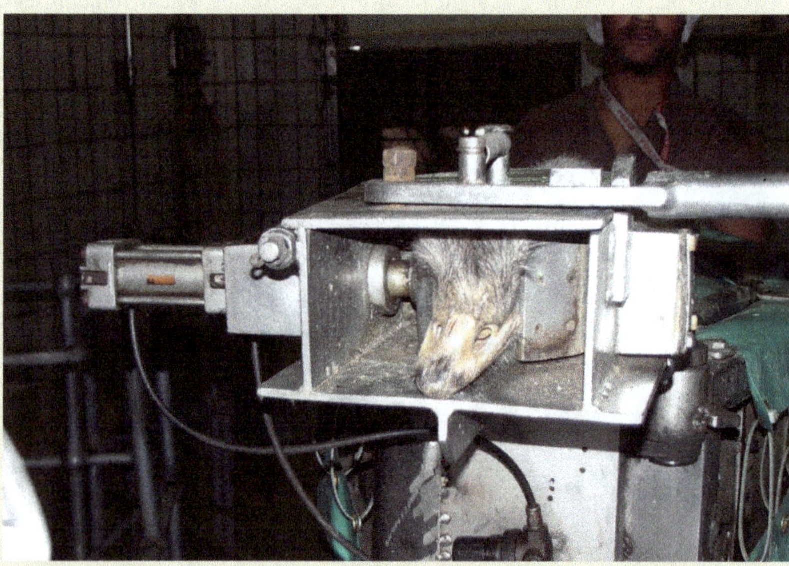

Figure 8.11. Ostrich's head in a stunning box (L. Hoffman, University of Queensland, Australia).

vs 3.3±0.34%). Furthermore, the additional thoracic stick had no effect on meat quality, in terms of drip and cooking loss, colour or pH and temperature readings, of the fillet (*M. Iliofibularis*), big drum (*M. Gastrocnemius, pars interna*) and inside loin (*M. Iliotibialis cranialis*). Nonetheless, the use of the thoracic stick would be recommended for ostrich slaughter as it may result in a faster death (L. Hoffman, personal observation).

8.5 Crocodiles

Crocodiles are classified as Appendix I animals through the Convention on International Trade in Endangered Species of Wild Fauna and Flora (https://cites.org/eng/app/appendices.php), with the aim of preserving the wild-stock and prevention of endangerment of fauna and flora. This has resulted in farming, or ranching, of these species becoming more favourable over wild harvest, as there was unpredictability and inconsistency of supplies associated with obtaining skins from wild crocodilians (Hoffman *et al.*, 2000). Today, most wild harvesting consists of the live capture of large adult animals as breeding stock to ensure genetic variability within the farmed stock. In Australia, salt-river crocodiles are still harvested from the wild, as are alligators in the Americas, mainly through a shot with a free bullet at night utilising spotlights (L. Hoffman, personal observation).

Crocodiles were farmed in Zimbabwe as early as 1965 (Hoffman *et al.*, 2000). The primary motivations for crocodile farming are harvesting of crocodilian skins, meat, and sale of live animals, while teeth are important by-products. Currently, the primary income from crocodiles is from their skin, with meat being a secondary product. Thus, size selection of slaughter animals depends on the surface area of their skin. There are three main cuts of skin: belly, back-strap and the horn-back skin. Skins are used in the manufacture of belts, wallets, purses, handbags, shoes, watch straps and clothing. Crocodilian skins undergo strict grading, from a grade scale ranging from 1 (absolutely no marks, scratches, or any other defect) to 5 (bad defects). The 1st grade would correspond to most highly prised skin, while the 5th grade would result in a revenue loss (Hoffman *et al.*, 2000). A large number of wild harvested skins fall within skin grades 4 and 5 (L. Hoffman, personal communication). Crocodiles destined to the skin market are slaughtered when they reach a specific length that meets the minimum size requirement for processing it into leather. Hatchlings arrive at 15-20 cm length from the breeding facility and are placed out in growth pens until they reach a length of 1.2-2 m (2-4 years old), with most slaughtered at a minimum length of 1.2 m. In the past, the crocodile's length would be judged by eye, and the selected crocodile would then be shot with a free bullet using a light calibre (0.22 mm) rifle, in the production pen (L. Hoffman, personal communication). However, this method had some disadvantages, including the fact that if the length of the crocodile was misjudged, then the producers suffered financial losses. Due to the shape of the head, it was also not uncommon for the bullet to ricochet off the head of the targeted crocodile (L. Hoffman, personal communication). The crocodiles would also bunch up into a corner as soon as the first bullet was fired, making it difficult to shoot the next animal.

Lately, producers are using electrical stunning by placing the V-shaped tongs behind the head. (135-165 V for 10 s) with the charge being produced from a backpack (Figure 8.12) to harvest the crocodiles (Pfitzer *et al.*, 2014). The electrical stunning procedure is also safer for the operator, and preferred over capturing with a noose (Pfitzer *et al.*, 2014).

Figure 8.12. Back-pack electrical stunner used to stun crocodiles (L. Hoffman, University of Queensland, Australia).

Figure 8.13. Crocodiles showing guillotine cut used to sever spinal column (L. Hoffman, University of Queensland, Australia).

After stunning, the crocodile typically lies unconscious for 3-5 min allowing sufficient time for the determination of its suitability (based on body length) for harvesting. Immobilisation is rapid following 3-10 s of rigor and tail twitching; the later tends to decrease in magnitude as the current continues to flow. The tongs are removed once the crocodile is immobile. The crocodiles should not be stunned whilst under water as this increases the risk of the water being inhaled, although the neck needs to be wet to ensure maximum efficiency of the stun (Isberg, 2016). If the crocodile is too short, it is left to regain consciousness, while, if it meets the required length, the crocodile's mouth is taped shut, it is removed from the pen, and moved to the abattoir for killing and further processing. Killing is performed by cutting with a guillotine behind the head (Figure 8.13) severing the spinal column and also allows it to bleed out.

8.6 Kangaroos

Macropods are harvested predominantly for their skin, under strict guidelines in Australia (AgriFutures, 2020), while their meat is only a secondary product (Spiegel and Greenwood, 2019). The method of harvesting of kangaroos is very similar to that followed for wildlife in Africa. Typically, registered teams/marksmen go out at night and using spotlights, and shoot the targeted kangaroos. Only males are allowed to be harvested, as most of the time the females will have joeys of different physiological development in their pouches (Spiegel and Wynn, 2014). There are a series of standard operating procedures on the euthanasia procedures to be followed for the different physiological developmental stages of the joeys (AgriFutures, 2020). The industry is only allowed to harvest six abundant species of kangaroo and wallaby, and their numbers are determined by annual government quotas set to between 10 and 15% of their populations corresponding to an average of around 4 million kangaroos per year. It is estimated that Australia has more than 40 million macropods, including kangaroos, wallabies, tree-kangaroos and pademelons (Wilson and Edwards, 2019). The more common species harvested include the red (*Macropus rufus*), eastern grey (*M. giganteus*), and western grey (*M. fuliginosus*) kangaroos, and a common wallaby, the common wallaroo or euro (*M. robustus*). There is a mandatory tagging system and recording of data (species, region harvested, etc.), similar to that utilised in other countries on different wildlife species ensuring full traceability.

The procedures and standards for the harvesting methods are outlined in the national code of practice for humane shooting of kangaroos and wallabies for commercial purposes (AgriFutures, 2020). The commercial harvesters are all marksmen who have to undergo an efficiency evaluation every five years. The code also stipulates rifle calibre, type of ammunition, methods of harvesting, competency of marksmen, etc. that are allowed to be used. Only head shots are permitted. No more than three animals are allowed to be shot in a group before the carcasses are checked and exsanguinated by the harvester. After bleeding which is performed via either a thoracic stick or transverse cut of the ventral region of the neck, thereby severing the veins and arteries, the carcass is placed on the outside of the hanging frame located on the back of the harvester's vehicle for evisceration.

8.7 Conclusions

While numerous non-traditional species may be used for meat production, there are several obstacles and requirements that their preslaughter handling and processing must overcome before their meat may enter the commercial market. Depending on the nature of their production and management, preslaughter handling and slaughter methods must be adjusted accordingly. While these may be discussed generally within this chapter, it is clear that each species is unique in their behaviour and stress tolerance for particular activities (herding, handling, transportation, and lairage). Some of these animals are becoming easier to handle preslaughter, after genetic section for temperament and habituation to handling, with such practices showing progress in both deer and antelope species. While livestock handling principles and, in some cases, equipment may be applied to these species, it is necessary to adjust them to accommodate for the specific behaviour of each of them, to ensure their utmost welfare. Despite the adjustment of these protocols, the effects of preslaughter handling and killing methods on the welfare of various non-traditional meat species still requires thorough investigation and further optimisation, particularly in African game species.

References

AgriFutures, 2020. National code of practice for the humane shooting of kangaroos and wallabies for commercial purposes. Available at: https://tinyurl.com/ru99ps8m.

Balthazart, J. and Taziaux, M., 2009.The underestimated role of olfaction in avian reproduction? Behaviour Brain Research 200: 248-259.

Barends-Jones, V. and Pienaar, L., 2020. The South African ostrich industry footprint. Western Cape Department of Agriculture, Elsenburg, South Africa. Available at: https://tinyurl.com/4putrv4e.

Bartos, L., 1985. Social activity and the antler cycle in red deer stags; In: Fennesy, P.F. and Drew, K.R. (eds.) Biology of deer production. The Royal Society of New Zealand, Wellington, New Zealand, pp. 269-272.

Bejaei, M. and Cheng, K.M., 2014. A survey of current ostrich handling and transport practices in North America with reference to ostrich welfare and transportation guidelines set up in other countries. Poultry Science 93: 296-306.

Blackmore, D.K., Cook, C.J., Devine, C.E., Gilbert, K.V., Tavener, A., Langdon, S., Isaacs, S. and Maasland, S.A., 1993. Electrical stunning of red deer (*Cervus elaphus*). New Zealand Veterinary Journal 41: 126-130.

Bothma, J. Du P., 1996. Harvesting game. In: Bothma, J. Du P. and Du Toit, J.G. (eds.) Game ranch management. Van Schaik Publisher, Pretoria, South Africa, pp. 246-270.

Brand, T.S., Kritzinger, W.J., Van der Merwe, D.A., Muller, A., Hoffman, L.C. and Niemann, G.J., 2018. Feather and skin development of ostriches *Struthio camelus*. Journal of the South African Veterinary Association 89: a1556. https://doi.org/10.4102/jsava.v89i0.1556

Ceacero, F., Landete-Castillejos, T., Bartosova, J., García, A.J., Bartos, L., Komarkova, M. and Gallego, L., 2014. Habituating to handling: factors affecting preorbital gland opening in red deer calves. Journal of Animal Science 92: 4130-4136.

Chakanya, C., Dokora, A., Muchenje, V. and Hoffman, L.C., 2016. The fallow deer (*Dama* spp.): endangered or not? Der Zoologische Garten 85: 160-172.

Clift, T.R., Challacombe, J. and Dyce, P.E., 1985. Electric fencing for fallow deer. Royal Society of New Zealand Bulletin 22: 363-365.

Conroy, A.M., 2005. The springbok. In: Bothma, J.Du P. and Van Rooyen, N. (eds.) Intensive wildlife production in Southern Africa. Van Schaik Publisher, Pretoria, South Africa, pp. 214-225.

DEFRA, 1985. Deer: code of welfare practice on abattoir slaughter of farmed deer. Ministry of Agriculture Fisheries and Food, London, UK. Available at: https://tinyurl.com/37un9389.

DEFRA, 1989. Deer: guidelines for the transport of farmed deer. Ministry of Agriculture Fisheries and Food, London, UK. Available at: https://tinyurl.com/3k5jf466.

Department of Agriculture, 2000. Regulation Gazette No. 8622, Meat Safety Act, 2000 (ACT No. 40 OF 2000): Ostrich regulations. Available at: https://tinyurl.com/4jpv5t8c.

Du Toit, J.G., 2005. The eland. In: Bothma, J. Du P. and Van Rooyen, N. (eds.) Intensive wildlife production in Southern Africa. Van Schaik Publ., Pretoria, South Africa, pp. 108-122.

Ebedes, H., Du Toit, J.G. and Van Rooyen, J., 1996. Game capture. In: Bothma, J. Du P. and Du Toit, J.G. (eds.) Game ranch management. Van Schaik Publishers, Pretoria, South Africa, pp. 271-333.

Encyclopedia Britannica, 2020. Domestication. Available at: https://www.britannica.com/science/domestication.

Fasone, V. and Priolo, A., 2005. Effect of stress on ostrich meat quality. Proceedings of the 3rd International Ratite Science Symposium, Madrid, Spain, pp. 393-396.

Furstenburg, D., 2005. The greater kudu. In: Bothma, J. Du P. and Van Rooyen, N. (eds.) Intensive wildlife production in southern Africa. Van Schaik Publ., Pretoria, South Africa, pp. 147-165.

Goble, K. 1991. Growing, yarding and harvesting of velvet. Proceedings of the 43rd Rukura Farmers' Conference. Ministry of Agriculture and Fisheries, Hamilton, New Zealand.

Goddard P., 2014. Deer handling and transport. In: Grandin, T. (ed.) Livestock handling and transport. CABI, London, UK, pp. 342-377.

Hattingh, J., Wright, P.G., De Vos, V., Mcnairan, I.S., Ganhao, M.F., Silove, M., Wolverson, G. and Cornelius, S.T., 1984. Blood composition in culled elephants and buffaloes. Journal of the South African Veterinary Association 55: 157-164.

Hoffman, L. C., and Wiklund, E., 2006. Game and venison – meat for the modern consumer. Meat Science 74: 197-208.

Hoffman, L.C. and Cawthorn, D.M., 2012. What is the role and contribution of meat from wildlife in providing high quality protein for consumption? Animal Frontiers 2: 40-53.

Hoffman, L.C. and Fisher, P., 2001. Comparison of meat quality characteristics between young and old ostriches. Meat Science 59: 335-337.

Hoffman, L.C. and Lambrechts, H., 2011. Bird handling, transportation, lairage and slaughter: implications for bird welfare and meat quality. In: Glatz, P., Lunam, C. and Malecki, I.A. (eds.) The welfare of farmed ratites. Springer, Heidelberg, Germany, pp. 195-235.

Hoffman, L.C. and Laubscher, L.L., 2009a. A comparison between the effects of day and night cropping on greater kudu (*Tragelaphus strepsiceros*) meat quality. South African Journal of Wildlife Research 39: 164-169.

Hoffman, L.C. and Laubscher, L.L., 2009b. Comparing the effects on meat quality of conventional hunting and night cropping of impala (*Aepyceros melampus*). South African Journal of Wildlife Research 39: 39-47.

Hoffman, L.C. and Laubscher, L.L., 2010. A comparison between the effects of day and night cropping on gemsbok (*Oryx gazella*) meat quality. Meat Science 85: 356-362.

Hoffman, L.C. and Laubscher, L.L., 2011. A comparison between the effects of day versus night cropping on the quality parameters of red hartebeest (*Alcelaphus buselaphus*) meat. South African Journal of Wildlife Research 41: 50-60.

Hoffman, L.C., 2001a. The effect of different culling methodologies on the physical meat quality attributes of various game species. Proceedings of the 5th International Wildlife Ranching Symposium. Pretoria, South Africa.

Hoffman, L.C., 2001b. The effect of daytime cropping on warthog (*Phacochoerus aethiopicus*) meat quality. Proceedings of the 47th International Congress of Meat Science and Technology, Krakow, Poland.

Hoffman, L.C., 2012. Advances in the electrical stunning of ostriches. Animal Welfare 21: 9-13.

Hoffman, L.C., Britz, T.J. and Schnetler, D.C., 2010. Bruising on ostrich carcasses and the implications on the microbiology and the losses in utilisable meat when removing them post-evisceration or post-chilling. Meat Science 86: 398-404.

Hoffman, L.C., Cloete, S.W..P, Van Schalkwyk, S.J. and Botha, S.S.C., 2009. Effect of bleeding method and low voltage electrical stimulation on meat quality of ostriches. South African Journal of Animal Sciences 39: 209-213.

Hoffman, L.C., Fisher, P.P. and Sales, J., 2000. Carcass and meat characteristics of the Nile crocodile (*Crocodylus niloticus*). Journal of the Science of Food and Agriculture 80: 390-396.

Hoffman, L.C., Muller, M., Schutte, D.W and Crafford, K., 2004. The retail of South African game meat: current trade and marketing trends. South African Journal of Wildlife Research 34: 123-134.

Hoffman, L.C., Wolmarans, W.J., Smith, C. and Brand, T.S., 2012. Effect of transportation on ostrich (*Struthio camelus*) weight loss and meat quality. Animal Production Science 52: 1153-1162.

HSA (Humane Slaughter Association), 2016. Captive-bolt stunning of livestock. Available at: https://tinyurl.com/4fp89r8p.

Hultgren, J., 2022. Avoiding live-animal transport to slaughter: mobile abattoirs. In: Faucitano, L. (ed.) Preslaughter handling and slaughter of meat animals. Wageningen Academic Publishers, Wageningen, the Netherlands, pp. 390-434.

Isberg, S., 2016. Farming dinosaur cousins: the unique welfare challenges of farming crocodiles. In: Hermesch, S. and Dominik, S. (eds.) Breeding focus 2016 – improving welfare. University of New England, Armidale, Australia, pp. 67-76.

Jago, J.G., Harcourt, R.G. and Matthews, L.R., 1997. The effect of road-type and distance transported on behaviour, physiology and carcass quality of farmed red deer (*Cervus elaphus*). Applied Animal Behaviour Science 51: 129-141.

Jago, J.G., Hargreaves, A.L., Harcourt, R.G. and Matthews, L.R., 1996. Risk factors associated with bruising in red deer at a commercial slaughter plant. Meat Science 44: 181-191.

Kelly, R.W., Fennessy, P.F., Moore, G.H., Drew, K.R. and Bray, A.R., 1984. Management, nutrition, and reproductive performance of farmed deer in New Zealand. In: Wemmer, C.M. (ed.) Biology and management of the cervidae. Smithsonian Institution Press, Washington, DC, USA, pp. 450-460.

Kuba, J., Landete-Castillejos, T. and Udała, J., 2015. Red deer farming: breeding practice, trends and potential in Poland – a review. Annals of Animal Science 15: 591-599.

Kudrnáčová, E., Bartoň, L., Bureš, D. and Hoffman, L.C., 2018. Carcass and meat characteristics from farm-raised and wild fallow deer (*Dama dama*) and red deer (*Cervus elaphus*): a review. Meat Science 141: 9-27.

La Grange, M., 2006. The capture, care, and management of wildlife: comprehensive studies on the management, capture, and translocation of wildlife from source to final release. Van Schaik Publishers, Pretoria, South Africa, 314 pp.

Langridge, M., 1992. Establishing a fallow deer farm: basic principles. In: Asher, G.W and Langridge, M. (eds.) Progressive fallow deer farming: farm development and management guide. Ruakura Agricultural Centre, Ruakura, New Zealand. p. 17-28.

Lincoln, G.A., Youngson, R.W. and Short, R.V., 1970. The social and sexual behaviour of the red deer stag. Journal of Reproduction and Fertility 11: 71-103.

MAFF, 1989. Guidelines for the transport of farmed deer. Ministry of Agriculture Fisheries and Food [now DEFRA], London, UK. Available at: https://tinyurl.com/332dp7x8.

Milner-Gulland, E.J. and Bennett, E.L., 2003. Wild meat: the bigger picture. Trends in Ecology and Evolution 18: 351-357.

Musa, A.S., Needham, T., Kotrba, R., Neradilova, S., Ganswindt, A. and Ceacero, F., 2021. Habituation of common eland (*Taurotragus oryx*) to intensive routine handling, and the effect of immunocastration thereon. Applied Animal Behaviour 237: 105294. https://doi.org/10.1016/j.applanim.2021.105294

NAWAC, 2018. Animal welfare act 1999, code of welfare: transport within New Zealand. Available at: https://www.mpi.govt.nz/dmsdocument/1407/direct.

Needham, T., Engels, R., Bureš, D., Kotrba, R., Janse van Rensburg, B. and Hoffman L., 2020b. Carcass yields and physiochemical meat quality of semi-extensive and intensively farmed impala (*Aepyceros melampus*). Foods 9: 418. https://doi.org/10.3390/foods9040418

Needham, T., Kotrba, R., Hoffman, L.C. and Bureš, D., 2020a. Ante-and post-mortem strategies to improve the meat quality of high-value muscles harvested from farmed male common eland (*Taurotragus oryx*). Meat Science 168: 108183. https://doi.org/10.1016/j.meatsci.2020.108183

O'Dell, 2016. Game capture methods used in game ranching. In: Oberem, P. and Oberem, P. (eds.) The new game rancher. Briza Publisher, Pretoria, South Africa, pp. 106-120.

Pfitzer, S., Ganswindt, A., Fosgate, G.T., Botha, P.J. and Myburgh, J.G., 2014 Capture of farmed Nile crocodiles (*Crocodylus niloticus*): comparison of physiological parameters after manual capture and after capture with electrical stunning. Veterinary Record 175: 304. https://doi.org/10.1136/vr.10243.

Price, S., Silby, R.M. and Davies, M.H., 1993. Effects of behaviour and handling on heart rate in farmed fallow red deer. Applied Animal Behaviour Science 37: 111-123.

Reiner, G., Seitz, K., Dzapo, V., 1996. A survey of farming environment and ostrich behaviour in Germany. In Deeming, D.C. (ed.) Improving our understanding of ratites in a farming environment. Ratite Conference, Oxfordshire, UK, pp. 23-24.

Schaefer, A.L., Jones, S.D.M., Robertson, W.M., Brereton, D.A. and Jeremiah, L.E., 1995. Carcass yield and meat quality of ostriches under two different ante mortem management regimes. Final Report for the Canadian Ostrich Association.

Spiegel, N.B. and Greenwood, P.L. 2019. Meat production from wild kangaroo: the species, industry, carcass characteristics and meat quality traits. In: Lorenzo, J.M., Munekataa, P., Barba, F.J. and Toldrá, F. (eds.) More than beef, pork and chicken – the production, processing, and quality traits of other sources of meat for human diet. Springer, Cham, Switzerland, pp. 347-383.

Spiegel, N.B. and Wynn, P.C., 2014. Promoting kangaroo as a sustainable option for meat production on the rangelands of Australia. Animal Frontiers 4: 38-45.

Taylor, W.A., Lindsey, P.A. and Davies-Mostert, H., 2015. An assessment of the economic, social and conservation value of the wildlife ranching industry and its potential to support the green economy in South Africa. The Endangered Wildlife Trust, Johannesburg, South Africa, 120 pp.

Taylor, W.A., Lindsey, P.A., Nicholson, S.K., Relton, C. and Davies-Mostert, H.T., 2020. Jobs, game meat and profits: the benefits of wildlife ranching on marginal lands in South Africa. Biological Conservation 245: 108561. https://doi.org/10.1016/j.biocon.2020.108561

Van Schalkwyk, D.L. and Hoffman, L., 2016. Guidelines for the harvesting & processing of wild game in Namibia 2016. Ministry of Environment & Tourism, Windhoek, Namibia. Available at: http://hdl.handle.net/10019.1/99655.

Van Schalkwyk, S.J., Hoffman, L.C., Cloete, S.W.P. and Mellett, F.D., 2005. The effect of feed withdrawal during lairage on meat quality characteristics in ostriches. Meat Science 69: 647-651.

Vercauteren, K.C., Anderson, C.W., Van Deelen, T.R., Drake, D.W, Walter, W.D., Vantassel, S.M. and Hygnstrom, S.E., 2011. Regulated commercial harvest to manage overabundant white-tailed deer: an idea to consider? Wildlife Society Bulletin 35: 185-194.

Wiklund, E., Rehbinder, C., Malmfors, G., Hansson, I. and Danielsson-Tham, M.L., 2001. Ultimate pH values and bacteriological condition of meat and stress metabolites in blood of transported reindeer bulls. Rangifer 21: 3-12.

Wilson, G.R. and Edwards, M., 2019. Professional kangaroo population control leads to better animal welfare, conservation outcomes and avoids waste. Australian Zoologist 40: 181-202.

Wirtu, G., Cole, A., Pope, C.E., Short, C.R., Godke, R.A. and Dresser, B.L., 2005. Behavioural training and hydraulic chute restraint enables handling of eland antelope (*Taurotragus oryx*) without general anaesthesia. Journal of Zoo and Wildlife Medicine 36: 1-11.

World Ostrich Association, 2007. WOA welfare codes for ostrich. Available at: http://www.ostrich.org.uk/industry/welfare.pdf.

Photos: Genaro Miranda-de La Lama, University of Zaragoza, Spain

Chapter 9

Species destined for non-traditional meat production: 2. Goats and South American domestic camelids

G.C. Miranda-de la Lama[1], W. Mamani-Linares[2] and L.X. Estévez-Moreno[1]*

[1]Department of Animal Production & Food Science, Faculty of Veterinary Medicine, Agri-Food Institute of Aragon (IA2), University of Zaragoza, Miguel Servet, 177, 50013 Zaragoza, Spain; [2]College of Veterinary Medicine, Public University of El Alto (UPEA), La Paz, Bolivia; genaro@unizar.es

Abstract

The ability of goats, llamas and alpacas to adapt to management and breeding, even in extreme climatic conditions, made these animals a valuable resource in ancient and contemporary human societies. The consumption of their meat is an important part of the culinary culture and agri-food heritage of many societies around the world. In addition, there is a growing demand for this type of meat from new consumer segments in developing and emerging countries seeking alternative sensory and culinary experiences to conventional ones. Therefore, traditional and alternative consumption largely determine the quality of preslaughter operations, slaughter infrastructure, logistics and transportation. In addition to the usual handling practices faced by goats and camelids during transport and preslaughter operations, there are a number of environmental stressors that are not directly dependent on handlers or, if dependent on handlers, are often unavoidable. These include aspects, such as journey time, inspection and/or rest stops, densities and location of animals in truck compartments, variations in air quality and temperature changes, vibration and noise. The effect of some of these factors on animal welfare has often been underestimated or overestimated. It is therefore important to have an integrated view of the phenomena associated with transport, identifying and prioritising, according to the particular conditions of each system, the most frequent risk points. This chapter presents the main environmental factors affecting the welfare of goats and camelids during transport and preslaughter operations, although there may be differences between countries depending on the type of trucks used, slaughter infrastructure, breed-related susceptibility to stress, environmental-geographical conditions, regulations and idiosyncrasies of each country. In addition, we present scientific and technical elements to consider in decision making, with the objective of preventing, mitigating or controlling the risk factors that cause suffering to goats, llamas and alpacas throughout the handling, transport and slaughter operations.

Keywords: alpaca, goat, llama, pre-slaughter stress, slaughter procedures, transport operations

9.1 Introduction

During preslaughter operations, even under favourable conditions, animals are exposed to a range of potential stressors, which may compromise their welfare, health and performance, including increased handling and human contact, loading and lairage, unfamiliar environments, food and water deprivation, alterations in weather conditions and also changes in social structure through separation, mixing and crowding, vibration, noise and environmental pollutants (Miranda-de la Lama *et al.*, 2014). Subsequently, animal welfare during transport depends greatly on the attitudes and training of hauliers and on appropriate facilities for small ruminants handling (Pulido *et al.*,

2018). Incorrect handling and transportation are also responsible for stress-induced carcass and meat quality problems, such as shrinkage of the carcass, higher pH, dark meat and skin bruises (Miranda-de la Lama *et al.*, 2010). Severely blemished carcasses are the main cause of goat carcass condemnation in Africa (Kusiluka *et al.*, 1995; Mandefro *et al.*, 2015; Regassa *et al.*, 2013).

In general, the international meat industry has experienced unprecedented growth and modernisation in terms of animal welfare, traceability and operational capacity, especially aimed at improving preslaughter systems. However, most of these technologies and improvements are applied to the species of greatest demand and consumption in national and international markets. In the case of goats and South American domestic camelids (SADC), many of these technologies and systems are not available or their use has not become widespread. Thus, goats and SADC can go directly to slaughterhouses, to livestock markets and to intermediate dealers, but the market is normally small and scarcely developed, meaning that there is a considerable amount of animals that are slaughtered for home consumption and do not pass through slaughterhouses. Moreover, this meat is rarely found for sale in butcher shops or supermarkets (Gallo and Huertas, 2016). This may be because the sale and consumption of meat from goats and SADC traditionally have had an important role in the diets and livelihoods of rural communities around the world, especially in many developing countries. Throughout history, meat from goats, llamas and alpacas has circulated in local and regional markets, with an important cultural component. In recent decades, their consumption is reaching urban or middle and high income environments, both in the breeding regions and in places far from the latter. Such dissemination has been linked to government programs, specialised commercial initiatives, and the dynamics of migration from the countryside to the cities (Markemann *et al.*, 2009).

9.2 Goats

The bezoar (*Capra aegagrus*) has been widely accepted as the common ancestor of the domestic goat (*Capra hircus*) (Amills *et al.*, 2017). This species was one of the first wild herbivores to be domesticated in the Neolithic, at least 10,000 years ago, in the area known as the Fertile Crescent, which corresponds to part of the territories of the Mediterranean Levant, Mesopotamia, and Persia (Daly *et al.*, 2018). The domestication of goats was a gradual process, with the animals first being hunted and later kept and managed by humans as a constant food source (Visser, 2019). Initially, goat domestication involved the selection of individuals that showed tolerance to human presence, including resistance to captivity or restricted grazing. Early selection criteria included the ability to breed in captivity, disease resistance, ability to survive better in arid environments, small horns, body size and coat pigmentation patterns (Scanes, 2018). Increasingly, goat farming became popular in Eurasia and Africa along with trade, warfare, thievery and human migration. However, the globalisation of the species took place between the 16th century when goats arrived in America and the 18th century in Japan and Australia, with the European colonisation of these territories (Gootwine, 2020).

During the last century, in the period of the great wars and the post-war period, goat breeding increased to alleviate meat and milk shortages. However, after these periods of food crisis, goat production was gradually replaced by cattle and sheep (Aréchiga *et al.*, 2008). Actually, goat farming can be commonly found in less productive areas, where production efficiency is not the main goal and inputs are low (i.e. areas where high-income crops cannot be cultivated, or high-income animals cannot be reared). In particular, goats do not require high inputs in terms of feed and housing, therefore, they are generally used in disadvantaged areas (Zucali *et al.*, 2020). However, in the last 30 years' intensive goat farming (especially for milk) has modernised and spread around the world. This is due to the increase in demand for goat products, mostly resulting from national and international migratory flows (Miranda-de la Lama, 2018), and a progressive popularity of goat meat and milk due to their unique nutritional properties, physicochemical composition, and sensory attributes, while having species-specific flavours compared to other red meats and dairy products (Watkins *et al.*, 2021). Goats have played an important role in modern agriculture, although they have differing functions in various cultures. On the one hand, they are a reliable source of meat, milk, skin and fibre, but they are also kept for their role in traditional and cultural ceremonies and are still regarded as almost sacred in many cultures (Visser, 2019). Goats are also valued for their productivity, adaptation capacity and disease resistance (Daskiran *et al.*, 2018).

Currently, there are an estimated 860 million goats in the world, with most goats being raised in Asia (55%) followed by Africa (28%), Americas (11%) and Europe (6%) (FAO, 2020). Devendra (2010) reported that globally there are 1,156 different goat breeds, which can be categorised as having three main purposes: meat (i.e. Boer and Kiko), milk (i.e. Saanen, Murciano-Granadina and Nubian) and fibre (i.e. Cashmere and Angora). Kids or suckling kids are also a valuable by-product of dairy, meat and hair production systems. It is important to note that 85% of these breeds originate from Europe (215 breeds or 37.2% of total), Asia (184 breeds or 32% of total) and Africa (95 breeds or 16.5% of total). However, 70% of the world's goat meat and 60% of the world's goat milk are produced in traditional, extensive and nomadic systems in Asia. In contrast, in Europe and America there is a trend towards intensification for dairy production, although traditional and self-subsistence production persists on both continents. Native goat breeds have an important socio-economic role in the livelihood strategies of poor farmers (Figure 9.1), especially those in rural and hard-to-reach areas (Daskiran *et al.*, 2018). Consequently, goats are exposed to a wide variety of systems for rearing, handling, transport and preslaughter operations, which can range from the most traditional and rudimentary to high-tech systems. Despite the fact that goats are one of the most popular and dispersed domestic species in agroecosystems around the world, the knowledge about its behaviour is generally based on sheep studies.

9.2.1 The goat meat market

The popularity of goat meat is subject to the common culture of communities and the forces of civilisation (Webb *et al.*, 2005). Consumers in the Western world do not favour goat meat, although their perception towards this meat is changing because of its healthy properties (Webb, 2014). Compared to other red meats, such as beef and lamb, chevon has lower fat, saturated fat and cholesterol content, but a higher polyunsaturated fatty acid (Mazhangara *et al.*, 2019). For marketing purposes, the categories usually slaughtered are kids or capretto, chevon, and cull goats for slaughter. Kids, meaning animals containing no permanent incisors and which are often still suckling or weaned, produce a light carcass weight (6 to 15 kg; Figure 9.2A) and pink meat. This production is popular in Latin America and the Mediterranean region, where the demand for capretto is high during Christmas and Easter holiday periods (Pophiwa *et al.*, 2020). Chevon meat is from mature

Figure 9.1. A herd of native goats of the Spanish 'Blanca Celtibérica' breed. Many of these breeds are part of extensive and sustainable production systems (photo courtesy of M. Campo, University of Zaragoza, Spain).

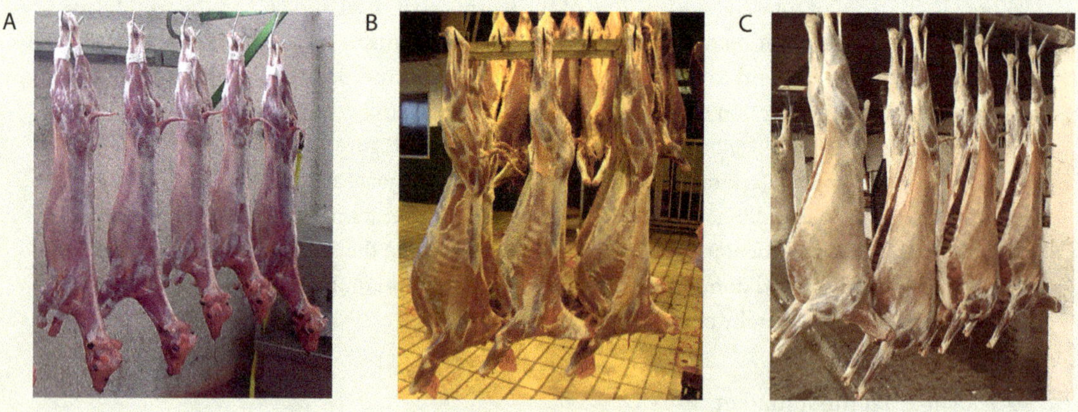

Figure 9.2. The three most common commercial categories in the world goat meat industry are: (A) kids (Spanish Moncaina breed); (B) chevon (Spanish Bermeya breed); (C) cull-goats (Alpine breed) (photo courtesy: A and B are courtesy of M. Campo, University of Zaragoza, Spain; C of G. Palacios, Abattoir of the Municipality of Calpulhuac, Mexico).

goats that present a carcass weight above 16 kg (Figure 9.2B) and produce a typically darker meat which is particularly appreciated in most parts of Africa and India (Dieters *et al.*, 2021). Cull goats usually come from dairy systems and present carcasses heavier than 20 kg (Figure 9.2C) and are particularly produced in most parts of Africa and Latin America.

The demand for goat meat obtained through religious slaughter is dependent on the festivities of each religion, both in their countries of origin and in those where those people have emigrated. This has forced the meat industry and government regulations to adjust to this new commercial trend. The Jewish consumer prefers kids between 9 and 18 kg live weight that are normally consumed during the New Year – Rosh Hashanah –, Passover – Pesach – and festival of lights – Chanukkah – (Ekanem *et al.*, 2013). In the case of Muslims, for the start of Ramadan and the festival of the breaking of the Ramadan fast (Id al Fitr), consumers prefer male, entire or castrated, but lean, and female kids with all their milk teeth (i.e. not older than 12 months). Optimum live weight is about 30 kg, but weaned kids from 20-55 kg are also accepted by different buyers. For the Islamic Festival of Sacrifice (Id al Adha), consumers prefer yearlings (i.e. animals with one set of adult teeth) that are blemish free. Animals with broken horns, open wounds, torn ears or physically unsound generally do not meet the criteria. In some cases, castrated animals are frowned upon (Luginbuhl, 2014). Finally, the consumption of goat meat obtained by the Jhatka slaughter method (see description in Section 9.5) in India is less dependent on festivities; it is highly appreciated, but is more expensive than other meats. Generally, consumers prefer animals between 6 and 9 months old, but in some regions older than one-year goats are also slaughtered (Sivachandiran *et al.*, 2020).

9.2.2 Principles of goat behaviour applied to handling

Goats, like other ungulates, have laterally positioned eyes, which gives them a monocular vision that provides a wide panoramic vision of more than 300°. However, their binocular vision, formed by the overlapping of the visual fields of the two eyes, is poor (Miranda-de la Lama, 2019). In contrast, goats have 55° of vertical vision compared to 140° in humans (Banks *et al.*, 2015), with rectangular and horizontal pupils helping them have a panoramic vision which is useful to detect potential predators (including humans), regardless of the direction from which they come (Heesy, 2004). In addition, this ocular adaptation prevents excessive light from entering from above, which could dazzle the animal and makes it more vulnerable to predation. The major axis of the pupil is always kept parallel to the ground through eyeball rotation in the eye socket (cyclovergence), whether the goat is looking at the horizon or grazing, allowing optimal panoramic vision to be maintained at all times (Banks *et al.*, 2015). Probably due to their lack of depth vision and limited ability to focus on objects, goats generally avoid abruptly crossing shadows cast on the ground and are susceptible to changes in textures or colours (Blakeman and Friend, 1986). Goat vision is dichromatic (two types of cones) which makes them sensitive to short or long wavelengths (bright purple and yellow-green colour range, respectively). However, the goat brain can combine eye signals and are thus able to distinguish between orange, yellow, green, blue and violet (Adamczyk *et al.*, 2015). In terms of visual abilities, although goats exhibit a marked interest in looking in mirrors, they are not able to recognise themselves in them, as it is the case in other mammals (Hals, 2016). This interest may be due to the innate need for socialisation of the species.

Olfaction is less important to goats in handling situations, although they have a strong olfactory apparatus and are able to make keen olfactory discriminations for social recognition. Their auditory range is from 78 to 37 kHz with a well-defined point of best sensitivity at 2 kHz (Heffner and Heffner, 1992). However, sound localisation acuity is relatively poor (Van der Staay *et al.*, 2011). The alarm call or alertness in goats is characterised by a kind of 'trumpeting' sound, accompanied by a sudden movement of the head and striking the ground with a foreleg. When this call is made, other goats quickly stand up and repeat these behaviours. Goats can differentiate between group members only on the basis of their body shape (Keil *et al.*, 2012) and can identify offspring by their vocalisations (Briefer *et al.*, 2012). They can also integrate visual and auditory cues to identify herd mates, especially for social partners (Pitcher *et al.*, 2017), indicating higher-order cognitive representations of individuals (Zobel *et al.*, 2019). These cognitive skills should be considered when preparing goats for transport as arbitrary social regrouping can trigger aggression in socially stable groups (Miranda-de la Lama *et al.*, 2011). Similar to other species, social mixing of familiar goats may be less detrimental than mixing groups of unfamiliar ones (Miranda-de la Lama *et al.*, 2012).

Human-goat relationships can be defined as the degree of association, distance and mutual perception between humans and animals. The establishment of an affinity between handler and animals is strategic for the welfare of the animals. In general, the lack of positive interactions with handlers can have negative repercussions on handling quality (Boivin and Braastad, 1996) as also showed by Jackson and Hackett (2007) who demonstrated that goats that have undergone negative or rough handling are more likely to have a larger flight zone or a greater latency time to approach people that those treated more positively or gently. Although not as sensitive as dogs, goats have developed certain abilities to respond to human body language, so they can predict potentially dangerous handling situations (Kaminski *et al.*, 2005). Nawroth *et al.* (2018) found that goats can discriminate between positive and negative human gestures, regardless of whether they are from man or woman, and between handlers by facial features and size. Moreover, like dogs, they can learn from humans to solve cognitive challenges, such as finding hidden food and transmitting that new knowledge to other herd members (Nawroth *et al.*, 2016). The memory of previous experience or knowledge acquisition can last for up to 4 months (Briefer *et al.*, 2014).

Specialised dairy breeds, e.g. Saanen, Alpine, Toggenburg, Anglo Nubian and Murciano-Granadina, tend to be tamer towards handlers, which is probably related to their milk production and the need to be milked to achieve mammary relief (Battini *et al.*, 2016). In native meat breeds or mixed breeds kept in pasture or transhumance, it is also possible to observe low reactivity to herders because of the frequent interaction with them making them accustomed to their presence and handling (Solaiman, 2010). On the contrary, kids of these breeds that are raised in complete stall feedlot systems tend to exhibit greater fear of handling compared to kids of dairy breeds destined for fattening (Ekiz *et al.*, 2020). Miranda de la Lama *et al.* (2013) found that social dominance influences the distance goats keep with their handlers and the time it takes to capture them, with higher-ranking goats tending to be more reactive compared to lower-ranking goats. However, goat responses to humans are highly conditioned by the quality of the relationship with their usual handlers. Although the process of domestication has substantially decreased their fear of humans, generally goats still exhibit ancestral anti-predatory behaviour (i.e. predatory fear) during capture and physical restraint, difficulty in adapting to unfamiliar environments and poor integration with unknown humans and animals. Kids, due to their post-weaning stress and the innate need to carry out the sucking behaviour, are particularly reactive to these factors. For this reason, goats may suffer more than other ruminant species during changes in housing and production practices, like regrouping, physical restriction, dehorning, claw trimming, castration, and changes in handlers (Sejian *et al.*, 2021). These responses may be exacerbated under novel and unpredictable management conditions, such as preslaughter handling. According to Underwood *et al.* (2015), signs of pain or discomfort in goats are usually bruxism or teeth grinding; other signs of discomfort, stress or disease are decreased time spent eating and chewing the cud, restlessness, prolonged lying with neck and head extended and hunched back when standing. Unhealthy goats may be thin, have external lumps or swollen joints, an unusual abdominal profile or a rough or dull coat.

9.2.3 Pre-transport handling and loading

This stage constitutes the first challenge for goats to be transported, and normally includes selecting the animals to be transported, handling and regrouping them, relocating them in pre-shipment pens, extreme ambient temperatures, overcrowding, fear to handlers, physical isolation or restriction and fasting. These activities, even under good conditions, can cause significant stress in goats, possibly due to their excitable nature (Casey and Webb, 2010), which may contribute to the unappealing dark colour and increased toughness of meat, often confounded with typical chevon meat characteristics (Pophiwa *et al.*, 2020). Kannan *et al.* (2002) in adult goats and Alcalde *et al.* (2017) and Zimerman *et al.* (2011) in kids found that pre-transport operations, such as herding and loading procedures, produced high levels of plasma creatine kinase (CK) even more than transport itself.

Fasting is another common pre-transport practice that consists of suspending the food supply, but maintaining the possibility of access to drinking water. Its benefits include reducing mortality and reducing the likelihood of contamination from spillage of ruminal content during slaughter (Pointon *et al.*, 2012). Although it is recognised that there is scarce information on goats with regard to carcass contamination at slaughter, in a study conducted in two Australian abattoirs, Duffy *et al.* (2009) found that rumen contents were contaminated with *Salmonella* in 72 out of 121 samples tested (59.5%), while faeces were contaminated 57% (68/120 sampled) of the time. However, research provided the evidence that goats should not be fasted for more than 12 h and that any extension of this period should include the provision of water (Greenwood *et al.*, 2010). Greenwood *et al.* (2008) found that three-month-old kids lost 5.9% of their initial live weight during the first 12 h of fasting, 7.5% at 24 h, 10.8% at 48 h and 14.1% at 72 h. In addition, these same authors found that as the hours of fasting increased, carcass yield decreased and the meat became darker (pHu ≥6.0).

Transporting goats showing pathological conditions or poor body conditions at loading can cause marked muscle weakness and the inability to move, increasing the frequency of non-ambulatory animals upon arrival (Miranda-de la Lama, 2019). According to Miranda-de la Lama (2018), when selecting animals to be transported, it is important to identify animals that are unfit for the journey, including: (1) goats close to the time of parturition, with the risk of parturition occurring during transport; (2) weak or emaciated goats (based on body condition score), which are not able to walk and/or stand during the journey; (3) totally blind animals; (4) kids with marked signs of gastrointestinal and/or respiratory disorders (diarrhoea and oculo-nasal secretions, respectively); (5) animals with exposed bone or horn fractures, or convalescent with recent surgical wounds; and (6) animals with obvious lameness. For example, dairy goats are especially prone to hoof overgrowth and lameness due to inadequate shelter and bare concrete floors (Ajuda *et al.*, 2019). In addition, it is important to detect stress-related behaviours and morphological traits that may indicate possible welfare and health problems that affects fitness for transport. Some of these behaviours may include excessive bleating, resistance or overreaction to touch or handling, hyperesthesia, obvious fear or aggression, frenzy, aimless running, compulsive walking, head pressing, constant chewing, teeth grinding, or fluttering of the eyelids (Smith and Sherman, 2009). Health can also be assessed by coat condition. For example, the percentage of goats with a coat described as 'a complete fur cover, even coat, presenting shiny, glossy and sheen hair, homogeneous and well adherent to the body' has been included in the EU-AWIN welfare assessment protocol for goats (Mattiello *et al.*, 2019). Particular consideration should be given at this stage to discarded or cull goats because they are more vulnerable than other commercial categories. According to Gautam *et al.* (2017), goats are discarded at the farm for health reasons (i.e. respiratory diseases, udder, locomotor and digestive disorders, and abortions) and because they are nervous, emaciated, old, infertile or low producers. A discarded goat will be more vulnerable to fall during the journey or to being stepped on by other animals.

Pre-shipment pens are recommended to help select, catch, weigh and loading goats. Many facilities include a gathering corral, holding pen, chute (alleyway), head catch or working box, and a loading area. Chain-link fences are dangerous because goats are curious and tend to stand on their hind legs against fencing or walls. Forelimbs and horns may be caught easily in the mesh (Underwood *et al.*, 2015). The walls of the pens should be at least 1.50 m high and solid to prevent goats from jumping over the fence (Vincent, 2005). In operations with a small number of goats, welded livestock panels can be used to create a circular corral to catch the goats. A major advantage of this type of corral is that it is relatively inexpensive and can be assembled and disassembled quickly (Solaiman, 2010). When handling horned goats in close quarters, handlers should use protective eye wear to prevent face and eye injuries (Smith and Sherman, 2009). While many farmers often use dogs to protect goats from predators, it is not recommended to use them for sorting, loading and unloading operations. This is because, compared to sheep, goats are more fearful of handling, resulting in jumping and injuries (Vincent, 2005). Animals that become very stressed in pre-loading pens may die during transport, especially when transport is also stressful (Taunde *et al.*, 2019). Thus, it is especially important that goats being sent to slaughter are handled as quietly and efficiently as possible. Isolation is highly associated with fear and discomfort in goats as showed by increased vocalisation, locomotor activity and escape attempts (Winblad von Walter *et al.*, 2021). Thus, this factor must be considered when selecting and preparing animals for transport.

Ideally, the slope of the loading ramp to the vehicle should be 20° or less (Miranda-de la Lama, 2018). Furthermore, there should be no gap between the ramp and the vehicle in order to prevent animals from getting stuck in between and injured. The direction of the slope of the ramp is another important factor as goats prefer moving upslope rather than downslope (Altinçekiç and Koyuncu, 2013). Although goats exhibit gregarious behaviour, they tend to show less social facilitation than sheep (Miranda-de la Lama and Mattiello, 2010). Therefore, it is expected adult goats to be more exploratory and individualistic, and not necessarily to get on the truck together immediately. Minka *et al.* (2009) found that loading took longer because the goats were especially reactive (slips, falls, and jumps) to handling. In kids, loading time can be shorter by using trained adult goats, because at an early age these animals tend to be behaviourally synchronised with adult ones (Miranda-de la Lama, personal observations). Unlike other domestic species, goats show a high motivation to explore the interior of the truck (Figure 9.3A). Considering this behavioural peculiarity, it is important to give them the possibility to explore and get familiar with the interior of the truck without the use of electric prods or blows. Therefore, it is particularly recommended to check the state of maintenance of the truck body, as well as the presence of gaps or holes where goats can be totally or partially trapped by the legs, head or horns. Once loaded and while awaiting the start of the journey (or in any stationary situation, i.e. truck stops), goats tend to compete for access to better ventilated and comfortable areas (Figure 9.3B). In these circumstances proper ventilation is critical. In case of poor ventilation, increased ammonia and other waste gases levels may induce respiratory problems (Underwood *et al.*, 2015).

9.2.4 Transport

The methods by which goats are transported are quite diverse due to their great territorial dispersion and can include journeys between farms, or from farm to market, feedlot, abattoir or even for export. In some developed countries, goats are herded for several days to reach local markets, abattoirs, or for export (Alarcon *et al.*, 2017). There is also a diverse range of vehicles that are used to transport goats, ranging from common small trucks to small trailers and specialised three-tier trailers (Figure 9.4). Goats are often transported in the same truck with other species, such as sheep. Smallholders with their own means usually transport small groups of animals for replacement or sale in markets or directly to the abattoir, although there is a trend towards using cooperatives that,

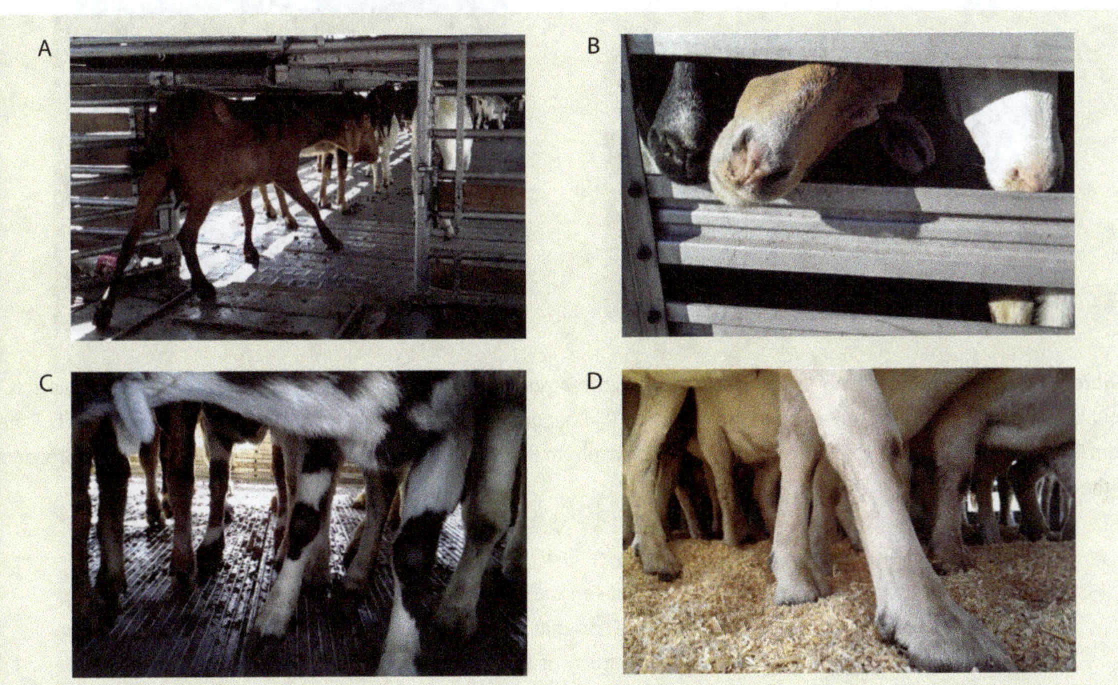

Figure 9.3. Important features of boarding the goats onto the truck: (A) goats boarding and exploring the compartment where they will journey; (B) when the truck is parked, animals tend to seek out and compete for the best ventilated areas; (C, D) the use of bedding material is important to control the humidity of the floor, but it is also a substrate that helps to enrich the environment during the journey (G. Miranda-de La Lama, University of Zaragoza, Spain).

Figure 9.4. Some examples of vehicles used in goat transport around the world: (A) pick-up trucks; (B) single-deck trailers; (C) double-deck trailers; (D) large trailers up to three-deckers (E) Boer goats loaded on a 4-tier pot-belly trailer (photos A, C, D, E are from G. Miranda-de La Lama, University of Zaragoza, Spain; photo B is courtesy of J.M. Micheo, Spanish Association of Breeders of the Malagueña Goat, Spain).

among other advantages, can offer a transport service (Miranda-de la Lama *et al.*, 2010). For cultural and religious reasons, there is a high demand for goat meat in some countries, which has stimulated an international trade that involves the transport of live goats, involving long-distance journey by road, boat, and even air. The use of straw or sawdust bedding for the journey is a common practice and is promoted since it can be an enriching and comforting stimulus while it is dry (Figure 9.3C,D). Goats are quite comfortable in colder temperatures if provided adequate amounts of dry dust-free bedding (Underwood *et al.*, 2015). However, the use of bedding in transport is also controversial, because it can favour falls by making the floor slippery and makes the trailer difficult to clean which may result in spread of pathogens. On the other hand, bedding material can favour the absorption of excrements.

Specialised trucks have compartments inside, which help to balance the load and control the movement of the animals during the journey, decreasing mostly longitudinal movements and to a lesser extent lateral movement (Miranda-de la Lama, 2018). The orientation and position an animal adopts during the journey can influence its ability to maintain postural control, stability and adapt to the vibrations of the truck. In this sense, Das *et al.* (2001) observed that goats transported 50 min by road spent most of the time parallel to the direction of vehicle, but with frequent diagonal and perpendicular changes in orientation to maintain balance. However, Minka and Ayo (2014) found that after the first hour of the journey goats changed their positions from parallel to perpendicular. It also seems that animals prefer a standing position during transport and only lie down when fatigued or exhausted, which occurred after 7 to 8 h of transportation. These authors also reported that goats that adopted the standing or lying down positions that were perpendicular or opposite to the direction of traffic had predominantly lower values of blood CK, alanine aminotransferase, aspartate aminotransferase, glucose and urea, which suggested less muscular damage and fatigue due to the choice of the best journey position.

Oxidative stress associated with transport occurs when the antioxidant defence system is overwhelmed by an increased oxidant burden or a reduced antioxidant supply resulting in cell damage and an increased susceptibility to disease if not adequately mitigated by antioxidant responses (Piccione *et al.*, 2013). Studies on oxidative stress have found that goats subjected to 7 h of transportation in hot and humid tropical conditions experience hematologic disturbance and oxidative stress, needing 3 to 16 days to recover (Polycarp *et al.*, 2016). Erythrocyte osmotic fragility in particular is a good indicator of transportation stress and represents an important index for quantifying oxidative stress in goats, which tends to decrease 24 h after the end of transportation (Fazio *et al.*, 2016). Also, oxidative stress can increase the expression of some of heat shock proteins (HSPs) in muscle and parenchymatous organs. Hu *et al.* (2020) found that HSP27 and HSP90 are preferentially increased during short journeys (2 h) and HSP70 is increased after longer time journeys (>6 h) in primary parenchymatous organs of goats. Perhaps the lung is the most vulnerable organ, while the heart may be the most resilient to transport stress. Zheng *et al.* (2021) confirmed these results and found that journeys longer than 6 h can cause a morphological impairment of lung, trachea and bronchia followed by the change of HSP70 expression. These results would explain why journeys in difficult conditions, such as high densities, long distances and poor vehicle design, can exacerbate respiratory conditions in goats. For example, Taunde *et al.* (2019) found that after long distance journeys (3,700 km) in tropical climates, goats presented clinical signs, such as dyspnoea, mucopurulent nasal discharge, and coughing, resulting from pleuropneumonia caused by non-haemolytic strains of *Mannheimia haemolytica*. Kannan *et al.* (2003) studied the effects of 2 h transport on physiological stress responses and meat quality characteristics in young goats (6 to 12 months of age) and old goats (24 to 30 months of age). These authors found that the muscle glycogen concentrations decreased due to transportation, with a more pronounced effect in young vs older goats, although the magnitudes of blood cortisol and glucose responses to stressor treatment were greater in older goats.

Goats are considered to be more tolerant to heat stress than dairy cows because they are more able to sweat and have a lower body weight. The heat tolerance limit in goats is between 35 and 40 °C (Salama *et al.*, 2014). However, transportation in hot climates causes stress responses in goats. Rajion *et al.* (2001) and Kannan *et al.* (2000, 2003) found that transporting goats for 1.5-2 h in humid tropical climates (25 to 30 °C) caused dramatic increases in plasma glucose concentrations and neutrophil-lymphocyte (N:L) ratios, with these ratios returning to pre-transport values about 12 h after transport. In goats transported for 3 h under hot humid tropical conditions (31-32 °C and 85-88% of relative humidity), Zulkifli *et al.* (2010) reported an increase in rectal temperature, leukocyte counts and circulating levels of cortisol. Also, Kadim *et al.* (2006) found that goats transported for 2 h in hot dry climates (35 °C) had higher plasma cortisol, adrenaline, noradrenaline and dopamine levels, increased body weight losses, and lower meat quality (i.e. darker and tougher meat). In a more recent study, meat from goats transported for 6 h in hot weather (42 °C) presented a greater pHu and shear force values compared to goats that were not transported (Kadim *et al.*, 2014). In 3.5 h transports under hot humid tropical conditions, Nikbin *et al.* (2016) found that goats transported at high densities (0.4 m^2 per animal) had darker meat and a higher pHu compared to goats transported at low density (0.2 m^2 per animal). In goats subjected to journeys of 6 h in temperate climates (range 7-14 °C), plasma cortisol and glucose levels decreased rapidly and returned to base values 6 h after the end of journey (Nwe *et al.*, 1996). These results show that at the start of transportation stress, the sympathetic nervous system is activated and then the adrenal medulla and cortex are simultaneously activated, followed by immediate physiological changes. It seems that the starting phase is the most important, and hence controlling the reaction of animals at the start of transportation may reduce their stress response.

The transport and slaughter of kids is a problem for the public image of the livestock industry, as public opinion often considers these practices to be cruel and unethical (López and Venegas, 2018). This is one of the reasons why some countries of north of Europe send goat kids to other Mediterranean countries to be slaughtered. For farmers and hauliers, transporting kids requires specialised transportation, precise handling, preventive medicine and logistics management. Hindle *et al.* (2013) studied the effects of suckling kid (8-10 kg) loading densities in a three-tier trailer (5, 7.5 or 10 animals per m^2) from the Netherlands to Spain (1,400 km journey). They reported that there were no significant differences in the behaviour and physiological parameters, although kids lost between 2.4 and 5.9% (or 0.2 and 0.5 kg) of their body weight. In weaned kids undergoing a long distance (>20 h) air and land journey from Australia to Malaysia, Shahrom and Zamri-Saad (2012) found a gradual increase in post-arrival mortality as early as week 1, with an average of 5% mortality reaching a peak at week 3 with 35% mortality, before it gradually decreased to 6% at week 6 and 3% at week 7. Shipping fever was the major cause of mortality reported in that study. Naive immune systems and lack of exposure to new environments in kids make them more vulnerable than adult goats (Miranda-de la Lama *et al.*, 2014). According to Malher and Noordhuizen (2008), the most common microbiological risks for kids that can pose problems before, during and after transportation are respiratory diseases (*Pasteurella* and *Mycoplasma* spp. after weaning) and digestive disorders, such as *Escherichia coli* diarrhoea (first week of life), *Cryptosporidium* diarrhoea (second and third weeks of life), ecthyma (up to 2 months old) and coccidiosis (from 1 to 5 months old). Before and after transportation kids need more water than adults due to their body structure, since they are growing faster, and their metabolism is also faster than older animals. Consequently, young animals are more prone to dehydration than adults. In general, pre-ruminants, such as suckling kids, are more susceptible to periods of food and water deprivation during transportation than weaning kids, because the rumen acts as a reservoir of potential nutrients and water (De la Fuente *et al.*, 2010).

9.2.5 Handling at the slaughter plant

Handling, loading, transportation and unloading at the abattoir adversely affect the electrolyte balance of goats and result in respiratory alkalosis, dehydration and muscular damage (Sejian *et al.*, 2021). Thus, a rest in lairage before slaughter is needed to help to reduce dehydration and regenerate muscle glycogen levels, favouring the most affected animals, and to prevent from production of dark cutters. These positive effects are obtained after a 3 h lairage (Kannan *et al.*, 2000). Extended lairage periods of more than 18 h without feed lead to economic losses due to increased stress and live weight shrinkage (Kannan *et al.*, 2002). From the logistics point of view, this interval is vital for slaughter planning per batch of animals (by destination and hours after admission), since it allows gastrointestinal emptying preventing possible contamination of carcasses and gives time for static and dynamic veterinary inspection (Miranda-de la Lama, 2013). However, similar to other species, animal response to lairage might be influenced by several factors (and their interactions), such as the concomitant fasting, environmental conditions, social mixing, previous preslaughter treatments (transport, unloading, mixing) and individual stress susceptibility. These stressors may lead to some changes in blood constituents, such as plasma concentration of CK, lactate dehydrogenase, glucose, cortisol and packed cell volume (Ekiz *et al.*, 2012).

One of the most common repercussions of transport and preslaughter operations on goats is fatigue. There are some factors that can exacerbate the presence of severely fatigued animals, such as the length of the journey, stressful handling, the position within the truck, very high or low densities, poor quality roads and the skills of the truck driver and individual predisposition (Miranda-de la Lama, 2018). Severely fatigued goats can be easily detected during unloading operations due to their inability to exit the truck or go down the ramp. Similar to sheep (Chulayo and Muchenje, 2013), affected goats may also show different clinical signs of fatigue, such as tachypnea with an abdominal component to breathing, muscle tremors, lameness, reluctance to move and prostration. To counteract the effects of fatigue and post-transport stress in goats upon arrival to the abattoir, several studies have assessed the efficiency of formulations containing vitamins, amino acids minerals and/or glycerine in the drinking water or feed. Administration of ascorbic acid (AA) to goats subjected to long distance journey (12 h) in hot and dry climates reduced the incidence of muscle damage as showed by lower blood CK levels at slaughter (Minka and Ayo, 2010). These effects were confirmed by Akinmoladun *et al.* (2020), who studied seven different degrees of water restriction (between 50 and 70% of the recommended daily intake) and AA supplementation in the water for 75 days prior to slaughter. The results of this study showed that, regardless of the degree of water restriction, goats that had been supplemented with AA had lower plasma cortisol concentrations, higher hot and cold carcass weights, and redder meat compared to unsupplemented animals. In contrast, the administration of AA did not have the same beneficial effect on kids transported for 3 h in controlled micro-climatic conditions as on lambs (Sarıbey and Karaca, 2019). However, better effects on the physiological response (i.e. lower plasma cortisol and urea levels) of kids transported for 3 h in hot and humid climates have been reported through the provision of crude glycerine in the feed 14 days prior to slaughter (Matos *et al.*, 2018). However, no effects on carcass and meat quality were found.

9.2.6 Stunning and slaughter

Goats are slaughtered using a wide variety of methods which can be classified as more humane, ritual or traditional. In order for the operator to properly stun a goat, an efficient physical restraint system is necessary. According to Small and Hewitt (2017), goats need to be restrained for stunning and bleeding in order to keep them in correct position for the procedure to be carried out accurately. Restraint in itself is stressful, however, when restraint is combined with isolation, as this situation can evoke more substantial behavioural responses, such as escape attempts by jumping and running away, vocalisations and aggressiveness to the operator. In horned goats, this state of reactivity can be especially dangerous for the operators (injuries to the face and abdomen) and for the animal itself (horn fractures) (Miranda-de la Lama, 2018). Many abattoirs have automated this part of the process through the installation of V-shaped conveyor restrainers, which deliver goats in a continuous flow. They are designed in such a way that animals remain in visual, auditory and tactile contact with each other, avoiding isolation of individual animals (Small and Hewitt, 2017).

Two technologies are commonly used to achieve effective stunning for goats: mechanical and electrical. The mechanical method is based on inducing loss of consciousness by administering a severe blow to the animal's head producing a cerebral concussion, which results in cessation of breathing and cardiac function (Mdletshe *et al.*, 2020). This unconsciousness condition is achieved using a penetrating or a non-penetrating captive-bolt. Both penetrating and non-penetrating devices are efficient for stunning of horned or hornless goats (Collins *et al.*, 2017). The penetrating captive bolt uses a pointed bolt which is propelled by pressurised air, spring mechanism, or a blank cartridge (0.22, 0.25, and 9 mm calibres). Poor shot placement has been identified as the most common cause of incomplete concussion, particularly in poorly restrained horned small ruminants (Cockram and Velarde, 2022 – Chapter 7 of this book; Gibson *et al.*, 2012). Recently, Plummer *et al.* (2018) have confirmed that effective damage to the brainstem and key adjacent structures may be accomplished by firing a penetrating bolt perpendicular to the skull over the anatomic site identified at the intersection of 2 lines, each drawn from the lateral canthus of one eye to the middle of the base of the opposite ear. Non-penetrating devices may be used for kids stunning. Grist *et al.* (2018) recommends the use of non-penetrating captive bolt delivering a calculated kinetic energy of 47 J for euthanasia of neonate kids when a specific shot position, on the midline, between the ears with the chin tucked into the neck, is applied in conjunction with a one grain 0.22" cartridge.

Electrical stunning is commonly used in small ruminants wherein an electric current is passed through the electrodes applied on the forehead of the animal to produce instantaneous insensibility by the induction of cardiac arrest and epileptic form of activity in the brain (Cockram and Velarde, 2022 – Chapter 7 of this book; Kiran *et al.*, 2019). Similar to sheep (Cockram and Velarde, 2022 – Chapter 7 of this book), two types of electric stunners are commonly used for goats: head-only (HO) and head-to-body (HB). Recommended parameters for HO stunning of goats are a voltage between 70 and 250 V and a minimum current of 1.0 A from 2 to 4 s with 50 Hz to achieve satisfactory stunning (Sabow *et al.*, 2018). While for HB stunning an electrical current of 50 Hz and 1 A applied for 3-4 s to both the brain and the heart of the animal is required (Sabow *et al.*, 2017). A major advantage of this technique is that animal welfare is maintained since it the level of unconsciousness is irreversible due to death of the animal by cardiac arrest (Farouk, 2013). According to the European Regulation (EC, 2009) on animal welfare at slaughter, a minimum current of 1.0 A (independently of the type of device) must be applied to obtain effective electrical stunning in adult goats. In kids, current intensities of 0.7 A are recommended by Llonch *et al.* (2015). Sabow *et al.* (2016a,b) examined the influence of low and high frequency HB electrical stunning (50 vs 850 Hz) on bleeding efficiency in goats. Results showed that blood loss was higher in goats stunned with 850 Hz compared to 50 Hz. The higher blood loss in high frequency stunned goats was confirmed by the lack of cardiac arrest. These authors tested the effectiveness of HO or HB electrical stunning with currents of 0.3, 0.5 and 0.7 A and found an immediate loss of consciousness, similar to that produced with 1.0 A, using the three current intensities. However, the percentage of animals that regained consciousness was lower when they were stunned with HB compared to HO.

Bleeding is the final part of the slaughter procedure performed immediately after stunning and consists of severing the main blood vessels of the animal to allow blood to drain from the body, resulting in death from cerebral anoxia. As for sheep slaughter (Cockram and Velarde, 2022 – Chapter 7 of this book)), thoracic stabbing and neck cutting are also used for goat slaughter (Chambers *et al.*, 2001). Neck cutting is inappropriate if the stab or cut is too small to completely sever both carotid arteries and jugular veins preventing a rapid bleeding and/or if bleed-out is further impeded by coagulation of blood in the stab wound (EFSA, 2013). Poor bleeding has at least two direct implications that affect animal welfare and meat safety. The first implication is related to the delay of brain death and the risk that the animal may regain consciousness and then experience pain, fear and distress (Terlouw *et al.*, 2016). Although the incidence of recovery of consciousness in goats due to poor bleeding is not known, this possibility cannot be ruled out (EFSA, 2013). The second implication is related to the fact that blood is an ideal medium for bacterial growth, and therefore a greater volume of residual blood in the carcass due to poor bleeding may reduce the sensorial quality and shelf-life of meat (Qekwana *et al.*, 2017).

Ritual slaughter is carried out for religious reasons and is regulated by norms and mandatory compliance requirements that are part of a specific method. The three ritual methods in goats that are most prevalent in the international sphere are shechitàh (Jewish), halal (Muslim), and jhatka (Sheik), with the latter being the least applied internationally. In general, both the shechitàh and halal methods are based on body restraining, i.e. hoisted before neck cutting, or placed manually on their sides, or upright (Cockram and Velarde, 2022 – Chapter 7 of this book; Gallo *et al.*, 2022 – Chapter 2 of this book; Velarde *et al.*, 2014), and a perfectly clean incision through the structures at the front of the neck: the trachea, oesophagus, carotid arteries and jugular veins (Rosen, 2004). If the cut is performed effectively, the animal will begin to lose consciousness within 10 to 15 s. The absence of signs of life has to be checked within 30-40 s (Cockram and Velarde, 2022 – Chapter 7 of this book; Farouk, 2013). The pain caused by the cut and the time required for loss of consciousness during bleeding are the two main criticisms for both methods. There is experimental evidence of the harmful stimulus of neck cutting, both in conscious and minimally anesthetised goats (Sabow *et al.*, 2016a). Also, Johnson *et al.* (2015) has reported that exsanguination and loss of consciousness may take up to 1 min or even more in some cases. The main difference between the halal and shechitàh methods is that prior to shechitàh/Kosher slaughter, the animal must be alive and conscious, while the Halal rite only requires the animal to be alive. Any form of stunning that renders the animal insensible to pain prior to slaughter is not acceptable for Kosher, while for Halal non-stunning is preferred, but reversible forms of stunning, such as HO electrical stunning, are widely accepted and practiced (Bozzo *et al.*, 2017; Farouk, 2013).

Jhatka slaughter is demanded by most Hindus meat eaters and is limited only to goats or sheep. During the procedure the head of the animal is attached to a stable column and the posterior legs are stretched out in the opposite direction. The animal is beheaded from the backbone side with a single stroke of a sword or axe (Aghwan, 2019). This will cause a sudden death of the animal because the spinal cord is cut and the circulation of blood to the brain is stopped, resulting in brain failure within seconds. For this reason, this procedure is considered to cause minimal pain and suffering to the animal compared to other ritual methods (Aghwan and Mac Regenstein, 2019). Because more blood is retained in the muscle, the meat is of high nutritional value, although its shelf life is shorter due to greater predisposition to microbial proliferation (Mahapatra, 2020). From the perspective of animal welfare, the success of this method is highly dependent on the sharpness of the cutting instrument and the experience/skills of the operator. Therefore, during the learning process it is possible that the method is not always accurate and fast. Animals that make noise are seen as a bad omen and indicates that the animal is suffering (McDonough and Holoyda, 2018).

The slaughter of goats by traditional methods is deeply rooted in agricultural societies around the world. In some developed countries these methods are tolerated (legally or politically), in others they are only allowed for self-consumption. These methods are varied and are frequently dependent on local traditions, although the most used practices are neck cutting and cardiac punctures with spears or daggers, and Spanish 'puntilla' (see description of the latter in Gallo *et al.*, 2022 – Chapter 2 of this book). The effectiveness of these methods is limited to the skill and experience of the person applying it. This would explain why in many communities this service is provided by specialised people. Research on the effect of these methods on goat welfare is very limited. Recently, Mdletshe *et al.* (2020) found that cardiac puncture (suprasternal position) is more effective in producing unconsciousness (based on the absence of breathing movements, threat, withdrawal, and corneal and eyelid reflex) compared to neck cutting method. However, meat from goats slaughtered by this method presents dark cutter characteristics (Mdletshe *et al.*, 2021).

9.3 South American domestic camelids

South American camelids are herbivores and pseudo-ruminants because they have only three stomach compartments analogous to the rumen, omasum and abomasum (Hund and Wittek, 2018). Four South American camelids are recognised today, two of which are wild species, the guanaco (*Lama guanicoe*) and the vicuña (*Vicugna vicugna*), two of which are domestic forms, the alpaca (*Lama pacos* L) and the llama (*Lama glama* L) (Vásquez *et al.*, 2020). There are two breeds of alpacas, the Huacaya (short and crimped fleece) and the Suri (longer and wavy fibres), and two breeds of llamas, the Chaku (woolly) and the Kara (tight hair with little fibre) (Antonini, 2010). Alpacas should be sheared annually, while llamas, depending on the wool type, every 1-2 years (woolly llama) or even less frequently (lightly haired llamas, such as the classic llama) (Neubert *et al.*, 2021). A recent phylogenomic study has shown that the llama was domesticated from the guanaco and the alpaca from the vicuña (Fan *et al.*, 2020). The llama is the larger of the two SADC measuring between 100-120 cm at the withers, with a weight of between 100 and 250 kg, while the alpaca measures only 80-90 cm at the withers and weighs between 55 and 90 kg (Björklund, 2014).

The archeo-zoological evidence places domestication of the alpaca in the wet plateau ecosystem of Peru's central Andes 6,000-7,000 years ago and that of the llama in the dry plateaus of southern Peru, Chile, and Argentina (Goñalons, 2008; Scanes, 2018). These animals were valued as transport in caravans facilitating the expansion of the Inca Empire, for resources, such as wool, leather, meat, bones, and dung, as well as for ritual and social aspects (Mader *et al.*, 2018). In contrast with many other domestic farm animals, there is no written history associated with the llama and alpaca. According to Kadwell *et al.* (2001), orally transmitted herding knowledge was largely lost during the Spanish conquest, and breeds disappeared as both the human and the native domestic livestock populations were reduced by 80-90% during the first 100 years of contact. The drastic reduction of the llama and alpaca populations could be due to: (1) a lack of understanding of the superior quality of their fibre for the textile industry; (2) the interest in the introduction and establishment of cattle and sheep for the production of wool that was in high demand in the international markets; and (3) as a measure to control local human populations by destabilising their means of sustenance (Diaz-Maroto *et al.*, 2021).

Currently, the llama has a broad geographical distribution, wider than the alpaca, and it is found in Ecuador, Peru, Bolivia, northern and central Chile, and northwest Argentina and its neighbouring areas. Comparatively, the alpaca has a more restricted distribution that includes the Titicaca basin, the plateau of northern Chile and central Peru, and coastal sites of northern and southern Peru (Wheeler, 1995). The largest herds of llamas and alpacas are raised by Andean families, playing a crucial role in their livelihoods (Figure 9.5), where alpaca husbandry has been directed mostly to fibre production, although recently there is interest in the commercial utilisation of their meat, while llama is considered a multi-producer animal, valued for meat, fibre, manure and transportation (Szpak *et al.*, 2014). Outside South America, the breeding of llamas and especially alpacas has spread in the last 40 years in North America, Europe and Oceania, especially as a valuable source of high-quality wool, recreation and trekking, farmers' pets, therapy animals and, recently, for meat production (McMullen, 2008).

On the Andean plateau where the llamas and alpacas are mainly concentrated, the climate is mainly dry with a short-wet season during winter months. The temperature often falls below zero at night resulting in frost (maximum nighttime temperatures of about -12 °C), but during the day the high altitude (3,500 m to 5,350 m) causes strong solar radiation to produce a milder climate with daytime temperatures reaching a maximum of 18 °C (Valenzuela *et al.*, 2021). Production systems of llamas and alpacas are based on the free grazing on vast pastures, mainly consisting of thick grasses of low nutritional quality as *Stipa*, *Festuca* and *Calamagrostis* genera (Reiner and Bryant, 1986). Both species are, in fact, adapted to areas where coarse forage is available, and the nutrients are diluted by structural carbohydrates that are difficult to digest (Sumar, 2010). In comparison to other species, such as cattle, sheep and goats, llamas and alpacas are less destructive for the fragile pasturelands, and their feeding habits can stimulate the growth of palatable plants, thus contributing

Figure 9.5. Alpacas and llamas are an important component of livelihoods in high Andean communities under adverse climatic and geographical conditions: (A) alpaca herd; and (B) llama herd (photo courtesy of C. Estévez-Moreno, Private Consultant, Colombia).

to the environmental sustainability of the production systems, and hence animal products (Popova *et al.*, 2021).

9.3.1 The SADC meat market

Historically, the consumption of llama or alpaca meat has been limited to the high Andean communities of Peru, Bolivia, Ecuador and Chile (García, 2013). While in the urban areas of these countries, consumption is less frequent, the main acceptability problems of SADC meat appear to be related to prejudices on the supply and demand, involving poor meat hygiene, eating quality and socio-cultural aspects (Franco *et al.*, 2018; Salvá *et al.*, 2009). This is mainly due to the fact that the SADC breeding is carried out under traditional management procedures and, often, poor sanitary conditions. One of the limitations for the production and commercialisation of SADC meat is the frequent finding of macroscopic cysts of the *Sarcocystis aucheniae* species resembling rice grains in skeletal muscles. Cysts are usually very abundant and give an unpleasant aspect to their meat, which leads to confiscation of carcasses by sanitary authorities or devaluation of their commercial value (Franco *et al.*, 2018). However, in the last 15 years, due to its reduced cholesterol content compared to beef, SADC meat is especially attractive for traditional and international gourmet cuisine (Mamani-Linares and Gallo, 2014a), which has not only increased its production and consumption in South America, but also in North America, Europe and Oceania (Popova *et al.*, 2021).

Based on productive performance data reported under traditional production conditions in South America, alpacas at birth (5.9 kg) have significantly lower live weights compared to llamas (7.9 kg). Starting from the 6 month of age this difference becomes more marked, with male llamas being heavier than alpacas (up to 79 vs 46.1 kg) at 25 months of age (Condori *et al.*, 2003; Cristofanelli *et al.*, 2004). Wurzinger *et al.* (2005) mention that in the environmental conditions of the High Andes, llamas reach maturity at the age of 5 years, registering weights of 93-96 kg for females. According to Claros *et al.* (2018), in Bolivia, the estimated production of camelid meat is 14,973 million tonnes.

Of this volume, 59-75% is destined for sale (live animals plus slaughtered animals) and 25-41% is destined for family consumption by producers. Of the total of meat produced, 60.25% is destined for fresh meat consumption (family consumption plus market sales) and 39.75% for processing into jerky and sausages. Of the total production 60% comes from animals over 3 years old, where the incidence of *Sarcocystosis* is higher than 70%, while the remaining 40% originates from younger animals (Romero *et al.*, 2017). In Peru, it is estimated that 0.5 to 1 million alpacas are slaughtered per year, with a production of 11,000 million tonnes of meat. According to Salvá (2009), alpaca slaughter is seasonal and takes place especially during the months of April and May, when the dry season begins in the highlands. Due to the scarcity of food, old males and females and defective animals are destined for this purpose to compensate for the rest of the herd, which is mostly composed of females or wombs that have just come out of the mating period (Borda *et al.*, 2007). Most of the animals that are slaughtered are between 7 and 8 years old to favour the utilisation of their fibre (Hack, 2001).

Data related to live weight and carcass yield of llamas are reported in different studies carried out in Chile, Peru and Bolivia. In all studies, the animals were raised and fed in grazing in an extensive system. In the work of Pérez *et al.* (2000) carried out in Chile, llamas were older than 3 years with an average live weight of 101 kg, carcass weight of 56.2 kg and carcass yield of 55.8%. In a work carried out in Peru (Cristofanelli *et al.*, 2004), the 25 month old male llamas had a carcass weight of 31.2 kg with a carcass yield of 52.4%. On the other hand, in studies carried out in Bolivia (Condori *et al.*, 2003), 24 and 36 months male llamas presented a carcass weight of 29.9 and 41 kg, respectively, and carcass yields of 52 to 54%. In alpacas, studies reported carcass weights ranging from 23 to 27 kg (Cristofanelli *et al.*, 2004; García, 1995; Salvá *et al.*, 2009). Carcass weight increase with slaughter age, varying from 22.3 kg at 18 months to 29 kg at more than 48 months of age, with carcass yield ranging from 52 to 55% (Quispe *et al.*, 2012; Smith *et al.*, 2015). Bustinza (2001), on the other hand, reported lower carcass yield in alpacas aged 2 and 6 years and higher in animals aged 3 and 4 years.

In Bolivia and Peru, the meat marketing circuit is divided into five markets defined according to the generation of added value: the market for live animals, the market for carcasses, the market for larger cuts, the market for smaller cuts and the market for jerky meat or charqui. In the camelid-producing area, of the total number of animals slaughtered, 90% corresponds to animals purchased by stockpilers, while the remaining 10% corresponds to animals directly marketed by producers near the production sites (Claros *et al.*, 2018). Etymologically, the word 'Charqui' comes from the Aimara language, which means 'dry' and refers to a traditional salted dried meat product of ancestral origin, currently made from the meat of alpacas, llamas or alpaca-llama hybrids in Peru, Bolivia, Argentina and Chile (Salvá *et al.*, 2012). Peru is the main producer of charqui with around 450,000 kg per year. This millenary processing technique of the High Andes consists of dehydrating lean meat saturated in salt, exposed to the sun and frost (-2 °C) in ventilated areas, for a period of 7-15 days, allowing the gradual and controlled dehydration of the surface of the tissues. Its durability is not determined, but some experiences indicate a durability of up to 20 years in cold environments and low humidity (Ayala *et al.*, 2019). Andean charqui composition requirements are defined in the Peruvian technical regulation 201.059 (INDECOPI, 2006), where minimum protein levels (45% dry matter) and maximum fat and moisture levels (12% dry matter and 20%, respectively) are established (Salvá *et al.*, 2012).

9.3.2 Principles of SADC behaviour applied to handling

In general terms, llamas and alpacas are more visually similar to goats than they are to other members of the *camalidae* family (Wang *et al.*, 2015). For example, the eye pupils of SADC are horizontal like many grazing animals, while dromedaries (*camelus*) have vertical pupils like many predatory mammals. This reason for this difference between members of the same family and their adaptive function has not been elucidated yet (Harman *et al.*, 2001). It has been postulated that the function of the horizontal pupil in most herbivores is to assist surveillance of predators guarding herds in open field, so that the animal does not require excessive eye and head movement (Wang *et al.*, 2015). The hearing of alpacas is typical of that of other artiodactyls (auditory range from 40 to 32.8 kHz with a well-defined point of best sensitivity at 8 kHz), having good sensitivity to sound, including at the low frequencies, but their sensitivity is not unique. At their highest sensitive point, they hear better than pigs and horses, but not as well as cattle, goats, and sheep. However, this difference is less than 10 dB (Heffner *et al.*, 2014). Similar to goats, alpacas (and probably llamas) also have relatively poor localisation acuity. Hearing capacity of SADC can be assessed by observing their reactions to their environment when unrestrained (Whitehead and Bedenice, 2009). Gauly *et al.* (2005) found that 7 out of 9 llamas and alpacas with pure white coat and blue iris pigmentation had bilateral congenital deafness. This anomaly should be considered during preslaughter handling to identify vulnerable animals during herding. Finally, camelids use visual and/or auditory information during handling and in general during their relationship with humans to determine their immediate behavioural responses, although in certain contexts they may use olfactory and tactile cues to confirm identification of their handlers or other animals (Bhattacharyya and Mukhopadhyay, 2014; Windschnurer *et al.*, 2020).

Alpacas are generally more vocal than llamas. Their most common sound has been described as a humming (bleating), which has been related to communication between group members. Other variations in intonation are described as a 'separation hum' or a 'distress hum', screams indicating extreme fear, and alarm calls expressed by high-pitched sounds described as 'whistles' or 'whinnies' (Fowler, 2008). Llamas, like goats, tend to have a closer interaction with people and show lower fear reactions to sudden movements. In contrast, alpacas accept human interaction to some extent, but flee if they perceive a threat (Anderson, 2013). Based on the innate territorial and leadership behaviour during herding, llamas and alpacas are often used to guard herds of goats and sheep (Cavalcanti and Knowlton, 1998). In Australia, castrated male alpacas are usually used joined to a flock of sheep at a ratio of one alpaca per 100 sheep. No more than three alpacas should be used for a single herd as they tend to flock together rather than with the livestock they are guarding (Matthews *et al.*, 2020). Pollard and Littlejohn (1995) found that when an individual alpaca was separated from a stable group, it experienced an increase in heart rate and performed behaviours, such as movement around the pen, vertical or horizontal head movements at the wall and door of the pen, nosing of the pen, and resting in ventral recumbence, as well as decreased eating. However, these responses disappeared when after being isolated from the group the animal was restrained by the handler. If an animal is isolated for any production procedure, it is recommended to ensure that these individuals always have herd mates nearby throughout the handling process.

Although SADC do not show visible pain, it is a myth that these animals do not feel pain or feel less pain compared to other livestock species (Grubb, 2014). In prey animals, such as SADC, a lower sensitivity to painful stimuli (analgesic response) has been reported when exposed to predators or situations related to predation (Kats and Dill, 1998). This would explain the difficulty of handlers or veterinarians to easily detect pain in llamas and alpacas when they are perceived as potential predators by the animals (Plummer and Schleining, 2013). According to Grubb (2014), common signs of general pain (regardless of intensity) in camelids are tachycardia, bruxism and tachypnoea, while signs of mild to moderate pain are anorexia, lethargy, excessive resting time, avoidance of contact and abnormal posture. More severe pain may result in a recognisable pain display, including violent movements, such as rolling and twisting. For example, kicking or staring at the abdomen and tenesmus are specific signs of abdominal pain (Grubb, 2014). Death from capture myopathy may occur in SADC in situations that combine pain from non-severe tissue injury, fear of predation and corresponding palliative medical care (Biasutti and Dart, 2019). This myopathy is characterised by decreases the flow of oxygen and nutrients in the blood to the tissues, which increases the production of lactic acid, and causes extensive necrosis of skeletal muscle tissue. Also, the myoglobin released from dead muscle cells can cause fatal renal failure (Carmanchahi *et al.*, 2011).

Human-animal interactions can be enhanced by accustoming animals to human contact, using the correct handling procedures, improving the skills of the handler and the design of the facilities in which the animals are handled (Turner, 2014). Llamas and possibly alpacas also have the ability to discriminate between known and unknown humans, responding negatively to the presence of strangers (Taylor and Davis, 1996). In a study including 20 animal holdings keeping alpacas, llamas, or both mixed, animal behaviour during handling by a familiar caretaker and a physical examination by a veterinarian was associated with handler attitudes and the amount of contact with the animals (Windschnurer *et al.*, 2020). Moreover, Windschnurer *et al.* (2021) found that the gentle headdress tactile, auditory and visual handling is related to calmer SADC as showed by no expression of signs of fear and easier handling. However, a scared SADC will avoid contact with the handler, and the establishment of a positive relationship between handler and animal without alleviation of fear is impossible (Kapustka and Budzyńska, 2021). The most frequent fear behaviours towards the humans in both species are spitting, vocalising (squealing and/or yelping, snorting and/or snapping, whining), biting, kicking, deliberate collapsing, i.e. abruptly lying on the belly on the ground, and freezing behaviours (Windschnurer *et al.*, 2020). According to Kapustka and Budzyńska (2021), when approaching alpacas and llamas it is important to take body language into account. For example, the position of the ears pinned back with the head raised, usually accompanied by the tail in the horizontal position, can be a warning signal before defensive aggression with spitting. Ears pinned back and a lightly raised tail are signals of anxiety and fear. On the contrary, behaviours like uplifted ears pointing to-ward the human, a neck and a body forming an angle of 90°, and a loosely lowered tail are signs of interest and curiosity. Such behaviours are most desirable in the human-animal relationship and reflect the positive attitude of the animal towards the person. Positive reinforcement training has been used successfully in alpacas and llamas for shearing, truck loading/unloading and medical check-up (Bennett, 2014; Waiblinger *et al.*, 2020).

9.3.3 Pre-transport handling and loading

The selection of animals for transport must be based on an evaluation to determine if the animals are fit for the journey. The criteria for this are similar to those for goats. In Argentina, fasting and shearing 24 h before transport is recommended so that on arrival at the slaughterhouse the animals can be sprayed with a shower to calm them, remove the dirt from the skin and favour the electrical stunning method (Aleu, 2010). To ease pre-transport loading and prevent stress-related abomasal ulcers, abortion and premature births, it is advised to handle animals in small groups (e.g. <10 alpacas or <5 llamas/group; Anderson, 2013; Hund and Wittek, 2018). SADC can be led into a narrow alley or catch pen by people with outstretched arms, however, arms should not be agitated aggressively as this may alarm the animals (Fowler, 2011). Ideally, it is recommended to use the farm facilities to create a series of enclosures, such as pens, so that the entire group or herd can be captured at the same time, and then separate it into smaller groups for loading. When only a single animal has to be captured within a herd, the entire group should be captured first. Llamas tend to accept halters better than alpacas, which facilitates their individual handling. Halters should be used for very specific management practices, such as physical restraint and loading/unloading onto a truck (Jones, 2013). Halters should not be used on a daily basis because they can cause significant damage to the nasal region, which is very sensitive in camelids. Additionally, as camelids are nasal breathers, care must be taken to avoid the obstruction of the nostrils because it can cause sudden death (Bennett, 2014).

To avoid sternal recumbence in response to handling and aggressiveness towards the handler, during loading restraining the animal may be necessary. In this case, neck protection should be considered. Some authors recommend the bandaging of the neck, so that during restraint the animals are less likely to get injured. Additionally, if the animal is very reactive, it is possible to place a hood to obstruct the field of vision (Anderson, 2013; Turner, 2014) or a halter and a towel fixed over the muzzle to prevent spitting (Jones, 2013). A useful technique to facilitate the loading of animals into the trailer proposed by Anderson (2013) is to use a rope to create a movable barrier or fence line. The rope can be anchored to a stationary post at the entryway to the containment area and then a single driver can close this rope around the group to create and ever-decreasing space until the animals move into the capture pen. Another technique to facilitate handling during loading is to train a camelid to get in and out of the trailer, so that other members of the group will follow (Bennett, 2014). Training, adapting, or habituating animals to handling and loading procedures may reduce fear and improve the camelids tolerance of novel situations, especially if this training is undertaken gradually using short sessions (Turner, 2014). For this, leading animals through the chute, haltering, handling and touching them on the neck and legs should be a routine activity during the production period (Marcoppido and Vila, 2013).

The effects of restraint on the stress response have not been studied in SADC, but have been studied in vicuñas and guanacos due to shearing for the sustainable use of their fleece (Arzamendia *et al.*, 2010). For example, in farmed vicuñas Bonacic *et al.* (2003) reported that, as a result of capture and restraint, a number of physiological parameters including rectal temperature, heart rate, respiratory rate, blood CK and cortisol concentration increased (by 41%) beyond the normal basal range seen in captive animals, suggesting that these procedures are stressful. In the case of wild guanacos,

Taraborelli *et al.* (2011) found that the rate of behavioural stress responses was negatively related to cortisol levels which may be due to animal fatigue or that behavioural stress responses may help mitigate physiological stress responses during physical restraint. These results suggest that both behavioural and physiological responses are important measures of stress in manipulated camelids, and that behavioural responses cannot be used as a simple surrogate for glucocorticoid levels.

A pre-transport practice to mitigate the effects of heat stress during transport in hot climates is shearing. This practice is supported by the results of Navarre *et al.* (2001) who found that non-sheared alpacas had higher rectal temperature and higher blood concentrations of sodium and blood urea nitrogen compared to sheared alpacas. However, shearing is not a common practice in llamas' production, which, together with a higher ratio of body volume to surface area, make these animals more susceptible to heat stress than alpacas (Whitehead and Bedenice, 2009). During loading operations SADC prefer to step up rather than walk up a ramp. It is unwise to pull or push a llama or alpaca into a step-up trailer. A leg may slip beneath the floor of the trailer and be fractured if the animal lunges forward (Fowler, 2011). SADC can tolerate and even habituate to the presence of known herding dogs, however, their use for herding is not recommended because they may be perceived by the camelids as a potential threat (Warbington, 2014). If the dogs are unfamiliar, the herd may adopt defensive behaviours in some individuals that can lead to stress myopathy (Biasutti and Dart, 2019).

9.3.4 Transport

Typical journeys are between farms (exchange or purchase of breeding stock), from breeding to fattening farms, from the farm to local and regional exhibition fairs, or to the slaughterhouse (Figure 9.6). Transport is either done in small vans, public buses and even cars, often in the trunks, or in specialised roofless trucks featuring wooden floors, a single compartment and no bedding (Mamani-Linares and Gallo, 2014b). During transport, camelids usually lie down and frequently extend their front legs forward when lying, so gaps below partitions should be high enough so that their legs are not trapped when the animals rise (Smith-Davila *et al.*, 2019). Keeping animals tethered during the journey should be avoided as they may hang themselves. If a covered horse trailer is used, the area above the rear doors should be blocked with netting to prevent the unlikely event of an animal attempting to jump out (Fowler, 2011). Studies on the effects of transport on SADC are still limited, and most of them refer to the journeys of alpaca breeders in North America or Europe. In this type of conditions, Anderson *et al.* (1999a,b) found that when alpacas were transported for 30 min, serum cortisol concentration was significantly higher after transport, but returned to baseline concentration after a 4 h recovery period. According to the British Alpaca Society (https://www.bas-uk.com/alpacapedia/alpaca-welfare/welfare-guide-for-alpacas/), good practices for the transport of domestic camelids should be based on the following recommendations: (1) camelids being held for transport should not be without water for periods longer than 3 h; (2) camelids should be kept off pasture for a minimum of 4 h prior to transport, but not more than 12 h. Clean hay from pastures may be fed during prolonged transports; (3) pregnant camelids should not be transported after 320 days of gestation. An exception may be made for short journeys (less than 20 min) with animals of calm temperament; (4) camelids often travel in the kush position (sternal decubitus), so cushioning in the form of rubber matting, carpet, straw or similar should be provided; and (5)

Figure 9.6. Llama transport operations in the Bolivian plateau: (A) auction market; (B) transport in small trucks; (C) loading of animals onto trucks; and (D) animals shipped in a truck with no interior partitions or roof (W. Mamani-Linares, Public University of El Alto, Bolivia).

females with calves less than 10 days old should not be transported (Turner, 2014). Due to the high value of these animals, these journeys are usually done in trailers or trucks like those used for equines, where they usually have access to bedding and at low densities that allow them to adopt the kush position (Gunsser, 2009).

The heat tolerance limit for llamas and alpacas is 30 °C, although they can withstand chronic heat stress for some time at the farm level (Gerken, 2010). However, the interaction of high temperatures with transport stress can lead to acute heat stress. High ambient temperature combined with high humidity further compromise SADC ability to thermoregulate as, without a temperature gradient between the animal and the environment, radiation, convection and conduction are diminished (Peng *et al.*, 2019). According to Whitehead and Bedenice (2009), SADC affected by heat stress, besides displaying signs of respiratory distress and tachycardia, present elevated body temperatures (often between 40-42 °C), are unable to rise as the forelimbs are weaker than the hindlimbs, and are depressed. Furthermore, they have greater blood concentrations of CK and aspartate aminotransferase, hypoproteinaemia due to cell damage despite dehydration (haemo-concentration), and possibly severe electrolyte imbalances, especially hypokalaemia and metabolic acidosis. In case studies conducted by Norton *et al.* (2014), necropsy results showed that SADC showing heat stress symptoms suffered from anaemia, hyperproteinaemia, mild spinal axonal degeneration, and skeletal muscle necrosis. Therefore, to increase evaporation and reduce body temperature during the journey the use of forced ventilation in the truck or the control of airflow in passively-ventilated equine trailers are recommended.

Cold conditions are also challenging factors for the welfare of SADC during transport. It has been reported that SADC are vulnerable to cold stress, especially when kept at sub-zero temperatures outdoors without shelter at night (Valenzuela *et al.*, 2021). Arias *et al.* (2016), in clinically healthy alpacas and llamas bred, transported and slaughtered in the Andean plateau, found increased interleukin-1 cytokine levels in the lung tissue, which indicated a degree of pneumonia. Streptococcosis (*Streptococcus equi* ssp. *zooepidemicus*) also known in Peru as 'alpaca fever' is the result of an interaction between cold stress during transport and exposure of SADC to carrier horses or other carrier species (Barrington *et al.*, 2006; Jones and Boileau, 2009). Clinical signs are high fever, depression, anorexia and recumbency which may result in death 4-8 days after onset of clinical signs (Fowler, 2001).

9.3.5 Handling at the slaughter plant

Currently in Peru and Bolivia, local and regional governments are making significant efforts to build or maintain specialised SADC slaughterhouses incorporating substantial improvements in design, operations and safety compared to older slaughterhouses or traditional home slaughter (Figure 9.7A). However, many layout designs are inspired by ruminant slaughterhouses and are not always functional for camelids. In addition, research in South America on welfare at the slaughterhouse level is scarce and insufficient. The necessary resting time in the slaughterhouse for animals to recover from transport has not been defined for camelids. In Peru and Bolivia, the recommended lairage time is between 12 and 16 h with access to water, but not food (Mamani-Linares and Gallo, 2014a; Smith-Davila *et al.*, 2019). However, time in lairage can frequently exceed 24 h (Cristofanelli *et al.*, 2005), which may result in greater proportion of severe lesions on the carcass (Smith-Davila *et al.*, 2019). This long lairage time can be explained by the fact that the animals are sheared during this period (Cortez *et al.*, 2011). In terms of the impact of lairage time on meat quality, Cristofanelli *et al.* (2004) measured the pH in the *Longissimus* muscle of 20 llamas and 40 male alpacas kept in lairage for 24 h, and reported no effect on pHu measured at 1, 6, 12, 24, 48 and 72 h post-mortem. In a recent study, Biffin *et al.* (2020) showed that alpacas with 15 h lairage had lower shear force values (greater tenderness) than alpacas that were rested for 7 days after transport, although no difference was found in *Longissimus* muscle pHu.

As in other livestock species, special care should be taken with the social mixing of animals of different origins (especially males) in the same lairage pen as this will increase the frequency of aggression and injury (Smith-Davila *et al.*, 2019). Another important aspect to consider during lairage is the monitoring of activity in the pens. Rest can provide valuable information on the status of the animals and can facilitate decision making on the order of slaughter or order an emergency slaughter (Mamani-Linares and Gallo, 2014b). Sternal decubitus is the most common position for resting and relaxing in llamas and alpacas. However, the sternal position with the head down suggests depression or, when combined with a hyperflexion of the neck and head towards the belly, suggests the presence of opisthotonos, depression or serious illness (Fowler, 2008). Another normal resting position is the lateral recumbence position.

Research has shown that the quality of ramps, corrals and handling areas design are the major risk factors for the increased prevalence and severity of bruising in SADC carcasses. In a study with 30 llamas transported for 3 h in Bolivia, a prevalence of 30% bruised carcasses was reported, with more contusions concentrated in the region of the back (58%), followed by the thorax (21%), shoulder (10%), leg (5%) and coxal tuberosity (5%). Bruises mostly consisted of irregular shaped lesions (81.2%; Mamani-Linares and Gallo, 2014b). In a more recent larger study on 203 carcasses of llamas and alpacas in Peru, Smith-Davila *et al.* (2019) reported a prevalence of 90% bruised carcasses, and the most affected anatomical regions were the thorax and abdomen, which concentrated 48% of bruises. The most common abuses reported in commercial camelid slaughterhouses during unloading and handling to stunning are ear pulling, pulling of the fleece on the sides of the thorax and abdomen, kicking and punching, as well as the use of objects, such as sticks and plastic tubes (Smith-Davila *et al.*, 2019).

9.3.6 Stunning and slaughter

In the slaughterhouses of the Andes the handling of the SADC is usually individualised, with the animals being captured in the holding pens and led one by one to the stunning-slaughter area (Figure 9.7B). The technique used for restraining camelids is to bring the animal's head and neck close to the chest with one hand, while the other hand rests on the upper back. Alternatively, the tail can be grasped with the second hand (Bennett, 2014; Fowler, 2011). During slaughter, the assistance of a second operator is required to hold the animal by the rump, while the main operator performs the stunning procedure and the animal is lifted to be exsanguinated. In these countries, the 'puntilla' is the most common stunning method, even in official slaughterhouses (Figure 9.7C,D). However, research in llamas has demonstrated that this method is often ineffective, with most llamas (95%) showing signs of brain and spinal cord function, suggesting that they were still sensible and able to experience pain and distress (Limon *et al.*, 2009, 2012).

As for sheep and goats, penetrating captive bolt and head-only (HO) electrical stunning, and free bullet are also in use at commercial slaughterhouses (BVA, 2020; Gibson *et al.*, 2015). The use of HO electrical stunning is increasing at SADC slaughterhouses. For example, in Argentina, HO electrical stunning is carried out with a 220 V discharge for 3 s (Zogbi and Frank, 2008). However, electrical stunning has some drawbacks in camelids due to the hairiness of the head which does not favour good electrical conduction, and in some cases can cause muscle spasms (Neijenhuis *et al.*, 2015), possibly resulting in injuries to the cervical vertebrae (Whitehead and Bedenice, 2009). For this reason, besides restraining the animal to ensure correct electrode positioning on its head and avoid neck injuries during the procedure, it is recommended to shear and bathe it to improve stunning efficiency (Aleu, 2010).

Gibson *et al.* (2015) studied the effectiveness of penetrating captive bolt at two anatomical sites (i.e. crown and frontal regions) in 96 alpacas. They found that the probability of an immediate loss of consciousness was increased when captive bolt penetrated at the highest point of the head (crown) and pointing in the direction of the base of the mandible. This position caused immediate damage to the thalamus and brainstem. In contrast, penetrative captive bolt applied in the frontal region, defined as the intersection of two lines drawn from the medial edge of each eye to the opposite ear,

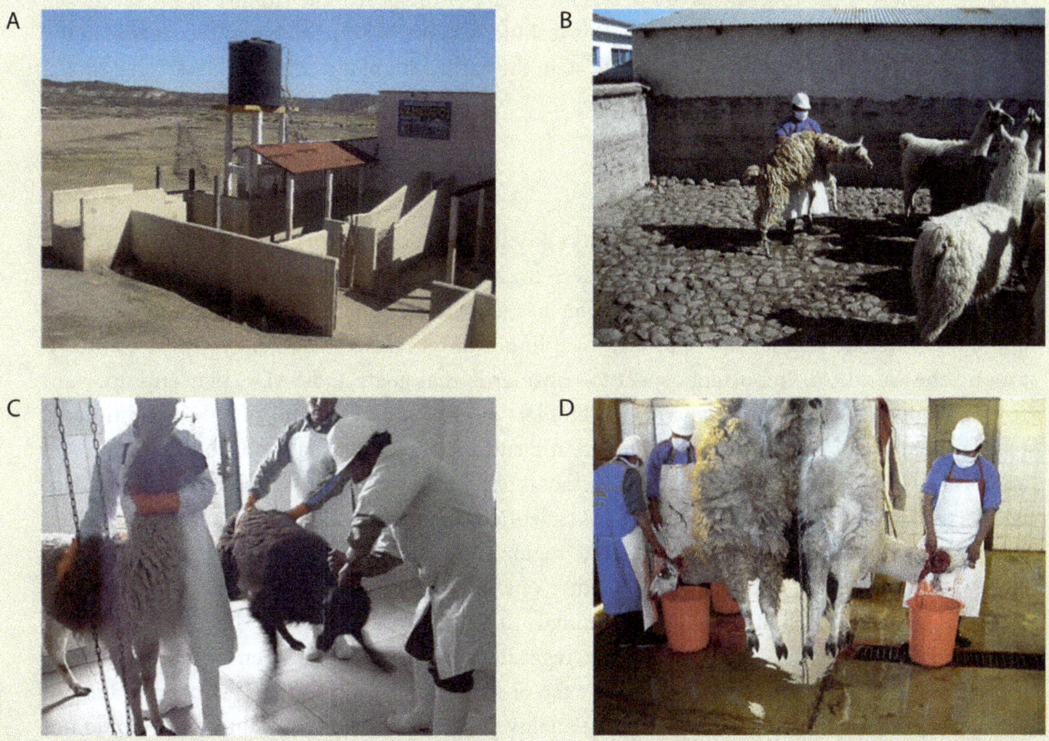

Figure 9.7. Example of slaughter conditions in modernised official slaughterhouses in Bolivia: (A) official slaughterhouse of SADC; (B) capture and restraint of the animals to be led to the slaughter area; (C) stunning with the 'puntilla' method; and (D) bleeding of the animals (W. Mamani-Linares, Public University of El Alto, Bolivia).

with the stunner pointed at the brainstem, was less effective because the animals moved their heads when the stunner entered their field of vision immediately before the shot. The authors of this study concluded that, although both regions are considered acceptable for captive bolt placement, for captive bolt stunning at the frontal region it is necessary to restrain the animal to prevent movement and incorrect placement.

Exsanguination in SADC can be performed by cutting the carotid arteries and jugular veins (ventral neck incision) at the level of the jaw line or by cutting the aorta in the thoracic cavity (chest stick or thoracic stab) (Gibson *et al.*, 2015). Traditional Andean ancestral methods as the 'degüello' and the 'ch'illa' are also used. The 'degüello' is used at the farm by the shepherds and at the slaughterhouse, sometimes after stunning with a Spanish 'puntilla' (Limon *et al.*, 2009). This type of slaughter is performed on the ground: the animal is tied by the legs and the large blood vessels are cut at the base of the neck, severing the trachea and oesophagus (Rodríguez-Loredo, 1998). The 'ch'illa' is used only by shepherds of the high Andean altitudes in their festivities and rituals, although the meat obtained is consumed (Rodríguez-Loredo, 1998). This slaughter technique consists of making a 10 cm incision in the right flank, behind the last rib, approaching the abdominal cavity, and then entering the thoracic cavity to reach the base of the heart. The rupture of the ascending aorta artery is performed with the hand. This technique takes 30 s and all blood is collected in the thoracic cavity without spilling, so as not to stain the earth with blood, which is considered sacred to the villagers (Aleu, 2010).

9.4 Conclusions

Goats, llamas and alpacas are part of humanity's oldest rural and culinary traditions. Therefore, their rearing, transport and slaughter are as diverse as the countries, regions and markets that appreciate and demand this type of meats. This chapter has shown the different scenarios, consequences and animal welfare risk factors related to handling, transport and slaughter of goats and SADC around the world. An important aspect to consider is that goats and SADC play an important role in the livelihoods of farmers and smallholders all over the world and that the conditions of rearing, handling, transport and slaughter are still rudimentary and usually not controlled by local authorities. Under these conditions, for a preslaughter animal welfare improvement programme to be effective, it must involve local stakeholders (scientists, farmers, middlemen, retailers, consumers) and consider the socio-cultural aspect of the region. A large variety of problems associated with stress during transport and other preslaughter operations can be solved by controlling the risk factors that affect the ability or fitness of animals to cope with the journey. This can be achieved through an operational quality program that, regardless of the conflict between animal welfare and productive efficiency, can become a strategic factor for the agri-food sector. The incorporation of animal welfare as the axis of an operational quality program has effects beyond the reduction of animal reactivity and suffering during transport and slaughter operations. It also has an indirect impact on other important aspects that influence product quality and the competitiveness of the agri-food sector, such as traceability, safety, reduction of losses, operational efficiency and risk prevention.

References

Adamczyk, K., Górecka-Bruzda, A., Nowicki, J., Gumułka, M., Molik, E., Schwarz, T., Earley, B. and Klocek, C., 2015. Perception of environment in farm animals – a review. Annals of Animal Science 15: 565-589.

Aghwan, Z.A., 2019. Ritual and traditional slaughter practices for meat production. Journal of Islamic, Social, Economics and Development 4: 224-30.

Aghwan, Z.A. and Mac Regenstein, J., 2019. Slaughter practices of different faiths in different countries. Journal of Animal Science and Technology 6: 111-121.

Ajuda, I.D.G.G., Battini, M. and Stilwell, G.T., 2019. The role of claw deformation and claw size on goat lameness. Veterinary and Animal Science 8: 100080. https://doi.org/10.1016/j.vas.2019.100080

Akinmoladun, O.F., Fon, F.N. and Mpendulo, C.T., 2020. Stress indicators, carcass characteristics and meat quality of Xhosa goats subjected to different watering regimen and vitamin C supplementation. Livestock Science 238: 104083. https://doi.org/10.1016/j.livsci.2020.104083

Alarcon, P., Fèvre, E.M., Murungi, M.K., Muinde, P., Akoko, J., Dominguez-Salas, P., Kiambi, S., Ahmed, S., Häsler, B. and Rushton, J., 2017. Mapping of beef, sheep and goat food systems in Nairobi – a framework for policy making and the identification of structural vulnerabilities and deficiencies. Agricultural Systems 152: 1-17.

Alcalde, M.J., Suárez, M.D., Rodero, E., Álvarez, R., Sáez, M.I. and Martínez, T.F., 2017. Effects of farm management practices and transport duration on stress response and meat quality traits of suckling goat kids. Animal 11: 1626-1635.

Aleu, G., 2010. Determinación de los aspectos tecnológicos y nutricionales de la carne de llama (*Lama glama*). MSc. Thesis, Universidad Católica de Córdoba, Córdoba, Argentina.

Altinçekiç, Ş.Ö. and Koyuncu, M., 2013. Goat welfare during transport. Macedonian Journal of Animal Science 3: 161-168.

Amills, M., Capote, J. and Tosser-Klopp, G., 2017. Goat domestication and breeding: a jigsaw of historical, biological and molecular data with missing pieces. Animal Genetics 48: 631-644.

Anderson, D.E., 2013. Behavior and capture techniques. In: Anderson, D.E., Jones, M.L., and Miesner, M.D. (eds.) Veterinary techniques for llamas and alpacas. John Wiley & Sons, Ames, IA, USA, pp. 2-7.

Anderson, D.E., Grubb, T. and Silveira, F., 1999a. The effect of short duration transportation on serum cortisol response in alpacas (*Llama pacos*). The Veterinary Journal 157: 189-191.

Anderson, D.E., Silveira, F. and Grubb, T., 1999b. Effects of venipuncture and correlation of plasma, serum and saliva cortisol concentration with transportation stress in camelids. Journal of Camel Practice and Research 6: 249-254.

Antonini, M., 2010. Hair follicle characteristics and fibre production in South American camelids. Animal 4: 1460-1471.

Aréchiga, C.F., Aguilera, J.I., Rincón, R.M., De Lara, S.M., Bañuelos, V.R. and Meza-Herrera, C.A., 2008. Situación actual y perspectivas de la producción caprina ante el reto de la globalización. Tropical and Subtropical Agroecosystems 9: 1-14.

Arias, N., Velapatino, B., Hung, A. and Cok, J., 2016. Cytokines expression in alpacas and llamas exposed to cold stress. Small Ruminant Research 141: 135-140.

Arzamendia, Y., Bonacic, C. and Vilá, B., 2010. Behavioural and physiological consequences of capture for shearing of vicunas in Argentina. Applied Animal Behaviour Science 125: 163-170.

Ayala, C., Condori, G., Renieri, C., Pilco, S. and Quispe, J.L., 2019. Carne y charqui de llama. In: Gerken, M., Renieri, C., Allain, D., Galbraith, H, Gutiérrez, J.P., McKenna, L., Niznikowski, R. and Wurzinger, M. (eds.) Advances in fibre production science in South American camelids and other fibre animals. Universitätsverlag Göttingen, Göttingen, Germany, pp. 279-303.

Banks, M.S., Sprague, W.W., Schmoll, J., Parnell, J.A. and Love, G.D., 2015. Why do animal eyes have pupils of different shapes? Science Advances 1: e1500391. https://doi.org/10.1126/sciadv.1500391

Battini, M., Barbieri, S., Vieira, A., Stilwell, G. and Mattiello, S., 2016. Results of testing the prototype of the AWIN welfare assessment protocol for dairy goats in 30 intensive farms in Northern Italy. Italian Journal of Animal Science 15: 283-293.

Barrington, G.M., Allen, A.J., Parish, S.M. and Tibary, A., 2006. Biosecurity and biocontainment in alpaca operations. Small Ruminant Research 61: 217-225.

Bhattacharyya, S. and Mukhopadhyay, G., 2014. A Review on a therapy with bonds between human and animals. International Journal of Pharmacy and Engineering 2: 519-531.

Bennett, M.M., 2014. Camelid management, handling techniques and facilities. In: Cebra, C., Anderson, D., Tibary, A., Van Saun, R. and Johnson Larue, W. (eds.) Llama and alpaca care: medicine, surgery, reproduction. Elsevier, St.-Louis, MI, USA, pp. 22-50.

Biasutti, S.A. and Dart, A.J., 2019. Suspected capture myopathy in an alpaca (*Vicugna pacos*) following a dog attack. New Zealand Veterinary Journal 67: 52-54.

Biffin, T.E., Hopkins, D.L., Bush, R.D., Hall, E. and Smith, M.A., 2020. The effects of season and post-transport rest on alpaca (*Vicunga pacos*) meat quality. Meat Science 159: 107935. https://doi.org/10.1016/j.meatsci.2019.107935

Björklund, C., 2014. Diseases and causes of death among camelids in Sweden: a retrospective study of necropsy cases 2001-2013. BSc. Thesis, Swedish University of Agricultural Sciences, Uppsala, Sweden.

Blakeman, N.E. and Friend, T.H., 1986. Visual discrimination at varying distances in Spanish goats. Applied Animal Behaviour Science 16: 279-283.

Boivin, X. and Braastad, B.O., 1996. Effects of handling during temporary isolation after early weaning on goat kids' later response to humans. Applied Animal Behaviour Science 48: 61-71.

Bonacic, C., Macdonald, D.W. and Villouta, G., 2003. Adrenocorticotrophin-induced stress response in captive vicugnas (*Vicugna vicugna*) in the Andes of Chile. Animal Welfare 12: 369-385.

Borda, A., Ottone, G. and Quicaño, I., 2007. No solo de fibra viven los alpaqueros. In: DESCO (eds.) Perú hoy: mercados globales y (des) articulaciones internas. Centro de Estudios y Promoción del Desarrollo DESCO, Lima, Perú, pp. 329-359.

Bozzo, G., Di Pinto, A., Bonerba, E., Ceci, E., Mottola, A., Roma, R., Capozza, P., Samoilis, G., Tantillo, G. and Celano, G.V., 2017. Kosher slaughter paradigms: evaluation of slaughterhouse inspection procedures. Meat Science 128: 30-33.

Briefer, E.F., Padilla de la Torre, M. and McElligott, A.G., 2012. Mother goats do not forget their kids' calls. Proceedings of the Royal Society B: Biological Sciences 279: 3749-3755.

Briefer, E.F., Haque, S., Baciadonna, L. and McElligott, A.G., 2014. Goats excel at learning and remembering a highly novel cognitive task. Frontiers in Zoology 11: 1-12.

British Veterinary Association (BVA), 2020. 'Policy statement – BVA position on the welfare of animals at slaughter' 2020. Available at: https://tinyurl.com/6z4apndz.

Bustinza, V. 2001. La alpaca. Conocimiento del gran potencial andino. Universidad Nacional del Altiplano, Puno, Perú, 495 pp.

Carmanchahi, P.D., Ovejero, R., Marull, C., Lopez, G.C., Schroeder, N., Jahn, G.A., Novaro, A.J., Gand, C. and Somoza, G. M., 2011. Physiological response of wild guanacos to capture for live shearing. Wildlife Research 38: 61-68.

Casey, N.H. and Webb, E.C., 2010. Managing goat production for meat quality. Small Ruminant Research 89: 218-224.

Cavalcanti, S.M. and Knowlton, F.F., 1998. Evaluation of physical and behavioral traits of llamas associated with aggressiveness toward sheep-threatening canids. Applied Animal Behaviour Science 61: 143-158.

Chambers, P.G., Grandin, T., Heinz, G. and Srisuvan, T., 2001. Slaughter of livestock. Guidelines for humane handling, transport and slaughter of livestock. RAP Publication 2001/4. Food and Agriculture Organization (FAO) and Humane Society International, Bangkok, Thailand, pp. 49-82.

Chulayo, A.Y. and Muchenje, V., 2013. The effects of pre-slaughter stress and season on the activity of plasma creatine kinase and mutton quality from different sheep breeds slaughtered at a smallholder abattoir. Asian-Australasian Journal of Animal Sciences 26: 1762-1772.

Claros, L., Quispe, J.L, Claros, G.A, Condori, G. and Ayala, C., 2018. Caracterización de la estructura y cuantificación de la cadena agroalimentaria de carne de llama estudio de caso: Lagunas, Papel Pampa y Sajama. Revista de Investigación e Innovación Agropecuaria y de Recursos Naturales 5: 185-192.

Cockram, M.S. and Velarde, A., 2022. Sheep. In: Faucitano, L. (ed.) Preslaughter handling and slaughter of meat animals. Wageningen Academic Publishers, Wageningen, the Netherlands, pp. 266-310.

Collins, S.L., Caldwell, M., Hecht, S. and Whitlock, B.K., 2017. Comparison of penetrating and nonpenetrating captive bolt methods in horned goats. American Journal of Veterinary Research 78: 151-157.

Condori, G., Ayala, C., Renieri, C., Rodriguez, T. and Martinez, Z., 2003. Evaluación biométrica de llama macho en crecimiento. Memoria del III Congreso Mundial sobre Camélidos, Potosi, Bolivia, pp. 611-617.

Cortez, M.A., Vides, H., Jurado, A. and Ruíz, M. 2011. Manual técnico de llamas. Protección del Medio Ambiente Tarija, Tarija, Bolivia, 63 pp.

Cristofanelli, S., Antonini, M., Torres, D., Polidori, P. and Renieri, C., 2004. Meat and carcass quality from Peruvian llama (*Lama glama*) and alpaca (*Lama pacos*). Meat Science 66: 589-593.

Cristofanelli, S., Antonini, M., Torres, D., Polidori, P. and Renieri, C., 2005. Carcass characteristics of Peruvian llama (*Lama glama*) and alpaca (*Lama pacos*) reared in the Andean highlands. Small Ruminant Research 58: 219-222.

Daly, K.G., Delser, P.M., Mullin, V.E., Scheu, A., Mattiangeli, V., Teasdale, M.D., Hare, A.J., Burger, J., Verdugo, M.P., Collins, M.J., Kehati, R., Erek, C.M., Bar-Oz, G., Pompanon, F., Cumer, T., Çakırlar, C, Mohaseb, A.F., Decruyenaere, D., Davoudi, H., Çevik, Ö., Rollefson, G., Vigne, J.D., Khazaeli, R., Fathi, H., Doost, S.B., Sorkhani, R.R., Vahdati, A.A., Sauer, E.W., Kharanaghi, H.A., Maziar, S., Gasparian, B., Pinhasi, R., Martin, L., Orton, D., Arbuckle, B.S., Benecke, N., Manica, A., Horwitz, L.K., Mashkour, M. and Bradley, D.G., 2018. Ancient goat genomes reveal mosaic domestication in the Fertile Crescent. Science 361: 85-88.

Das, K.S., Srivastava, B.B. and Das, N., 2001. Standing orientation and behaviour of goats during short-haul road transportation. Small Ruminant Research 41: 91-94.

Daskiran, I., Savas, T., Koyuncu, M., Koluman, N., Keskin, M., Esenbuga, N., Konyali, A., Cemal, I., Gül, S., Elmaz, O., Kosum, N., Della, G. and Bingöl, M., 2018. Goat production systems of Turkey: nomadic to industrial. Small Ruminant Research 163: 15-20.

De la Fuente, J., Sánchez, M., Pérez, C., Lauzurica, S., Vieira, C., De Chávarri, E.G. and Díaz, M.T., 2010. Physiological response and carcass and meat quality of suckling lambs in relation to transport time and stocking density during transport by road. Animal 4: 250-258.

Devendra, C., 2010. Concluding synthesis and the future for sustainable goat production. Small Ruminant Research 89: 125-130.

Diaz-Maroto, P., Rey-Iglesia, A., Cartajena, I., Núñez, L., Westbury, M.V., Varas, V., Moraga, M., Campo, P.F., Orozco-terWengel, P., Marín, J.C. and Hansen, A.J., 2021. Ancient DNA reveals the lost domestication history of South American camelids in Northern Chile and across the Andes. Elife 10: e63390. https://doi.org/10.7554/eLife.63390

Dieters, L.S., Meale, S.J., Quigley, S.P. and Hoffman, L.C., 2021. Meat quality characteristics of lot-fed Australian Rangeland goats are unaffected by live weight at slaughter. Meat Science 175: 108437. https://doi.org/10.1016/j.meatsci.2021.108437

Duffy, L., Barlow, R., Fegan, N. and Vanderlinde, P., 2009. Prevalence and serotypes of *Salmonella* associated with goats at two Australian abattoirs. Letters in applied Microbiology 48: 193-197.

Ekanem, E., Mafuyai-Ekanem, M., Tagegne, F., Singh, S. and Favors, D., 2013. Goat meat consumer preferences: implications for goat meat marketing in metropolitan Nashville, Tennessee area. Journal of Food Distribution Research 44: 1-9.

Ekiz, B., Ekiz, E.E., Kocak, O., Yalcintan, H. and Yilmaz, A., 2012. Effect of pre-slaughter management regarding transportation and time in lairage on certain stress parameters, carcass and meat quality characteristics in Kivircik lambs. Meat Science 90: 967-976.

Ekiz, E.E., Yalcintan, H. and Ekiz, B., 2020. Haematological stress parameters and behavioural characteristics of dairy type goat kids compared to indigenous breeds during an intensive fattening programme. Archives Animal Breeding 63: 441-449.

European Commission (EC), 2009. Council Regulation (EC) No 1099/2009 of 24 September 2009 on the protection of animals at the time of killing. Official Journal of the European Union L303: 1-30. Available at: http://data.europa.eu/eli/reg/2009/1099/oj.

European Food Safety Authority (EFSA), 2013. Scientific opinion on monitoring procedures at slaughterhouses for sheep and goats. EFSA Journal 11: 3522. https://doi.org/10.2903/j.efsa.2013.3522

Fan, R., Gu, Z., Guang, X., Marín, J. C., Varas, V., González, B.A., Wheeler, J.C., Hu, Y., Li, E., Sun, X., Yang, X., Zhang, C., Gao, W., He, J., Munch, K., Corbett-Detig, R., Barbato, M., Pan, S., Zhan, X, Bruford, M.D. and Dong, C., 2020. Genomic analysis of the domestication and post-Spanish conquest evolution of the llama and alpaca. Genome Biology 21: 1-26.

FAO, 2020. Total production of goat in all countries around the world; Food and Agriculture Organization, Rome, Italy. Available at: http://www.fao.org/faostat/en/#data/QA/visualize.

Farouk, M.M., 2013. Advances in the industrial production of halal and kosher red meat. Meat Science 95: 805-820.

Fazio, F., Casella, S., Giannetto, C., Giudice, E. and Piccione, G., 2016. Erythrocyte osmotic fragility in response to a short road transport in cattle, horses, and goats. Journal of Veterinary Behaviour: Clinical Applications and Research 12: 82-84.

Fowler, M.E., 2001. Selected diseases of South American camelids. Journal of Camel Practice and Research 8: 99-112.

Fowler, M.E., 2008. Behavioral clues for detection of illness in wild animals: models in camelids and elephants. In: Fowler, M.E. and Miller, E.R. (eds.) Zoo and wild animal medicine current therapy-e-book. Elsevier, St.-Louis, MI, USA, pp. 33-49.

Fowler, M.E., 2011. Restraint and handling. Medicine and surgery of camelids. John Wiley & Sons. Ames, IA, USA, pp. 59-87.

Franco, C.D., Romero, S., Ferrari, A., Schnittger, L. and Florin-Christensen, M., 2018. Detection of *Sarcocystis auchenia*e in blood of llama using a duplex semi-nested PCR assay and its association with cyst infestation. Heliyon 4: e00928. https://doi.org/10.1016/j.heliyon.2018.e00928

Gallo, C.B. and Huertas, S.M., 2016. Main animal welfare problems in ruminant livestock during preslaughter operations: a South American view. Animal 10: 357-364.

Gallo, C., Schwartzkopf-Genswein, K. and Gibson, T., 2022. Cattle. In: Faucitano, L. (ed.) Preslaughter handling and slaughter of meat animals. Wageningen Academic Publishers, Wageningen, the Netherlands, pp. 62-116.

García, C.A., 1995. Contribución al conocimiento del consumo de carne de camélidos sudamericanos (CSA) en Chile. BSc. Thesis. Universidad Austral de Chile, Valdivia, Chile.

García, M.E., 2013. The taste of conquest: colonialism, cosmopolitics, and the dark side of Peru's gastronomic boom. Journal of Latin American and Caribbean Anthropology 18: 505-524.

Gauly, M., Vaughan, J., Hogreve, S. K. and Erhardt, G., 2005. Brainstem auditory-evoked potential assessment of auditory function and congenital deafness in llamas (*Lama glama*) and alpacas (*Lama pacos*). Journal of Veterinary Internal Medicine 19: 756-760.

Gautam, M., Stevenson, M.A., Lopez-Villalobos, N. and McLean, V., 2017. Risk factors for culling, sales and deaths in New Zealand dairy goat herds, 2000-2009. Frontiers in Veterinary Science 4: 191. https://doi.org/10.3389/fvets.2017.00191

Gerken, M., 2010. Relationships between integumental characteristics and thermoregulation in South American camelids. Animal 4: 1451-1459.

Gibson, T.J., Ridler, A.L., Lamb, C.R., Williams, A., Giles, S. and Gregory, N.G., 2012. Preliminary evaluation of the effectiveness of captive-bolt guns as a killing method without exsanguination for horned and unhorned sheep. Animal Welfare 21: 35-42.

Gibson, T.J., Whitehead, C., Taylor, R., Sykes, O., Chancellor, N.M. and Limon, G., 2015. Pathophysiology of penetrating captive bolt stunning in alpacas (*Vicugna pacos*). Meat Science 100: 227-231.

Goñalons, G.L.M., 2008. Camelids in ancient Andean societies: a review of the zooarchaeological evidence. Quaternary International 185: 59-68.

Gootwine, E., 2020. Genetics and breeding of sheep and goats. In: Bazer, F.W., Lamb, G.C. and Wu, G. (eds.) Animal agriculture, sustainability, challenges and innovations. Academic Press, London, UK, pp. 183-198.

Greenwood, P.L., Finn, J.A., May, T.J., and Nicholls, P.J., 2008. Preslaughter management practices influence carcass characteristics of young goats. Australian Journal of Experimental Agriculture 48: 910-915.

Greenwood, P.L., Finn, J.A., May, T.J. and Nicholls, P.J., 2010. Management of young goats during prolonged fasting affects carcass characteristics but not pre-slaughter liveweight or cortisol. Animal Production Science 50: 533-540.

Grist, A., Lines, J.A., Knowles, T.G., Mason, C.W. and Wotton, S.B., 2018. Use of a non-penetrating captive bolt for euthanasia of neonate goats. Animals 8: 58. https://doi.org/10.3390/ani8040058

Grubb, T., 2014. Pain management in camelids: systemic techniques. In: Cebra, C., Anderson, D., Tibary, A., Van Saun, R. and Johnson Larue, W. (eds.) Llama and alpaca care: medicine, surgery, reproduction, nutrition, and herd health. Elsevier, St. Louis, MI, USA, pp. 618-624.

Gunsser, I., 2009. Problems in keeping and in transporting of llamas and alpacas. Deutsche Tierärztliche Wochenschrift 116: 121-124.

Hack, W., 2001. The Peruvian alpaca meat and hide industries – a travel report, Publication No. 01/19 (11 pp.). Rural Industries Research & Development Corporation (RIRDC), Barton, Australia.

Hals, S.K., 2016. Responses to mirrors in domestic goats (*Capra aegagrus hircus*): assessing mirror use to solve a problem and in self-recognition. MSc. Thesis, Norwegian University of Life Sciences, Oslo, Norway.

Harman, A., Dann, J., Ahmat, A., Macuda, T., Johnston, K. and Timneyb, B., 2001. The retinal ganglion cell layer and visual acuity of the camel. Brain, Behavior and Evolution 58: 15-27.

Heesy, C.P., 2004. On the relationship between orbit orientation and binocular visual field overlap in mammals. Anatomical Record Part A 281: 1104-1110.

Heffner, R. S. and Heffner, H. E., 1992. Hearing in large mammals: sound-localization acuity in cattle (*Bos taurus*) and goats (*Capra hircus*). Journal of Comparative Psychology 106: 107-113.

Heffner, R.S., Koay, G. and Heffner, H.E., 2014. Hearing in alpacas (*Vicugna pacos*): audiogram, localization acuity, and use of binaural locus cues. The Journal of the Acoustical Society of America 135: 778-788.

Hindle, V.A., Reimert, H.G.M., Van der Werf, J.T.N. and Lambooij, E., 2013. Loading density and welfare of goat kids during long distance road transport. Animal Welfare 22: 345-356.

Hu, W., Fang, M., Yang, Y., Ye, T., Liu, B. and Zheng, W., 2020. Detection of heat shock protein 27, 70, 90 expressions in primary parenchymatous organs of goats after transport stress by real-time PCR and ELISA. Veterinary Medicine and Science 6: 788-795.

Hund, A. and Wittek, T., 2018. Abomasal and third compartment ulcers in ruminants and South American camelids. Veterinary Clinics: Food Animal Practice 34: 35-54.

INDECOPI, 2006. Norma Técnica Peruana 201.059: 2006. Carne y productos cárnicos: Charqui. Instituto Nacional de Defensa de la Competencia y de la Protección de la Propiedad Intelectual, Lima, Perú.

Jackson, K.M.A. and Hackett, D., 2007. The effects of human handling on heart girth, behaviour and milk quality in dairy goats. Applied Animal Behaviour Science 108: 332-336.

Johnson, C.B., Mellor, D.J., Hemsworth, P.H. and Fisher, A.D., 2015. A scientific comment on the welfare of domesticated ruminants slaughtered without stunning. New Zealand Veterinary Journal 63: 58-65.

Jones, M., 2013. Haltering. In: Anderson, D.E., Jones, M.L. and Miesner, M.D. (eds.) Veterinary techniques for llamas and alpacas. John Wiley & Sons, Ames, IA, USA, pp. 11-13.

Jones, M. and Boileau, M., 2009. Camelid herd health. Veterinary Clinics: Food Animal Practice 25: 239-263.

Kadim, I.T., Mahgoub, O., Al-Kindi, A., Al-Marzooqi, W. and Al-Saqri, N., 2006. Effects of transportation at high ambient temperatures on physiological responses, carcass and meat quality characteristics of three breeds of Omani goats. Meat Science 73: 626-634.

Kadim, I.T., Mahgoub, O. and Khalaf, S., 2014. Effects of the transportation during hot season and electrical stimulation on meat quality characteristics of goat Longissimus dorsi muscle. Small Ruminant Research 121: 120-124.

Kadwell, M., Fernandez, M., Stanley, H.F., Baldi, R., Wheeler, J.C., Rosadio, R. and Bruford, M.W., 2001. Genetic analysis reveals the wild ancestors of the llama and the alpaca. Proceedings of the Royal Society of London. Series B: Biological Sciences 268: 2575-2584.

Kaminski, J., Riedel, J., Call, J. and Tomasello, M., 2005. Domestic goats, *Capra hircus*, follow gaze direction and use social cues in an object choice task. Animal Behaviour 69: 11-18.

Kannan, G., Terrill, T.H., Kouakou, B., Gazal, O.S., Gelaye, S., Amoah, E.A. and Samake, S., 2000. Transportation of goats: effects on physiological stress responses and live weight loss. Journal of Animal Science 78: 1450-1457.

Kannan, G., Terrill, T.H., Kouakou, B., Gelaye, S. and Amoah, E.A., 2002. Simulated preslaughter holding and isolation effects on stress responses and live weight shrinkage in meat goats. Journal of Animal Science 80: 1771-1780.

Kannan, G., Kouakou, B., Terrill, T.H. and Gelaye, S., 2003. Endocrine, blood metabolite, and meat quality changes in goats as influenced by short-term, preslaughter stress. Journal of Animal Science 81: 1499-1507.

Kapustka, J. and Budzyńska, M., 2021. Human ability to interpret alpaca body language. Journal of Veterinary Behavior 42: 16-21.

Kats, L.B. and Dill, L.M. 1998. The scent of death: chemosensory assessment of predation risk by prey animals. Écoscience 5: 361-394.

Keil, N.M., Imfeld-Mueller, S., Aschwanden, J. and Wechsler, B., 2012. Are head cues necessary for goats (*Capra hircus*) in recognising group members? Animal Cognition 15: 913-921.

Kiran, M., Naveena, B.M., Smrutirekha, M., Reddy, P.B., Rituprna, B., Kumar, Y.P., Venkatesh, Ch. and Rapole, S., 2019. Traditional halal slaughter without stunning versus slaughter with electrical stunning of sheep (*Ovis aries*). Meat Science 148: 127-136.

Kusiluka, L.J.M., Kambarage, D.M., Daborn, C.J., Matthewman, R.W. and Harrison, L.J. S., 1995. Causes of condemnation of carcasses and organs in goats in Tanzania. Journal of Applied Animal Research 8: 185-189.

Llonch, P., Rodríguez, P., Casal, N., Carreras, R., Muñoz, I., Dalmau, A. and Velarde, A., 2015. Electrical stunning effectiveness with current levels lower than 1 A in lambs and kid goats. Research in Veterinary Science 98: 154-161.

Limon, G., Guitian, J. and Gregory, N.G., 2009. A note on the slaughter of llamas in Bolivia by the puntilla method. Meat science 82: 405-406.

Limon, G., Guitian, J. and Gregory, N.G., 2012. A review of the humaneness of puntilla as a slaughter method. Animal Welfare 21: 3-8.

López, Á.L. and Venegas, G.J.Q., 2018. The (unethical) consumption of a newborn animal: cabrito as a tourist and recreational dish in Monterrey, Mexico. In: Kline, K. (ed.) Animals, food, and tourism. Routledge, New York, NY, USA, pp. 36-51.

Luginbuhl, J., 2014. Consider breeding your does to target ethnic holidays where goat meat is part of the traditional holiday feast. Technical Report. College of Agriculture and Life Science, North Carolina State University, Raleigh, NC, USA.

Mader, C., Hölzl, S., Heck, K., Reindel, M. and Isla, J., 2018. The llama's share: highland origins of camelids during the Late Paracas period (370 to 200 BCE) in south Peru demonstrated by strontium isotope analysis. Journal of Archaeological Science: Reports 20: 257-270.

Mahapatra, M.S., 2020. Unit-21 Welfare issues in animal slaughter. Notes of MAW-002 Animal Welfare Issues. Indira Gandhi National Open University, New Delhi, India.

Malher, X. and Noordhuizen, J.P.T.M., 2008. Applying the HACCP principles to selected hazards during goat kids rearing on milking goat farms in western France. Revue de Médecine Vétérinaire 159: 38-48.

Mamani-Linares, L.W. and Gallo, C.B., 2014a. Meat quality, proximate composition and muscle fatty acid profile of young llamas (*Lama glama*) supplemented with hay or concentrate during the dry season. Meat Science 96: 394-399.

Mamani-Linares, L.W. and Gallo, C.B., 2014b. Una nota sobre el efecto de las operaciones presacrificio en llamas (*Lama glama*) sobre las concentraciones de algunos componentes sanguíneos relacionados al estrés y la calidad de la canal. Archivos de Medicina Veterinaria 46: 463-469.

Mandefro, A., Aragaw, K., Hailu, B., Alemayehu, G. and Chala, G., 2015. Major cause of organ and carcass condemnation and its financial loss at Bishoftu Elfora export abattoir. International Journal of Nutrition and Food Sciences 4: 364-372.

Marcoppido, G.A. and Vila, B.L., 2013. Manejo de llamas extra-andinas: observaciones que contribuyen a su bienestar en un contexto no originario. Revista Argentina de Producción Animal 33: 139-154.

Markemann, A., Stemmer, A., Siegmund-Schultze, M., Piepho, H.P. and Zárate, A.V., 2009. Stated preferences of llama keeping functions in Bolivia. Livestock Science 124: 119-125.

Matos, A.M., Silva, F.V., Oliveira, L.L.S., Borges, I., Ruas, J.R.M., Novaes, C.G., Novais, F.C., Alves, D.D. and Oliveira, A.M.F., 2018. Strategies for the supply of crude glycerin for kids in pre-slaughter lairage. Small Ruminant Research 167: 87-91.

Matthews, P.T., Barwick, J., Doughty, A.K., Doyle, E.K., Morton, C.L. and Brown, W.Y. 2020. Alpaca field behaviour when cohabitating with lambing ewes. Animals 10: 1605. https://doi.org/10.3390/ani10091605

Mattiello, S., Battini, M., De Rosa, G., Napolitano, F. and Dwyer, C., 2019. How can we assess positive welfare in ruminants? Animals 9: 758. https://doi.org/10.3390/ani9100758

Mazhangara, I.R., Chivandi, E., Mupangwa, J.F. and Muchenje, V., 2019. The potential of goat meat in the red meat industry. Sustainability 11: 3671. https://doi.org/10.3390/su11133671

McDonough, S.P. and Holoyda, B., 2018. Ritualistic animal killing. In: Brooks, J.W. (ed.) Veterinary forensic pathology. Springer, Cham, Switzerland, pp. 129-138.

McMullen, C., 2008. Romancing the alpaca: passionate consumption, collection, and companionship. Journal of Business Research 61: 502-508.

Mdletshe, Z.M., Marufu, M.C., and Chimonyo, M., 2020. Effect of indigenous slaughter methods on the behavioural response, bleeding efficiency and cardiac arrest of Nguni goats. Animals 10: 247. https://doi.org/10.3390/ani10020247

Mdletshe, Z.M., Chimonyo, M. and Mapiye, C., 2021. Indigenous slaughter techniques: effects on meat physico-chemical characteristics of Nguni goats. Animals 11: 858. https://doi.org/10.3390/ani11030858

Minka, N.S. and Ayo, O.J., 2010. Serum biochemical activities and muscular soreness in transported goats administered with ascorbic acid during the hot-dry season. European Journal of Translational Myology 20: 193-203.

Minka, N.S. and Ayo, J.O., 2014. Effects of transportation positions and orientations on muscular damage of goats transported by road for 12 h during the hot-dry conditions. Journal of Veterinary Medicine and Animal Health 6: 75-82.

Minka, N.S., Ayo, J.O., Sackey, A.K.B. and Adelaiye, A.B., 2009. Assessment and scoring of stresses imposed on goats during handling, loading, road transportation and unloading, and the effect of pretreatment with ascorbic acid. Livestock Science 125: 275-282.

Miranda-de la Lama, G.C., 2013. Transport and pre-slaughter logistics: definitions and current tendencies in animal welfare and meat quality. Veterinaria Mexico 44: 31-56.

Miranda-de La Lama, G.C., 2018. Transporte y bienestar animal: un enfoque integrador. Editorial Servet, Grupo Asís Biomedia S.L., Zaragoza, Spain, 160 pp.

Miranda-de la Lama, G.C., 2019. Goat handling and transport. In: Grandin, T. (ed.) Livestock handling and transport. CABI, Wallingford, UK, pp. 271-289.

Miranda-de la Lama, G.C. and Mattiello, S., 2010. The importance of social behaviour for goat welfare in livestock farming. Small Ruminant Research 90: 1-10.

Miranda-de La Lama, G.C., Villarroel, M., Liste, G., Escós, J. and María, G.A., 2010. Critical points in the pre-slaughter logistic chain of lambs in Spain that may compromise the animal's welfare. Small Ruminant Research 90: 174-178.

Miranda-de la Lama, G.C., Sepúlveda, W.S., Montaldo, H.H., María, G.A. and Galindo, F., 2011. Social strategies associated with identity profiles in dairy goats. Applied Animal Behaviour Science 134: 48-55.

Miranda-de La Lama, G.C., Villarroel, M. and María, G.A., 2012. Behavioural and physiological profiles following exposure to novel environment and social mixing in lambs. Small Ruminant Research 103: 158-163.

Miranda-de la Lama, G.C., Pinal, R., Fuchs, K., Ducoing, A. and Galindo, F. 2013. Environmental enrichment and social rank affects the fear response to human of dairy goats. Journal of Veterinary Behaviour: Clinical Applications and Research 8: 342-348.

Miranda-de la Lama, G.C., Villarroel, M. and María, G.A., 2014. Livestock transport from the perspective of the pre-slaughter logistic chain: a review. Meat Science 98: 9-20.

Navarre, C.B., Heath, A. M., Wenzel, J., Simpkins, A., Blair, E., Belknap, E. and Pugh, D.G., 2001. A comparison of physical examination and clinicopathologic parameters between sheared and nonsheared alpacas (*Lama pacos*). Small Ruminant Research 39: 11-17.

Nawroth, C., Baciadonna, L. and McElligott, A.G., 2016. Goats learn socially from humans in a spatial problem-solving task. Animal Behaviour 121: 123-129.

Nawroth, C., Albuquerque, N., Savalli, C., Single, M.S. and McElligott, A.G., 2018. Goats prefer positive human emotional facial expressions. Royal Society Open Science 5: 180491. https://doi.org/10.1098/rsos.180491

Neijenhuis, F., Ruis, M. and Bokma, M., 2015. Advies over de geschiktheid van de alpaca als productiedier (No. 928). Wageningen Livestock Research, Wageningen, the Netherlands, 30 pp.

Neubert, S., von Altrock, A., Wendt, M. and Wagener, M.G., 2021. Llama and alpaca management in Germany – results of an online survey among owners on farm structure, health problems and self-reflection. Animals 11: 102. https://doi.org/10.3390/ani11010102

Nikbin, S., Panandam, J.M. and Sazili, A.Q., 2016. Influence of pre-slaughter transportation and stocking density on carcass and meat quality characteristics of Boer goats. Italian Journal of Animal Science 15: 504-511.

Norton, P.L., Gold, J.R., Russell, K.E., Schulz, K.L. and Porter, B.F., 2014. Camelid heat stress: 15 cases (2003-2011). The Canadian Veterinary Journal 55: 992-996.

Nwe, T.M., Hori, E., Manda, M. and Watanabe, S., 1996. Significance of catecholamines and cortisol levels in blood during transportation stress in goats. Small Ruminant Research 20: 129-135.

Peng, S., Fu, E., Lee, S. and Tsai, S. 2019. Methods to ameliorate heat stress in non-native alpaca, *Vicugna pacos*. Small Ruminant Research 175: 90-95.

Pérez, P., Maino, M., Guzman, R., Vaquero, C., Kobrich, C. and Pokniak, J., 2000. Carcass characteristics of llamas (*Lama glama*) reared in central Chile. Small Ruminant Research 37: 93-97.

Piccione, G., Casella, S., Giannetto, C., Bazzano, M., Giudice, E. and Fazio, F., 2013. Oxidative stress associated with road transportation in ewes. Small Ruminant Research 112: 235-238.

Pitcher, B.J., Briefer, E.F., Baciadonna, L. and McElligott, A.G., 2017. Cross-modal recognition of familiar conspecifics in goats. Royal Society Open Science 4: 160346. https://doi.org/10.1098/rsos.160346

Plummer, P.J. and Schleining, J.A., 2013. Assessment and management of pain in small ruminants and camelids. Veterinary Clinics of North America: Food Animal Practice, 29: 185-208.

Plummer, P.J., Shearer, J.K., Kleinhenz, K.E. and Shearer, L.C., 2018. Determination of anatomic landmarks for optimal placement in captive-bolt euthanasia of goats. American Journal of Veterinary Research 79: 276-281.

Pointon, A., Kiermeier, A. and Fegan, N., 2012. Review of the impact of pre-slaughter feed curfews of cattle, sheep and goats on food safety and carcase hygiene in Australia. Food Control 26: 313-321.

Pollard, J.C. and Littlejohn, R.P., 1995. Effects of social isolation and restraint on heart rate and behaviour of alpacas. Applied Animal Behaviour Science 45: 165-174.

Polycarp, T.N., Obukowho, E.B. and Yusoff, S.M., 2016. Changes in haematological parameters and oxidative stress response of goats subjected to road transport stress in a hot humid tropical environment. Comparative Clinical Pathology 25: 285-293.

Pophiwa, P., Webb, E.C. and Frylinck, L., 2020. A review of factors affecting goat meat quality and mitigating strategies. Small Ruminant Research 183: 106035. https://doi.org/10.1016/j.smallrumres.2019.106035

Popova, T., Tejeda, L., Peñarrieta, J.M., Smith, M.A., Bush, R.D. and Hopkins, D.L., 2021. Meat of south American camelids – sensory quality and nutritional composition. Meat Science 171: 108285. https://doi.org/10.1016/j.meatsci.2020.108285

Pulido, M.A., Mariezcurrena-Berasain, M.A., Sepúlveda, W., Rayas-Amor, A.A., Salem, A.Z. and Miranda-de la Lama, G.C., 2018. Hauliers' perceptions and attitudes towards farm animal welfare could influence the operational and logistics practices in sheep transport. Journal of Veterinary Behaviour 23: 25-32.

Qekwana, D.N., McCrindle, C.M., Oguttu, J.W. and Grace, D., 2017. Assessment of the occupational health and food safety risks associated with the traditional slaughter and consumption of goats in Gauteng, South Africa. International Journal of Environmental Research and Public Health 14: 420. https://doi.org/10.3390/ijerph14040420

Quispe, E., Poma, A., Siguas, O., Berain, MJ. and Purroy, A., 2012. Estudio de la carcasa de alpacas (*Vicugna pocos*) en relacion al peso y clasificacion carnica. Revista de Investigaciónes Veterinarias del Perú 23: 43-51.

Rajion, M.A., Mohamed, I., Zulkifli, I and Goh, Y.M., 2001. The effects of road transportation on some physiological stress measures in goats. Asian-Australasian Journal of Animal Sciences 14: 1250-1252.

Regassa, A., Moje, N., Megersa, B., Beyene, D., Sheferaw, D., Debela, E., Abunna, F. and Skjerve, E., 2013. Major causes of organs and carcass condemnation in small ruminants slaughtered at Luna export abattoir, Oromia Regional State, Ethiopia. Preventive Veterinary Medicine 110: 139-148.

Reiner, R.J. and Bryant, F.C., 1986. Botanical composition and nutritional quality of alpaca diets in two Andean rangeland communities. Rangeland Ecology & Management/Journal of Range Management Archives 39: 424-427.

Rodríguez-Loredo, C., 1998. Estudio arqueozoológico del sitio Inca Potrero-Chaquiago, Barrios La Solana y Retambay, Andalgalá, Pcia. de Catamarca (Argentina). Relaciones de la Sociedad Argentina de Antropología 22: 203-245.

Romero, S., Carletti, T., Franco, C.D., Moré, G., Schnittger, L. and Florin-Christensen, M., 2017. Seropositivity to *Sarcocystis* infection of llamas correlates with breeding practices. Veterinary Parasitology: Regional Studies and Reports 10: 65-70.

Rosen, S.D., 2004. Physiological insights into shechita. Veterinary Record 154: 759-765.

Sabow, A.B., Goh, Y.M., Zulkifli, I., Sazili, A.Q., Kaka, U., Ab Kadi, M.Z.A., Ebrahimi, M., Nakyinsige, K. and Adeyemi, K.D., 2016a. Blood parameters and electroencephalographic responses of goats to slaughter without stunning. Meat Science 121: 148-155.

Sabow, A.B., Zulkifli, I., Goh, Y.M., Ab Kadir, M.Z.A., Kaka, U., Imlan, J.C.,Abubakar, A.A., Adeyemi, K.D. and Sazili, A.Q., 2016b. Bleeding efficiency, microbiological quality and oxidative stability of meat from goats subjected to slaughter without stunning in comparison with different methods of pre-slaughter electrical stunning. Plos One 11: e0152661. https://doi.org/10.1371/journal.pone.0152661

Sabow, A.B., Nakyinsige, K., Adeyemi, K.D., Sazili, A.Q., Johnson, C.B., Webster, J. and Farouk, M.M., 2017. High frequency pre-slaughter electrical stunning in ruminants and poultry for halal meat production: a review. Livestock Science 202: 124-134.

Sabow, A.B., Goh, Y.M., Zulkifli, I., Ab Kadir, M.Z., Kaka, U., Adeyemi, K.D., Abubakar, A.A., Imlan, A.J.C., Ebrahimi, M. and Sazili, A.Q., 2018. Electroencephalographic and blood parameters changes in anaesthetised goats subjected to slaughter without stunning and slaughter following different electrical stunning methods. Animal Production Science 59: 849-860.

Salama, A.A.K., Caja, G., Hamzaoui, S., Badaoui, B., Castro-Costa, A., Facanha, D.A.E., Guilhermino, M.M. and Bozzi, R., 2014. Different levels of response to heat stress in dairy goats. Small Ruminant Research 121: 73-79.

Salvá, B.K., 2009. Caracterización de la carne y Charqui de Alpaca (*Vicugna pacos*). PhD. Thesis, Norwegian University of Leon, Leon, Spain.

Salvá, B.K., Zumalacárregui, J.M., Figueira, A.C., Osorio, M.T. and Mateo, J., 2009. Nutrient composition and technological quality of meat from alpacas reared in Peru. Meat Science 82: 450-455.

Salvá, B.K., Fernández-Diez, A., Ramos, D.D., Caro, I. and Mateo, J. 2012. Chemical composition of alpaca (*Vicugna pacos*) charqui. Food chemistry 130: 329-334.

Sarıbey, M. and Karaca, S., 2019. Effects of pre-slaughter ascorbic acid administration on some physiological stress response and meat quality traits of lambs and kids subjected to road transport. Animal Production Science 59: 954-964.

Scanes, C.G., 2018. The neolithic revolution, animal domestication, and early forms of animal agriculture. In: Scanes, C.G. and Toukhsati, S.R. (eds.) Animals and human society. Academic Press, London, UK, pp. 103-131.

Sejian, V., Silpa, M.V., Reshma Nair, M.R., Devaraj, C., Krishnan, G., Bagath, M., Chauhan, S.S., Suganthi, R.U., Fonseca, V.F., König, S., Gaughan, J.B., Dunshea, F.R. and Bhatta, R., 2021. Heat stress and goat welfare: adaptation and production considerations. Animals 11: 1021. https://doi.org/10.3390/ani11041021

Shahrom, M.S. and Zamri-Saad, M., 2012. A retrospective study on post-arrival mortality rate of Australian Boer goats in a breeder farm in Malaysia. Pertanika Journal of Tropical Agricultural Science 35: 737-742.

Sivachandiran, R., Bhanu Rekha, V., Natchimuthu, K. and Ajay Kumar, V.J., 2020. Challenges and mitigation strategies in goat meat production for assuring food security through goat farming in Puducherry. Pharma Innovation Journal 9: 57-59.

Small, A. and Hewitt, L., 2017. Transport and pre-slaughter management. In: Ferguson, D.M., Lee, C. and Fisher, A. (eds.) Advances in sheep welfare. Woodhead Publishing, Duxford, UK, pp. 227-243.

Smith, M.C. and Sherman, D.M., 2009. Nervous system. In: Smith, M.C. and Sherman, D.M. (eds.) Goat medicine. John Wiley and Sons, Ames, IA, USA, pp.163-256.

Smith, M.A., Bush, R.D., Thomson, P.C. and Hopkins, D.L., 2015. Carcass traits and saleable meat yield of alpacas (*Vicugna pacos*) in Australia. Meat Science 107: 1-11.

Smith-Davila, C.E., Mendoza Torres, G.J., Barbeito, C.G. and Ghezzi, M.D., 2019. Evaluation of animal welfare conditions of South American camelids admitted to the Huancavelica municipal slaughterhouse, Peru. Revista Mexicana de Ciencias Pecuarias 10: 379-390.

Solaiman, S.G., 2010. Housing requirements. In: Solaiman, S.G. (ed.) Goat science and production. John Wiley and Sons, Ames, IA, USA, pp. 323-337.

Sumar, J., 2010. Nutrition in camelids. Proceedings of the XXVI World Buiatric Congress, Santiago, Chile, pp. 343-357.

Szpak, P., Millaire, J.F., White, C.D. and Longstaffe, F.J., 2014. Small scale camelid husbandry on the north coast of Peru (Virú Valley): insight from stable isotope analysis. Journal of Anthropological Archaeology 36: 110-129.

Taylor, A.A. and Davis, H., 1996. The Response of llamas (*Lama glama*) to familiar and unfamilar humans. International Journal of Comparative Psychology 9: 43-50.

Taraborelli, P., Ovejero, R., Schroeder, N., Moreno, P., Gregorio, P. and Carmanchahi, P., 2011. Behavioural and physiological stress responses to handling in wild guanacos. Journal for Nature Conservation 19: 356-362.

Taunde, P.A., Argenta, F.F., Bianchi, R.M., Cecco, B.S.D., Vielmo, A., Lopes, B.C., Siqueira, F.M., De Andrade, C.P., Snel, G.G.M., De Barros, C.S.L., Sonne, L., Pavarini, S.P. and Driemeier, D., 2019. Mannheimia haemolytica pleuropneumonia in goats associated with shipping stress. Ciência Rural 49: e20180621. https://doi.org/10.1590/0103-8478cr20180621

Terlouw, C., Bourguet, C. and Deiss, V., 2016. Consciousness, unconsciousness and death in the context of slaughter. Part II. Evaluation methods. Meat Science 118: 147-156.

Turner, A.A., 2014. Alpacas, llamas & guanaco. Welfare guide issued to form secondary legislation attached to the Animal Welfare Act 2006. British Alpaca Society, Exeter, UK, 60 pp.

Underwood, W.J., Blauwiekel, R., Delano, M.L., Gillesby, R., Mischler, S.A. and Schoell, A., 2015. Biology and diseases of ruminants (sheep, goats, and cattle). In: Fox, J.G. (ed.). Laboratory animal medicine. Elsevier, St.-Louis, IA, USA, pp. 623-694.

Van der Staay, F.J., Joosse, M., Van Dijk, H., Schuurman, T., and Van der Meulen, J., 2011. Physiological and behavioral reactions elicited by simulated and real-life visual and acoustic helicopter stimuli in dairy goats. BMC Veterinary Research 7: 1-12.

Valenzuela, S., Benites, R.M., Moscoso-Muñoz, J.E., Youngs, C.R. and Gómez-Quispe, O.E., 2021. Impact of cria protection strategy on post-natal survival and growth of alpacas (*Vicugna pacos*). Veterinary and Animal Science 11: 100162. https://doi.org/10.1016/j.vas.2020.100162

Vásquez, V.F., Redondo, R., Rosales, T.E., Dorado, G. and Peiró, V., 2020. Osteometric and isotopic (δ13C and δ15N) evidence of Pre-Hispanic camelid-herd breeding in Moche site of 'Huaca de la Luna' (North coast of Peru). Journal of Archaeological Science: Reports 29: 102083. https://doi.org/10.1016/j.jasrep.2019.102083

Velarde, A., Rodriguez, P., Dalmau, A., Fuentes, C., Llonch, P., Von Holleben, K.V., Anil, M.H., Lambooij, J.B., Pleiter, H., Yesildere, T.Y. and Cenci-Goga, B.T., 2014. Religious slaughter: evaluation of current practices in selected countries. Meat Science 96: 278-287.

Vincent, B., 2005. Farming meat goats: breeding, production and marketing. Landlinks Press, Collingwood, Australia, 330 pp.

Visser, C., 2019. A review on goats in southern Africa: an untapped genetic resource. Small Ruminant Research 176: 11-16.

Waiblinger, S., Hajek, F., Lambacher, B. and Wittek, T., 2020. Effects of the method of restraint for shearing on behaviour and heart rate variability in alpacas. Applied Animal Behaviour Science 223: 104918. https://doi.org/10.1016/j.applanim.2019.104918

Wang, H.H., Gallagher, S.K., Byers, S.R., Madl, J. E. and Gionfriddo, J.R., 2015. Retinal ganglion cell distribution and visual acuity in alpacas (*Vicugna pacos*). Veterinary Ophthalmology 18: 35-42.

Warbington, M.C., 2014. Protection of camelids from predators. In: Cebra, C., Anderson, D., Tibary, A., Van Saun, R. and Johnson Larue, W. (eds.) Llama and alpaca care: medicine, surgery, reproduction. nutrition, and herd health. Elsevier: St.-Louis, MI, USA, pp. 5-8.

Watkins, P.J., Jaborek, J.R., Teng, F., Day, L., Castada, H.Z., Baringer, S. and Wick, M., 2021. Branched chain fatty acids in the flavour of sheep and goat milk and meat: a review. Small Ruminant Research 200: 106398. https://doi.org/10.1016/j.smallrumres.2021.106398

Webb, E.C., 2014. Goat meat production, composition, and quality. Animal Frontiers 4: 33-37.

Webb, E.C., Casey, N.H. and Simela, L., 2005. Goat meat quality. Small Ruminant Research 60: 153-166.

Wheeler, J.C., 1995. Evolution and present situation of the South American Camelidae. Biological Journal of the Linnean Society 54: 271-295.

Whitehead, C.E. and Bedenice, D., 2009. Neurologic diseases in llamas and alpacas. Veterinary Clinics of North America: Food Animal Practice 25: 385-405.

Winblad von Walter, L., Forkman, B., Högberg, M. and Hydbring-Sandberg, E., 2021. The effect of mother goat presence during rearing on kids' response to isolation and to an arena test. Animals 11: 575. https://doi.org/10.3390/ani11020575

Windschnurer, I., Eibl, C., Franz, S., Gilhofer, E.M. and Waiblinger, S., 2020. Alpaca and llama behaviour during handling and its associations with caretaker attitudes and human-animal contact. Applied Animal Behaviour Science 226: 104989. https://doi.org/10.1016/j.applanim.2020.104989

Windschnurer, I., Fischer, L., Yanagida, T., Eibl, C., Franz, S. and Waiblinger, S., 2021. Caretaker attitudes and animal training are associated with alpaca behaviour towards humans – an online survey. Applied Animal Behaviour Science 236: 105224. https://doi.org/10.1016/j.applanim.2021.105224

Wurzinger, M., Delgado, J., Nürnberg, M., Zárate, A.V., Stemmer, A., Ugarte, G. and Sölkner, J., 2005. Growth curves and genetic parameters for growth traits in Bolivian llamas. Livestock Production Science 95: 73-81.

Zheng, W., Liu, B., Hu, W. and Cui, Y., 2021. Effects of transport stress on pathological injury and main heat shock protein expression in the respiratory system of goats. Journal of Animal Physiology and Animal Nutrition 105: 1-13.

Zimerman, M., Grigioni, G., Taddeo, H. and Domingo, E., 2011. Physiological stress responses and meat quality traits of kids subjected to different pre-slaughter stressors. Small Ruminant Research 100: 137-142.

Zobel, G., Neave, H.W. and Webster, J., 2019. Understanding natural behaviour to improve dairy goat (*Capra hircus*) management systems. Translational Animal Science 3: 212-224.

Zogbi, A.P. and Frank, E.N., 2008. Procesos tecnológicos y nutricionales de la carne de llama (*Lama glama*). In: Pérez Álvarez, J.A., Fernández López, J. and Sayas Barbera, M.E. (eds.) Industrialización de productos de origen animal. Gráficas Limencop S.L., Elche, Spain.

Zucali, M., Lovarelli, D., Celozzi, S., Bacenetti, J., Sandrucci, A. and Bava, L., 2020. Management options to reduce the environmental impact of dairy goat milk production. Livestock Science 231: 103888. https://doi.org/10.1016/j.livsci.2019.103888

Zulkifli, I., Bahyuddin, N., Wai, C.Y., Farjam, A.S., Sazili, A.Q., Rajion, M.A. and Goh, Y.M., 2010. Physiological responses in goats subjected to road transportation under the hot, humid tropical conditions. International Journal of Agriculture and Biology 12: 840-844.

Photo: Torill Malmstrøm, Mobilslakt AS, Vestby, Norway

Chapter 10

Avoiding live-animal transport to slaughter: mobile abattoirs

J. Hultgren

Department of Animal Environment and Health, Swedish University of Agricultural Sciences, P.O. Box 234, 532 23 Skara, Sweden; jan.hultgren@slu.se

Abstract

Large-scale, automation, high process line speeds and a cheap and easily replaceable workforce enabled the industrialisation of slaughter during the 19th and 20th centuries, and continued consolidation, vertical integration and high-tech inventions took the development to the present day. The first mention of organised mobile slaughter may have been in the United States towards the end of the 20th century. Since then, the development has to some extent been determined by regulations on official meat inspections. At present, fish, poultry, livestock, reindeer and bison are slaughtered in various types of mobile establishments across many countries, but the total numbers are not known. This chapter examines the implications of mobile slaughter for animal welfare, followed by a discussion on its sustainability and feasibility. Finally, reference is made to many training resources that provide guidance for those considering setting up a mobile slaughterhouse, and summarise experiences from already started or completed projects in several countries. By eliminating live-animal transport and increasing farmers' control over the food chain, mobile slaughter contributes to rural development and potentially improves animal welfare and meat quality. The success of mobile abattoirs is affected by many factors and profitability largely depends on the opportunities to create added value and get premium retail prices for end products. Practical experience and further research can provide a basis for a slightly more liberal view of local meat and mobile slaughter.

Keywords: animal welfare, food system, meat processing, on-farm slaughter, small-scale slaughter

10.1 Introduction

Meat from a variety of animals has always been part of the human diet, with variations due to the availability of different species, and the possibilities to catch and kill edible animals and prepare the meat. It is only during the last 50-60 years or so that concerns for universal moral obligations, animal wellbeing, human health and the environment have had any major impact on meat consumption. Thousands of years ago, wild fish were trapped and kept in lagoons, ponds or shallow lakes, so that they would be available for consumption at all times. Freshwater aquaculture developed empirically before 1000 B.C. in China (Rabanal, 1988). The Romans fattened fish in specially designed basins or tanks (Busana, 2018), and during the Middle Ages, pond farming techniques developed in Europe, for example in monasteries where Christian faith prescribed non-meat food during Lent (Bonow *et al.*, 2016). In difficult times, for example during war, livestock and poultry have been slaughtered and processed in more or less temporary and mobile facilities to obtain food under primitive conditions. Even today, the slaughter of farmed animals or game for domestic consumption on farm, including in emergencies, is often carried out with very simple methods, some of which can be described as mobile. For a long time, animal farming was the primary source of farm income, rural employment and subsistence. Today, the role of meat production varies greatly from country to country, depending on conditions for production, standard of living and traditions. It still plays a significant role in rural livelihoods in many low-income countries (Herrero *et al.*, 2012; Long, 1993).

Worldwide, the meat industry is dominated by farmed fishes, poultry, pigs, sheep, goats and cattle. Other species are slaughtered in smaller numbers, but may be important in certain regions or to specific groups of producers or consumers (see later in this chapter and Miranda-de la Lama *et al.*, 2022 and Needham and Hoffman, 2022 – Chapters 9 and 8 of this book). In each of the last few years, slaughter took place of approximately 51-167 billion farmed fishes (FAO, 2020; Fishcount, 2017), 73.2 billion birds, 1.5 billion pigs, 1.1 billion sheep and goats, 302 million heads of cattle, 28 million buffaloes and 5.0 million horses (FAO, 2018) and 200,000 heads of semi-domestic reindeer in Norway, Sweden, Finland and Russia (Jernsletten and Klokov, 2002; Reinbase, 2019; Reindeer Herders' Association, 2019; Sámediggi, 2020). Birds included 68.8 billion chickens, 3.0 billion ducks and 656 million turkeys in 2018 (FAO, 2020). A total of about 27 million buffaloes per year (FAO, 2020) included 71,000 head of farmed bison in North America (Canadian Bison Association, 2020; USDA, 2020b). Thus, most animals slaughtered are fishes, followed by birds and only in third place are different mammals.

Meat is increasingly produced industrially for the international market by large, vertically integrated companies. Infrastructure and technological innovations have made it possible to transport animals over long distances, and slaughter and process them into meat which is marketed, distributed and sold in an efficient way under hygienically acceptable conditions. Large-scale operations have proven to have major economic benefits, here as in many other industries. However, there are clearly also significant disadvantages. There is a risk of distancing slaughterhouse staff from the animals they handle and kill and of separating consumers from farmers and animals, with potentially negative consequences for animal health and welfare, product quality and occupational health, not to mention interpersonal attitudes and actions. This has created public opinion for alternative production methods and a demand for sustainably produced meat, and even an interest in alternative protein sources such as synthetic or cell-based meat (Bekker *et al.*, 2017; Valente *et al.*, 2019).

As described in previous chapters in this book (e.g. Gallo *et al.*, 2022; Miranda-de la Lama *et al.*, 2022; Needham and Hoffman, 2022 – Chapters 2, 9 and 8 of this book), livestock can be accustomed to human handling and transport (Lürzel *et al.*, 2015; Waiblinger *et al.*, 2004), and gentle touching can reduce flight and stress levels in cattle and pigs (Hemsworth and Barnett, 1991; Hutson, 2014; Lürzel *et al.*, 2016). It is possible to create a positive relationship between humans and animals through stimuli, such as brushing, stroking with the hands, and a calm human voice (Boivin and Braastad, 1996; Lansade *et al.*, 2004; Ligout *et al.*, 2008; Shahin, 2018). Frequent rotation between pastures in combination with human handling has been shown to reduce reactivity in cattle (Ceballos *et al.*, 2018). In contrast, cattle kept under extensive rangeland conditions with little experience of human handling can become heavily stressed when handled before slaughter (Boissy and Bouissou, 1988; Boivin *et al.*, 1998; Waiblinger *et al.*, 2006).

Mobile slaughter means that killing and possibly further processing is done where the animals have been reared by taking the slaughterhouse to the animals and not the other way around. Consequently, at least the first part of the slaughter process takes place on the farm. Mobile slaughter can take different forms, depending on the animal species and local conditions, but it is usually small-scale. Processing at the mobile slaughterhouse can theoretically include cutting, deboning, further processing, packing and labelling, but in most cases ends with dressing and chilling of the carcasses. Under some circumstances, it only includes stunning and bleeding while the remaining process is done in a stationary slaughterhouse.

In a sense, mobile slaughter has been practised as long as animals have been consumed. Livestock have long been slaughtered and the meat prepared on the same farm for household needs, and wild animals have always been killed in their natural environment. Nevertheless, the interest in small-scale and mobile slaughter that is seen today is to some extent a reaction to large-scale and an attempt to create better conditions for sustainable production and consumption, and possibly more stable and resilient food chains locally and globally (Caldwell *et al.*, 2014). The question is whether and under what circumstances added value can make small-scale and mobile slaughter profitable. There is therefore a need to study these alternative food systems in more detail. In the UK, the Farm Animal Welfare Committee (FAWC, 2019) recommended further research on the feasibility of the economics, design and use of mobile slaughter facilities, to reduce the need to transport animals over long distances.

On-farm slaughter may undoubtedly have consequences for food safety, but also animal welfare, environmental protection, rural development and sustainability. To understand current trends and future perspectives, it is necessary to know how the slaughter industry developed. It is also important to recognise the characteristics of different species of animals and the specific conditions of different geographical regions, which are relevant to slaughter strategies. Large-scale industrialised slaughter, including the transportation of animals from farms to slaughterhouses, has proven to be a winning concept for most commercial species, at least in the short term. During the 20th century, it spread to most parts of the world, including Europe, North and South America, China, India, Southeast Asia, and many African countries. However, the massive livestock and poultry slaughterhouses of the previous century were not built as places for respectful animal handling and artisanal butchering, but mainly as meatpacking plants with an easily replaceable manual work force and cheap products in focus (Brantz, 2001).

In this chapter, we will take a closer look at small-scale and, more specifically, mobile slaughter of different species of farmed fishes, birds and mammals. We start by reviewing the history of industrialised slaughter. In terms of preslaughter handling, the perhaps most striking consequence of mobile slaughter is the avoidance of animal transports. Nevertheless, the prospects for mobile slaughter depend on the ability to meet legal provisions and expectations from farmers, consumers and the society as a whole.

10.2 Industrialised slaughter

The early history of large-scale slaughterhouses is about the production of meat from livestock, mainly pigs, cattle, sheep and goats. Already in the 18th century, reformers advocated public livestock slaughterhouses to move their activities out of public view, regulate 'morally dangerous' work, contain pollution and facilitate national surveillance, while ensuring sufficient volumes of meat

(Brantz, 2001; Fitzgerald, 2010; MacLachlan, 2007, 2008; Otter, 2008). Until then, animals had been slaughtered and marketed in various places, including backyards, streets or right beside the stand from which the meat was sold (Brantz, 2001). The public slaughterhouse emerged as a unique institution in the early 19th century as part of a larger transition from an agrarian to industrial system, accompanied by increased urbanisation and technological achievements. Despite the alleged benefits, it has been argued that this development distanced people from the animals they consumed, the act of killing and the natural environment (Cronon, 1991).

After years of revolutionary unrest, Napoleon Bonaparte ordered the construction of five municipal livestock slaughterhouses in a ring around Paris. When they opened in 1818, they were the first public slaughterhouses in Europe (Brantz, 2001), and probably worldwide. The French word 'abattoir' was introduced to denote a place where animals are slaughtered for human consumption (Vialles, 1994). Almost 50 years later, new livestock markets and slaughterhouses opened at *La Villette* in Paris, during the World Exposition in 1867, and were considered a monument of the new industrial design of the city (Brantz, 2001). Trains delivered livestock right to the markets, where animals were traded and sent directly to the abattoir. Slaughterhouses had become objects of city pride. *La Villette* continued to operate until 1974. Similar centralised and industrial slaughterhouses were erected in other European cities, as part of the massive transformation of medieval towns into modern metropoles (Brantz, 2001; Fitzgerald, 2010). In the UK, livestock slaughter was identified as a nuisance in the mid-18th century. Edinburgh's municipal slaughterhouse was built in 1852, but private slaughter persisted well into the 20th century in London, executed by hundreds of small private establishments, typically located behind or beneath a retail butcher shop and scattered throughout the capital (MacLachlan, 2007). Later, in the second half of the 20th century, there was a substantial drop in the number of stationary livestock slaughterhouses in Europe and a further increase in size of the remaining plants.

The large-scale livestock slaughter industry in the United States developed quite differently. Contrary to Europe, private interests took the initiative. The first large-scale slaughter plant opened in 1818 in Cincinnati for production of salted pork, and a couple of decades later, the city got its nickname, 'Porkopolis' (Lisa, 2020). Beef grew in popularity and ranchers started grazing enormous cattle herds on endless land in the West. In 1865, following the invention of mechanical refrigeration and the arrival of railroads which made Chicago an important transportation hub (Azzam, 1998; Hill, 1923), the Union Stockyards and Transit Company was founded on the South Side of the city. Around 1860, Chicago had only 220,000 inhabitants and there were but a couple of small livestock dealers, mainly satisfying local demands. However, at the start of the 20th century, Chicago had grown to the second largest city in the US with a population of 2.9 million (CensusViewer, 2021), and turned into the largest producer of meat (Brantz, 2001). The conveyor belt was introduced to increase production speed and efficiency, and slaughter became centralised and industrialised. US livestock slaughterhouses and meat-packing facilities ended up being much larger and operating at much higher line speeds than in Europe.

The privately owned Chicago Stock Yards forever transformed the production of meat in the US (Cronon, 1991; Jablonsky, 1993). The complex was huge with 2,300 separate livestock pens accommodating 75,000 pigs, 21,000 cattle and 22,000 sheep (Halpern, 1997; Figure 10.1). By the

Figure 10.1. The maze of livestock pens and walkways at Chicago's Union Stock Yards around 1947 (Wikimedia Commons by J.F. Vachon, Farm Security Administration, Washington, DC, USA).

beginning of the 1890s, it could hold more than 400,000 live animals at a time, and between 1865 and 1900, about 400 million animals, mostly pigs, were killed within the confines of the yards (Wilson, 2004). They also provided housing for many of the workers in the Back of the Yards community, which was notorious for its extreme poverty, criminality and environmental pollution. By the time of the First World War, the yards employed almost 50,000 workers (Pacyga, 2008) and their community had become home to nearly 60,000 people (Jablonsky, 1993). The yards closed down in 1971.

As the US meat industry continued to grow in the 20th century due to increasing demand and improved distribution possibilities, large meat-packing companies started to move slaughterhouses out from the large cities to small communities, nearer to the farms and feedlots (Broadway, 2000; Whittaker, 2006). An additional driver for this migration may have been a quest for cheap labour, away from the urban unions. The change was enabled by developments in refrigeration techniques and the introduction of vacuum-packed cutting details ('boxed beef') which gained in popularity. Instead of hanging and transporting sides of carcasses, the meat was cut and boxed up, which made distribution more efficient (Azzam, 1998; Brueggemann and Brown, 2003). Between 1963 and 1992, the percentage of meat from US processing plants delivered as cut-up products increased five-fold for beef and almost doubled for pork (MacDonald *et al.*, 2000). Unfortunately, the combined decentralisation and consolidation of the meat industry led to the shutdown of small and old nationally inspected abattoirs and slaughter facilities (Broadway and Ward, 1990) resulting in a host of environmental and social problems to small communities (Broadway, 2000; Whittaker, 2006).

Commercial production of meat chickens and the genetic modification of stocks started in the United States in the 1920s, when the production was characterised by small backyard flocks kept mainly for egg production (Mozdziak, 2014; Thiruvenkadani *et al.*, 2011). Since then, average growth rate has increased by at least 400% and the weight of the birds at the time of slaughter has more than doubled (Barbut, 2015; Thiruvenkadani *et al.*, 2011). As the demand for white meat increased in the 1950s, enterprises ('integrators') embarked on a process of vertical integration that included hatcheries, feed mills, chicken suppliers, slaughterhouses and packing plants (Mozdziak, 2014; Josephson, 2020). Concurrently, poultry companies began to produce domesticated chickens specifically bred for meat production (Tallentire *et al.*, 2016). By the end of the century, this production model had spread worldwide (Josephson, 2020) and contributes to the vast majority of poultry meat production today.

Large-scale aquaculture is a later phenomenon. For example, industrialised slaughter of farmed finfish did not develop until after the Second World War, towards the end of the 20th century. Since the 1980s, the trend has been a rise in the contribution of aquaculture relative to fishery (Barton and Staniford, 1998; FAO, 2020) and in 2018, aquaculture accounted for 52% of all fish for human consumption in the world (FAO, 2020).

In 2018, China was the largest producer of livestock and poultry for slaughter (FAO, 2018). Chinese and Brazilian slaughterhouses each processed 13% of all cattle, followed by the US and Argentina. In the same year, China slaughtered 27% of all sheep, while Australia, New Zealand and Turkey followed at considerably lower numbers (4-5%). For goats, Asia and Africa dominated even more, with China again as the main producer, with 32%. For pigs, China produced 47%, followed by the US, Germany and Spain at much lower figures (4-8%). Chinese slaughterhouses produced 15% of all slaughter chickens, followed by the US, Brazil and Indonesia at 13, 8 and 7%, respectively.

10.3 Transportation of meat versus animals

The opportunities to produce and market meat products are determined by a wide range of factors, such as the location and size of animal suppliers, the efficiency of transportation systems, the availability of adequate labour, processes and technical aids, and considerations regarding risks to animals, humans and the environment. Regulations vary between regions and exist at different levels and with different interests in focus (Anonyous, 2005; Australian Government, 2012; CARC, 2001; EC, 2019; OIE, 2019). Before the expansion of the railroads, the only way to get beef to consumers was to drive the cattle on foot to slaughterhouses located near the cities. Later, increased demand for vacuum-packed cutting details made distribution more efficient, which facilitated the relocation of slaughterhouses nearer to the farms.

Fish, poultry and livestock are often transported from their birthplaces or fattening farms to a central slaughterhouse, where they are killed and processed into meat (see the species-related chapters of this book: Cockram and Velarde, 2022; Driessen *et al.*, 2022; Faucitano and Raj, 2022;

Gallo *et al.*, 2022; Gerritzen *et al.*, 2022; Miranda-de la Lama *et al.*, 2022; Needham and Hoffman, 2022; Villarroel and Lambooij, 2022). Most mammals and birds are loaded on trucks or trailers and shipped by road, while fish are transported either on road or at sea. The animals are usually slaughtered in the country where they were farmed, but the abattoirs are sometimes scattered and situated far away, so some animals have to travel over long distances to be slaughtered. The major trend in high-income countries is continued consolidation of the slaughter industry towards larger processing facilities, which results in longer slaughter transports, at least for some animals. Due to slaughterhouse location, capacity and policies, as well as market issues, pricing and farmer preferences, animals are not always delivered to the nearest or even the second or third nearest slaughterhouse. Recently, the 2020 COVID-19 pandemic caused longer transport distances and a surplus of slaughter volumes with negative animal welfare consequences (Marchant-Forde and Boyle, 2020). According to FAO (2018), 12 million cattle, 44 million pigs, 20 million sheep and goats, 320,000 horses and 1.8 billion chickens were traded between countries in 2018, by road, railway, boat or air. The only available report on livestock trading within the US territory informed that more than 50 million live cattle, sheep and pigs were traded annually long distances between US states (Shields and Mathews, 2003).

Small-scale production and on-farm slaughter play a significant role in many low-income countries, where centralised slaughter is scarce or non-existent (Alders *et al.*, 2019a,b; McLeod *et al.*, 2009; Robinson *et al.*, 2011; Wong *et al.*, 2017). Even in industrialised countries, on-farm slaughter can be a viable alternative if, for example, the producer wants to maintain control of the value chain, ensure certain product qualities and market the meat locally. Producers may thus decide to administer slaughter and/or product distribution themselves to gain more direct access to local markets and capture more of the total product value (Carlsson *et al.*, 2007; FAO, 2019; Gwin and Thiboumery, 2014; Johnson *et al.*, 2012). Additional reasons for on-farm slaughter might be sanitary restrictions, or that animals are too dangerous to handle or unfit for lawful transport due to, for example, lameness (Woods *et al.*, 2009). Small slaughter facilities that operate on farm are sometimes called 'micro abattoirs' or 'very small processors'.

Although consolidation and standardised mass production has dominated since the mid- to late 20th century, many retailers are now trying to serve a more diversified and sophisticated market with value-added sales of meat products characterised by zero transport, good animal welfare, excellent taste, health-promoting properties and high ethical standards for the discerning consumer (Adams and Salois, 2010; Kirwan, 2006). This shift may suggest new opportunities for local food flows as it makes more sense to transport meat instead of animals (Baltussen *et al.*, 2009, 2017). In a sense, livestock 'butchering' has retained or regained its status as an artisan craft, distinguishing it from 'slaughter' in the sense of merely killing and processing the animals in a standardised way (Brantz, 2001; Kennard and Young, 2018; MacLachlan, 2008). Sonnino and Marsden (2006) argued that alternative food networks that fall outside the conventional model of agriculture short-circuit the conventional food chain through the development of new relationships between producers and consumers. A key feature of the new systems of food provision is their capacity to define food by its locality or even the specific farm where it was produced. Caldwell *et al.* (2014) argued that governments must adapt to changing needs and facilitate a transition to alternative food systems and networks.

However, large agricultural and food companies try to take advantage of consumer preferences, thus maintaining a dominant role and making local food markets vulnerable, with serious implications for small farmers, consumers and many other stakeholders (Guptill and Wilkins, 2002). In their quest for a maximised profit, transnational retailing corporations could quite easily use their vast resources to offer local foods as a symbol or gesture, thus undermining the potential of small-scale food production to create more meaningful and equitable social relations in rural communities (Guptill and Wilkins, 2002).

The terms 'local food' and 'local meat' have been associated not only with geographic proximity of producer and consumer, but also social and supply-chain characteristics, such as local farm ownership and natural, organic or other special foods. In contrast to the mainstream food chain, the products are marketed through direct-to-consumer outlets (farmers' markets, on-farm stores, roadside stands) or intermediate marketing channels, such as 'regional food hubs', grocers and restaurants, as well as food-service providers in institutions like schools, universities and hospitals (Low *et al.*, 2015; Martinez *et al.*, 2010). Regional food hubs are collaborative enterprises for taking local foods to larger mainstream markets (Barham *et al.*, 2012). These hubs typically work with farmers to preserve the characteristics of the food's origin and any special circumstances under which it was produced (Low *et al.*, 2015). By transparency of differentiated products throughout the local food supply chain, food hubs attempt to provide premium prices to farmers. Food that meets certain standards is often given priority. In the US, the share of livestock farms with direct-to-customer sales of locally produced meat rose from 7.5 to 9.3% between 2007 and 2012 (Low *et al.*, 2015). The number of farmers' markets increased by 180% between 2006-2007 and 2014, and the number of regional food hubs by 288%. Hence consumers were given more opportunities to buy food directly from producers.

10.4 Mobile slaughter versus transportation

Private initiatives for small-scale meat production, sometimes supported by authorities, have been seen in many industrialised countries, e.g. the United States (Coleman, 2008; Dickenson *et al.*, 2013; Gwin *et al.*, 2013; McCann, 2011; USDA, 2020a), Canada (Lees, 2015), the UK (All-Party Parliamentary Group for Animal Welfare, 2020; Kennard and Young, 2018), Sweden (Morchain, 2009) and the Baltic countries (Seppänen, 2004). In a resolution on the implementation of the European Union (EU) Council Regulation (EC) 1/2005 on the protection of animals during transport and related operations, the European Parliament (EC, 2019) called on the Member States and the European Commission to encourage and support options for on-farm and mobile slaughter, so that animals are slaughtered as close as possible to their place of rearing, thus promoting short distribution circuits and direct sales. The Parliament acknowledged that this would be in the interest of high animal welfare standards, low emissions and maintained employment in rural areas.

Due to the massive consolidation and re-structuration of the meat industry over the past decades, the time that stockpersons spend with the animals on farm has diminished (Raussi, 2013), which can make livestock less accustomed to human contact and lower their tolerance to preslaughter handling and transport (Bunzel-Drüke *et al.*, 2009; Waiblinger *et al.*, 2006). For this reason, some animals become severely stressed when being transported (Terlouw and Bourguet, 2022 – Chapter 1 and species-related chapters in this book). The European Food Safety Authority (EFSA, 2004) argued that fearful animals, such as ratites and deer, should be killed on farm using appropriate methods, e.g. accurate shooting or a well-designed stunning enclosure (Needham and Hoffman, 2022 – Chapter 8 of this book) and a mobile abattoir. Rearing of beef cattle on pasture may have positive effects on animal health and welfare, production economy, and sustainable landscape management, and is the most feasible type of meat production in certain regions (Bunzel-Drüke *et al.*, 2009; Le Neindre *et al.*, 1996). However, extensively kept livestock may be difficult to handle at the time of transport and slaughter (Gallo *et al.*, 2022 – Chapter 2 of this book; Jayme *et al.*, 2018; Petherick, 2005), with severe consequences for animal welfare, meat quality, food safety and occupational safety (Gallo *et al.*, 2022 – Chapter 2 of this book; Hultgren, 2018; Le Neindre *et al.*, 1996; Lürzel *et al.*, 2015, 2016; Probst *et al.*, 2012; Terlouw *et al.*, 2012). Furthermore, preslaughter handling of wild or semi-domesticated animals, like reindeer or bison, should be kept to a minimum to reduce animal stress (Church *et al.*, 2016; Galbraith *et al.*, 2014; Malmfors and Wiklund, 1996; Needham and Hoffman, 2022 – Chapter 8 of this book). On-farm slaughter of ranched bison helps reduce the risk of muscle damage and trim losses, and prevents injuries (McCorkell *et al.*, 2013).

Instead of transporting the animals to a central stationary slaughterhouse or building a fixed on-farm plant, the abattoir can be accommodated in a vehicle and moved to the farm. This practice has been termed 'mobile slaughter'. If the abattoir is used on the farm where the animals are kept, it prevents on-road transportation completely. At the time of slaughter, large livestock are taken on foot from housing or outdoor facilities to the stunning area. Poultry and small livestock may have to be collected in a container or carried by hand. In contrast, farmed fish are subjected to much the same preslaughter procedures, regardless of where the slaughter takes place (Villarroel and Lambooij, 2022 – Chapter 3 of this book). However, stunning and bleeding procedures for fish can still differ between mobile and stationary slaughter.

At on-farm slaughter of livestock, mobile or not, animals may originate from other farms in the neighbourhood, and be transported the short distance some time before slaughter. For example, if a large abattoir vehicle cannot be fit into the farm yard, it may be possible to park it on a nearby, larger farm and transport the animals a short distance. The transported animals can be kept for one or more nights on pasture or in confinement on the farm where they are finally slaughtered, and it will then be a matter of definition which is their 'home farm'. As described in the last section, the mobile abattoir can also be parked at a 'docking station' with adequate infrastructure for slaughter. Although short hauls may be stressful to pigs because the animals have little time to accommodate (Faucitano and Raj, 2022 – Chapter 5 of this book), short transports between farms are likely to be less demanding to most animals and they will not be as exhausted on arrival, as long as appropriate environmental conditions and handling are provided. Farmers are often allowed to transport their animals for short distances without special requirements for transport conditions. For example within the EU, farmers transporting their own animals in their own vehicle up to a maximum of 65 km are excluded from the main provisions of Council Regulation (EC) 1/2005 (EC, 2005).

Road transports of livestock are often administered by haulers who have been contracted by the slaughterhouses. Hitherto, many haulers have planned the routes without sophisticated logistics software, although trucks may be equipped with GPS technology, which has resulted in unnecessarily long and time-consuming transports. Swedish researchers developed an animal-transport simulation model and showed that strategic transportation planning and redirection of journeys could reduce the total slaughter transport distance in Sweden by 40% for cattle and 25% for pigs (Håkansson *et al.*, 2016). They also found that it would be possible to shut down approximately 50% of the Swedish livestock slaughterhouses without serious effects on the total transport distance, if all transports were optimised. Later, Frisk *et al.* (2018) reported that a more narrow time window for transport planning increases the animal transport time, the total driving time and the total distance driven.

Håkansson *et al.* (2016) also estimated that five additional mobile or small-scale abattoirs could reduce the number of Swedish slaughter transports of more than 4 h by 78% for cattle and 91% for pigs, compared to the 2008 situation with 46 and 40 existing Swedish slaughterhouses, respectively. Still, new small-scale abattoirs do not seem to shorten total mileage if they are not strategically located. Several authors have argued that mobile slaughter of livestock and poultry has the potential to serve farmers in remote places locally, create job opportunities through small-scale companies and provide high efficiency in local meat distribution (Díaz Charquero, 2017; Gwin *et al.*, 2013; Hedberg and Gebresenbet, 1999; Johnson *et al.*, 2012).

The equipment required to carry out adequate slaughter without the animals suffering unnecessarily may be technically advanced and bulky, making it difficult to fit in a vehicle, which limits the possibilities for mobile slaughter. Particularly expensive and bulky stunning equipment and machinery, such as pneumatic guns, paternoster or dip-lift gas systems for livestock, low atmosphere pressure systems or even electric water baths for poultry, are less likely to be found in mobile abattoirs, although they are not completely excluded.

Legislation on mobile slaughter varies between countries and is in many cases complicated and confusing. Basically, the requirements for mobile abattoirs are the same as for stationary slaughterhouses, and licensing requires that the same basic conditions be met. For example EU, US, Canadian and Australian regulations do not explicitly ban mobile slaughter of domestic animals, which indirectly makes it permissible. EU Council Regulation (EC) 1099/2009 (EC, 2009) on the protection of animals at the time of killing also allows EU Member States to apply national legislation for derogations from general provisions on the layout, construction and equipment of mobile slaughterhouses. Some countries, states or territories have adopted legislation on official control, in particular meat inspection, at mobile slaughter. For example, in the US, producers can to some extent opt for different types of inspection for mobile slaughter, which affects labelling and the opportunities to market the meat.

10.5 Mobile slaughter of fish

In Europe and the United States towards the end of the 19th century, the interest in salmonid culturing increased. Pollution had caused wild fish stocks to diminish and dams and irrigation canals obstructed the migration of salmon and other species, which motivated the establishment of marine laboratories and fish hatcheries in many countries (Isabella and Hunt, 2020). After Second World War, large-scale manufacturing of plastics revolutionised aquaculture through the introduction of light, durable and inexpensive floating cages (Isabella and Hunt, 2020). Today, commercial fish farming takes place outdoors in cages, closed containers, ponds or irrigation ditches, or indoors in tanks or integrated recycling systems. Floating or submerged cages or containers are placed in lakes, ponds, rivers or the sea (Villarroel and Lambooij, 2022 – Chapter 3 of this book). The industry is characterised by large companies, with a high degree of vertical integration from feed manufacturing and hatching to marketing, and many companies belong to international consortia with their own production sites in several countries.

In 2018, the 20 most common species made up 83.6% of the global production of farmed finfish, by weight (FAO, 2020). In 2018, China was by far the largest producer of cultured fish, accounting for almost half of the global production, but Norway was almost equal when it came to marine and coastal aquaculture (FAO, 2020). Developing countries dominate inland aquaculture of finfish, while a small number of industrialised countries, like Norway, Chile, Japan, the United Kingdom, Canada and Greece, are major producers of marine species, especially cold-water salmonids (FAO, 2020). Inland aquaculture represents 86.4% of the total production and various carp species are most common (34.1%), while Atlantic salmon represents 4.5% of the total production and rainbow trout a mere 1.6%. However, salmonid aquaculture is the fastest growing food production system globally, accounting for most of the market. Salmon was once a luxury food, but is now among the most popular fish species in the US, Europe and Japan. The world's first successful salmon farm was established in Norway in 1970 (Isabella and Hunt, 2020), and Norway is presently the largest producer of salmonids, followed by Chile (FAO, 2020; Iversen *et al.*, 2020).

Despite the fact that the industry is dominated by a relatively small number of species or groups of species, each one has its own characteristics and requirements that need to be considered during handling and slaughter (Villarroel and Lambooij, 2022 – Chapter 3 of this book). The suitability of the stunning/slaughter system depends on the species (Villarroel and Lambooij, 2022) and it is therefore not possible to design systems that work for mobile slaughter of all different kinds of farmed fish.

At some fish farms, slaughter is still practised in a traditional way, often involving mobile methods. Fish are stunned, bled and chilled on farm, on the shore close to the holding pen. They can then be further processed locally or shipped refrigerated in containers to a central plant. However, technological advances have led to the use of increasingly large and advanced facilities. Some facilities operate offshore, while others are intended for coastal areas. In 2018, China's first deep-sea salmon farm (Shenlan 1) was placed 240 km offshore in the Yellow Sea, claimed to be the world's largest fully submersible fish cage (Evans, 2018). The Norwegian Directorate of Fisheries (2021) listed a number of development projects. The Norwegian company Nordlaks has designed a 'dynamic' mobile farm for salmon and rainbow trout, with propulsion to move or maintain position, as well as possibilities for mooring at rest (the Havfarm project; https://www.nordlaks.no/english). The farm can thus be moved between areas depending on season, weather, environmental conditions and other commercial or leisure interests. Recently, China has also constructed a mobile marine fish farm (Guoxin 1), the first in a fleet of 50 vessels (Anonymous, 2020). Similarly, the Chilean venture Ocean Arks Tech has introduced a vessel with eight cages that can produce fish of several commercial species (http://oatech.cl/).

With the objective to preserve fish in good condition and retain its usability and quality, it can instead be kept alive in water tanks during long transports to slaughter facilities. Transports can be made in trucks on road or by boat at sea (Villarroel and Lambooij, 2022 – Chapter 3 of this book).

On-farm stunning and killing could be done in a fixed construction, a stationary abattoir, or in some kind of mobile unit. The European Commission (EC, 2017) reported on mobile slaughter of fish in Norway, the United Kingdom and Ireland. In January 2021, Norway was estimated to have 17 boats that slaughtered farmed finfish, mainly salmon, with some of them also processing fish further, while others carried them to a central processing plant (I. Fyllingen, Norwegian Food Safety Authority, personal communication; Figure 10.2).

Ferrada Schulz (2009) described different systems for transport and slaughter of salmonids in Chile. In traditional systems, fish are stunned and bled in a floating facility on pontoons directly next to the holding cages or on the deck of a vessel, after which they are transported at sea by a so-called 'tenderboat' or on road to a central plant for degutting and further processing. Boat and truck transportation are sometimes combined, with reloading on the way. In well-boat systems, on the other hand, fish are pumped from the holding cages and transported alive to a collection centre and

Figure 10.2. Crowding and pressure-vacuum pumping of salmon from a farming cage onto a slaughter and processing boat in Norway (Photo courtesy of J. Hultgren, Swedish University of Agricultural Sciences, Sweden).

slaughterhouse, or directly to a central processing plant. In October 2020, there were 53 authorised companies for the processing of salmonids in Chile, of which 20 carried out slaughter (National Fisheries and Aquaculture Service, 2020), and at the beginning of 2021, there were 63 authorised well-boats for aquaculture services (National Fisheries and Aquaculture Service, 2021).

A Swedish producer carries out electrical stunning and bleeding of Arctic char in a pontoon-based facility, which is placed at the shore during summer, and next to the cages during winter when the water is lower due to hydropower regulation of the lake where the farm is (J. Nygren, Umlax AB, personal communication). Fish are then transported on ice in containers on road to the processing plant. Furthermore, Nordic Trout AB, the largest producer of farmed rainbow trout in Sweden and Finland, has developed a mobile slaughter trailer which has been used in Sweden since 2018 (J. Leed, Nordic Trout AB, personal communication). It slaughters and processes approximately 20-22 tonnes of fish daily between January and October of each year. Fish are pumped or brailed to the dry electrical stunning equipment, bled and then transported on road in refrigerated containers to a stationary slaughterhouse for further processing.

10.6 Mobile slaughter of birds

For smallholders in developing countries, family poultry represents one of the few opportunities for saving and investment. In some of these countries, family poultry accounts for most of the poultry production. Small numbers of meat chickens can be fairly easily slaughtered and processed in basic facilities. Compared to livestock, birds are small and easily handled, and small-scale chicken production is not as strictly regulated as for livestock.

For decades, however, development of the industry has brought about an adaptation to large-scale food chains, a substantial drop in the number of stationary poultry abattoirs, and further increase in size of the remaining plants, especially in industrialised countries. As for fish, large slaughterhouses are typically part of vertically integrated multinational companies with their own contracted feed manufacturers, hatcheries, chicken farmers, laboratories and advisory services, where small-scale backyard producers have no role.

Yet, the demand for local poultry meat from small outdoor flocks is increasing and in some regions of the US, these systems have gained immense popularity in response to consumer demands and, as a consequence, the shortage of slaughterhouses has resulted in an insufficient capacity to effectively process all birds (McLeod *et al.*, 2009; Van Loo *et al.*, 2013), which has stimulated the interest in mobile slaughter (Niche Meat Processor Assistance Network, 2021; O'Bryan *et al.*, 2014). The total number of mobile abattoirs is unknown, but In June 2021, the Niche Meat Processor Assistance Network listed 11 inspected mobile poultry abattoirs in operation in the US, although the list may be incomplete as not all facilities are inspected by federal or state authorities. The slaughter volumes are generally small. According to Johnson *et al.* (2012), a mobile slaughter facility typically slaughters up to 500 chickens per day. However, mobile poultry abattoirs nowadays come in many different designs and can be housed in vans or trucks, or even be mounted on a small trailer (O'Bryan *et al.*, 2014; Figure 10.3). A mobile facility can be owned or rented by one or several producers and shared between farms.

Figure 10.3. Outdoor mobile poultry abattoir during preparation for turkey slaughter; stainless steel construction covered with a custom tarpaulin during processing (Photo courtesy of A. Fanatico, Appalachian State University, NC, USA).

In contrast to the EU and many other countries, US legislation does not require stunning before bleeding of poultry, because they are not covered by the Humane Methods of Slaughter Act (USC, 1958). Stunning is standard practice in large US slaughterhouses, but is only sometimes used in small plants and typically not at on-fam slaughter, where it is considered as impractical. Due to their size, turkeys and geese are more often stunned before slaughter compared to chickens. When slaughtered on farm, birds are placed head down in funnel-shaped killing cones and their neck vessels are cut immediately (Fanatico, 2003a,b). In countries where stunning is prescribed, such as the EU, head-only electrical stunning using dry electrodes or mechanical stunning with a cartridge-driven weapon appears to be appropriate for on-farm slaughter.

Some of the mobile poultry slaughter units mentioned above are said to be used not only for chickens, but also turkeys, pheasants and quail. Considering their sensitivity to transportation stress (Needham and Hoffman, 2022 – Chapter 8 in this book), the development of mobile abattoirs has been recommended for the slaughter of ostriches and other ratites (Bejaei, 2013; EFSA, 2004; Glatz and Miao, 2008; Wotton and Hewitt, 1999). Adams and Revell (1998) reported on a UK company operating a mobile abattoir for ostriches. Barter (2013) and Caldwell *et al.* (2014) described a mobile abattoir in Quebec where ostriches and emus were processed. Mobile ratite slaughter has also been reported in the US (https://www.provisioneronline.com/articles/106341-the-necessity-of-the-slaughterhouse), Australia (https://www.theland.com.au/story/5819588/mini-abattoir-within-reach-for-paddock-to-plate-producers/) and New Zealand (https://www.stuff.co.nz/business/industries/123979104/meatworks-are-gory-and-messy-and-nasty-but-the-slaughterings-humane).

Apart from the mentioned slaughter for household purposes, no information has been found about mobile slaughter of ducks or geese, but technically, there should be no obstacle to it.

10.7 Mobile slaughter of mammals

One of the first mentions of a commercial mobile abattoir was in Texas (USA) in the late 20th century, where exotic game species (e.g. various types of deer) were slaughtered under the supervision of state inspectors (Pitts, 2021). Later, a Sioux tribe in South Dakota set up a mobile abattoir to slaughter American bison in a traditional and respectful way (Pitts, 2021). In 2001, the Lopez Community Land Trust on Lopez Island in Washington State (USA) designed a mobile abattoir that later became the first mobile facility to receive a USDA license for the slaughter of beef, sheep, goats and pigs (Stevenson and Fukumoto, 2011; https://www.lopezclt.org/; http://www.mobileslaughter.com/index.htm). Canada's first mobile abattoir arrived ready for business in Yukon in 2006 and was intended for slaughter of pigs, cattle, sheep, goats, bison and elk (Yukon, 2007). Mobile slaughter of pigs, sheep and goats was introduced in Argentina, Brazil, Chile and Uruguay in 2012 and the following years (Díaz Charquero, 2017).

The total number of mobile abattoirs worldwide is unknown. Based on audits in 13 EU Member States between 2013 and 2015, the European Commission (EC, 2015) found that mobile slaughterhouses existed in Denmark and Germany, although their use was not extensive. In June 2021, the Niche Meat Processor Assistance Network (2021) listed nine mobile red meat abattoirs in operation in the US, of which six for livestock under federal inspection, but the list may not have been exhaustive. Nevertheless, according to news media during the last decade, mobile abattoirs of livestock have been introduced or are underway in a large number of countries across the globe, including the United States, Canada, Brazil, Chile, Argentina, Uruguay, the United Kingdom, Ireland, France, Germany, Austria, the Netherlands, Sweden, Latvia, Russia, China, Australia, New Zealand, South Africa and Namibia. In many countries, equipment manufacturers and meat suppliers advertise mobile livestock abattoirs and meat from such facilities on the internet.

As for poultry, the slaughter volumes are generally small. According to Johnson *et al.* (2012), a mobile slaughter facility could typically slaughter 5-10 cattle, 10-25 pigs or 10-40 sheep per day. Since then, the technique has evolved and the line speed has increased. Today, mobile abattoirs for mammals come in many different shapes and designs, ranging from basic systems mounted on a standard van for small livestock to a fully-equipped cattle abattoir housed in an 18 to 21 m long truck or trailer. Stunning can take place in an outdoor pen or enclosure or in a specially designed box inside the abattoir. Mechanical stunning with a cartridge-driven captive bolt weapon or a rifle appears to be appropriate for large and small livestock. Electrical stunning is an option for pigs and small ruminants.

Sometimes two trucks or trucks with trailers are used, so that one can take chilled carcasses to a hanging and cutting facility, while slaughter continues on the farm. An example is given by a mobile abattoir introduced by a Swedish company in 2014 (Hultgren, 2018). The unit consisted of a portable inspection pen and two trucks with trailers, with one being used for stunning with a penetrating captive bolt gun in a stun box and dressing, and the other for carcass chilling, storage and transport for further processing in a cutting plant (Figure 10.4). A maximum of four animals per hour and 30-35 animals per day were processed, usually three days a week, leaving the remaining weekdays for travel and maintenance. During 2017, 4,630 animals or 1.2% of the Swedish commercial slaughter of large cattle were processed in this mobile abattoir (Swedish Board of Agriculture, 2018). A sophisticated digital tracking system provided detailed information to the consumer about the animal and the farm through text and a QR code on the meat package in the store. In 2019, the company ceased operations due to low profitability. Nevertheless, the concept has been adopted in other countries, including France (Hälsingestintan, 2016) and Australia (Simmons, 2019).

As mentioned above, livestock with limited experience of human interaction can be easily stressed and difficult to handle at the time of slaughter, even in connection with mobile slaughter on the farm. Considering risks for animal welfare, food safety and occupational safety, it might be favourable to stun such animals without restraint using a rifle in the field at close range (Gallo *et al.*, 2022 and Needham and Hoffman, 2022 – Chapter 2 and 8, respectively, in this book). This has been called 'the gunshot method' (Schiffer, 2015; Figure 10.5). This way, the animal will remain in a familiar environment among familiar herd mates until it loses consciousness and dies. Immediately after a successful shot, the stun quality can be checked, the animal hoisted with a loader and bled, and the body taken to a nearby slaughterhouse for processing, optionally a mobile abattoir parked in the farmyard. As always, it is important to hoist and bleed the animal shortly following the shot and to take care of the carcass in a hygienic manner without delay. Practical experience shows that it in most cases is possible to stick the animals within 180 s after the shot (Hultgren *et al.*, unpublished data; Schiffer, 2015). Provided the method is applied in an orderly and safe way under acceptable

Figure 10.4. Holstein bulls in portable inspection pen waiting to be slaughtered in a mobile abattoir; entrance to the stun box to the left (Photo courtesy of A. Larsen, Swedish University of Agricultural Sciences, Sweden).

Figure 10.5. The gunshot method for slaughter of a beef steer; open trailer transport from the shooting pen to the abattoir (Photo courtesy of A. Larsen, Swedish University of Agricultural Sciences, Sweden).

sanitary conditions, and the meat can be successfully processed and marketed, it might provide a feasible business opportunity.

Generally speaking, a firearm with free bullet can be used for the stunning of all species of livestock. For example, this method has been approved in the US (Humane Slaughter of Livestock Regulations, 9 CFR 313, 2011) and the EU (Council Regulation (EC) 1099/2009), and by the World Organisation for Animal Health (OIE, 2019). The appropriate firearm and ammunition should be used, to achieve penetration and destruction of brain tissue and large blood vessels in the brainstem without passing through the head (Blackmore, 1985; Finnie, 1997; Woods *et al.*, 2009). Aged bulls, boars and rams have especially thick frontal skull bones, which requires an extra powerful weapon. The firearm should be aimed so that the bullet enters the brain and causes immediate unconsciousness, accounting for differences in skull conformation and brain position between animal species, sexes and ages (Gilliam *et al.*, 2012; Kohlen, 2011).

The choice of ammunition is a trade-off between adequate stunning and safety. Full metal jacket and solid-point bullets are designed to easily penetrate the target without much deformation or fragmentation. In comparison, semi-jacketed or hollow-point bullets expand and fragment rapidly upon impact, which transfers more energy without a risk of over-penetration, but they may also fail to penetrate at all. Hollow-point bullets are less likely to ricochet, but if a free bullet hits a person, it becomes more dangerous than more solid bullets. Muzzle velocity and energy of a bullet are generally higher when fired from a rifle instead of a handgun. High-velocity weapons tend to perforate whereas low-velocity bullets produce penetrating injury and retention of the bullets within the skull (Finnie, 2016). The most appropriate firearm and ammunition may thus depend on the circumstances. The American Veterinary Medical Association (AVMA, 2020) argued that bullet selection is perhaps the most important consideration for slaughter of livestock by gunshot. On the other hand, in the study of Schiffer *et al.* (2017), correct shot placement appeared to be more important than calibre. Rimfire cartridges are most common for killing livestock and small game hunting, but a more powerful centrefire ammunition may be required for heavier animals.

The 0.22 Long Rifle (0.22 LR, indicating a diameter of 0.22 inches, rimfire cartridge) may be one of the most frequently used ammunitions to shoot livestock. However, shooting success varies. Schiffer *et al.* (2017) applied the shotgun method to free-range Galloway and German Angus cattle at a distance of 2.5-20 m, using different types of ammunition. The authors concluded that the method with great certainty leads to immediate unconsciousness and death, provided that the shot is accurate. It was considered important that the bullet remained within the skull, which was achieved only with 0.22 Winchester Magnum rimfire cartridge (Retz *et al.*, 2014; Schiffer *et al.*, 2014). Hence, for purposes of outdoor slaughter with the shotgun method, the semi-jacketed 0.22 Magnum calibre seemed to be the most favourable, at least for relatively small cattle breeds, like the Galloway and German Angus. Swedish researchers used semi-jacketed 0.22 Magnum for Hereford steers and confirmed that the stun quality was acceptable (Hultgren *et al.*, unpublished data). Other researchers found that standard and high-velocity 0.22 LR bullets failed to penetrate the skulls of steers and heifers at a distance of 25 m (Baker and Scrimgeour, 1995).

According to the US Humane Slaughter of Livestock Regulations (https://ecfr.federalregister.gov/current/title-9/chapter-III/subchapter-A/part-313), small-bore firearms (calibre 0.32 or less) are usually effective for stunning cattle, sheep, swine, and goats, and calves, horses, and mules, except mature bulls, boars and rams. Canadian regulations (Safe Food for Canadians Regulations, SOR/2018-108; Government of Canada, 2019) prescribe 0.22 Magnum solid point bullets for steers, heifers, cows and bulls, as well as 0.30 Remington Carbine centerfire cartridges for large bulls. The same Canadian regulations recommend 0.22 Magnum over long-rifle ammunition to kill livestock, and advise against hollow-point bullets.

Several authors have reviewed the literature on mechanical stunning of livestock with a free bullet, the type of ammunition and the location of the shot (Gallo *et al.*, 2022 – Chapter 2 in this book; Grandin, 1994; Lambooij and Algers, 2016; Shearer, 2018; Woods *et al.*, 2009). Numerous guidelines have judged a properly placed gunshot suitable for on-farm stunning and killing of cattle (AVMA, 2016, 2020; FRSC, 2007; HSA, 2016; OIE, 2019; Shearer and Ramirez, 2013; UFAW, 1988). In some cases, gunshot may be the only possible stunning method (HSA, 2016; Hultgren, 2018).

The gunshot method with subsequent processing at a slaughter plant is permitted in several countries, including the United States (https://ecfr.federalregister.gov/current/title-9/chapter-III/subchapter-A/part-313), Canada (Safe Food for Canadians Regulations, SOR/2018-108; Government of Canada, 2019), Australia (FRSC, 2007) and New Zealand (New Zealand Government, 2018). Until recently, EU Regulation (EC) 853/2004 (EC, 2004) on specific hygiene rules for food of animal origin required that domestic ungulates (including all common livestock) intended for sale be brought into the slaughterhouse alive, with the exception of emergency slaughter. Despite this, the method has been used regularly for commercial purposes in Germany, where it is allowed for free-range cattle since 2011 if the transport time to the slaughterhouse does not exceed 1 h and permission is granted by the responsible regional public authorities (Tier-LMHV, 2018). In 2021, following public consultation, the European Commission (EC, 2021) adopted a delegated regulation which amended EU Regulation (EC) 853/2004, allowing the on-farm slaughter and bleeding of a limited number of domestic animals, provided that a high level of food safety for the meat is maintained and that the animals are accompanied by an official certificate that the hygiene requirements for slaughter have been met. New research may facilitate the implementation of the changed regulation. Switzerland has also passed legislation to allow the method (VSFK, 2020). Interestingly, farmed game intended for human consumption is usually shot with a rifle in the field before being taken to a slaughterhouse, in accordance with legislation in, e.g. the EU (Needham and Hoffman, 2022 – Chapter 8 in this book).

Semi-domestic reindeer have long been herded and used for meat production by the Sámi in Norway, Sweden, Finland and Russia, the Evenki in Russia, China and Mongolia, the Yakuts in Russia, and other ethnic groups in the United States (Alaska), Canada, China, Greenland, Mongolia, Russia and Scotland (https://reindeerherding.org/). Only since the 1990s, reindeer slaughter is regulated in EU Member States (i.e. Sweden and Finland) and for the most part in Norway. The slaughter is seasonal and almost all of it takes place between September and March, starting with the bulls in early autumn. In winter, the castrated males are in good condition and the cold weather makes meat refrigeration unnecessary. Traditionally, the Sámi slaughtered outdoors at the sites where they selected the reindeer, at various locations around the herding districts (Malmfors and Wiklund, 1996).

In Sweden, new regulations regarding reindeer meat inspection came into force in 1994 (SLVFS, 1994), stipulating that reindeer should be processed in a slaughterhouse and that meat inspection should normally be carried out. Many former outdoor sites for reindeer slaughter were then closed, larger numbers of animals were instead transported to slaughter and the development of specialised mobile reindeer abattoirs was initiated (Wiklund and Malmfors, 2004; Wiklund *et al.*, 2014). A Swedish company designed the mobile 'SANMO Multi-Species' unit, mainly for slaughter of reindeer, but also pigs (Benfalk *et al.*, 2005; Figure 10.6). About 10 such units were operating during the 1990's (Wikberg, 2001), with a capacity of 60-70 reindeer per hour. Stunning was done with a cartridge-driven captive bolt gun. The abattoirs operated exclusively in the cold season, without the need for forced chilling. They were often parked close to the selection corrals and pens and slaughter often took place at the same occasion as the sorting of animals, which resulted in short waiting times, but relatively unorganised conditions (Malmfors and Wiklund, 1996).

Many of the Swedish mobile reindeer abattoirs were later converted into stationary slaughterhouses (EFSA, 2006) and approved according to EU standards (Wiklund *et al.*, 2014). There is currently no mobile slaughter of reindeer in Sweden (A.H. Kautto, Swedish Food Agency, personal communication). In December 2020, 10 mobile abattoirs for livestock were registered in Norway, all of which were reported to slaughter reindeer, while two slaughtered sheep, goats and potentially even finishing pigs and calves (H.B. Bysheim, Norwegian Food Safety Authority, personal communication; Figure 10.7 and 10.8). The companies reported capacities of approximately 500 to 200 heads per year, resulting in between 15 and 300 tonnes of meat per year.

Figure 10.6. Reindeer slaughter in the mobile 'SANMO Multi-Species' abattoir (Photo courtesy of B. Haarstad, Selbu, Norway).

Figure 10.7. Mobile on-farm household slaughter of sheep (Photo courtesy of T. Malmstrøm, Mobilslakt AS, Norway).

Figure 10.8. Mobile on-farm household slaughter of pigs (Photo courtesy of T. Malmstrøm, Mobilslakt AS, Norway).

In the United States and Canada, there is a fairly large population of farmed American bison (Statistics Canada, 2017; USDA, 2019). In the US, two of the mobile abattoirs listed by the Niche Meat Processor Assistance Network in June 2021 were said to slaughter bison (Figure 10.9), and one to slaughter reindeer. There are no mobile slaughter facilities for bison operating commercially in Canada at present (T. Kremeniuk, Canadian Bison Association, personal communication).

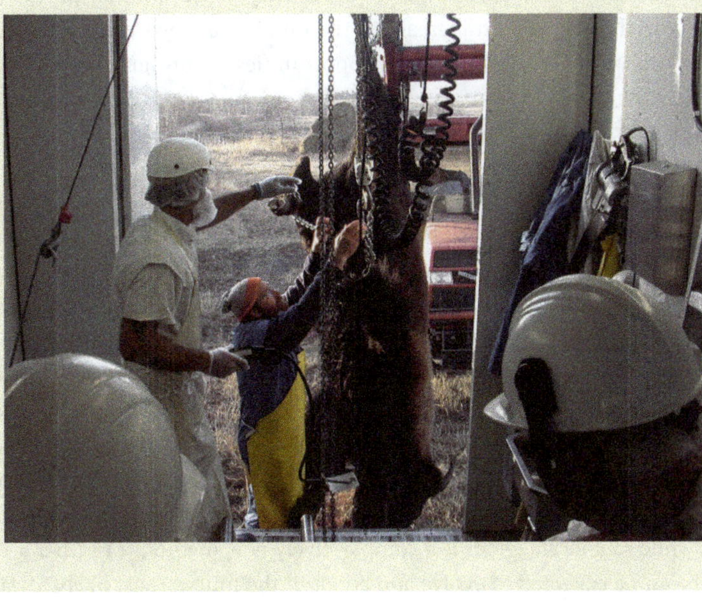

Figure 10.9. Bison being hoisted into mobile abattoir (Photo courtesy of J. Galbraith, Alberta Agriculture and Forestry, Canada).

However, mobile abattoirs for American bison have been tested to eliminate transportation in Canada (Galbraith, 2011; Galbraith *et al.*, 2014; McCorkell *et al.*, 2013).

10.8 Animal welfare

Traditional animal transport and transfer to a large slaughter plant, including handling during loading and unloading, as well as overnight lairage at the plant, present a large number of well-known risks to animal health and welfare (see other species-related chapters in this book: Cockram and Velarde, 2022; Driessen *et al.*, 2022; Faucitano and Raj, 2022; Gallo *et al.*, 2022; Gerritzen *et al.*, 2022; Miranda-de la Lama *et al.*, 2022; Needham and Hoffman, 2022; Villarroel and Lambooij, 2022). The EU Council Regulation (EC) 1099/2009 recognises that mobile slaughter can reduce the need to transport animals over long distances and can therefore help to protect animal welfare. Moreover, small-scale slaughter usually implies a low line speed, without serious time constraints for the personnel, which reduces the risk for inadequate animal handling and stunning (Gregory *et al.*, 2007; Stocchi *et al.*, 2014). This is of course important to bear in mind when comparing mobile and stationary slaughter methods.

Transport of so-called unfit animals is considered to be one of the most serious animal welfare issues in connection with industrialised slaughter. Only healthy animals may be slaughtered for human consumption and preslaughter inspection should establish whether the animal was healthy before the event that led to the need for emergency slaughter. EU Council Regulation (EC) 1/2005 provides details where animals are to be considered unfit for transport. The European Commission (EC, 2015) found indications of fewer animals arriving unfit at slaughterhouses when many animals were slaughtered on farm. For example, in a region of Poland, the incidence of unfit animals transported fell by 60% between 2013 and 2014 when the number of animals emergency-slaughtered on farm in this region increased by 235%. A regional project in Italy to prevent transport of unfit cows for slaughter was followed by a five-fold increase in the number of cows emergency-slaughtered on farm in less than two years (EC, 2015). Hence opportunities for on-farm slaughter may decrease the risk of transporting them.

Research evaluating animal welfare at mobile slaughterhouses or comparing the conditions for animals at mobile with stationary slaughter is uncommon, and scientific studies of animal welfare in mobile slaughter of fish or poultry are virtually non-existent. Direct comparisons of the two methods are complicated by the fact that the number of slaughterhouses (especially mobile, but also stationary) is limited and the conditions at different facilities vary considerably in many respects (Hultgren, 2018; Hultgren *et al.*, 2018, 2020). Interior design, equipment and routines may differ, regardless of the type of slaughter or size of the premises. The animal material may differ, too, either because farmers prefer one of the systems instead of the other or because the slaughterhouse prioritises certain animal categories. Similarly, the skills of the staff at a mobile abattoir may be different from a stationary plant, for a number of reasons not related to the method *per se*. It is also important to realise that stationary slaughter has a long history of gradual development and improvements, while mobile slaughter is a comparatively new method (or rather a set of different methods) which can be expected to evolve and adapt in the future, most likely towards increased functionality. Varying conditions on farms where mobile slaughter is carried out can contribute to variation in studied characteristics and in that way further complicate evaluations of the method (Hultgren *et al.*, 2020). In a sense, each farm is almost unique in terms of its breeding strategy, animal housing conditions, pastures, feeding and management routines. The transfer of livestock from housing facilities to the point of stunning at mobile slaughter may be demanding if the driveways are not adequately designed or the farm stockpersons are insufficiently trained (Benfalk *et al.*, 2005; Hultgren *et al.*, 2018, 2020).

Nine experts in lamb slaughter assessed individually the risk of poor animal welfare at small-scale lamb slaughter in Norway, Iceland, Sweden and Finland, comparing with large-scale industrialised slaughter, and found lower risk estimates for most hazards in small-scale slaughter (Hultgren *et al.*, 2016). However, the reverse was true for the hazards denoted 'splitting of groups' and 'separation of one sheep from the group'. The authors concluded that small-scale slaughter has a potential for improved sheep welfare, but overnight lairage at the slaughterhouse and prolonged fasting should be avoided.

Compared with a large stationary slaughterhouse, Skog Eriksen *et al.* (2013) found signs of reduced stress in lambs at a Norwegian mobile abattoir, as indicated by lower frequencies of vocalisation and aggressive behaviour and lower blood cortisol and glucose levels at exsanguination. In addition, meat (as assessed in the *Longissimus* muscle) from the mobile slaughterhouse had a lower pH and was more tender.

Hultgren *et al.* (2020) compared a large stationary plant with a Swedish mobile abattoir during one year, observing the behaviour of 298 cattle and stockpersons in the driveway to the point of stunning, and analysing blood samples for cortisol, glucose and lactate. Animals displayed backing and violent behaviours less frequently when moving them into the stun box at the mobile than at the stationary abattoir. Furthermore, blood samples indicated low stress levels (Hultgren *et al.*, unpublished data). In the end, the differences between the two slaughter systems were relatively few and mainly explained by factors specific to the studied facilities, such as design details and manning, rather than mobile and stationary slaughter in general. The big difference between the systems was the transport and lairage required for stationary slaughter.

On-farm mobile slaughter of bison at a maximum line speed of 10 animals per day was shown to result in fewer cases of dark cutting, less bruising, and improved meat tenderness, compared to transport and subsequent slaughter at a research abattoir or a conventional slaughterhouse (McCorkell *et al.*, 2013). However, the level of animal stress varied depending on the amount of handling before stunning.

Narrow spaces in small facilities can easily result in compromises and suboptimal solutions. For example, in an inadequately designed stun box, it may be difficult to stun large animals without head restraint and to hoist them after stunning, possibly resulting in delayed stunning and/or sticking (Grandin and Shivley, 2015; Hultgren *et al.*, 2020).

High noise levels are common in slaughterhouses, which can cause animal stress, complicate handling and poor meat quality (Faucitano and Raj, 2022 – Chapter 5 of this book; Weeks *et al.*,

2009). At mobile livestock slaughter, the absence of a constant, rather high noise level from many different activities inside a plant can make the animals more sensitive to occasional sharp sound impressions. Hultgren *et al.* (2020) observed that sudden loud noises from inside the mobile abattoir sometimes disturbed the cattle that waited outside, which may have contributed to handling difficulties. If the slaughter is done outdoors, the animals can be exposed to a variety of stressful sensory stimuli, including moving objects, direct sunlight, rough weather and slippery ground.

The gunshot method implies that animals are not only killed on the home farm, but also in a familiar environment among familiar herd-mates and totally relaxed. Studies from Germany (Schiffer, 2015; Schiffer *et al.*, 2017) and Sweden (Hultgren *et al.*, unpublished data) have shown that the gunshot method has potential to secure good animal welfare at slaughter, especially for animals that are difficult to handle. However, several authors underline that the gunshot method should only be performed by highly skilled personnel trained in the use of firearms, and that the safety of other animals that are nearby should be considered (AVMA, 2016; Schiffer, 2015; Schiffer *et al.*, 2017). The conditions should be as calm as possible, so that only one animal should be shot at a time, while being kept in an enclosure together with a few herd-mates. The animals must be accustomed to the enclosure. To ensure a clean shot, the marksman should have a free choice to shoot the animal standing in the most appropriate position. At the time of the shot, the animals next to the fallen individual usually display only a slight behavioural reaction (indicative of surprise and attention, sometimes followed by curiosity) and no sign of strong fear or stress (i.e. no increased blood lactate levels; Hultgren *et al.*, unpublished data; Schiffer, 2015).

Ideally, firearms and captive-bolt guns should be fired perpendicular to the front of the head and placed in such a way that the projectile travels through the brain and ricochet is avoided (AVMA, 2016). However, depending on the height above the ground where the shooter is, it may be difficult to achieve a shot at a right angle. The shooter must therefore be aware of the angle and plan the trajectory so that the projectile travels through the midbrain and brainstem (Government of Canada, 2019). Consequently, when the marksman is on the ground or on a low platform, the ideal point of aim might be more to the front (rostrally) than with a perpendicular shot. For safety reasons, firearm weapons must be used with caution or not at all in confined spaces, for example a mobile abattoir with a contained stun box (AVMA, 2016).

10.9 Feasibility and sustainability

It is commonly believed that slaughter itself is a break-even business, i.e. without profit, and that further processing or other value-adding activities are necessary to create profitability. Although mobile abattoirs are typically less expensive to build than stationary plants of the same size, small-scale slaughter implies a lower line speed than in industrial slaughter, which results in a higher cost per kilogram of meat under most circumstances. Especially cutting and further processing are more expensive when performed on farm and the product range will be reduced because some equipment are simply not available on farm (Lambooij *et al.*, 2011). In general, low throughput is likely a more limiting factor than a low processing capacity in on-farm slaughter (Gwin *et al.*, 2013). On the other hand, the capacity of a mobile abattoir may sometimes be limiting. For example, large livestock farms and feedlots are likely to send many animals to slaughter at the same time, which requires high-capacity processing facilities.

Lack of availability of large-scale stationary slaughter makes mobile facilities more economically advantageous. However, the distance between farms and the availability of animals for slaughter affect the degree of utilisation of the facility, which influences profitability. There is a lower limit to the number of animals to be slaughtered at the same time to make it worthwhile take a mobile abattoir to the farm. In regions where the farms are very small and scattered, too much time will be spent travelling between farms and setting up the mobile abattoir to make it profitable (Benfalk *et al.*, 2005). If alternative tasks for slaughter personnel, such as meat cutting or facility maintenance, can be found on days with few animals to slaughter, the costs will decrease. As a rule, the total costs for processing, marketing and distribution of locally marketed products are considerably higher than for commodity products from large-scale processing facilities (Gwin *et al.*, 2013). Therefore, profitability in mobile slaughter presupposes that the products can be sold at a premium price and/or that farmers are willing to pay extra for the services that a mobile slaughterhouse can offer. Consequently, when assessing the viability of small-scale and mobile slaughter, much depends on local conditions and how the business is organised.

The cost of slaughter differs between animal species. For example, for several reasons, it is more difficult to achieve good profitability in the slaughter and processing of poultry compared to livestock. Most farmers cannot commit to bringing birds regularly because production is seasonal, poultry are less flexible than livestock in terms of scheduling because they grow quickly, and the processing cost per weight is very high because the technology needed to decrease cost and increase throughput is expensive (Gwin *et al.*, 2013). As a result, companies that offer slaughter and processing services to poultry farmers typically must maintain high prices and require farmers to bring many birds at a time, possibly a minimum of 2,000 birds per day (Gwin *et al.*, 2013). A simple mobile unit owned by one or several farmers may still be feasible, as experiences in the US have shown (Gwin *et al.*, 2013).

Benfalk *et al.* (2005) estimated the total cost per slaughter weight at a mobile unit to between 6.4 and 70% higher for cattle and between 31% lower and 9.6% higher for pigs compared to the common stationary slaughter in Sweden. The calculations assumed a mobile abattoir capacity of 24 cattle or 120 finishing pigs during 5 h of slaughter per day, excluding travelling, preparations, cleaning, etc. Mobile slaughter profitability depends on the geographical region of Sweden, appearing more economically favourable in the more sparsely populated north. If the capacity of the mobile abattoir can be increased, it will lower the costs significantly. Comparing to stationary slaughter, processing (including cutting), Lambooij *et al.* (2011) estimated the operating costs for (1) mobile slaughter and processing, (2) mobile slaughter with processing in a separate on-farm facility and (3) mobile slaughter with processing at a central slaughterhouse to be 54, 19 and 31% higher, respectively, by weight of final products from cattle. Corresponding figures for pigs were 64, 41 and 28%, respectively, and for sheep 282, 64 and 24%, respectively. At mobile slaughter and central processing, which seemed most realistic, cattle were estimated to have the highest cost per kg of final products, followed by pigs and sheep. Furthermore, coordination with animal suppliers can be logistically intensive and demanding (Lambooij *et al.*, 2011). Gwin *et al.* (2013) calculated the cost for local (on-farm) slaughter, processing, marketing, distribution and sale of cattle and commodity (large-scale) production, and found the local product costs in local slaughter to be nearly twice as high at the end of the supply chain. Yet, processing accounted for only 13% of the final price of local beef, compared to 44% for marketing, distribution, and retail.

McCann (2011) provided financial advice to small and very small abattoirs and concluded that there are often significant opportunities to improve profitability with little or no capital investment, by changing the weekly slaughter schedule, keeping the workforce stable and adapting the product range. Still, Hedberg and Gebresenbet (1999) identified a risk for low profitability of mobile livestock and poultry abattoirs due to low line speed, unless the slaughter is integrated with further processing and marketing of meat products. Worker skill and experience with animals and equipment were judged to be crucial. Cartoni Mancinelli *et al.* (2018) estimated the costs for different approaches to chicken slaughter. The analysis indicated that the costs for mobile slaughter are, on average, lower than the costs for centralised large-scale slaughter, and on the same scale as the costs for on-farm stationary slaughter, provided that the number of processed animals is higher than 10,000 per year.

A way to reduce investment costs may be the so-called 'modular abattoir' approach (Coppin, 2010), where modified cargo containers are utilised to serve different processing purposes. A complete abattoir will include modules specially designed for all the operations involved in slaughter, processing, chilling, administration and accommodation. The approach is flexible to individual situations and containers can be stacked on top of each other to save space or increase vertical space for slaughter activities. The modules can also be modified and new modules added as the operation expands over time.

Animal by-products are usually an important source of significant income for large slaughterhouses (Gwin *et al.*, 2013). Jayathilakan *et al.* (2012) and Mora *et al.* (2019) reviewed the literature on possible uses of slaughter by-products. Products that can be processed for usage in human food, known as edible coproducts, are valued on the international market because they are part of the culinary tradition in different countries (Nollet and Toldrá, 2011). They can also be used as feedstuffs or to produce processing aids or bioactive peptides and proteins (Chernukha *et al.*, 2015; Martínez-Álvarez *et al.*, 2015). By-products can also be used to produce fertilisers, chemical or pharmaceutical substances, biodiesel or biogas. Small abattoirs primarily provide processing services and may have difficulty utilising these assets, resulting in additional costs for proper by-product disposal (Marti *et al.*, 2011). Yet, animal by-products may constitute a hitherto undervalued and unused resource in small-scale slaughter facilities.

Thus, added value of meat from small-scale and mobile slaughter may increase competitiveness and create profit. According to the theory of planned behaviour (Ajzen, 1991, 2006), consumers are probably more likely to take such action if they have a positive attitude towards mobile slaughter, feel morally compelled to buy meat from a mobile abattoir, perceive supportive social norms and believe in their own ability (Hoeksma *et al.*, 2017; López-Mosquera and Sánchez, 2012). This is the case of Swedish consumers who seem willing to pay extra for beef and pork meat from mobile abattoirs (Carlsson *et al.*, 2007; Liljenstolpe, 2005).

In most industrial countries, despite the fact that huge numbers of farm animals exist to be killed for food and that most people in the world actually eat them from time to time, slaughter is largely hidden away in large plants that are seldom or never visited by ordinary people (Pachirat, 2011). For this reason, many citizens do not even want to know how cattle turn into burgers, pigs into hot-dogs and chickens into nuggets (Cudworth, 2015). This contrasts with traditional household slaughter and even the proud city enterprises of early-day industrialised slaughter described above. The mere knowledge of livestock being shot on farm for food production may be perceived by the public as offensive and unethical (Gullone, 2014). Although public interest and outrage may change over time, these issues require careful consideration when performing mobile slaughter outdoors, for example with the gunshot method.

Although a low line speed, such as that of a butcher at the Chicago Stock Yards (two cattle per hour) or at a Swedish mobile abattoir (four cattle per hour), reduces the company's revenues, it may also contribute to work satisfaction and occupational safety, which may indirectly affect animal welfare. Parts of the large-scale meat processing are increasingly being robotised, but the high line speeds of industrialised meat processing facilities in the US, with rates ranging to 400 cattle or more than 10,000 chickens per hour (Andrews, 2015; Barbut, 2015; Kelloway, 2019; Schlosser, 2001), are considered to be a major cause of injuries among the workers (Human Rights Watch, 2004; Schlosser, 2001). However, when animals are stunned with a free bullet, as with the gunshot method, the safety of personnel and other people in the vicinity must be also taken into account. The method requires an enclosed shooting area with restricted public access, a protective backstop, such as a sand wall behind the animals and a raised platform for the marksman, preferably out of sight of the animals (Hultgren, 2018; Shiffer *et al.*, 2014).

Industrialised slaughterhouses can have extremely high employee turnover rates, sometimes exceeding 100% per year (GAO, 2005). Kandel (2009) reviewed reports from 1988 to 2007 on worker turnover in the US meat-processing industry and found annual rates between 40 and 400%. Rapid turnover results in a high percentage of inexperienced workers, which jeopardises the safety of all workers. Moreover, the work in a small-scale or mobile abattoir is probably more varied, compared to a large slaughter plant, and repetitive tasks are probably less common, which can help promote the health of workers (Finkel, 1985; Sherman, 2010). On the other hand, as also discussed above, the working conditions in a mobile abattoir can be demanding due to, for example, limited or unreliable access to water, electricity and computer networks, poor sanitary conditions, restricted working space, long journeys and irregular working hours with short notice.

Compared to stationary slaughter, and especially industrial slaughter at large-scale plants, mobile slaughter offers flexibility. In theory, operations can take place on almost any farm, provided there is enough space to park the abattoir. In practice, especially in remote areas, scarce supporting resources and poor opportunities for official inspections limit. As mentioned above, stationary slaughter has developed gradually over a long period of time, whereas mobile slaughter is a fairly new phenomenon, which in some respects can give an unfavourable impression. Furthermore, as Hultgren *et al.* (2020) and Sherman (2010) pointed out, the establishment and operation of mobile abattoirs probably require a great deal of creative thinking and problem solving. Pinkney (2014) argued that mobile abattoirs can contribute to the development of more robust local food systems.

Provisions, processing codes and technical standards are often based on large-scale conditions, making them less suitable and sometimes causing unreasonable costs for small farmers or processors with no dedicated staff and lower volumes over which to spread the costs (Adams and Salois, 2010; Gwin *et al.*, 2013; Khatri and Collins, 2007; Losito *et al.*, 2011; Morchain, 2009; Muth *et al.*, 2007; Raynolds, 2004). Existing production structures can thus hinder the development of small-scale rearing systems (Cartoni Mancinelli *et al.*, 2018). The European Parliament (EC, 2019) acknowledged that legal requirements, as well as changing market conditions and policy decisions, had made small slaughterhouses economically unprofitable, resulting in a decline in their numbers.

Most industrial countries require that slaughter of livestock is preceded by official ante-mortem inspection of the animals, mainly to ensure that they are fit for human consumption, but also to secure good animal welfare. Inspection before slaughter has also been recommended in resource-poor communities (Joshi *et al.*, 2003). However, in some countries, ante-mortem inspection is not statutory, or there are exemptions for small-scale or on-farm slaughter, and certain species. Emergency slaughter is commonly exempted, and inspection regulations are generally less strict for poultry than for livestock. For ante-mortem inspection in connection with mobile slaughter, it is required that the animals are available for inspection in approved premises on-farm shortly before slaughter begins. Inspection can be done in the existing rearing system indoors, although a pre-approved portable inspection pen outdoors may be advisable (Hultgren, 2018). Lambooij *et al.* (2011) pointed out that the inspection is complicated if the mobile abattoir slaughters several species.

In the United States, processors of meat and poultry are to some extent dependent on state agreements with federal authorities. The federal Food Safety and Inspection Service has cooperative agreements with a number of states (currently 27; USDA, 2015b) that allow state inspectors to enforce requirements 'at least equal to' those imposed under the federal acts for state-inspected facilities. Products produced under such conditions must not be sold outside the state. However, states operating a cooperative agreement may also opt into the Cooperative Interstate Shipment Program (currently 9 states; USDA, 2015a), allowing plants with 25 or fewer employees to conduct interstate trade (NASDA, 2021). Canada has similar provisions for commerce between provinces.

Exemptions for home or 'custom' slaughter for personal, household, guest or employee consumption exist in many countries. Facilities may charge less for custom processing than commercial processing and meat customers who purchase live animals prior to slaughter can sometimes take advantage of this (Johnson *et al.*, 2012; Low *et al.*, 2015). In some regions, 'mobile butchers' provide custom slaughter and processing on farm for the owner's personal consumption (Figure 10.7 and 10.8). Provisions may also allow the sale of poultry from small abattoirs without inspection under certain conditions. However, such meat cannot be marketed as inspected and may therefore provide a smaller profit (Angioloni *et al.*, 2015; O'Bryan *et al.*, 2014).

10.10 Guidance and educational resources

A large number of guidelines, recommendations and other resources on mobile slaughter are readily available as publications or websites. A non-exhaustive list is given in Table 10.1. Several guides have been based on experiences and expertise of existing facilities, mainly in the US, but also

Table 10.1. Readily available guidance and educational resources on small-scale and mobile slaughter.

Subject	Country	Reference
Killing of livestock using firearms	Unspecified	HSA, 2016
Design, operation and marketing	Unspecified	FAO, 1988
	USA	Anderson and Hashley, 2012; Bliven *et al.*, 2012; Fanatico, 2003b; Hashley and Gillan, 2012a,b; Niche Meat Processor Assistance Network, 2011; Thiboumery, 2009, 2010; Tum Tum's Black Gilt Meats, 2020; USDA, 2010; Yukon, 2007
	Sweden	Benfalk *et al.*, 2005
Evaluation of existing operations	USA/Australia	Coppin, 2010; Jonas, 2017
	USA	Galbraith *et al.*, 2014; Gwin *et al.*, 2013
	Sweden	Hultgren *et al.*, 2020
Prospective feasibility	USA	Center for Innovative Food Technology, 2009; Cowee and Harris, 2011; Dickenson *et al.*, 2013; Methow Conservancy, 2018; Mills, 2007; Sleeping Lion Associates, Inc., 2005
	Canada	Babb and Kennedy, 2013; Gibson and Epprechat, 2017; Melanson, 2005
	Australia	Australian Government, 2015
	UK	All-Party Parliamentary Group for Animal Welfare, 2020; Menzies *et al.*, 2020

other countries, covering HACCP plans, Standard Operating Procedures and recommendations on humane animal handling.

Coppin (2010) reviewed eight mobile red-meat facilities in the US to draw conclusions about prospects for the Australian livestock industry, and acknowledged the ability to process livestock in isolated regions, the opportunity for producers to vertically integrate their businesses, the capacity to avoid pressure from radical animal rights groups on live transport and the ability to move with the seasonal flow and supply of animals that are raised on natural pastures. On the other hand, the author also mentioned the lack of skills and knowledge, vulnerability to harsh environmental conditions, low throughput, lack of storage capacity and the need for large initial investment as weaknesses. The Australian Department of Industry and Science (Australian Government, 2015) was in favour of introducing mobile slaughter, but expressed concern about national regulations, which did not explicitly support the method.

To varying degrees, prospective feasibility reports have assessed the needs, designs, furnishing, manning, organisation, regulations, marketing and costs of projects to develop meat processing in local communities. Although most reports acknowledged obvious benefits of mobile slaughter, concerns were expressed regarding, for example, the need for long-term commitment of farmers, degree of utilisation, distances to cover, waste disposal, sanitation, maintenance and repair, as well as possible negative consequences for other small-scale producers in the area. A couple of reports considered the possible benefits of fixed docking stations instead of a fully mobile unit.

In the United Kingdom, the All-Party Parliamentary Group for Animal Welfare (2020) found that mobile abattoirs may provide opportunities in some areas, under certain circumstances, to help support the wider meat production chain and improve financial viability of small or rural businesses where there is a low geographical density of stationary abattoirs. At the same time, the report stressed that the abattoirs must provide robust business plans, can demonstrate need and that there is no impact on existing abattoir businesses, as well as show that they can meet current legislative requirements in terms of animal welfare and food hygiene before considering to introduce them. Menzies *et al.* (2020) gathered information from a few existing mobile facilities in the UK, units operating in 10 foreign countries and miscellaneous international experience. A docking station approach was considered most practical and preferred, with the mobile abattoir connecting at central locations, such as auction marts, industrial units and farms, with existing infrastructure for electricity, water, drains and chilling. The docking stations would act as hubs for a number of farmers. Such a system would facilitate the planning of slaughter on scheduled days. A cost-benefit analysis revealed that payback could be achieved without grant funding, in a time period of 11 to 15 years, provided premium prices from the sale of meat and offal (Menzies *et al.*, 2020).

10.11 Concluding remarks

Mobile slaughter usually takes place on a small scale, but in many different forms, depending on the species, specific business objectives and local conditions. Examples of currently viable systems are units with stunning equipment mounted on stationary pontoons or boats next to marine salmon cages, farmer-owned outdoor chicken abattoirs mounted on trailers, fully equipped cattle abattoirs for dressing and subsequent transportation of chilled carcasses to a cutting plant, and arrangements for on-farm rifle stunning and bleeding of pastured cattle or bison. Mobile slaughter of fish, poultry and livestock is carried out in many countries around the world, but the total volumes are not known.

Arguments for applying mobile slaughter methods can be respect for the animals' wellbeing, local demand from discerning consumers and a desire to increase farmers' control over the food chain by securing slaughter and process services and marketing high-quality meat products locally. At best, by avoiding unnecessary handling and transport of live animals to slaughter, mobile abattoirs can contribute to improved animal welfare, prevention of infectious diseases and superior meat quality. They may even contribute to environmental protection, rural development and the sustainability of food systems.

Nonetheless, the viability and success of mobile abattoirs are affected by many factors, such as animal species, accessibility and distance to centralised stationary slaughter, distance between farms, long-term commitment of farmers, the availability of trained and skilled slaughterhouse staff, inspection services and retail prices. Profitability largely depends on the opportunities to create added value. Legislation and standards are poorly adapted for mobile slaughter, making them less suitable and sometimes causing unreasonable costs. Practical experience from ongoing initiatives and continued research can provide a basis for a somewhat more liberal view of local meat and mobile slaughter.

Acknowledgements

The author acknowledges valuable assistance that was offered by a number of people. Professor Emeritus Bo Algers at the Swedish University of Agricultural Sciences contributed ideas and views on text structure during an initial collaboration. Associate Professor Albin Gräns at the Swedish University of Agricultural Sciences provided information about Swedish fish farming and Dr José Miguel Burgos González at the University of Chile about Chilean aquaculture. Dr Jason Galbraith at Alberta Agriculture and Forestry, Canada contributed with information about bison production and senior Veterinary Inspector Arja Helena Kautto at the Swedish Food Agency helped with clarifications about EU legislation and reindeer slaughter in Sweden. Photographs were kindly provided by Associate Professor Anne Fanatico at the Appalachian State University, North Carolina, USA, Ms Anne Larsen at the Swedish University of Agricultural Sciences, Sweden, Ms Barbro Söderlund, Ms Torill Malmstrøm at Mobilslakt AS, Norway, and Dr Jayson Galbraith at Alberta Agriculture and Forestry, Canada.

References

Adams, D.C. and Salois, M.J., 2010. Local versus organic: a turn in consumer preferences and willingness to pay. Renewable Agriculture and Food Systems 25: 331-341.

Adams, J. and Revell, B.J., 1998. Ostrich farming: a review and feasibility study of opportunities in the EU. A feasibility study undertaken for the European Network for Livestock Systems in Integrated Rural Development. School of Management, Harper Adams University College, Newport, UK. Available at: https://macaulay.webarchive.hutton.ac.uk/livestocksystems/feasibility/osr.pdf.

Ajzen, I., 1991. The theory of planned behavior. Organizational Behavior and Human Decision Processes 50: 179-211.

Ajzen, I., 2006. Behavioral interventions based on the theory of planned behavior. Available at: http://people.umass.edu/~aizen/pdf/tpb.intervention.pdf.

Alders, R., Costa, R., Gallardo, R.A., Sparks, N. and Zhou, H., 2019a. Smallholder poultry: contributions to food and nutrition security. In: Ferranti, P., Berry, E.M. and Anderson, J.R. (eds.) Encyclopedia of food security and sustainability. Elsevier Science Publishing Co., Inc., San Diego, CA, USA, pp. 292-298.

Alders, R., Costa, R., Gallardo, R.A., Sparks, N. and Zhoue, H., 2019b. Smallholder poultry: leveraging for sustainable food and nutrition security. In: Ferranti, P., Berry, E.M. and Anderson, J.R. (eds.) Encyclopedia of food security and sustainability. Elsevier Science Publishing Co., Inc., San Diego, CA, USA, pp. 340-346.

All-Party Parliamentary Group for Animal Welfare, 2020. The future for small abattoirs in the UK: report on an inquiry into small red meat abattoir provision. UK Parliament. Available at: https://tinyurl.com/7n7n54jf.

Anderson, S. and Hashley, J., 2012. Building an on-farm poultry processing facility: a guide to planning and constructing a mobile poultry processing unit (MPPU) or stationary poultry processing facility on your farm. New Entry Sustainable Farming Project Report. Available at: https://www.sare.org/resources/building-an-on-farm-poultry-processing-facility/.

Andrews, J., 2015. Q&A with Ted Genoways on food safety in pork processing, Part 1. Food safety news, 5 February, 2015. Available at: https://tinyurl.com/n7mpmjb5.

Angioloni, S., Kostandini, G., Alali, W.Q. and O'Bryan, C.A., 2015. Economic feasibility of mobile processing units for small-scale pasture poultry farmers. Renewable Agriculture and Food Systems 31: 387-401.

Anonymous, 2005. Meat packing industry criticized on human rights grounds. New York Times, 25 January, 2005. Available at: https://tinyurl.com/v45365d5.

Anonymous, 2020. Construction begins on world's first 100,000-tonne intelligent fish farming vessel in Qingdao. Global Times, 20 December, 2020. Available at: https://www.globaltimes.cn/content/1210431.shtml.

Australian Government, 2012. Australian animal welfare standards and guidelines – land transport of livestock. Animal Health Australia, Canberra, Australia. Available at: http://www.animalwelfarestandards.net.au/land-transport/.

Australian Government, 2015. Effect of market consolidation on the red meat processing sector. Department of Industry and Science. Available at: https://tinyurl.com/kn2x65mr.

AVMA, 2016. Guidelines for the humane slaughter of animals. American Veterinary Medical Association, Schaumburg, IL. Available at: https://tinyurl.com/4p6yc5mu.

AVMA, 2020. Guidelines for the euthanasia of animals. American Veterinary Medical Association, Schaumburg, IL. Available at: https://tinyurl.com/fb6t4m5t.

Azzam, A., 1998. Competition in the US meatpacking industry: is it history? Agricultural Economics 18: 107-126.

Babb, A. and Kennedy, E., 2013. Mobile slaughter unit feasibility study. Department of Natural Resources Forestry and Agrifoods Agency, Corner Brook, Canada. Available at: https://tinyurl.com/2xsywttp.

Baker, H.J and Scrimgeour, H.J., 1995. Evaluation of methods for the euthanasia of cattle in a foreign animal disease outbreak. Canadian Veterinary Journal 36: 160-165.

Baltussen, W., Nuhoff-Isakhanyan, G. and Van Wagenberg, C., 2017. Transport of live animals versus meat. Case studies of spent hens and lambs, using newly developed calculation model. Wageningen University & Research, Wageningen, the Netherlands. Available at: https://library.wur.nl/WebQuery/wurpubs/fulltext/420339.

Baltussen, W.H.M, Spoolder, H.A.M., Lambooij, E. and Backus, G.B.C., 2009. Sustainable production: transporting animals or meat? In: Guither, H.H., Merry, J.L. Merry, C.E. (eds.) Agriculture: food, fiber and energy for the future. Proceedings of the 17th International Farm Management Congress, Bloomington, IL, USA, pp. 635-648.

Barbut, S., 2015. The science of poultry and meat processing. University of Guelph, Guelph, Canada. Available at: http://www.poultryandmeatprocessing.com/.

Barham, J., Tropp, D., Enterline, K., Farbman, J., Fisk, J. and Kiraly, S., 2012. Regional food hub resource guide. United States Department of Agriculture, Agricultural Marketing Service, Washington, DC, USA. Available at: https://tinyurl.com/5wycuzep.

Barter, H., 2013. Mobile abattoirs: lessons from Quebec. Sustain Ontario, Ottawa, Canada. Available at: https://tinyurl.com/d3mjbdh8.

Barton, J.R and Staniford, D., 1998. Net deficits and the case for aquacultural geography. Area 30: 145-155.

Bejaei, M., 2013. Effects of pre-slaughter handling, transportation, and nutrient supplementation on ostrich welfare and product quality. PhD. Thesis, University of British Columbia, Vancouver, Canada.

Bekker, G., Fischer, A., Tobi, H. and Van Trijp, H., 2017. Explicit and implicit attitude toward an emerging food technology: the case of cultured meat. Appetite 108: 245-254.

Benfalk, C., Lindgren, K., Edström, M., Qiuqing, G. and Nordberg, Å., 2005. Mobile slaughter of cattle and pigs. Lantbruk & Industry Report 339. Swedish Institute of Agricultural and Environmental Engineering – JTI, Uppsala, Sweden. Available at: https://tinyurl.com/2m3rusf7.

Blackmore, D.K., 1985. Energy requirements for the penetration of heads of domestic stock due to development of a multiple projectile. Veterinary Record 116: 36-40.

Bliven, L., Stanton, T. and Frenay, E., 2012. On-farm poultry slaughter guidelines. College of Agriculture and Life Sciences, Cornell University, Ithaca, NY, USA. Available at: https://tinyurl.com/3fd2bunj.

Boissy, A. and Bouissou, A.-M., 1988. Effects of early handling on heifers' subsequent reactivity to humans and to unfamiliar situations. Applied Animal Behaviour Science 20: 259-273.

Boivin, X. and Braastad, B.O., 1996. Effects of handling during temporary isolation after early weaning on goat kids' later response to humans. Applied Animal Behaviour Science 48: 61-71.

Boivin, X., Garel, J.P., Mante, A. and Le Neindre, P., 1998. Beef calves react differently to different handlers according to the test situation and their previous interactions with their caretaker. Applied Animal Behaviour Science 55: 245-257.

Bonow, M., Olsén, H., and Svanberg, I., 2016. Historical aquaculture in Northern Europe. Södertörn University, Huddinge, Sweden. Available at: https://tinyurl.com/yk5cer3t.

Brantz, D., 2001. Recollecting the slaughterhouse. Cabinet, No. 4. Available at: https://www.cabinetmagazine.org/issues/4/brantz.php.

Broadway, M., 2000. Planning for change in small towns or trying to avoid the slaughterhouse blues. Journal of Rural Studies 16: 37-46.

Broadway, M.J. and Ward, T., 1990. Recent changes in the structure and location of the US meatpacking industry. Geography 75: 76-79.

Brueggemann, J. and Brown, C., 2003. The decline of industrial unionism in the meatpacking industry: event-structure analyses of labor unrest, 1946-1987. Work and Occupations 30: 327-360.

Bunzel-Drüke, M., Böhm, C., Finck, P., Kämmer, G., Luick, R., Reisinger, E., Riecken, U., Riedl, J., Scharf, M. and Zimball, O., 2009. 'Wilde Weiden'. Praxisleitfaden für ganzjahresbeweidung in naturschutz und landschaftsentwicklung. Working Group on biological environmental protection in the Soest e.V. district (Federal Agency for Nature Conservation; Naturschutz Schleswig-Holstein Foundation; Free State of Thuringia; North Rhine-Westphalia Foundation Nature Conservation, Homeland and Cultural Care), Bad Sassendorf-Lohne, Germany.

Busana, M.S., 2018. Fishing, fish farming and fish processing during the Roman age in the Northern Adriatic: literary sources and archaeological data. Regional Studies in Marine Science 21: 7-16.

Caldwell, W., Pinkney, D., Robson, D. and Chalel, I., 2014. Mobile processing: opportunities in the local meat sector. School of environmental design and rural development, University of Guelph, Guelph, Canada. Available at: https://tinyurl.com/bp8px7jy.

Canadian Agri-Food Research Council (CARC), 2001. Recommended code of practice for the care and handling of farm animals: transportation. Canadian Agri-Food Research Council, Ottawa, Canada.

Canadian Bison Association, 2020. Slaughter numbers. Available at: https://tinyurl.com/3zcczfxv.

Carlsson, F., Frykblom, P. and Lagerkvist, C.J., 2007. Consumer willingness to pay for farm animal welfare: mobile abattoirs versus transportation to slaughter. European Review of Agricultural Economics 34: 321-344.

Cartoni Mancinelli, A., Dal Bosco, A., Mattioli, S., Ranucci, D. and Castellini, C., 2018. Mobile poultry processing unit as a resource for small poultry farms: planning and economic efficiency, animal welfare, meat quality and sanitary implications. Animals 8: 229.

Ceballos, M.C., Góis, K.C.R., Sant'Anna, A.C. and Paranhos da Costa, M.J.R., 2018. Frequent handling of grazing beef cattle maintained under the rotational stocking method improves temperament over time. Animal Production Science 58: 307-313.

CensusViewer, 2021. Chicago, Illinois population: census 2010 and 2000 Interactive map, demographics, statistics, quick facts. Available at: http://45.79.181.212:8080/city/IL/Chicago.

Center for Innovative Food Technology, 2009. Assessing the feasibility and sustainability of small-scale poultry processing in Ohio. Ohio Department of Agriculture. Available at: https://tinyurl.com/krzjr8mn.

CFR, 2011. Humane slaughter of livestock regulations. Code of Federal Regulations, 9 CFR 313, USA, 1 January 2011. Available at: https://tinyurl.com/a5drcxtu.

Chernukha, I.M., Fedulovaa, L.V. and Kotenkovaa, E.A., 2015. Meat by-product is a source of tissue-specific bioactive proteins and peptides against cardio-vascular diseases. Procedia Food Science 5: 50-53.

Church, J., Galbraith, J., McCorkell, R., Rioja-Lang, F.C. and Silzer, M., 2016. Code of practice for the care and handling of bison: review of scientific research on priority issues. Bison Code of Practice Scientists' Committee, National Farm Animal Care Council, Saskatoon, Canada. Available at: https://tinyurl.com/58pa2k27.

Cockram, M.S. and Velarde, A., 2022. Sheep. In: Faucitano, L. (ed.) Preslaughter handling and slaughter of meat animals. Wageningen Academic Publishers, Wageningen, the Netherlands, pp. 266-310.

Coleman, K., 2008. Demand and options for local meat processing: finding the way from pasture to market in the CT River Valley. Community Involved in Sustaining Agriculture, South Deerfield, MA. Available at: https://www.buylocalfood.org/upload/resource/Slaughter.FeasibilityStudy2008.pdf.

Coppin, A., 2010. Mobile processing systems for the Australian livestock industry. International Specialised Skills Institute, Melbourne, Australia. Available at: https://tinyurl.com/4bckavn4.

Cowee, M. and Harris, T., 2011. Mobile slaughter potential for Nevada, 2011. University of Nevada Cooperative Extension, University of Nevada, Reno, NV. Available at: https://extension.unr.edu/publication.aspx?PubID=2201.

Cronon, W., 1991. Nature's metropolis: Chicago and the Great West. W.W. Norton and Company, New York, NY, USA, 592 pp.

Cudworth, E., 2015. Killing animals: sociology, species relations and institutionalized violence. The Sociological Review 63: 1-18.

Díaz Charquero, J., 2017. Diseño e implementación de un sistema de faena móvil como herramienta de apoyo al desarrollo sustenable de la agricultura familiar en Uruguay. MSc. Thesis. University of the Republic, Montevideo, Uruguay.

Dickenson, E., Joseph, S. and Ward, J., 2013. Local meat processing study. Confronting challenges in the local meat industry: focus on the pioneer valley of Western Massachusetts. Isenberg School of Management, University of Massachusetts, Amherst, MA, USA. Available at: https://www.buylocalfood.org/local-meat-processing-study/.

Driessen, B., Marlin, D. and Buyse, J., 2022. Horses. In: Faucitano, L. (ed.) Preslaughter handling and slaughter of meat animals. Wageningen Academic Publishers, Wageningen, the Netherlands, pp. 150-177.

European Commission (EC), 2004. Regulation (EC) No 853/2004 of the European Parliament and of the Council of 29 April 2004 laying down specific hygiene rules for food of animal origin. Official Journal of the European Union L139: 55-205. Available at: http://data.europa.eu/eli/reg/2004/853/oj.

European Commission (EC), 2005. Council Regulation (EC) No 1/2005 of 22 December 2004 on the protection of animals during transport and related operations and amending Directives 64/432/EEC and 93/119/EC and Regulation (EC) No 1255/97. Official Journal of the European Union L3: 1-44. Available at: http://data.europa.eu/eli/reg/2005/1/oj.

European Commission (EC), 2009. Council Regulation (EC) No 1099/2009 of 24 September 2009 on the protection of animals at the time of killing. Official Journal of the European Union L303: 1-30. Available at: http://data.europa.eu/eli/reg/2009/1099/oj.

European Commission (EC), 2015. Overview report: systems to prevent the transport of unfit animals in the EU. Directorate General of Health and Food Safety. Available at: https://tinyurl.com/pkkf22xh.

European Commission (EC), 2017. Welfare of farmed fish: common practices during transport and at slaughter. Final Report. Directorate-General for Health and Food Safety. Available at: https://tinyurl.com/3n7pz5re.

European Commission (EC), 2019. European Parliament resolution of 14 February 2019 on the implementation of Council Regulation (EC) No 1/2005 on the protection of animals during transport within and outside the EU Official Journal of the European Union C 449: 157-169. Available at: https://tinyurl.com/emekjnzu.

European Commission (EC), 2021. Commission Delegated Regulation (EU) 2021/1374 of 12.4.2021 amending Annex III to Regulation (EC) No 853/2004 of the European Parliament and of the Council on specific hygiene requirements for food of animal origin. C(2021) 199 final. Official Journal of the European Union L297, 20.8.2021: 1-15.

European Food Safety Authority (EFSA), 2004. The welfare of animals during transport. Opinion of the Scientific Panel on Animal Health and Welfare on a request from the Commission related to the welfare of animals during transport. Question EFSA-Q-2003-094. EFSA Journal 2: 44.

European Food Safety Authority (EFSA), 2006. Opinion of the Scientific Panel on Animal Health and Welfare on a request from the Commission related to the welfare aspects of the main systems of stunning and killing applied to commercially farmed deer, goats, rabbits, ostriches, ducks, geese and quail. Question EFSA-Q-2005-005. EFSA Journal 4: 326.

Evans, O., 2018. China gets ready to harvest first batch of farmed salmon from huge, deep sea fully-submersible fish cage. Salmon Business, 21 November, 2018. Available at: https://tinyurl.com/47ew8hfr.

Fanatico, A., 2003a. Small-scale poultry processing. National Center for Appropriate Technology, Butte, MT, USA. Available at: https://sd.appstate.edu/sites/default/files/poultryprocess.pdf.

Fanatico, A., 2003b. Small-scale poultry processing. National Center for Appropriate Technology, Butte, MT, USA (updated by K. Ellis, 2017). Available at: https://attra.ncat.org/product/small-scale-poultry-processing/.

FAO, 1988. Standard design for small-scale modular slaughterhouses. Paper No. 73. Food and Agriculture Organization of the United Nations, Rome, Italy. Available at: http://www.fao.org/3/t0034e/T0034E00.htm.

FAO, 2018. FAOSTAT. Food and agriculture data. Food and Agriculture Organization of the United Nations, Rome, Italy. Available at: http://www.fao.org/faostat/.

FAO, 2019. Developing sustainable value chains for small-scale livestock producers. FAO Animal Production and Health Guidelines No. 21. Food and Agriculture Organization of the United Nations, Rome, Italy. Available at: http://www.fao.org/3/ca5717en/ca5717en.pdf.

FAO, 2020. The State of world fisheries and aquaculture 2020. Sustainability in action. Food and Agriculture Organization of the United Nations, Rome, Italy. Available at: https://doi.org/10.4060/ca9229en.

Faucitano, L. and Raj, M., 2022. Pigs. In: Faucitano, L. (ed.) Preslaughter handling and slaughter of meat animals. Wageningen Academic Publishers, Wageningen, the Netherlands, pp. 178-230.

FAWC, 2019. Opinion on the welfare of animals during transport. Farm Animal Welfare Committee, London, UK. Available at: https://tinyurl.com/38axxhy5.

Ferrada Schulz, M.V., 2009. Análisis de alternativas para la operación de cosecha de salmónidos. MSc. Thesis, University of Chile, Santiago de Chile, Chile.

Finkel, M.L., 1985. The effects of repeated mechanical trauma in the meat industry. American Journal of Industrial Medicine 8: 375-379.

Finnie, I.W., 1997. Traumatic head injury in ruminant livestock. Australian Veterinary Journal 75: 204-208.

Finnie, J.W., 2016. Forensic pathology of traumatic brain injury. Veterinary Pathology 53: 962-978.

Fishcount, 2017. Fish count estimates. Available at: http://fishcount.org.uk/fish-count-estimates-2#farmedestimate.

Fitzgerald, A.J., 2010. A social history of the slaughterhouse: from inception to contemporary implications. Human Ecology Review 17: 58-69.

Frisk, M., Jonsson, A., Sellman, S., Flisberg, P., Rönnqvist, M. and Wennergren, U., 2018. Route optimization as an instrument to improve animal welfare and economics in pre-slaughter logistics. PLoS ONE 13: e0193223.

FRSC, 2007. Australian standard for the hygienic production and transportation of meat and meat products for human consumption. Food Regulation Standing Committee, Technical Report 3, AS 4696:2007. Available at: https://www.publish.csiro.au/book/5553/.

Galbraith, J., Rodas-González, A., López-Campos, O., Juárez, M. and Aalhus, J., 2014. Bison meat: characteristics, challenges, and opportunities. Animal Frontiers 4: 68-73.

Galbraith, J.K., 2011. Meat characteristics and stress of bison slaughtered in a mobile or stationary abattoir. PhD Thesis, University of Alberta, Calgary, Canada.

Gallo, C., Schwartzkopf-Genswein, K. and Gibson, T., 2022. Cattle. In: Faucitano, L. (ed.) Preslaughter handling and slaughter of meat animals. Wageningen Academic Publishers, Wageningen, the Netherlands, pp. 62-116.

GAO, 2005. Workplace safety and health: safety in the meat and poultry industry, while improving, could be further strengthened. Report to the Ranking Minority Member, Committee on Health, Education, Labor, and Pensions, US Senate, January 2005. United States Government Accountability Office, Report GAO-05-96. Available at: https://www.gao.gov/products/GAO-05-96.

Gerritzen, M., Shynkaruk, T., Buchynski, K., Schwean-Lardner, K. and Crowe, T.G., 2022. Poultry. In: Faucitano, L. (ed.) Preslaughter handling and slaughter of meat animals. Wageningen Academic Publishers, Wageningen, the Netherlands, pp. 232-265.

Gibson, S.D. and Epprechat, G., 2017. Meat on wheels: the potential for a mobile slaughterhouse unit in Haliburton County. Journal of Undergraduate Studies 5:1. Available at: https://ojs.trentu.ca/ojs/index.php/just/article/view/85.

Gilliam, J.N., Shearer, J.K., Woods, J., Hill, J., Reynolds, J., Taylor, J.D., Bahr, R.J., Crochik, S. and Snider, T.A., 2012. Captive-bolt euthanasia of cattle: determination of optimal-shot placement and evaluation of the Cash Special Euthanizer Kit® for euthanasia of cattle. Animal Welfare 21 (Suppl. 2): 99-102.

Glatz, P.C. and Miao, Z.H., 2008. Husbandry of ratites and potential welfare issues: a review. Australian Journal of Experimental Agriculture 48: 1257-1265.

Government of Canada, 2019. Guidelines for stunning techniques of mammalian food animals. Available at: https://tinyurl.com/pm83cndb.

Grandin, T. and Shivley, C., 2015. How farm animals react and perceive stressful situations such as handling, restraint, and transport. Animals 5: 1233-1251.

Grandin, T., 1994. Euthanasia and slaughter of livestock. Journal of the American Veterinary Medical Association 204: 1354-1360.

Gregory, N.G., Lee, C. and Widdecomb, J.P., 2007. Depth of concussion in cattle shot by penetrating captive bolt. Meat Science 77: 499-503.

Gullone, E., 2014. Risk factors for the development of animal cruelty. Journal of Animal Ethics 4: 61-79.

Guptill, A. and Wilkins, J.L., 2002. Buying into the food system: trends in food retailing in the US and implications for local foods. Agriculture and Human Values 19: 39-51.

Gwin, L. and Thiboumery, A., 2014. Beyond the farmer and the butcher: institutional entrepreneurship and local meat. Journal of Agriculture, Food Systems, and Community Development 4: 81-96.

Gwin, L., Thiboumery, A. and Stillman, R., 2013. Local meat and poultry processing: the importance of business commitments for long-term viability. Economic Research Service, United States Department of Agriculture. Economic Research Report 150. Available at: https://tinyurl.com/s2cjaxhc.

Håkansson, N., Flisberg, P., Algers, B., Jonsson, A., Rönnqvist, M. and Wennergren, U., 2016. Improvement of animal welfare by strategic analysis and logistic optimization of animal slaughter transportation. Animal Welfare 25: 255-263.

Halpern, R., 1997. Down on the killing floor: black and white workers in Chicago's packinghouses, 1904-54. University of Illinois Press, Chicago, IL, USA, 336 pp.

Hälsingestintan, 2016. Mobile abattoir and digital traceability to France. Press Release, 21 September, 2016. Available at: https://tinyurl.com/ynjnjw7p.

Hashley, J. and Gillan, J., 2012a. Handbook for small-scale poultry producer-processors. New England Small Farm Institute & New Entry Sustainable Farming Project Report. Available at: https://tinyurl.com/768ryxbx.

Hashley, J. and Gillan, J., 2012b. Mobile poultry processing unit: farm & food safety management guide for small-scale poultry producers and processors using a Massachusetts-inspected mobile poultry processing unit or stationary on-farm facility. New England Small Farm Institute & New Entry Sustainable Farming Project Report. Available at: https://tinyurl.com/rtuhvynu.

Hedberg, E. and Gebresenbet, G., 1999. Mobila och semi-mobila slakterier som alternativa slaktsystem. Department of Agricultural Engineering, Swedish University of Agricultural Sciences, Uppsala, Sweden. Report 238. Available at: https://tinyurl.com/ejs8t65y.

Hemsworth, P.H. and Barnett, J.L., 1991. The effects of aversively handling pigs, either individually or in groups, on their behaviour, growth and corticosteroids. Applied Animal Behaviour Science 30: 61-72.

Herrero, M., Grace, D., Njuki, J., Johnson, N., Enahoro, D., Silvestri, S. and Rufino, M.C., 2012. The roles of livestock in developing countries. Animal 7: 3-18.

Hill, H.C., 1923. The development of Chicago as a center of the meat packing industry. The Mississippi Valley Historical Review 10: 253-273.

Hoeksma, D.L., Gerritzen, M.A., Lokhorst, A.M. and Poortvliet, P.M., 2017. An extended theory of planned behavior to predict consumers' willingness to buy mobile slaughter unit meat. Meat Science 128: 15-23.

Hultgren, J., 2018. Is livestock transport a necessary practice? Mobile slaughter and on-farm stunning and killing before transport to slaughter. CAB Reviews 13: 1-15.

Hultgren, J., Algers, B., Atkinson, S., Ellingesen, K., Eriksson, S., Hreinsson, K., Nordensten, L. and Valtari, H. and Mejdell, C.M., 2016. Risk assessment of sheep welfare at small-scale slaughter in Nordic countries, comparing with large-scale slaughter. Acta Veterinaria Scandinavica 58: 34.

Hultgren, J., Arvidsson Segerkvist, K., Berg, C., Karlsson, A.H. and Algers, B., 2020. Animal handling and stress-related behaviour at mobile slaughter of cattle. Preventive Veterinary Medicine 177C: 104959.

Hultgren, J., Berg, C., Karlsson, A.H., Schiffer, K.J. and Algers, B., 2018. On-farm slaughter – ethical implications and prospects. In: Springer, S. and Grimm, H. (eds.) Professionals in food chains. Wageningen Academic Publishers, Wageningen, the Netherlands, pp. 146-151.

Human Rights Watch, 2004. Blood, sweat, and fear: workers' rights in U.S. Meat and Poultry Plants. New York, NY, USA. Available at: https://www.hrw.org/sites/default/files/reports/usa0105.pdf.

Humane Slaughter Association (HAS), 2016. Humane killing of livestock using firearms. Humane Slaughter Association, Wheathampstead, UK. Available at: https://tinyurl.com/tt4k4tvp.

Hutson, G.D., 2014. Behavioural principles of sheep handling. In: Grandin, T. (ed.) Livestock handling and transport. CAB International, Wallingford, UK, pp. 193-217.

Isabella, J. and Hunt, S., 2020. A short history of aquaculture innovation. Kakai Magazine, 24 August, 2020. Available at: https://www.hakaimagazine.com/features/a-short-history-of-aquaculture-innovation/.

Iversen, A., Asche, F., Hermansen, Ø. and Nystøyl, R., 2020. Production cost and competitiveness in major salmon farming countries 2003-2018. Aquaculture 522: 735089.

Jablonsky, T.T., 1993. Pride in the jungle: community and everyday life in back of the Yards Chicago. Johns Hopkins University Press, Baltimore, MD, USA, 216 pp.

Jayathilakan, K., Sultana, K., Radhakrishna, K. and Bawa, A.S., 2012. Utilization of byproducts and waste materials from meat, poultry and fish processing industries: a review. Journal of Food Science and Technology 49: 278-293.

Jayme, D.G., Oliveira, N.D., Jayme, C.G., Gonçalves, L.C., Glória, J.R., Martins, T., Borges, A.L.C.C., Silva, R.R. and Raposo, V.S., 2018. Bovine natural taming. In: Abubakar, M. and Manzood, S. (eds.) Animal welfare. IntechOpen, London, UK, pp. 87-97.

Jernsletten, J.-L.L. and Klokov, K., 2002. Sustainable reindeer husbandry. Arctic Council 2000-2002 University of Tromsø, Norway. Available at: http://www.reindeer-husbandry.uit.no/online/Final_Report/final_report.pdf.

Johnson, R.J., Marti, D.L. and Gwin, L., 2012. Slaughter and processing options and issues for locally sourced meat. Report No. LDP-M-216-01. United States Department of Agriculture, Economic Research Service, Washington DC, USA. Available at: https://www.ers.usda.gov/publications/pub-details/?pubid=37460.

Jonas, T., 2017. Dead local meat: building and operating a small-scale abattoir. Lessons learnt from seven US abattoirs. Jonai Farms Meatsmiths & Australian Food Sovereignty Alliance, Australia. Available at: https://tinyurl.com/v9vzzrw2.

Josephson, P.R., 2020. Chicken: a history from farmyard to factory. Wiley & Sons, New York, NY, USA, 252 pp.

Joshi, D.D., Maharjan, M., Johansen, M.V., Willingham, A.L. and Sharma, M., 2003. Improving meat inspection and control in resource-poor communities: the Nepal example. Acta Tropica 87: 119-127.

Kandel, W., 2009. Recent trends in rural-based meat processing. Conference presentation. Immigration reform: implications for farmers, farm workers, and communities, 21-22 May, 2009, Washington DC, USA. Available at: https://tinyurl.com/a7m7whhu.

Kelloway, C., 2019. Meat processing is a dangerous job. It's about to get a lot worse. Talk Poverty, 24 April, 2019. Available at: https://tinyurl.com/ymac9awm.

Kennard, B. and Young, R., 2018. A Good life and a good death: re-localising farm animal slaughter. The Sustainable Food Trust, Bristol, UK. Available at: https://tinyurl.com/htvcbtdc.

Khatri, Y. and Collins, R., 2007. Impact and status of HACCP in the Australian meat industry. British Food Journal 109: 343-354.

Kirwan, J., 2006. The interpersonal world of direct marketing: examining conventions of quality at UK farmers' market. Journal of Rural Studies 22: 301-312.

Kohlen, S., 2011. Untersuchungen zum korrekten treffpunkt für den bolzenschuss bei der betäubung von rindern bei der schlachtung. PhD Thesis, Ludwig-Maximilians-Universität München, Munich, Germany.

Lambooij, B. and Algers, B., 2016. Mechanical stunning and killing methods. In: Velarde, A. and Raj, M. (eds.) Animal welfare at slaughter. 5M Publishing Ltd., Sheffield, UK, pp. 91-110.

Lambooij, E., Puister-Jansen, L., Graven, W., Bemelman, Y. and Hoste, R., 2011. Haalbaarheid mobiel slachthuis en uitsnijderij. Report No. 460. Wageningen UR Livestock Research, Lelystad, the Netherlands. Available at: https://tinyurl.com/zhd77rb9.

Lansade, L., Bertrand, M., Boivin, X. and Bouissou, M.F., 2004. Effects of handling at weaning on manageability and reactivity of foals. Applied Animal Behaviour Science 87: 131-149.

Le Neindre, P., Boivin, X. and Boissy, A., 1996. Handling of extensively kept animals. Applied Animal Behaviour Science 49: 73-81.

Lees, W., 2015. Advancing the small scale, local food sector in Manitoba: a path forward. Small Scale Food Manitoba working group, Winnipeg, Canada. Available at: https://tinyurl.com/2fhhb5wf.

Ligout, S., Bouissou, M.F. and Boivin, X., 2008. Comparison of the effects of two different handling methods on the subsequent behaviour of Anglo-Arabian foals toward humans and handling. Applied Animal Behaviour Science 113: 175-188.

Liljenstolpe, C., 2005. Valuing animal welfare with choice experiments: an application to Swedish pig production. Proceedings of the 11th Congress of the European Association of Agricultural Economists. Copenhagen, Denmark. Available at: https://tinyurl.com/48bnc7c2.

Lisa, A., 2020. History of America's meat-processing industry. Stacker. Available at: https://tinyurl.com/2z69z5d8.

Long, D., 1993. The role of state sponsored and co-operatively organized support services in meat production, processing and marketing in developing countries. In: Mack, S. (ed.) Strategies for sustainable animal agriculture in developing countries. Proceedings of FAO Expert Consultation, Rome, Italy.

López-Mosquera, N. and Sánchez, M., 2012. Theory of planned behavior and the value-belief-norm theory explaining willingness to pay for a suburban park. Journal of Environmental Management 113: 251-262.

Losito, P., Visciano, P., Genualdo, M. and Cardone, G., 2011. Food supplier qualification by an Italian large-scale-distributor: Auditing system and non-conformances. Food Control 22: 2047-2051.

Low, S.A., Adalja, A., Beaulieu, E., Key, N., Martinez, S., Melton, A., Perez, A., Ralston, K., Stewart, H., Suttles, S., Vogel, S. and Jablonski, B.B.R., 2015. Trends in U.S. local and regional food systems: a report to congress. Administrative Publication AP-068. U.S. Department of Agriculture, Economic Research Service, Washington DC, USA. Available at: https://www.ers.usda.gov/publications/pub-details/?pubid=42807.

Lürzel, S., Münsch, C., Windschnurer, I., Futschik, A., Palme, R. and Waiblinger, S., 2015. The influence of gentle interactions on avoidance distance towards humans, weight gain and physiological parameters in group-housed dairy calves. Applied Animal Behaviour Science 172: 9-16.

Lürzel, S., Windschnurer, I., Futschik, A. and Waiblinger, S., 2016. Gentle interactions decrease the fear of humans in dairy heifers independently of early experience of stroking. Applied Animal Behaviour Science 178: 16-22.

MacDonald, J.M., Ollinger, M.E., Nelson, K.E. and Handy, C.R., 2000. Consolidation in U.S. meatpacking. Agricultural Economic Report No. 785. Food and Rural Economics Division, Economic Research Service, U.S. Department of Agriculture. Available at: https://tinyurl.com/69zcf52u.

MacLachlan, I., 2007. A bloody offal nuisance: the persistence of private slaughter-houses in nineteenth century London. Urban History 34: 227-254.

MacLachlan, I., 2008. Humanitarian reform, slaughter technology, and butcher resistance in nineteenth-century Britain. In: Young Lee, P. (ed.) Meat, modernity and the rise of the slaughterhouse. University of New Hampshire Press, Durham, NH, USA, pp. 107-126.

Malmfors, G. and Wiklund, E., 1996. Pre-slaughter handling of reindeer – effects on meat quality. Meat Science 43: 257-264.

Marchant-Forde, J.N. and Boyle, L.A., 2020. COVID-19 effects on livestock production: a one welfare issue. Frontiers in Veterinary Science 7: 585787.

Marti, D.L., Johnson, R.J. and Mathews, Jr., K.H., 2011. Where's the (Not) meat? Byproducts from beef and pork production. Report LDP-M-209-01. Economic Research Service, US Department of Agriculture. Washington DC, USA. Available at: https://tinyurl.com/75hzjner.

Martinez, M., Hand, M., Da Pra, M., Pollack, S., Ralston, K., Smith, T., Vogel, S., Clark, S., Lohr, L., Low, S. and Newman. C., 2010. Local food systems: concepts, impacts, and issues. Economic Research Report ERR-97. US Department of Agriculture, Economic Research Service, Washington DC, USA. Available at: https://www.ers.usda.gov/publications/pub-details/?pubid=46395.

Martínez-Álvarez, O., Chamorro, S. and Brenes, A., 2015. Protein hydrolysates from animal processing by-products as a source of bioactive molecules with interest in animal feeding: a review. Food Research International 73: 204-212.

McCann, N., 2011. Meat plants: improving profitability in small and very small operations. National Sustainable Agriculture Information Service, National Center for Appropriate Technology, Butte, MT, USA. Available at: https://tinyurl.com/3u3zprc2.

McCorkell, R., Wynne-Edwards, K., Galbraith, J., Schaefer, A., Caulkett, N., Boysen, S. and Pajor, E., 2013. Transport versus on-farm slaughter of bison: physiological stress, animal welfare, and avoidable trim losses. Canadian Veterinary Journal 54: 769-774.

McLeod, A., Thieme, O. and Mack, S.D., 2009. Structural changes in the poultry sector: will there be smallholder poultry development in 2030? World's Poultry Science Journal 65: 191-199.

Melanson, J., 2005. Restrictions, needs, markets and opportunities associated with a mobile slaughter facility in Nova Scotia. Undergraduate project report, Nova Scotia Agricultural College, Dalhousie University, Truro, Canada.

Menzies, B., Wood, D. and Dimambro, M., 2020. Assessing the viability and sustainability of mobile abattoirs in Scotland. Director-General Economy, Scottish Government, Edinburgh, UK. Available at: https://tinyurl.com/5bycykbf.

Methow Conservancy, 2018. Assessing needs and opportunities: a feasibility study for USDA-inspected livestock slaughter in Okanogan County. Washington. Winthrop, WA, USA. Available at: https://tinyurl.com/463pdxp8.

Mills, S., 2007. Report on the feasibility of a small-scale small-animal slaughter facility for independent meat producers in North Carolina. North Carolina Department of Agriculture & Consumer Services. Available at: https://www.smithsonmills.com/animalreport.pdf.

Miranda-de la Lama, G.C., Mamani-Linares, W. and Estévez-Moreno, L.X., 2022. Species destined for non-traditional meat production: 2. Goats and South American domestic camelids. In: Faucitano, L. (ed.) Preslaughter handling and slaughter of meat animals. Wageningen Academic Publishers, Wageningen, the Netherlands, pp. 348-389.

Mora, L., Toldrá-Reig, F., Reig, M. and Toldrá, F., 2019. Possible uses of processed slaughter byproducts. In: Galanakis, C.H. (ed.) Sustainable meat production and processing. Academic Press, Cambridge, MA, USA, pp. 145-160.

Morchain, D., 2009. Factors affecting the small-scale production of artisan meat. A study of livestock farming in Sweden. MSc Thesis, The International Institute for Industrial Environmental Economics, Lund University, Lund, Sweden.

Mozdziak, P., 2014. Species of meat animals. Poultry. In: Dikeman, M. and Devine, C. (eds.) Encyclopedia of meat sciences. Elsevier Science Publishing Co., Inc., San Diego, CA, USA, pp. 369-373.

Muth, M.K., Wohlgenant, M.K. and Karns, S.A., 2007. Did the pathogen reduction and HACCP regulation cause slaughter plants to exit? Review of Agricultural Economics 29: 596-611.

NASDA, 2021. Interstate meat sales background. National Association of State Departments of Agriculture Available at: https://tinyurl.com/frb7vzps.

National Fisheries and Aquaculture Service, 2020. Listado de plantas autorizadas para faenar y/o procesar salmónidos. Ministry of Economy, Development and Tourism, Government of Chile, Santiago de Chile, Chile. Available at: https://tinyurl.com/hrtu9363.

National Fisheries and Aquaculture Service, 2021. Naves prestadoras de servicios de acuicultura. Ministry of Economy, Development and Tourism, Government of Chile, Santiago de Chile, Chile. Available at: https://tinyurl.com/ymdz53cv.

Needham, T. and Hoffman, L.C., 2022. Species destined for non-traditional meat production: 1. African game species, cervids, ostriches, crocodiles and kangaroos. In: Faucitano, L. (ed.) Preslaughter handling and slaughter of meat animals. Wageningen Academic Publishers, Wageningen, the Netherlands, pp. 312-347.

New Zealand Government, 2018. Code of welfare: commercial slaughter. Minister of agriculture, Hamilton, New Zealand. Available at: https://www.mpi.govt.nz/dmsdocument/46018/direct.

Niche Meat Processor Assistance Network, 2011. Mobile slaughter unit manual. Available at: https://www.nichemeatprocessing.org/mobile-slaughter-unit-manual/.

Niche Meat Processor Assistance Network, 2021. Mobile slaughter and processing. Oregon State University, Corvallis, OR, USA. Available at: http://www.nichemeatprocessing.org/mobile-unit-overview.

Nollet, L.M.L. and Toldrá, F., 2011. Introduction – offal meat: definitions, regions, cultures and generalities. In: Nollet, L.M.L. and Toldrá, F. (eds.) Handbook of analysis of edible animal by-products. CRC Press, Boca Raton, FL, USA, pp. 3-11.

Norwegian Directorate of Fisheries, 2021. Kunnskap fra utviklingsprosjektene. Available at: https://tinyurl.com/3wtzfat8.

O'Bryan, C.A., Crandall, P.G., Davis, M.L., Kostadini, G., Gibson, K.E., Alali, W.Q., Jaroni, D., Ricke, S.C. and Marcy, J.A., 2014. Mobile poultry processing units: a safe and cost-effective poultry processing option for the small-scale farmer in the United States. World's Poultry Science Journal 70: 787-802.

OIE, 2019. Terrestrial animal health code. World Organisation for Animal Health, Paris, France. Available at: https://tinyurl.com/43ratf8e.

Otter, C., 2008. Civilizing slaughter: the development of the British public abattoir, 1850-1910. In: Young Lee, P. (ed.) Meat, modernity and the rise of the slaughterhouse. University of New Hampshire Press, Durham, NH, USA, pp. 89-106.

Pachirat, T., 2011. Every twelve seconds: industrialized slaughter and the politics of sight. Yale University Press, New Haven, CT, USA, 320 pp

Pacyga, D., 2008. Chicago: slaughterhouse to the world. In: Young Lee, P. (ed.) Meat, modernity and the rise of the slaughterhouse. University of New Hampshire Press, Durham, NH, USA, pp. 153-166.

Petherick, J.C., 2005. Animal welfare issues associated with extensive livestock production: the northern Australian beef cattle industry. Applied Animal Behaviour Science 92: 211-234.

Pinkney, D., 2014. Mobile abattoirs in the context of Canadian local food system development: niches and the potential for local food system resiliency. MSc. Thesis. University of Guelph, Guelph, Canada.

Pitts, L., 2021. Meals on wheels. Triple A Livestock Report, New Mexico Stockman and Livestock Market Digest, Albuquerque, NM, USA. Available at: https://www.aaalivestock.com/xtra/meals-on-wheels/.

Probst, J.K., Spengler Neff, A., Leiber, F., Kreuzer, M. and Hillmann, E., 2012. Gentle touching in early life reduces avoidance distance and slaughter stress in beef cattle. Applied Animal Behaviour Science 139: 42-49.

Rabanal, H.R., 1988. History of aquaculture. FAO/UNDP Network of Aquaculture Centers in Asia Training Programme for Senior Aquaculturists. 24 March, 1988. Southeast Asian Fisheries Development Center, Tigbauan, Iloilo, Philippines. Available at: http://www.fao.org/3/ag158e/AG158E01.htm.

Raussi, S., 2013. Human-cattle interactions in group housing. Applied Animal Behaviour Science 80: 245-262.

Raynolds, L.T., 2004. The globalization of organic agro-food networks. World Development 32: 725-743.

Reinbase, 2019. Slakteuttak. Norway. Available at: https://www.reinbase.no/nb-no/Studer-reindriften/Slakteuttak.

Reindeer Herders' Association, 2019. Reindeer meat production. Rovaniemi, Finland. Available at: https://tinyurl.com/khyz4tv8.

Retz, S., Schiffer, K.J., von Wenzlawowicz, M. and Hensel, O., 2014. Stunning effect of different rifle-bullets for slaughter of outdoor cattle. Landtechnik 69: 296-300.

Robinson, T.P., Thornton, P.K., Franceschini, G., Kruska, R.L., Chiozza, F., Notenbaert, A., Cecchi, G., Herrero, M., Epprecht, M., Fritz, S., You, L., Conchedda, G. and See, L., 2011. Global livestock production systems. Food and Agriculture Organization of the United Nations and International Livestock Research Institute, Rome, Italy. Available at: http://www.fao.org/3/i2414e/i2414e00.htm.

Sámediggi, 2020. Renslakt. Sámi Parliament. Available at: https://www.sametinget.se/statistik/renslakt.

Schiffer, K.J., 2015. On-farm slaughter of cattle via gunshot method. PhD Thesis, University of Kassel, Witzenhausen, Germany.

Schiffer, K.J., Retz, S.K., Algers, B. and Hensel, O., 2017. Assessment of stun quality after gunshot used on cattle: a pilot study on effects of diverse ammunition on physical signs displayed after the shot, brain tissue damage and brain haemorrhages. Animal Welfare 26: 95-109.

Schiffer, K.J., Retz, S.K., Richter, U., Algers, B. and Hensel, O., 2014. Assessment of key parameters for gunshot used on cattle: a pilot study on shot placement and effects of diverse ammunition on isolated cattle heads. Animal Welfare 23: 479-489.

Schlosser, E., 2001. Fast food nation: the dark side of the all-American meal. Houghton Mifflin Company, Boston, MA, USA, 356 pp.

Seppänen, L. 2004. Local and organic food and farming around the Baltic Sea. Report No. 40, Centre for Sustainable Agriculture, Swedish University of Agricultural Sciences, Uppsala, Sweden. Available at: https://tinyurl.com/ntfz62k8.

Shahin, M., 2018. The effects of positive human contact by tactile stimulation on dairy cows with different personalities. Applied Animal Behaviour Science 204: 23-28.

Shearer, J.K. and Ramirez, A., 2013. Humane euthanasia of sick, injured and/or debilitated livestock. Iowa State University, Ames, IA, USA. Available at: https://tinyurl.com/8ws8j23y.

Shearer, J.K., 2018. Euthanasia of cattle: practical considerations and application. Animals 8: 57.

Sherman, G.L., 2010. Considerations unique to mobile red meat slaughter. Webinar presentation to the Food Safety and Inspection Service, United States Department of Agriculture, Available at: https://tinyurl.com/2tkcnsnk.

Shields, D.A. and Mathews, K.H., Jr., 2003. Interstate livestock movements. Electronic Outlook Report LDP-M-108-01. United States Department of Agriculture, Economic Research Service, Washington, DC, USA. Available at: https://www.ers.usda.gov/publications/pub-details/?pubid=37686.

Simmons, D., 2019. Provenir's mobile abattoir set to revolutionise ag-tech and animal welfare. Business News Australia. Available at: https://tinyurl.com/4pz4h7fz.

Skog Eriksen, M., Rødbotten, R., Grønahl, A.M., Friestad, M., Andersen, I.L. and Mejdell, C.M., 2013. Mobile abattoir versus conventional slaughterhouse – impact on stress parameters and meat quality characteristics in Norwegian lambs. Applied Animal Behaviour Science 149: 21-29.

Sleeping Lion Associates, 2005. Slaughterhouse feasibility report. Sleeping Lion Associates, Inc., Montpelier, VT, USA. Available at: https://ucanr.edu/sites/placernevadasmallfarms/files/107066.pdf.

SLVFS, 1994. Statens Livsmedelsverks kungörelse med föreskrifter och allmänna råd om slakt och annan hantering av färskt kött av ren. SLVFS 1994:12 (H 197). Swedish Food Agency regulations.

Sonnino, R. and Marsden, T., 2006. Beyond the divide: rethinking relationships between alternative and conventional food networks in Europe. Journal of Economic Geography 6: 181-199.

Statistics Canada, 2017. 2016 census of agriculture. Farm and Farm Operator Data, Livestock in Canada. Available at: https://www.statcan.gc.ca/eng/ca2016.

Stevenson, M. and Fukumoto, G., 2011. A Review of mobile/modular slaughter and processing technology. Food Safety and Technology, FST-44. University of Hawaii at Mānoa. Available at: https://www.ctahr.hawaii.edu/oc/freepubs/pdf/FST-44.pdf.

Stocchi, R., Mandolini, N.A., Marinsalti, M., Cammertoni, M., Loschi, A.R. and Rea, S., 2014. Animal welfare evaluation at a slaughterhouse for heavy pigs intended for processing. Italian Journal of Food Safety 3: 54-56.

Swedish Board of Agriculture, 2018. Godkänd slakt, antal djur: storboskap. Jönköping, Sweden. Available at: https://tinyurl.com/fdjwj88m.

Tallentire, C.W. Leinonen, I. and Kyriazakis, I., 2016. Breeding for efficiency in the broiler chicken: a review. Agronomy for Sustainable Development 36: 66.

Terlouw, C. and Bourguet, C., 2022. Quantifying animal welfare preslaughter using behavioural, physiological and carcass and meat quality measures. In: Faucitano, L. (ed.) Preslaughter handling and slaughter of meat animals. Wageningen Academic Publishers, Wageningen, the Netherlands, pp. 12-61.

Terlouw, E.M.C., Bourguet, C. and Deiss, V., 2012. Stress at slaughter in cattle: role of reactivity profile and environmental factors. Animal Welfare 21: 43-49.

Thiboumery, A., 2009. Guide to designing a small red meat plant with two sizes of model designs. Iowa State University, Ames, IA, USA. Available at: https://www.sare.org/resources/guide-to-designing-a-small-red-meat-plant/.

Thiboumery, A., 2010. Iowa meat processors' resource guidebook. Iowa State University, Ames, IA, USA. Available at: https://www.sare.org/resources/iowa-meat-processors-resource-guidebook/.

Thiruvenkadani, A.K., Prabakaran, R. and Panneerselvam, S., 2011. Broiler breeding strategies over the decades: an overview. World's Poultry Science Journal 67: 309-336.

Tier-LMHV, 2018. Verordnung über anforderungen an die hygiene beim herstellen, behandeln und inverkehrbringen von bestimmten lebensmitteln tierischen ursprungs. Animal food hygiene regulation BGBl. I S. 480 (619), Tier-LMHV. Available at: https://www.gesetze-im-internet.de/tier-lmhv/BJNR182800007.html.

Tum Tum's Black Gilt Meats, 2020. Yukon mobile abattoir procedures manual. Tum Tum's Black Gilt Meats Ltd, Whitehorse, Canada. Available at: https://tinyurl.com/supjbksc.

UFAW, 1988. Humane killing of animals. Universities Federation for Animal Welfare, Potters Bar, UK, 28 pp.

USC, 1958. Humane methods of livestock slaughter act. United States Code, Title 7 of the Agriculture Code, Chapter 48. Available at: https://tinyurl.com/th6eah98.

USDA, 2010. Mobile slaughter unit compliance guide. Food Safety and Inspection Service, United States Department of Agriculture, Washington, DC, USA. Available at: https://tinyurl.com/kmpjspeb.

USDA, 2015a. Cooperative interstate shipping program. Food Safety and Inspection Service. United States Department of Agriculture, Washington, DC, USA. Available at: https://tinyurl.com/2vtnwxfr.

USDA, 2015b. States with and without inspection programs. Food Safety and Inspection Service, United States Department of Agriculture, Washington, DC, USA. Available at: https://tinyurl.com/bn5fkprx.

USDA, 2019. Census of agriculture. United States Summary and State Data, Volume 1. Geographic Area Series, Part 51, AC-17-A-51. United States Department of Agriculture, Washington, DC, USA. Available at: https://www.nass.usda.gov/Publications/AgCensus/2017/#full_report.

USDA, 2020a. Grants, loans and other support. Agricultural Marketing Service. United States Department of Agriculture, Washington, DC, USA. Available at: https://www.ams.usda.gov/services/local-regional/food-sector/grants.

USDA, 2020b. Quick stats agricultural database. National Agricultural Statistics Service, United States Department of Agriculture, Washington, DC, USA. Available at: https://catalog.data.gov/dataset/quick-stats-agricultural-database.

Valente, J.deP.S., Fiedler, R.A., Sucha Heidemann, M. and Molento, C.F.M., 2019. First glimpse on attitudes of highly educated consumers towards cell-based meat and related issues in Brazil. PLoS ONE 14: e0221129.

Van Loo, E.J., Alali, W.Q., Welander, S., O'Bryan, C.A., Crandall, P.G. and Ricke, S.C., 2013. Independent poultry processing in Georgia: survey of producers' perspective. Agriculture, Food and Analytical Bacteriology 3: 70-77.

VFSK, 2020. Verordnung über das Schlachten und die Fleischkontrolle. Ordinance 817.190, Germany, 30 July, 2020. Available at: https://www.fedlex.admin.ch/eli/cc/2017/66/de#a22.

Vialles, N., 1994. Animal to edible. Cambridge University Press, Cambridge, UK, 158 pp.

Villarroel, M. and Lambooij, E., 2022. Fish. In: Faucitano, L. (ed.) Preslaughter handling and slaughter of meat animals. Wageningen Academic Publishers, Wageningen, the Netherlands, pp. 118-149.

Waiblinger, S., Boivin, X., Pedersen, V., Tosi, M.-V., Janczak, A.M., Visser, E.K. and Jones, R.B., 2006. Assessing the human-animal relationship in farmed species: a critical review. Applied Animal Behaviour Science 101: 185-242.

Waiblinger, S., Menke, C., Korff, J. and Bucher, A., 2004. Previous handling and gentle interactions affect behaviour and heart rate of dairy cows during a veterinary procedure. Applied Animal Behaviour Science 85: 31-42.

Weeks, C.A., Brown, S.N., Lane, S., Heasman, L., Benson, T. and Warriss, P.D., 2009. Noise levels in lairages for cattle, sheep and pigs in abattoirs in England and Wales. Veterinary Record 165: 308-314.

Whittaker, W.G., 2006. Labor practices in the meat packing and poultry processing industry: an overview. CRS Report for the Unites States Congress, Washington, DC, USA. Available at: https://nationalaglawcenter.org/wp-content/uploads/assets/crs/RL33002.pdf.

Wikberg, P., 2001. Allt fler vill ha mobila slakterier: Svensk kunskap kan bli intressant inom hela EU. Available at: https://www.miljomagasinet.se/artiklar/slakt12%20_01.html.

Wiklund, E. and Malmfors, G., 2004. The effects of pre-slaughter handling on reindeer meat quality – a review. Animal Breeding Abstracts 72: 1N-6N.

Wiklund, E., Farouk, M. and Finstad, G., 2014. Venison: meat from red deer (*Cervus elaphus*) and reindeer (*Rangifer tarandus tarandus*). Animal Frontiers 4: 55-61.

Wilson, M.R., 2004. Union stock yard & transit Co. In: Grossman, J.R., Keating, A.D. and Ruff, J.L. (eds.) Encyclopedia of Chicago. University of Chicago Press, Chicago, IL, USA. Available at: http://www.encyclopedia.chicagohistory.org/pages/2883.html.

Wong, J.T., De Bruyn, J., Bagnol, B., Grieve, H., Li, M., Pym, R. and Alders, R.G., 2017. Small-scale poultry and food security in resource-poor settings: a review. Global Food Security 15: 43-52.

Woods, J., Shearer, J.K. and Hill, J., 2009. Recommended on-farm euthanasia practices. In: Grandin T. (ed.) Improving animal welfare: a practical approach. CABI Publishing Wallingford, UK, pp. 186-213.

Wotton, S.B. and Hewitt, L., 1999. Transportation of ostriches – a review. Veterinary Record 145: 725-731.

Yukon, 2007. Mobile abattoir procedures manual. 40056 Yukon Inc., Whitehorse, Canada. Available at: https://tinyurl.com/db8jydea.

www.ingramcontent.com/pod-product-compliance
Lightning Source LLC
LaVergne TN
LVHW082053011026
849162LV00020B/1634
* 9 7 8 9 0 8 6 8 6 3 7 2 3 *